Vortex Laser Beams

Vortex Laser Beams

V. V. Kotlyar

A. A. Kovalev

A. P. Porfirev

CRC Press is an imprint of the
Taylor & Francis Group, an **informa** business

CRC Press
Taylor & Francis Group
6000 Broken Sound Parkway NW, Suite 300
Boca Raton, FL 33487-2742

First issued in paperback 2023

ISBN-13: 978-1-138-54211-2 (hbk)
ISBN-13: 978-1-03-265334-1 (pbk)
ISBN 13: 978-1-351-00960-7 (ebk)

DOI: 10.1201/9781351009607

Publisher's Note
The publisher has gone to great lengths to ensure the quality of this reprint but points out that some imperfections in the original copies may be apparent.

Library of Congress Cataloging-in-Publication Data

Names: Kotlyar, Victor, author. | Alexey, Kovalev, author. | Alexey, Porfirev, author.
Title: Vortex laser beams / Victor Kotlyar, Kovalev Alexey, and Porfirev Alexey.
Description: Boca Raton : Taylor & Francis, a CRC title, part of the Taylor & Francis imprint, a member of the Taylor & Francis Group, the academic division of T&F Informa, plc, 2018. | Includes bibliographical references and index.
Identifiers: LCCN 2018009767 | ISBN 9781138542112 (hardback : acid-free paper) | ISBN 9781351009607 (e-book)
Subjects: LCSH: Laser beams. | Vortex-motion. | Nonlinear optics. | Quantum optics.
Classification: LCC TA1675 .K68 2018 | DDC 621.36/6--dc23
LC record available at https://lccn.loc.gov/2018009767

Visit the Taylor & Francis Web site at
http://www.taylorandfrancis.com

and the CRC Press Web site at
http://www.crcpress.com

Contents

Preface

Derivation of analytical solutions of Maxwell's equations, as well as other mathematical tools widely employed in optics—such as a Helmholtz equation or paraxial Schrodinger-type equations—has always been in the focus of interest of optical researchers. The said solutions describe in detail properties of coherent electromagnetic light fields and laser beams, which have found numerous practical applications. In particular, widely known light fields that can be described analytically include plane or spherical waves and Gaussian and Bessel beams [1], to name just a few. Recently, new promising light beams that can be described by exact analytical relations have been proposed. These include Hermite-Gaussian and Laguerre-Gaussian modal beams [2], Hermite-Laguerre-Gauss beams [3], elliptic Mathieu and Ince beams [4,5], hypergeometric beams [6], accelerating Airy beams [7], and self-focusing Pearcey beams [8]. Further research of elegant Laguerre-Gaussian and Hermite-Gaussian beams is currently underway [9,10], with their behavior being described using polynomials with complex argument. The elliptic Laguerre-Gaussian beams have been studied using a number of approaches [11,12]. Recent years have seen an increase of interest in deriving exact solutions of paraxial Schrodinger-type equations in cylindrical coordinates. More recently, hypergeometric Gaussian beams [13] and circular beams [14] have been proposed. A number of well-studied light beams, such as conventional and elegant Laguerre-Gaussian modes, quadratic Bessel-Gaussian beams [15], and Gaussian optical vortices [16], have been shown to be a particular case of the circular beams [14].

Light fields can be grouped into two classes: those that carry orbital angular momentum (OAM) [17] and those devoid of OAM. Beams that carry OAM are termed as vortex or singular beams. The vortex laser beams are characterized by a helical or spiral phase, wavefront dislocations, and isolated intensity nulls.

Currently, vortex laser beams have been put to many practical uses, including turbulent atmosphere probing [18], wireless optical communications [19], fiber optic communication channel multiplexing [20], astronomy [21], quantum informatics [22], and micromanipulation [23].

It has not been our goal to offer a comprehensive review of all exact solutions of Maxwell's, Helmholtz, and Schrodinger equations that are currently utilized in optics. Instead, our book covers in detail only the vortex laser beams that were personally proposed by the current authors over years of research, which are quite many. For instance, these include hypergeometric beams, Hankel-Bessel beams, half Pearcey beams, asymmetric Bessel modes, Lommel modes, asymmetric Laguerre-Gaussian beams, vortex Hermite-Gaussian beams, vector Hankel beams, and others.

It is worth noting that a vortex laser beam is most easy to generate via illuminating a spiral phase plate (SPP) by a conventional Gaussian beam. Hence, diffraction of light by a SPP is given particular attention in this book.

The book contains research results financially supported by the Russian Science Foundation grant No. 17-19-01186.

Authors

V. V. Kotlyar is a Head of Laboratory at the Image Processing Systems Institute (Samara) of the Russian Academy of Sciences and a professor in the Computer Science department at Samara National Research University. He received his MS, PhD, and DrSc degrees in physics and mathematics from Samara State University (1979), Saratov State University (1988), and Moscow Central Design Institute of Unique Instrumentation, the Russian Academy of Sciences (1992). He is a co-author of 300 scientific papers, 5 books, and 7 inventions. His current interests are diffractive optics, gradient optics, nanophotonics, and optical vortices.

A. A. Kovalev (b. 1979) graduated (2002) from Samara National Research University, majoring in Applied Mathematics. He received his Doctor in Physics & Maths degree in 2012. He is a senior researcher in the Laser Measurements laboratory at Image Processing Systems Institute of the Russian Academy of Sciences, a branch of the Federal Scientific Research Centre "Crystallography and Photonics" RAS. He is a co-author of more than 150 scientific papers. His current research interests are mathematical diffraction theory and photonic crystal devices.

A. P. Porfirev (b. 1987) graduated (2010) from Samara National Research University, majoring in Applied Physics and Mathematics. He is a candidate in Physics and Mathematics (2013). Currently he is an associate professor in the Technical Cybernetics department of Samara National Research University and a researcher in the Micro- and Nanotechnologies laboratory of the Image Processing Systems Institute of the Russian Academy of Sciences, a branch of the Federal Scientific Research Centre "Crystallography and Photonics" RAS. His current research interests include diffractive optics, optical manipulation, and structured laser beams.

1 A Spiral Phase Plate for an Optical Vortices Generation

1.1 DIFFRACTION OF A PLANE, FINITE-RADIUS WAVE BY A SPIRAL PHASE PLATE

For the first time, spiral phase plates (SPPs) were fabricated by optical lithography and analyzed as optical elements whose transmittance is proportional to $\exp(in\varphi)$ (where φ is the polar angle and n is the integer) in Ref. [24]. In Ref. [25] the SPP was implemented by index matching a microscopic structure in an optical immersion. The recent years have seen an increase of interest in the SPPs, especially due to the opportunities the SPPs offer for optical micromanipulation [26–31]. The SPPs have also gained importance in quantum information studies [32]. Thus, further research into diffraction of light by the SPP is highly relevant. In Ref. [33] the diffraction of an infinite plane wave by an SPP with an arbitrary integer n was studied theoretically. Diffraction of the same wave by a fraction-numbered SPP was dealt with in Ref. [34]. Diffraction of a Gaussian beam by the SPP was reported in Ref. [35]. The Fraunhofer diffraction of a plane wave in a circular aperture was discussed in Refs. [24,26,35–37]. Note that in Refs. [24,35] identical expressions for the complex amplitude were obtained only for an SPP of the first order. Expressions for the complex amplitude for diffraction by an SPP of an arbitrary order were deduced in Refs. [26,36]. However, the relations presented (Eq. (1.4) in Ref. [26] and Eq. (1.1) in Ref. [36]) are not quite accurate. In Ref. [37] it was indicated that the radius of the first ring of the nth order vortex is proportional to the first root of the nth order Bessel function.

In this section, we derive analytical relations for the Fresnel and Fraunhofer diffraction of a plane wave with circular and ring-like cross-section by an SPP of arbitrary integer order. An accurate formula for the intensity on the vortex's first ring and an approximate formula for the radius of the vortex's first ring are derived. Experimental results on diffraction of a plane, finite-radius wave by SPPs with $n = 2$ and 3 are also discussed. The high-quality, 32-level SPPs are fabricated (with 1.5% accuracy for the second-order and 4.3% accuracy for the third-order SPP) using direct e-beam writing in resist. It is shown that the experimental and theoretical diffraction patterns are in good agreement.

The Fraunhofer Diffraction of a Finite-Radius Plane Wave

Let us first consider the Fraunhofer diffraction of a plane, finite-radius wave by an SPP. The plane wave of unit intensity, radius R, and wavenumber $k = 2\pi/\lambda$ (λ is wavelength), propagating along the z-axis, is described by its complex amplitude at $z = 0$:

$$E_0(r) = \text{circl}\left(\frac{r}{R}\right), \tag{1.1}$$

where:

$$\text{circl}(x) = \begin{cases} 1, & |x| \le 1, \\ 0, & |x| > 1. \end{cases} \tag{1.2}$$

Assume that the plane wave of Eq. (1.1) falls on the SPP of transmittance:

$$\tau(\phi) = \exp(in\phi), \quad n = 0, \pm 1, \pm 2, \ldots, \tag{1.3}$$

where (r,φ) are the polar coordinates at $z = 0$. The Fraunhofer diffraction pattern for the plane wave of Eq. (1.1) by the SPP of Eq. (1.3) is generated in the rear focal plane of a spherical Fourier-lens of focal length f and described by the Fourier transform:

$$\begin{aligned} E_n(\rho,\theta) &= \frac{(-i)^{n+1}k}{f}\exp(in\theta)\int_0^R J_n\left(\frac{k}{f}r\rho\right)r\,\mathrm{d}r \\ &= \frac{(-i)^{n+1}\exp(in\theta)}{(n+2)n!}\left(\frac{kR^2}{f}\right)\left(\frac{kR\rho}{2f}\right)^n {}_1F_2\left[\frac{n+2}{2},\frac{n+4}{2},n+1;-\left(\frac{kR\rho}{2f}\right)^2\right], \end{aligned} \tag{1.4}$$

where:

(ρ,θ) are the polar coordinates in the Fourier plane
$J_n(x)$ is the Bessel function of the nth order and first kind
${}_1F_2(a,b,c;x)$ is a hypergeometric function:

$${}_1F_2(a,b,c;x) = \sum_{m=0}^{\infty}\frac{(a)_m x^m}{(b)_m(c)_m m!}, \tag{1.5}$$

where $(a)_m = a(a+1)(a+2)\ldots(a+m-1)$ is Puckhammer symbol, $(a)_0 = 1$. Equation (1.4) was derived using a reference integral given in Ref. [38]. Equation (1.4) is similar to the previously presented relations (Eq. (1.4) in Ref. [26] and Eq. (1.1) in Ref. [36]). Note, however, that there are differences. The complex amplitudes presented in Refs. [26,36] have the dimensions of $[\text{mm}]^2$, being different from the dimensionless complex amplitude in Eq. (1.4) of the present paper by a constant $-ik(2\pi f)^{-1}$. In addition, in Ref. [26] the power of the radial variable is indicated inaccurately, whereas

Eq. (1.4) rigorously describes the Fraunhofer diffraction of a plane wave with a circular diaphragm by the SPP (up to an insignificant phase factor of $\exp(ikz)$).

Simulation Results

From Eq. (1.4) it can be seen that at $n \neq 0$ the amplitude at the center of the Fourier plane ($\rho = 0$) is zero: $E_n(\rho = 0,\theta) = 0$. From Eqs. (1.4) and (1.5) it also follows that at small ρ $(\rho \ll 2f/kR)$ ${}_1F_2\,(a,b,c;x) \approx 1$ and $E_n(\rho \to 0,\theta) \sim (kR^2/f)[kR\rho/(2f)]^n$, where ~ symbolizes proportionality.

Figure 1.1 shows plots for the intensity function $I_n(\rho) = \left|E_n(\rho,\theta)\right|^2$ derived from Eq. (1.4). In the series in Eq. (1.5), 110 terms were retained. The calculation parameters are $\lambda = 0.633$ µm, $f = 100$ mm, and $R = 1$ mm. From Eq. (1.4) it follows that, because the amplitude $E_n(\rho,\theta)$ is proportional to ρ^n with increasing number n, the radius of the diffraction pattern's first ring will also increase (Figure 1.1). Figure 1.2 depicts radial intensity distributions of the diffraction pattern of a plane wave with different radii R diffracted by an SPP with $n = 2$.

From Figure 1.2 it is seen that with increasing radius R the radius and width of the first ring of the diffraction pattern are decreased, whereas the maximal intensity on the ring is increased. From Eq. (1.4) it follows that the intensity is proportional to

$$I_n(\xi) \sim R^4 \xi^{2n} \left| {}_1F_2\left[\frac{n+2}{2}, \frac{n+4}{2}, n+1; -\left(\frac{k\xi}{2f}\right)^2\right]\right|^2, \tag{1.6}$$

where $\xi = R\rho$. From Eq. (1.6) it follows that with increasing R the diffraction pattern changes only in scale, with the rings' radii decreasing as many times as the radius R has increased. The maximal intensity on the ring increases in direct proportion to R^4. For example, for curves 1 and 3 of Figure 1.2 the ratio of the maximal intensity values is $3^4/2^4 \approx 5$. It can be shown that the intensity of the vortex first ring is given by

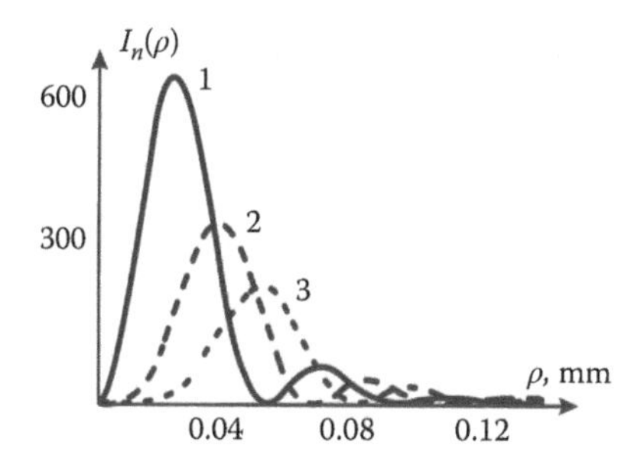

FIGURE 1.1 Radial intensity distributions of the Fraunhofer diffraction pattern for a plane wave of radius $R = 1$ mm diffracted by an SPP with $n = 1$ (curve 1), $n = 2$ (curve 2), and $n = 3$ (curve 3).

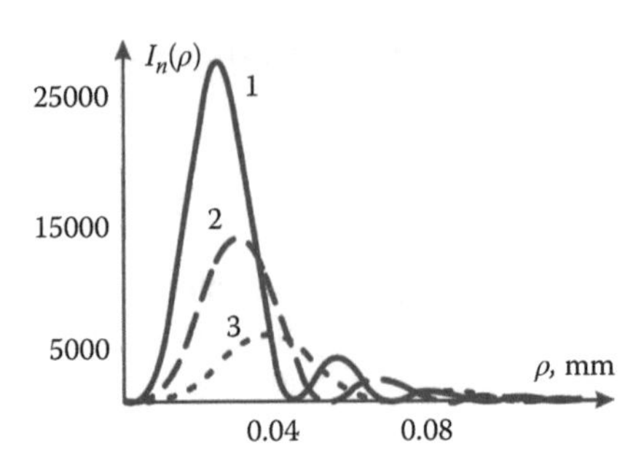

FIGURE 1.2 Radial intensity distribution of the Fraunhofer diffraction pattern of a plane wave diffracted by an SPP with $n = 2$ at $R = 3$ mm (curve 1), $R = 2.5$ mm (curve 2), and $R = 2$ mm (curve 3).

$$I_n(\rho_n) = \left(\frac{kR^2}{2f}\right)^2 J_n^2\left(\frac{kR\rho_n}{f}\right), \tag{1.7}$$

where ρ_n is the radius of the vortex first ring. From Eq. (1.4) it follows that the radius of the vortex first ring is approximated by

$$\rho_n \approx \frac{\gamma_{n-1,1}\lambda f}{2\pi R}, \tag{1.8}$$

where $\gamma_{n-1,1}$ is the first root of the Bessel function of the $(n - 1)$th order, $J_{n-1}(\gamma_{n-1,1}) = 0$. In Ref. [37] it was indicated that the radius of the first ring of the n-order vortex is proportional to the first root of the same-order Bessel function. Equation (1.8) is in good agreement with numerical simulation (Table 1.1). From the Table, the relative error is seen not to exceed 2%.

Equation (1.4) can easily be generalized onto the case of a ring-like diaphragm:

TABLE 1.1
Comparison of Vortex Radii Derived Numerically and by Use of Eq. (1.8) at $\lambda = 0.633$ µm, $f = 100$ mm, and $R = 1$ mm

Vortex Number, n	1	2	3	4	5	6	7
Vortex radii by (1.8), $\rho_n \times 10^{-2}$, mm	2.42	3.86	5.18	6.43	7.65	8.84	10.01
Vortex radii by (1.4), $\rho_n \times 10^{-2}$, mm	2.44	3.86	5.15	6.37	7.56	8.71	9.83

$$E'_n(\rho,\theta)=\frac{(-i)^{n+1}k}{f}\exp(in\theta)\int_{R_1}^{R}J_n\left(\frac{k}{f}r\rho\right)r\,\mathrm{d}r=\frac{2(-i)^{n+1}\exp(in\theta)}{(n+2)n!}\left(\frac{k}{2f}\right)^{n+1}\rho^n$$
$$\times\left\{R^{n+2}\,{}_1F_2\left[\frac{n+2}{2},\frac{n+4}{2},n+1;-\left(\frac{kR\rho}{2f}\right)^2\right]-R_1^{n+2}\,{}_1F_2\left[\frac{n+2}{2},\frac{n+4}{2},n+1;-\left(\frac{kR_1\rho}{2f}\right)^2\right]\right\},\tag{1.9}$$

where R and R_1 are the inner and outer radii of the ring-like diaphragm. From Eqs. (1.9) and (1.8) and from Figure 1.2 it follows that by fitting the radius R_1 the second ring of the vortex can be effectively "suppressed," as it was suggested in Ref. [36].

The Fresnel Diffraction of a Finite-Radius Plane Wave

Let us next consider the Fresnel diffraction of a finite-aperture plane wave by an SPP. The paraxial diffraction of the wave of Eq. (1.1) by the SPP of Eq. (1.3) is described by the Fresnel transform:

$$E_n(\rho,\theta,z)=\frac{(-i)^{n+1}k}{z}\exp\left(\frac{ik\rho^2}{2z}+in\theta\right)\int_0^R\exp\left(\frac{ikr^2}{2z}\right)J_n\left(\frac{k}{z}r\rho\right)r\,\mathrm{d}r$$
$$=2\exp\left(\frac{iz_0\bar{\rho}^2}{z}+in\theta\right)\frac{\left(\frac{-iz_0}{z}\right)^{n+1}\bar{\rho}^n}{n!}\tag{1.10}$$
$$\times\sum_{m=0}^{\infty}\frac{\left(\frac{iz_0}{z}\right)^m}{(2m+n+2)m!}\,{}_1F_2\left[\frac{2m+n+2}{2},\frac{2m+n+4}{2},n+1;-\left(\frac{z_0\bar{\rho}}{z}\right)^2\right],$$

where:

$z_0=kR^2/2$ is Rayleigh range

$\bar{\rho}=\rho/R$

The difference between Eqs. (1.10) and (1.4) is that the hypergeometric functions of Eq. (1.5) are found in the former as terms of a series. From Eq. (1.10) it is seen that at $n\neq 0$ in the beam center ($\rho=0$) the amplitude is zero, $E_n(\rho=0,\theta,z)=0$, at any z, except at $z=0$. From Eq. (1.10) it can also be seen that with increasing z only several initial terms of the series of hypergeometric functions give their contribution, whereas at $z \to \infty$ ($z \gg z_0$, far field) the contribution is only from the first term at $m = 0$, which coincides with the right-hand side of Eq. (1.4).

Experiment

Figure 1.3 gives comparison between experimental results and a theoretical prediction. Shown in Figure 1.3a is the surface profile of an SPP with $n = 3$ and diameter 2.5 mm produced on the NewView 5000 Zygo interferometer (200x magnification). The SPP profile deviates from an ideal shape by 4.3%. The SPP with 32 microrelief levels was fabricated in a low-contrast negative resist XAR-N7220 via direct e-beam writing on the Leica LION LV1 lithographer of a 5 μm pixel size. Figure 1.3b and c depicts experimental and theoretical patterns of the SPP-diffracted plane wave of radius $R = 1.25$ mm and wavelength $\lambda = 0.633$ μm at distance $z = 80$ mm. Both diffraction patterns feature the same number of diffraction rings (8 rings). A spiral seen in Figure 1.3a instead of concentric rings, as in Figure 1.3b, is due to a small displacement (about 10% of R) between the SPP center and the center of the plane wave with circular cross-section. Shown in Figure 1.3d is the calculated diffraction pattern at the same parameters as in Figure 1.3a and b, but with a $0.1R$ displacement between the centers of the beam and the SPP. It is noteworthy that the spiral spin does not depend on the direction in which the displacement occurs, but only on the sign of the SPP number n.

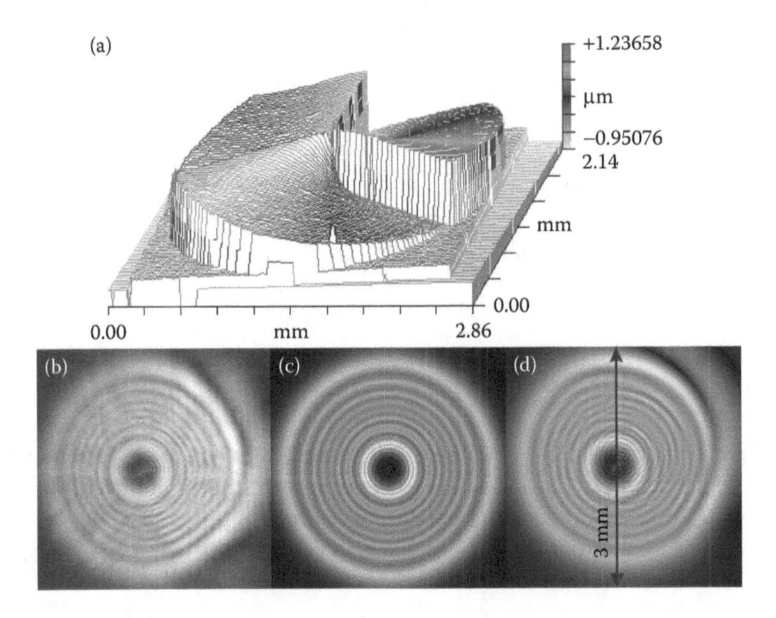

FIGURE 1.3 (a) The SPP surface profile ($n = 3$), the Fresnel diffraction pattern for a plane wave of radius $R = 1.25$ mm and wavelength $\lambda = 0.633$ μm at distance $z = 80$ mm from the SPP; (b) experiment and theory; (c) without; and (d) with displacement ($0.1R$) between the centers of the beam and the SPP.

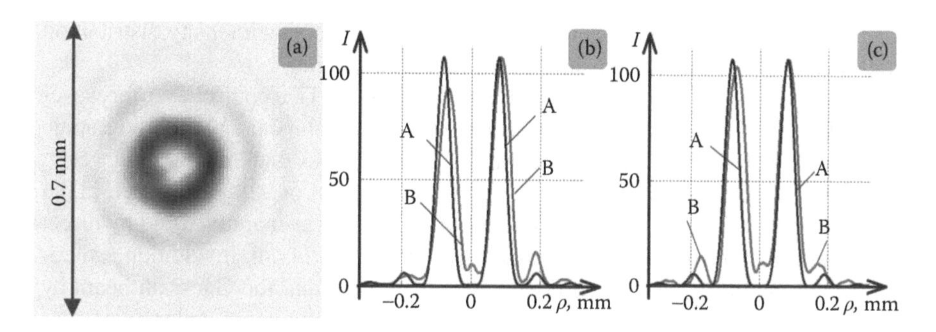

FIGURE 1.4 The Fraunhofer diffraction (negative) of a plane wave of radius 1.25 mm and wavelength 0.633 μm by an SPP with $n = 3$, generated at the focal plane of a Fourier lens of focus 150 mm: (a) intensity distribution (negative) and intensity cross-sections; (b) horizontal, and (c) vertical (A – theory, B – experiment).

Figure 1.4 shows the Fraunhofer diffraction pattern registered with a CCD camera at a lens focus ($f = 150$ mm) when a plane wave of radius 1.25 mm and wavelength 0.633 μm is diffracted by an SPP with $n = 3$. The root-mean-square deviation of the theoretical and experimental curves in Figure 1.4b, c was found to be 14.3%.

In conclusion, we have derived an analytical expression to describe the paraxial diffraction of a finite with circular and ring-like diaphragms plane wave by an SPP. We have also derived formulae for the radius and intensity of the vortex first ring. The Fresnel and Fraunhofer diffraction patterns have been produced experimentally using a high-quality SPP with $n = 3$. The root-mean-square deviation of the theory and experiment was no more than 15%.

1.2 DIFFRACTION OF CONIC AND GAUSSIAN BEAMS BY A SPIRAL PHASE PLATE

The SPP as an optical element with its transmittance being proportional to $\exp(in\varphi)$, where φ is the polar angle and n is the integer, was manufactured and analyzed for the first time in Ref. [24]. Currently, an interest in SPPs has increased, especially because of new opportunities for optical microparticle manipulation [27–30,33,37]. In Ref. [27], a SPP with $n = 3$ of several millimeters in diameter and maximal microrelief height of 5 μm was manufactured using mould technology and characterized for wavelength of 831 nm. The fabrication accuracy of the SPP surface relief on the polymer was very high (the error ~3%). In Ref. [28], with the conventional SEM converted into an electronic lithographer a SPP with $n = 1$, continuous relief profile (500 μm in diameter) and maximal step height of 1.04 μm for a He-Ne laser of wavelength 633 nm was recorded by direct e-beam writing mode on a negative photoresist SU-8. The SSP-produced Gaussian beam diffraction pattern was found to differ from the ideal "doughnut" shape by only 10%. The SPP manufactured in this way was used for simultaneous optical trapping of 6 latex microspheres, each of which has a diameter of 3 μm and the refractive index $n' = 1.59$. The same authors showed in Ref. [29] that when the SPP center was displaced from the Gaussian beam axis then an off-axis vortex was formed, with the cross-section intensity distribution rotating around the optical axis upon the beam propagation.

In Ref. [30] a 16-level and 16-sector SPP for impulse Ti sapphire laser for wavelength 789 nm was manufactured and studied using a standard photolithographic technique with four binary amplitude masks. The SPP was manufactured on a SiO_2 substrate of diameter 100 mm and a maximal step height of 928 nm. In Ref. [33] a SPP of diameter 2.5 mm and a relief depth of 1082 nm was manufactured by direct e-beam writing on a negative photoresist for wavelength 514 nm. In addition, a theoretical analysis of Fresnel diffraction of the plane wave and the Gaussian beam by SPP was done in Ref. [33]. In Ref. [37] a SPP with large singularity order $n = 80$ was formed using a liquid-crystal spatial light modulator (SLM) and double-frequency Neodymium laser of wavelength 532 nm. Also, in Ref. [37] analytical expressions for the Gaussian beam Fraunhofer diffraction by SPPs with large-order singularities, $n \gg 1$ were obtained.

The SPP can be used as a spatial filter to implement the generalized Hilbert transform [24] and to increase contrast when viewing phase objects under microscope [39]. In Ref. [40] the results of computer simulation of the vector diffraction of linearly polarized light by a SPP were reported. In particular, the intensity distribution in the focus of a spherical lens of near-unit numerical aperture was numerically calculated. It was shown that the intensity distribution on the ring becomes asymmetrical because of polarization effects; this is indeed a general property of high-numerical-aperture focusing systems.

Alongside studies concerning diffraction of a plane wave and a Gaussian beam by the SPP, the diffraction of a conic wave by a SPP was also considered. In Ref. [41] an optical element of transmittance proportional to the product of the transmittances of an axicon and a SPP was manufactured on a low-contrast photoresist using grayscale photolithography. This element has been called a Trochoson (i.e., an element forming a light tube) in Ref. [41]. Such an optical element is also referred to as a helical axicon (HA) [42]. Plane-wave diffraction by such an element is equivalent to the conic wave diffraction by the SPP. In Ref. [31] optical trapping and rotation (with a period of two seconds) of yeast particles and polystyrene beads of diameter 5 μm was performed with a He-Ne laser and a 16-level HA of the fifth order and diameter 6 mm, manufactured by direct e-beam writing. In Ref. [43] helico-conical optical beams with helical (instead of ring-like) intensity distributions were generated with a liquid-crystal SLM.

In this section, an analytical expression is derived which describes the wave field at the focal plane of an ideal diffraction-unlimited positive Fourier transform lens illuminated by a monochromatic helico-conical wave. Analysis of the derived expression shows that a diffraction pattern in the form of a light ring of infinite intensity is formed when the plane wave is diffracted by the HA. The radial intensity distribution near the diffraction pattern center is proportional to r^{2n}, where r is the radial coordinate and n is the SPP's singularity order. When r is large, the intensity beyond the ring decreases proportionally to n^2r^{-4}. An analytical expression for the Gaussian-beam Fraunhofer diffraction by the SPP was obtained in Refs. [33,35]. In this section, we show that the Gaussian beam also forms a ring in the far-field diffraction pattern after diffraction by the SPP. It is also shown that when r is small (inside the ring) then the intensity function changes proportionally to $(wr)^{2n}$, where w is the Gaussian beam waist radius, and when r is large (outside the ring) the intensity decreases similarly to the case of a HA, proportionally to n^2r^{-4}.

Also, in this section we conduct experimental studies on SPP-aided generation of a high-quality tube-shaped beam with ring-like intensity distribution. The SPP was manufactured on a resist by direct e-beam writing. Distinctly from Ref. [28], the technology used does not impose strong restrictions on the SPP diameter. Also, distinctly from Ref. [28], a four-order DOE was manufactured for simultaneous manipulation of several microparticles. This DOE forms simultaneously four light beams with ring intensity distributions and phase singularities of the ±3 and ±7 orders, propagating at different angles with respect to the optical axis. The transmittance of such a DOE is proportional to a linear combination of four angular harmonics, each of which is the transmittance of the SPP of the corresponding order.

Fourier Transform of Helico-Conical-Wave

The transmittance of the HA [41] has the following form:

$$f_n(r,\theta) = \exp(i\alpha r + in\theta), \quad n = 0, \pm1, \pm2, ..., \tag{1.11}$$

where α is an axicon's parameter, which is related to the angle v at the cone apex, the refractive index n', and the wavelength λ by the following equation:

$$\alpha = \frac{2\pi(n'-1)}{\lambda \tan(v/2)}. \tag{1.12}$$

Equation (1.12) is valid for the paraxial approximation. When the SPP is illuminated by a conic wave with apex angle of 2β, the function (1.11) can also be considered as the complex amplitude of light immediately after the SPP, with the transmittance given by

$$g_n(r,\theta) = \exp(in\theta), \quad n = 0, \pm1, \pm2, ... \tag{1.13}$$

In this case, the conic wave parameter α will be given by

$$\alpha = \frac{2\pi \sin\beta}{\lambda}. \tag{1.14}$$

The spatial spectrum of an unlimited plane wave diffracted by a HA or the diffraction of a conic wave by a SPP is described by the Fourier transform of the function (1.11). In polar coordinates, the far-field complex light amplitude will take the form

$$F_n(\rho,\phi) = \frac{-ik}{2\pi f}\int_0^{\infty}\int_0^{2\pi} \exp\left[i\alpha r + in\theta - \frac{ik}{f} r\rho\cos(\theta - \phi)\right] r dr d\theta, \tag{1.15}$$

where:

$k = 2\pi/\lambda$ is the wave number

f is the focal length of the spherical lens, in the rear focal plane in which the spatial spectrum pattern is generated

Transforming Eq. (1.15) yields

$$F_n(\rho,\phi) = \frac{(-i)^{n+2}k}{f}\exp(in\phi)\frac{d}{d\alpha}\int_0^\infty \exp(i\alpha r)J_n\left(\frac{k}{f}r\rho\right)dr, \tag{1.16}$$

where $J_n(x)$ is a Bessel function of the nth order and first kind. Using the reference integral [38] the following expression can be obtained:

$$\int_0^\infty \exp(ibx)J_n(cx)dx = \begin{cases} \dfrac{\exp\left[in\arcsin(b/c)\right]}{\left(c^2-b^2\right)^{1/2}}, & b < c, \\ \dfrac{c^n i^{n+1}}{\left[b+(b^2-c^2)^{1/2}\right]^n (b^2-c^2)^{1/2}}, & b > c. \end{cases} \tag{1.17}$$

By applying Eqs. (1.17) to (1.16) we obtain an explicit expression for the complex amplitude of spatial spectrum of the conic wave diffracted by the SPP:

$$F_n(\rho,\phi) = \frac{-k}{f}\exp(in\phi)\begin{cases} \dfrac{\bar{\rho}^n\left(\alpha + in\sqrt{\bar{\rho}^2-\alpha^2}\right)}{\left(\bar{\rho}^2-\alpha^2\right)^{3/2}\left(\alpha + i\sqrt{\bar{\rho}^2-\alpha^2}\right)^n}, & \alpha < \bar{\rho}, \\ \dfrac{-i\bar{\rho}^n\left(\alpha + n\sqrt{\alpha^2-\bar{\rho}^2}\right)}{\left(\alpha^2-\bar{\rho}^2\right)^{3/2}\left(\alpha + \sqrt{\alpha^2-\bar{\rho}^2}\right)^n}, & \alpha > \bar{\rho}, \end{cases} \tag{1.18}$$

where $\bar{\rho} = k\rho/f$. From Eq. (1.18) it follows that when $n = 0$, the expression for spatial spectrum of a plane wave diffracted by a simple axicon takes the form

$$F_0(\rho) = \frac{-k}{f}\begin{cases} \dfrac{\alpha}{\left(\bar{\rho}^2-\alpha^2\right)^{3/2}}, & \alpha < \bar{\rho}, \\ \dfrac{-i\alpha}{\left(\alpha^2-\bar{\rho}^2\right)^{3/2}}, & \alpha > \bar{\rho}. \end{cases} \tag{1.19}$$

From Eqs. (1.18) and (1.19) it can be seen that the amplitude $F_n(\rho, \varphi)$ has a singularity when $\alpha = \bar{\rho}$. This means that when the distance from the Fourier plane center equals

$\rho_0 = \alpha f/k$, a light ring with infinite energy density but finite width is being formed. The width of the ring with infinite intensity means the radial distance between two points at which the intensity equals, for example, one (if the intensity is dimensionless). From Eq. (1.18), the intensity is given by

$$\bar{I}_n(\rho) = |F_n(\rho,\phi)|^2 = \left(\frac{k}{f}\right)^2 \begin{cases} \dfrac{\alpha^2 + n^2(\bar{\rho}^2 - \alpha^2)}{\left(\bar{\rho}^2 - \alpha^2\right)^3}, & \alpha < \bar{\rho}, \\[2ex] \dfrac{\bar{\rho}^n\left(\alpha + n\sqrt{\alpha^2 - \bar{\rho}^2}\right)^2}{\left(\alpha^2 - \bar{\rho}^2\right)^3\left(\alpha + \sqrt{\alpha^2 - \bar{\rho}^2}\right)^{2n}}, & \alpha > \bar{\rho}. \end{cases} \tag{1.20}$$

To obtain the width of the ring, we assume that the intensity equals $\bar{I}_0$ and find from Eq. (1.20) (when $n = 0$) two numbers ρ_1 and ρ_2, which are the internal and external ring radii:

$$\rho_{1,2} = \frac{f}{k}\left[\alpha^2 \pm \left(\frac{\alpha k}{\sqrt{\bar{I}_0} f}\right)^{2/3}\right]^{1/2}. \tag{1.21}$$

The full width of the ring with respect to the intensity drop $\bar{I}_0(\rho) = \bar{I}_0$ equals $\rho_1 - \rho_2$ and increases with the growth of the axicon parameter α. Let us note that when $\alpha < (k/f)^{1/2}$ the internal ring radius becomes zero or negative. This is because on the optical axis, where $\rho = 0$, the intensity becomes larger than one ($\bar{I}_0(\rho = 0) > \bar{I}_0$). Indeed, when $\rho = 0$, it follows from Eq. (1.20) that

$$\bar{I}_n(\rho = 0) = \begin{cases} 0, & n \neq 0, \\[1ex] \left(\dfrac{k}{f\alpha^2}\right)^2, & n = 0. \end{cases} \tag{1.22}$$

It is seen from Eq. (1.22) that for a HA ($n \neq 0$) a point of zero intensity, $\bar{I}_n(0) = 0$, will be found in the center of the Fourier plane ($\rho = 0$) at any α.

Let us find a function describing the intensity on the internal side of the ring when ρ tends to zero. From Eq. (1.20) at $\alpha \gg \bar{\rho}$ we obtain the following expression:

$$\bar{I}_n(\rho) = \left(\frac{k}{f}\right)^2 \left(\frac{\bar{\rho}}{\alpha}\right)^{2n} \frac{(n+1)^2}{4^n \alpha^4}. \tag{1.23}$$

It is seen from Eq. (1.23) that if both α and $\bar{\rho}$ tend to zero, while their ratio $\alpha/\bar{\rho}$ remains constant, we obtain

$$\bar{I}_n(\bar{\rho} \to 0, \alpha \to 0) \approx \alpha^{-4}. \tag{1.24}$$

From Eq. (1.24) it follows that near the center of the Fourier plane $\bar{\rho} = 0$ the intensity tends to infinity when $\alpha \to 0$, but in the central point itself ($\rho = 0$) the intensity will remain zero, $\bar{I}_n(\bar{\rho} = 0) = 0$, at any small value of α.

It is also seen from Eq. (1.23) that the near-center intensity in the Fourier plane will tend to infinity proportionally to the 2nth power of the radius:

$$\bar{I}_n(\bar{\rho}) \approx \left(\frac{\bar{\rho}}{\alpha}\right)^{2n}, \quad \bar{\rho} \ll \alpha. \tag{1.25}$$

Furthermore, it follows from Eq. (1.20) that on the outer side of the ring, when $\alpha \ll \bar{\rho}$, the intensity function decreases with growing radial variable ρ as

$$\bar{I}_n(\rho) \approx \left(\frac{fn}{k\rho^2}\right)^2. \tag{1.26}$$

It is seen from Eq. (1.26) that when ρ is large the behavior of the intensity function is independent on α. In particular, when $\alpha = 0$ and a plane wave illuminates the SPP, it follows from Eq. (1.26) that

$$\bar{I}_n(\bar{\rho}) = \begin{cases} 0, & n = 0, \\ \left(\dfrac{kn}{f\bar{\rho}^2}\right)^2, & n \neq 0. \end{cases} \tag{1.27}$$

The intensity distributions on the ring's internal and external sides for some different values of index n are illustrated in Figure 1.5a.

Shown in Figure 1.5b, c are simulated patterns of Fraunhofer diffraction of a conic wave of finite radius R by an SPP. The intensity distributions I calculated using the Fourier transform are plotted against the radial variable $\bar{\rho}$. It is seen from Figure 1.5b that with increasing SPP number n there takes place an insignificant decrease in the maximal intensity value in the main ring (at $\bar{\rho} = 2$ mm^{-1}) due to an insignificant widening of the ring, and an insignificant shift of intensity maximum toward larger values of the radial variable $\bar{\rho}$. With n changing from 0 to 10, the ring radius is increased by about 7%. This might be suggested by plots in Figure 1.5a, which show that for an infinite-radius conic beam an increase in number n also leads to a shift of the ring's external-intensity curves toward larger values of $\bar{\rho}$.

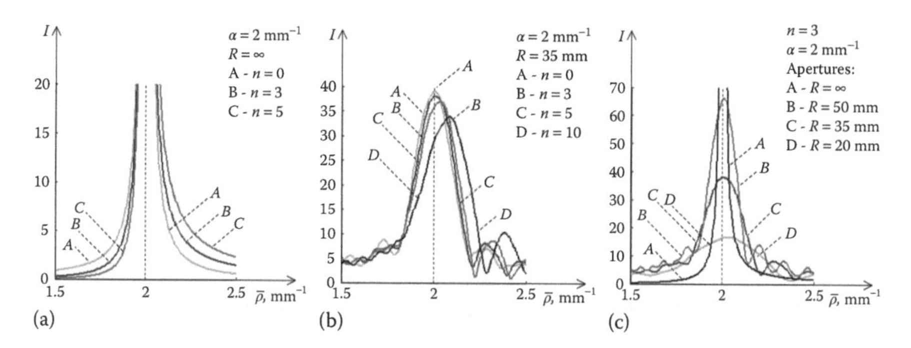

FIGURE 1.5 Radial intensity distribution in spatial spectrum plane at conic wave diffraction on SPP: (a) aperture is infinite, different orders of singularities n: 0 (curve A), 3 (curve B), and 5 (curve C); (b) aperture is finite ($R=35$ mm), different orders of singularities n: 0, 3, 5, 10 (curves A, B, C, and D); (c) singularity order $n=3$, different apertures—infinite (curve A), $R=50$ mm, 35 mm, 20 mm (curves B, C, and D).

From Figure 1.5c it is seen that with n held constant, a decrease in radius R of the conic wave causes the main ring of diffraction pattern to widen and the maximal value of the ring intensity to decrease. Note that the ring radius, that is, the distance of the ring maximum to the center, which equals $\bar{\rho}=2$ mm^{-1}, remains unchanged. It is noteworthy that the narrowest intensity distribution is formed when the conic wave's radius is infinite.

Thus, the analytical expressions derived for diffraction of an unlimited conic wave by an SPP prove to be useful for analysis of a real situation when the conic wave's radius is limited.

In conclusion of this section, we will consider in detail the amplitude (1.19) as a function of the second-kind discontinuity at $\alpha=\bar{\rho}$. Inside the ring, when $\alpha>\bar{\rho}$, $F_0(\rho)$ in Eq. (1.19) is a real function and outside the ring, when $\alpha<\bar{\rho}$, the $F_0(\rho)$ function is purely imaginary. Thus, when the radius ρ intersects the point $\alpha=\bar{\rho}$ the $F_0(\rho)$ function acquires a phase shift of $-\pi/2$ radians. When $\alpha=\bar{\rho}$, the $F_0(\rho)$ function tends to infinity, and if we assume α to be equal to zero, $F_0(\rho)$ becomes the δ-function. It follows from Eq. (1.15), when $n=0$ and $\alpha=0$, that

$$F_0(\rho)=\frac{-ik}{f}\int_0^{\infty}J_0(r\bar{\rho})rdr=\frac{-ik\delta(\bar{\rho})}{f\bar{\rho}}. \tag{1.28}$$

Since the integral of the δ-function is equal to one, that is, $\int_{-\infty}^{+\infty}\delta(x)dx=1$, the function (1.28) yields

$$\int_0^{\infty}F_0(\rho)\rho d\rho=\frac{-if}{2k}. \tag{1.29}$$

To obtain an expression analogous to Eq. (1.29) from the explicit form of the $F_0(\rho)$ function in Eq. (1.29), we need to temporarily get rid of the divergence at $\alpha = \bar{\rho}$. This can be done by using the well-known integral [38]:

$$\int_0^\infty \exp\left[-(i\alpha+\beta)r\right]J_0(\bar{\rho}r)rdr = \frac{i\alpha+\beta}{(\bar{\rho}^2-\alpha^2+\beta^2+2i\alpha\beta)^{3/2}}. \tag{1.30}$$

It can be seen that the function in the right-hand side of Eq. (1.30) coincides with the right-hand side of Eq. (1.19) when $\beta=0$. The function in the right-hand side of Eq. (1.30) cannot have a zero in the denominator at any real values of α and $\bar{\rho}$. Using this idea lets us consider the following function:

$$F_0(\rho,\beta) = \frac{-\alpha k}{f\left(\bar{\rho}^2-\alpha^2+2i\alpha\beta\right)^{3/2}}. \tag{1.31}$$

When $\beta=0$ the function $F_0(\rho,\beta=0)$ coincides with the function (1.19). Let us take the integral of this function:

$$\int_0^\infty F_0(\rho,\beta)\rho d\rho = \frac{-if}{2k(1-2i\beta/\alpha)^{1/2}}. \tag{1.32}$$

Equation (1.32) clearly coincides with Eq. (1.29) when $\beta=0$. Note that Eq. (1.32) is correct for any α, even for $\alpha=0$. This proves that the function (1.31) tends to the δ-function when α approaches zero and when $\beta=0$. Although the integral of the amplitude (1.32) is finite, despite the second-kind discontinuity at the point $\alpha = \bar{\rho}$, the integral of the light field intensity (full energy) will be infinite. Actually, it follows from Eq. (1.31) that

$$\bar{I}_0(\rho,\beta) = \left|F_0(\rho,\beta)\right|^2 = \left(\frac{\alpha k}{f}\right)^2\left[(\bar{\rho}^2-\alpha^2)^2+4\alpha^2\beta^2\right]^{-3/2}. \tag{1.33}$$

The integral of the function (1.33) can be evaluated analytically and the result is

$$\int_0^\infty \bar{I}_0(\rho,\beta)\rho d\rho = \frac{1}{8\beta^2}\left[1+\left(1+4\frac{\beta^2}{\alpha^2}\right)^{-1/2}\right]. \tag{1.34}$$

It follows from Eq. (1.34) that when $\beta \to 0$ the integral of the intensity in Eq. (1.33) will tend to infinity independently of α.

Fraunhofer Diffraction of a Gaussian Beam on Spiral Phase Plate

Explicit analytical expressions that describe Fresnel diffraction of the Gaussian beam by a SPP were obtained in Refs. [33,35]. The expression for the Gaussian beam's Fraunhofer diffraction by the SPP was also obtained in Ref. [33] by passage to the limit from the Fresnel diffraction to the far-field diffraction. In this section, an analytical formula for description of the Gaussian beam's Fraunhofer diffraction by a SPP, formed in the focal plane of a spherical lens, is obtained. Some results reported in Refs. [33,35] are briefly recalled in order to compare the diffraction by the SPP of conical and Gaussian beams.

Let us consider the following source function, instead of Eq. (1.11):

$$f_n'(r,\theta) = \exp\left(-\frac{r^2}{w^2} + in\theta\right), \tag{1.35}$$

where w is the waist radius of the incident Gaussian beam. Then, the complex amplitude of the Fraunhofer diffraction of the Gaussian beam by the SPP is given by

$$F_n'(\rho,\phi) = \frac{(-i)^{n+1}k}{f}\exp(in\phi)\int_0^\infty \exp\left(-\frac{r^2}{w^2}\right)J_n\left(\frac{k}{f}r\rho\right)rdr. \tag{1.36}$$

The following reference integral is well-known [35]:

$$\int_0^\infty \exp(-px^2)J_n(cx)xdx = \frac{c\sqrt{\pi}}{8p^{3/2}}\exp\left(-\frac{c^2}{8p}\right)\left[I_{(n-1)/2}\left(\frac{c^2}{8p}\right) - I_{(n+1)/2}\left(\frac{c^2}{8p}\right)\right]. \tag{1.37}$$

Here $I_\nu(x)$ is a modified Bessel function or a Bessel function of the second kind. Considering Eq. (1.37), the expression (1.36) can be rewritten as

$$F_n'(\rho,\phi) = (-i)^{n+1}\exp(in\phi)\left(\frac{kw^2}{4f}\right)\sqrt{2\pi x}\exp(-x)\left[I_{(n-1)/2}(x) - I_{(n+1)/2}(x)\right], \tag{1.38}$$

where $x = [kw\rho/(2f)]^2/2$. The intensity function of the Fraunhofer diffraction pattern of the Gaussian beam by the SPP takes the form

$$\bar{I}_n'(\rho) = \left|F_n'(\rho,\phi)\right|^2 = 2\pi\left(\frac{kw^2}{4f}\right)^2 x\exp(-2x)\left[I_{(n-1)/2}(x) - I_{(n+1)/2}\right]^2. \tag{1.39}$$

It is seen from Eq. (1.39) that when $x = 0$ the intensity equals zero (if $n \neq 0$) in the center of the Fourier plane: $\bar{I}_n'(0) = 0$. The factors $x\exp(-2x)$ in Eq. (1.39) show that for the far-field diffraction a circular intensity distribution is formed. The radius of the ring can be obtained from [33]:

$$(n-4x)I_{(n-1)/2}(x)+(n+4x)I_{(n+1)/2}(x)=0. \tag{1.40}$$

Let us find the intensity function on the external side of the ring when $\rho \to \infty$ (or $x \to \infty$). To do this we use the asymptotic form of the Bessel function:

$$I_\nu(x) \approx \frac{\exp(x)}{\sqrt{2\pi x}}\left(1-\frac{4\nu^2-1}{8x}\right), \quad x \gg 1. \tag{1.41}$$

So, when $x \to \infty$, we obtain instead of Eq. (1.39) the result

$$\bar{I}_n'(\rho) \approx \left(\frac{nf}{k\rho^2}\right)^2. \tag{1.42}$$

It is interesting that Eq. (1.42) does not depend on the Gaussian beam's waist and that it coincides with Eq. (1.26). From this coincidence it can be concluded that for $\rho \to \infty$, the asymptotics of the intensity function are determined only by the SPP number, the focal length of the spherical lens, and the light wavelength; they do not depend on the amplitude-phase parameters α and w of the beam illuminating the SPP.

Let us note that the expression (1.42) can be obtained from Eq. (1.39) by extending the Gaussian beam's waist radius to infinity ($w \to \infty$), while ρ remains fixed.

Let us next find the intensity function inside the ring. When ρ tends to zero (with w fixed), the Bessel function argument x tends to zero also, and the first members of the cylinder function's decomposition into a series can be used:

$$I_\nu(x) \approx \left(\frac{x}{2}\right)^\nu \Gamma^{-1}(\nu+1), \quad x \ll 1, \tag{1.43}$$

where $\Gamma(x)$ is a gamma-function. So, instead of Eq. (1.39) we obtain, when $\rho \to 0$,

$$\bar{I}_n'(\rho) \approx \pi\Gamma^{-2}\left(\tfrac{n+1}{2}\right)\left(\frac{kw^2}{f}\right)\left(\frac{kw\rho}{4f}\right)^{2n}. \tag{1.44}$$

It is seen from Eq. (1.44) that the intensity near the center of the Fourier plane increases as the $2n$th power of the radial coordinate:

$$\bar{I}_n'(\rho) \approx (w\rho)^{2n}, \quad \rho \ll 1 \tag{1.45}$$

The form of the asymptotics (1.45) coincides with Eq. (1.25) for the diffraction of the conic wave by the SPP, if w is replaced by $1/\alpha$.

If ρ tends to zero, while the Gaussian beam's radius w tends to infinity, with their product $w\rho$ remaining constant, Eq. (1.44) suggests that the intensity near the center of the Fourier plane will tend to infinity proportionally to the squared waist radius:

$$\overline{I}_n'(\rho \to 0, w \to \infty) \approx w^2, \quad \rho w = \text{const}, \tag{1.46}$$

but at the central point itself (when $\rho = 0$) the intensity will be zero, $\overline{I}_n'(\rho = 0) = 0$, at any value of w.

Experimental Studies of Light Diffraction by a Spiral Phase Plate

Fabrication and experiments with SPPs obtained via modern microlithography technology have been reported elsewhere [24,27–30]. However, all these technologies have drawbacks of their own. For example, in Ref. [28] a SPP was fabricated using a conventional scanning electron microscope, specially adapted for direct writing on a resist. However, the control system of the microscope e-beam does not allow one to write on the resist high-quality patterns of diameter larger than 500 μm. The photolithography technology with use of several photomasks, which was used in Ref. [30] for SPP fabrication, involves a critical operation of photomask alignment, resulting in additional errors in microrelief fabrication. Use of a conventional e-beam lithographer in the direct e-beam writing mode on a resist for SPP fabrication is free of the listed drawbacks.

In this section, the light field with the second-order singularity was generated with a 32-level SPP. The element size is 2.5×2.5 mm^2 and the pixel size is 5×5 μm^2. The SPPs were intended to operate at wavelength $\lambda = 633$ nm. The microrelief depth measured with a contact profilometer is 1320 nm. The optimal depth of the 32-level microrelief is 1341 nm if we assume that the refractive index of the resist is $n_r = 1.457$ (the exact value is not known). Thus, the deviation from the intended depth is only about 1%. In Ref. [33] a SPP was fabricated using the same technology but for wavelength of 514 nm, with wavelength 543 nm used in experiments. This led to a low quality of tubular beam generation. In the present paper, the same wavelength 633 nm of a He-Ne laser was used both for designing the SPP and in experiments. Because of this, the intensity distribution of the generated beam really possesses a ring symmetry. Figure 1.6 shows the experimental results of generating the laser field with the second-order phase singularity. Figure 1.6a illustrates the theoretical phase distribution (white denotes the zero phase and black shows the phase $2\pi(1 - 1/N)$, N being the number of quantization levels). Figure 1.6b shows the SPP microrelief derived using the NewView 5000 Zygo interferometer (200× magnification, tilted view). Figure 1.6c–e show the intensity distributions registered with a CCD camera at different distances from the SPP. Figure 1.6f depicts the Fraunhofer diffraction pattern obtained with use of a spherical lens of focus $f = 135$ mm when a plane wave is diffracted by the above-specified SPP.

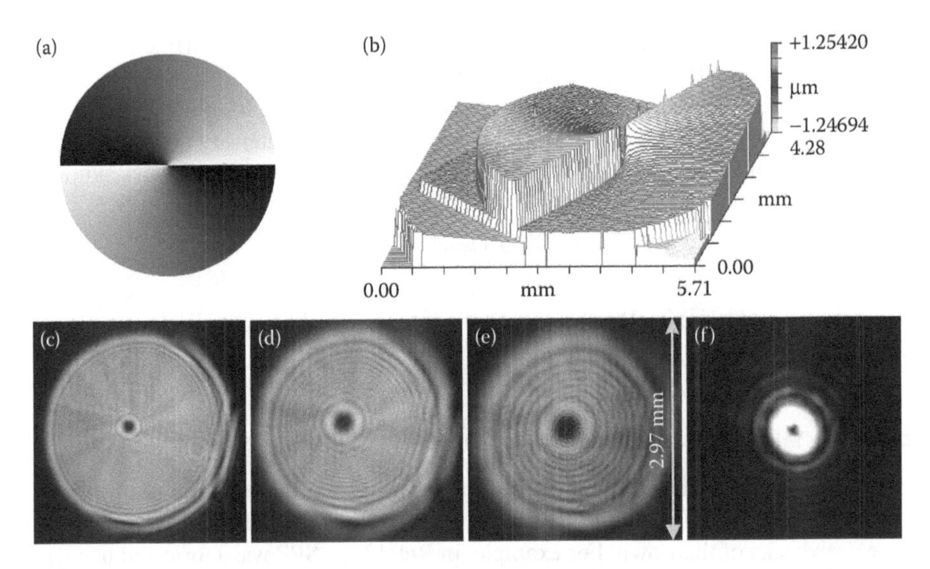

FIGURE 1.6 Generation of a laser field with a second-order phase singularity. (a) Desired phase distribution. (b) Central part of the SPP microrelief. Shown also are the field intensity distributions registered by a CCD camera at different distances from the SPP: (c) $z = 10$ mm, (d) $z = 40$ mm, (e) $z = 80$ mm, and (f) at the focal plane of the lens.

Figure 1.7 shows experimental intensity distributions in the focal plane of a spherical lens of focus $f = 135$ mm. The annular intensity distribution in Figure 1.7a was obtained when a plane wave of radius $R = 2.5$ mm was diffracted by a diffraction axicon with parameter $\alpha = 44.5\ \text{mm}^{-1}$ and an SPP with number $n = 5$. The wavelength of light was $\lambda = 0.633$ μm of a 1 mW laser beam. A second, larger ring (Figure 1.7a) appears because the maximal relief height fails to produce an exact phase delay of $2\pi m$. The annular intensity distribution in Figure 1.7b was derived when a Gaussian beam of waist radius $\sigma = 0.8$ mm was diffracted by an SPP with number $n = 2$. Some ellipticity of the diffraction pattern is due to inaccurate alignment of the Gaussian

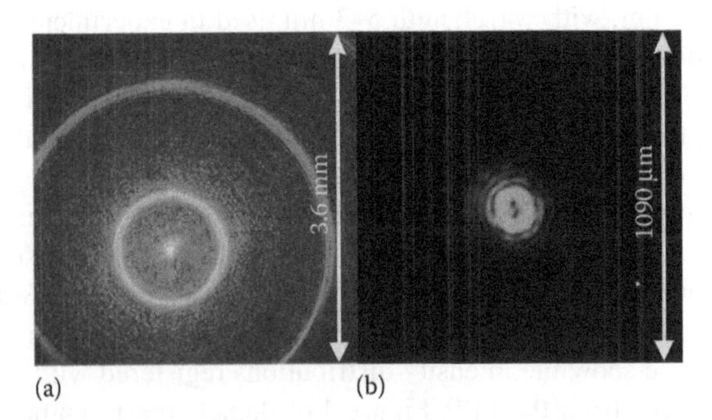

FIGURE 1.7 Fraunhofer diffraction pattern for (a) a conic wave (diffraction axicon's parameter: $\alpha = 44.5\ \text{mm}^{-1}$) and (b) a Gaussian beam (waist radius: $\sigma = 0.8$ mm) diffracted by an SPP with number (a) $n = 5$ and (b) $n = 2$.

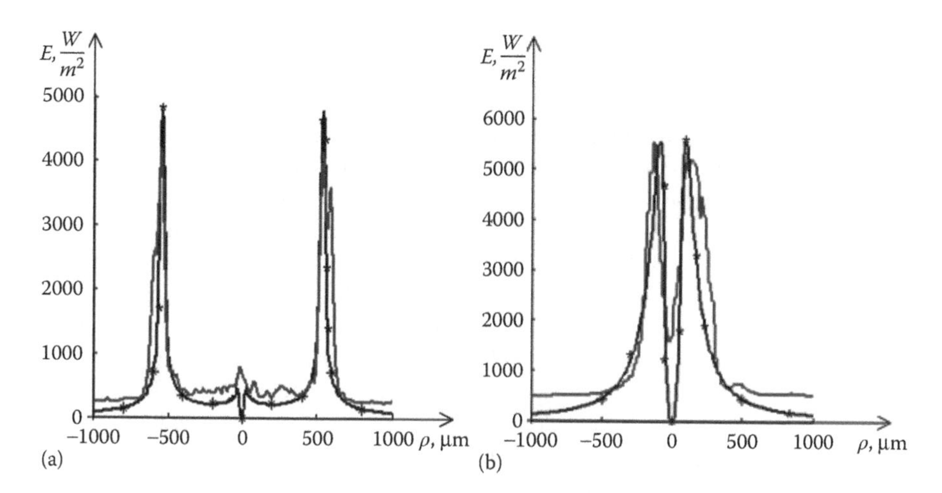

FIGURE 1.8 Diffraction of (a) a conic wave and (b) a Gaussian beam by an SPP: ---- experimental and —*—* theoretical intensity distribution.

beam and the SPP center. Note that there is no such problem when a plane wave of limited radius is diffracted by the SPP (Figure 1.6f).

Figure 1.8 compares theoretical and experimental profiles of the annular intensity distributions shown in Figure 1.7. The ring's radius in Figure 1.8a approximately equals the estimate in Eq. (1.18), $\rho_0 = \alpha f/k = 605$ μm, while the ring width can be derived from $\Delta\rho = 2\lambda f/R \approx 68$ μm. The radius of the ring in Figure 1.8b can be derived from a relation in Ref. [33]: $\rho_2 = 0.46\,\lambda f/\sigma = 45$ μm. It is seen from Figure 1.8a and b that the experimental and theoretical curves agree fairly well.

In this section the following results have been obtained. An analytical expression for spatial spectrum of the conic wave diffracted by a SPP with arbitrary integer singularity order n has been obtained. The diffraction of the conic wave by SPP is equivalent to the diffraction of the plane wave by a HA. An analytical expression for the Gaussian beam's Fraunhofer diffraction by a SPP located in the beam's waist has been obtained. Diffraction of the conic wave and that of the Gaussian beam by the SPP have been compared analytically. It has been shown that in both cases a light ring is formed, with the intensity function increasing proportionally to ρ^{2n} at small values of the radial variable ρ and decreasing as $n^2\rho^{-4}$ at large values of ρ.

1.3 DIFFRACTION OF A FINITE-RADIUS PLANE WAVE AND A GAUSSIAN BEAM BY A HELICAL AXICON AND A SPIRAL PHASE PLATE

The first time, the HA [41] and the SPP [24] were fabricated by photolithography and experimentally studied in 1992. The HA is used to generate diffraction-free laser Bessel beams, whereas the SPP can generate optical vortices, perform the radial Hilbert transform [24,44,45], and be used for phase microscopy [46,47] and astronomy [21]. Recent years have seen an increase of interest in the HA and

SPP [26,28,33,48–63]. This is due to the fact that the improved quality of fabrication of SLMs has made them suitable for generating diffractive optical elements (DOEs), including HA and SPP. For example, in Ref. [26] the higher-order SPPs ($n > 30$) were generated using the SLM and the higher-order optical vortices were studied. Using the SLM enables producing a composite SPP to generate a laser beam composed of several coaxial optical vortices [48]. Also, the diffraction-free Bessel beams [49–53,63], elliptic Bessel beams [54], Ince-Gaussian beams [55], and hollow beams [56] were generated using the SLM. SLMs are used for dynamic transformation of laser beams in order to eliminate aberrations and equalize intensity [64–67].

On the other hand, studies of the HA and SPP fabricated through the traditional e-beam lithography have been in progress [33,57–59,68]. Diffraction of a plane wave by a second- and third-order SPP was experimentally studied in Refs. [33,57]. In Ref. [58] a fifth-order HA was studied, and a double axicon to generate two conical light beams that mutually interfere to produce zero axial intensity was presented in Ref. [59].

The theoretical analysis of the paraxial Fresnel and Fraunhofer diffraction by the SPP was conducted for the incident Gaussian beam [16], unbounded plane wave [24,33], bounded plane wave [26,57], and elliptical beam [60]. Diffraction by the HA was theoretically studied for the unbounded plane wave [58] and the Gaussian beam [61]. In Ref. [69] diffraction of polychromatic light on SPP has been studied. Interest in the studies of the HA and SPP is also due to the potential for optical micromanipulation they show [26,28,58,66,68].

In this section, we conduct theoretical analysis of the paraxial Fraunhofer diffraction of a finite-radius plane wave by the HA and the SPP, as well as of the paraxial Fraunhofer and Fresnel diffraction of a Gaussian beam by the HA and the SPP. For the analysis of the diffraction of a plane wave new analytical expressions for the light complex amplitude have been derived as a series of the Bessel functions (for the HA) and a finite sum of the Bessel functions (for the SPP). The number of terms in this sum is $(n+1)$, where n is the SPP number. A characteristic feature of these relationships is that they describe Fraunhofer diffraction pattern of a plane, finite-radius wave by the HA using different analytical relations inside and outside the bright ring of a certain radius, which equals the maximal-intensity ring of lower-order SPPs. It is also interesting that in the diffraction pattern the spacing between the inside sub-rings found within the major bright ring is almost half as large as that between the outside rings. The Fraunhofer diffraction pattern of a plane, finite-radius wave by the SPP is described by different relationships for the even and odd numbers of n. For the analysis of the diffraction of a Gaussian beam new analytical expressions for the light complex amplitude have been derived as a series of the hypergeometric functions.

Also, we have fabricated by photolithography a binary DOE (a HA with number $n = 10$) able to produce in the focal plane of a spherical lens two bright rings (optical vortices), each having 40% efficiency. One of the rings was then used to perform rotation of several polystyrene beads of diameter 5 μm.

Description of the Fraunhofer Diffraction of the Limited Plane Wave by a Helical Axicon

Let us consider a scalar paraxial diffraction of a plane wave of finite radius R and complex amplitude given by

$$E_0(r) = \mathrm{circl}\left(\frac{r}{R}\right) = \begin{cases} 1, r \le R, \\ 0, r > R, \end{cases} \tag{1.47}$$

by a HA, whose transmittance in the thin transparency approximation is given by

$$\tau_n(r,\phi) = \exp(i\alpha r + in\phi), \tag{1.48}$$

where:

(r,ϕ) are the polar coordinates in the HA plane at $z = 0$
z is the optical axis
α is the axicon parameter
$n = 0, \pm 1, \pm 2, \ldots$ is the SPP number

Then, the complex amplitude of light $F_n(\rho,\theta)$ in the focal plane of an ideal spherical lens of a focus f takes the form (omitting the insignificant pre-factor $\exp(ikf)$):

$$\begin{aligned} E_n(\rho,\theta) &= -\frac{ik}{2\pi f}\int_0^R\int_0^{2\pi} \exp\left[i\alpha r + in\phi - \frac{ik}{f} r\rho\cos(\phi-\theta)\right] r\mathrm{d}r\mathrm{d}\phi \\ &= (-i)^{n+1}\frac{k}{f}\exp(in\theta)\sum_{m=0}^{\infty}\frac{(i\alpha)^m}{m!}\int_0^R r^{m+1} J_n\left(\frac{k}{f}\rho r\right)\mathrm{d}r, \end{aligned} \tag{1.49}$$

where:

(ρ,θ) are the polar coordinates in the Fraunhofer diffraction plane
$k = 2\pi/\lambda$ is the wavenumber
$J_n(x)$ is the Bessel function.

Using the reference integral [38]

$$\int_0^R x^\lambda J_\nu(ax)\,dx = \frac{(aR)^{\nu+\lambda+1}}{a^{\lambda+1}2^\nu(\nu+\lambda+1)\nu!}\,{}_1F_2\left[\frac{\nu+\lambda+1}{2},\frac{\nu+\lambda+3}{2},\nu+1,-\left(\frac{aR}{2}\right)^2\right], \quad (1.50)$$

we obtain, instead of Eq. (1.49):

$$\begin{aligned} E_n(\rho,\theta) = {} & \frac{(-i)^{n+1}\exp(in\theta)}{n!}\left(\frac{kR^2}{f}\right)\left(\frac{kR\rho}{2f}\right)^n \\ & \times\sum_{m=0}^{\infty}\frac{(i\alpha R)^m}{(m+n+2)m!}\,{}_1F_2\left[\frac{m+n+2}{2},\frac{m+n+4}{2},n+1,-\left(\frac{kR\rho}{2f}\right)^2\right], \end{aligned} \quad (1.51)$$

where ${}_1F_2(a,b,c,x)$ is a hypergeometric function:

$${}_1F_2(a,b,c,x) = \sum_{m=0}^{\infty}\frac{(a)_m x^m}{(b)_m (c)_m m!}, \quad (1.52)$$

$(a)_m = \Gamma(a+m)/\Gamma(a)$, $(a)_0 = 1$, and $\Gamma(x)$ is the Gamma-function.

From Eq. (1.51) it follows that the diffraction pattern represents a set of concentric rings. At $\rho = 0$ and any $n \neq 0$, there will be zero intensity in the center of the diffraction pattern. Because the complex amplitude in Eq. (1.51) depends on the combination of variables $kR\rho/(2f)$, the radii ρ_l of the diffraction pattern local maxima and minima should satisfy the relation:

$$\rho_l = \frac{\lambda f \gamma_l}{\pi R}, \quad (1.53)$$

where γ_l are constants that depend only on the ring number $l = 1,2, \ldots$ of the diffraction pattern and the values of n and α.

At $\alpha = 0$ (no axicon), from Eq. (1.51) follows a relationship for the complex amplitude of Fraunhofer diffraction of a plane, finite-radius wave by the SPP [57]:

$$E_n(\rho,\theta) = \frac{(-i)^{n+1}\exp(in\theta)}{(n+2)n!}\left(\frac{kR^2}{f}\right)\left(\frac{kR\rho}{2f}\right)^n {}_1F_2\left[\frac{n+2}{2},\frac{n+4}{2},n+1,-\left(\frac{kR\rho}{2f}\right)^2\right]. \quad (1.54)$$

It would be interesting to compare the expression in Eq. (1.51) with the complex amplitude of the Fresnel diffraction of a plane, finite-radius wave by the SPP [57]:

$$E_n(\rho,\theta,z)=2\exp\left(\frac{iz_0\bar{\rho}^2}{z}+in\theta\right)\frac{\bar{\rho}^n}{n!}\left(\frac{-iz_0}{z}\right)^{n+1}$$
$$\times\sum_{m=0}^{\infty}\frac{\left(\frac{iz_0}{z}\right)^m}{(2m+n+2)m!}{}_1F_2\left[\frac{2m+n+2}{2},\frac{2m+n+4}{2},n+1,-\left(\frac{z_0\bar{\rho}}{z}\right)^2\right], \quad (1.55)$$

where:

$z_0=kR^2/2$

$\bar{\rho}=\rho/R$

The expressions in Eqs. (1.51) and (1.55) are similar in structure, describing a diffraction pattern composed of a set of concentric bright rings. Note that, just as in Eq. (1.51), out of the entire sum the first term is only retained when α tends to zero, likewise in Eq. (1.55) the only term is retained when z tends to infinity, being equal to the relation in Eq. (1.54) at $z=f$.

Shown in Figure 1.9 is the calculated amplitude distribution $|E_n(\rho,\theta)|$ in relative units as a function of the radial variable. This curve presents the radial profile of the Fraunhofer diffraction pattern when a plane wave of radius $R=2$ mm and wavelength $\lambda=633$ nm is diffracted by a HA with the parameter $\alpha=30$ mm^{-1} and $n=4$. The lens focus is $f=140$ mm. The vertical line is drawn at the point $\rho=\alpha f/k$. The maximal value of the amplitude $|E_n(\rho,\theta)|$ is near this line.

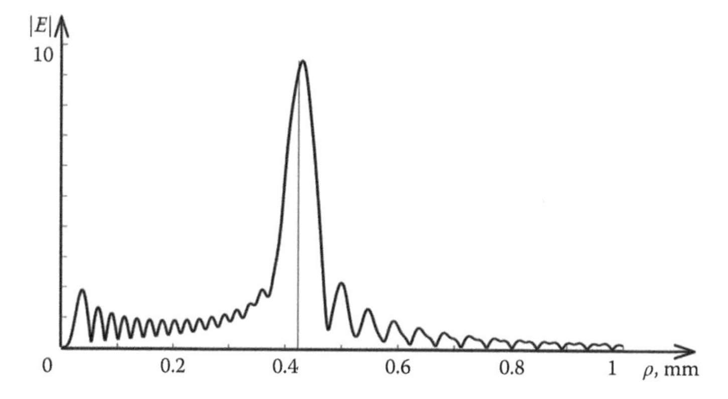

FIGURE 1.9 Radial profile of the Fraunhofer diffraction pattern (amplitude $|E_n(\rho,\theta)|$) for a plane, finite-radius wave diffracted by an HA.

Description of the Fraunhofer Diffraction by a Helical Axicon Using a Bessel Function Series

Instead of the series in Eq. (1.51) composed of the hypergeometric functions, the complex amplitude that describes the Fraunhofer diffraction of a plane wave of radius R by an HA can be derived as a series composed of the Bessel functions:

$$E_n(\rho,\theta)=$$
$$\frac{(-i)^{n+1}k\exp(in\theta)}{2\pi f}\left[(-i)^n\frac{\partial I_1^n}{\partial\alpha}-\exp(i\alpha R)\sum_{m=-\infty}^{+\infty}i^m\left(iRI_1^m+\frac{\partial I_1^m}{\partial\alpha}\right)J_{m+n}(R\bar{\rho})\right], \tag{1.56}$$

where:

$$I_1^n=\int_0^{2\pi}\frac{\exp(in\varphi)d\varphi}{\alpha+\bar{\rho}\cos\varphi}, \tag{1.57}$$

$\bar{\rho}=k\rho/f$, and $J_{m+n}(x)$ is the Bessel function. The integral I_1^n and the derivative $\partial I_1^n/\partial\alpha$ of Eq. (1.56) can be derived in the explicit form:

$$I_1^n=\begin{cases}\dfrac{2\pi\,\mathrm{sgn}\,\alpha}{\sqrt{\alpha^2-\bar{\rho}^2}}\chi^{|n|},0<\bar{\rho}<|\alpha|,\\[2ex]\dfrac{\pi i\left(\beta^{*|n|}-\beta^{|n|}\right)}{\sqrt{\bar{\rho}^2-\alpha^2}},\bar{\rho}>|\alpha|,\end{cases} \tag{1.58}$$

where:

$$\begin{cases}\beta=\dfrac{-\alpha+i\sqrt{\bar{\rho}^2-\alpha^2}}{\bar{\rho}},\\[2ex]\chi=\dfrac{-\alpha+\mathrm{sgn}\,\alpha\sqrt{\alpha^2-\bar{\rho}^2}}{\bar{\rho}},\end{cases} \tag{1.59}$$

$$\mathrm{sgn}\,\alpha=\begin{cases}1,\alpha>0,\\-1,\alpha<0,\end{cases}$$

$$\frac{\partial I_1^n}{\partial \alpha} = \begin{cases} -2\pi\chi^{|n|} \cdot \dfrac{\alpha \operatorname{sgn}\alpha + |n|\sqrt{\alpha^2 - \bar{\rho}^2}}{\left(\alpha^2 - \bar{\rho}^2\right)^{3/2}}, 0 < \bar{\rho} < |\alpha|, \\ \pi i \left[\dfrac{\alpha\left(\beta^{*|n|} - \beta^{|n|}\right)}{\left(\bar{\rho}^2 - \alpha^2\right)^{3/2}} - \dfrac{i|n|\left(\beta^{*|n|} + \beta^{|n|}\right)}{\bar{\rho}^2 - \alpha^2} \right], \bar{\rho} > \alpha. \end{cases} \tag{1.60}$$

As $R \to \infty$ the expression in Eq. (1.56) changes to equations that describe the diffraction of an unbounded plane wave by the HA, deduced in Ref. [56]. Equations (1.58) and (1.60) also suggest that different relationships describe the field $E_n(\rho,\theta)$ at $0 < \bar{\rho} < |\alpha|$ and at $\bar{\rho} > |\alpha|$. Differently looking diffraction patterns as $\bar{\rho}$ is changing from 0 to $|\alpha|$ and beyond $|\alpha|$ can also be seen in Figure 1.9. It should be noted that the spatial frequency of the diffraction pattern in Figure 1.9 on the interval $0.1 \text{ mm} < \rho < 0.3 \text{ mm}$ is approximately twice as large as that on the interval $0.5 \text{ mm} < \rho < 0.7 \text{ mm}$. It is also noteworthy that the global intensity maximum $I_n(\rho,\theta) = |E_n(\rho,\theta)|^2$ occurs at $\bar{\rho} > |\alpha|$.

Description of the Fraunhofer Diffraction by a Spiral Phase Plate Using a Finite Sum of the Bessel Functions

To derive an expression for the Fraunhofer diffraction of a plane wave of radius R by an SPP with integer n, we must put $\alpha = 0$ in Eqs. (1.56) through (1.60). Then, the integrals in Eqs. (1.56) and (1.60) will be essentially simpler:

$$I_1^n = \begin{cases} 0, n = 2m, \\ -\dfrac{2\pi i^{|n|+1}}{\bar{\rho}}, n = 2m+1; \end{cases} \tag{1.61}$$

$$\frac{\partial I_1^n}{\partial \alpha} = \begin{cases} \dfrac{2\pi i^{|n|}|n|}{\bar{\rho}^2}, n = 2m, \\ 0, n = 2m+1. \end{cases} \tag{1.62}$$

Substituting Eqs. (1.61) and (1.62) into Eq. (1.56) produces an expression for the complex amplitude of Fraunhofer diffraction of a plane, finite-radius wave by the SPP:

$$E_n(\rho,\theta)=\frac{(-i)^{n+1}k\exp(in\theta)}{f\,\bar{\rho}^2}\cdot\begin{cases}n\left[1-J_0(y)-2\sum_{m=1}^{(n-2)/2}J_{2m}(y)\right]-yJ_{n-1}(y), n=2m,\\ n\left[\int_0^y J_0(t)dt-2\sum_{m=1}^{(n-1)/2}J_{2m-1}(y)\right]-yJ_{n-1}(y), n=2m+1,\end{cases} \tag{1.63}$$

where $y=R\bar{\rho}=kR\rho/f$,

$$\int_0^y J_0(t)dt=\frac{y}{2}\{\pi J_1(y)H_0(y)+J_0(y)[2-\pi H_1(y)]\}, \tag{1.64}$$

$H_{0,1}(y)$ is the Struve function of zero and first order.

The expression in Eq. (1.63) describes the diffraction by the SPP in the form of finite sums of the Bessel functions. For the even and odd SPP numbers the diffraction pattern is described by different expressions. For the first three numbers n = 0, 1, 2, Eq. (1.64) reduces to simple expressions:

$$E_0(\rho,\theta)=-\frac{ikR^2}{f}\cdot\frac{J_1(y)}{y}, \tag{1.65}$$

$$E_1(\rho,\theta)=-\frac{k\exp(i\theta)}{f\,\bar{\rho}^2}\left[\int_0^y J_0(t)dt-yJ_0(y)\right], \tag{1.66}$$

$$E_2(\rho,\theta)=\frac{ik\exp(i2\theta)}{f\,\bar{\rho}^2}\left[2-2J_0(y)-yJ_1(y)\right]. \tag{1.67}$$

Equation (1.65) describes a conventional pattern of Fraunhofer diffraction of the plane wave by a circular aperture of radius R. Figure 1.10 shows the calculated amplitude modules $|E_n(\rho,\theta)|$ for the radial profiles of the Fraunhofer diffraction (f = 100 mm) of the plane wave of radius R = 1 mm and wavelength $\lambda = 633$ nm by SPPs with $n=2$, 16, 50.

From Figure 1.10, the side lobes' contrast on the amplitude module plot is seen to decrease with increasing n. The side lobes' amplitude does not reach zero values.

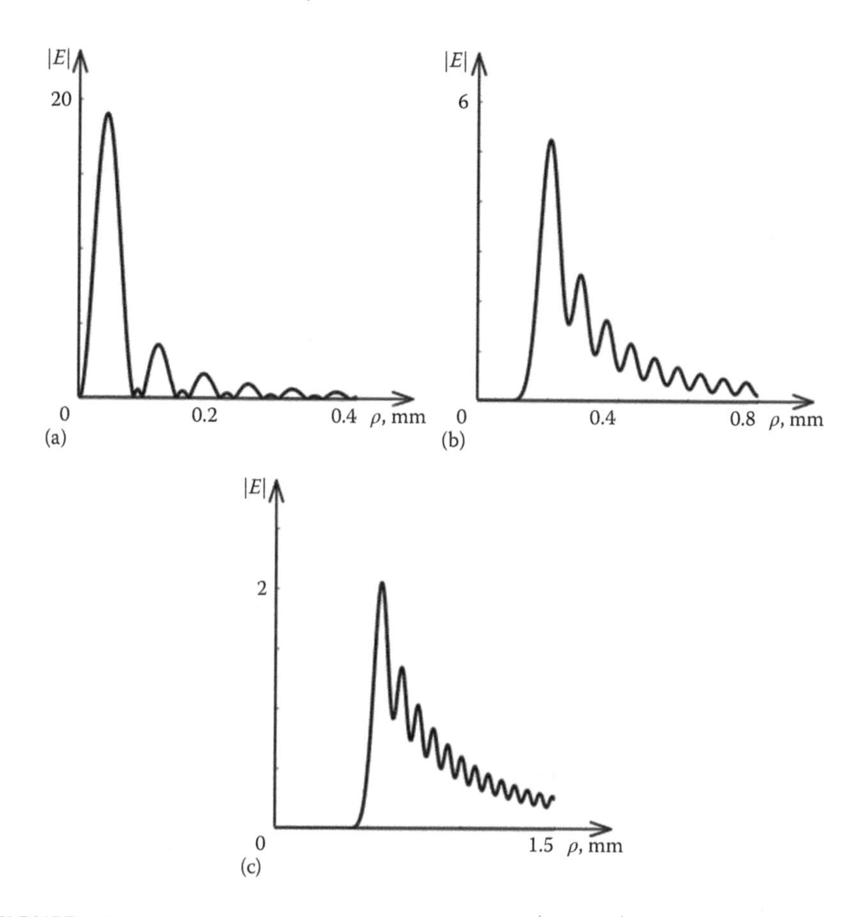

FIGURE 1.10 Calculated complex amplitude modules $|E_n(\rho,\theta)|$ versus the radial coordinate for different numbers of the SPP: (a) $n = 2$, (b) $n = 16$, and (c) $n = 50$.

From Eq. (1.63) we can asymptotically estimate the radius of the first ring where the module of the field amplitude equals zero. When $\rho \to \infty$ ($n = 2m$), from Eq. (1.63) the following approximate equation can be derived:

$$\left|E_n(\rho,\theta)\right| \approx \frac{k\left[n - yJ_n(y)\right]}{f\,\bar{\rho}^2} \approx \frac{k}{f\,\bar{\rho}^2}\left[n - \sqrt{\frac{2y}{\pi}}\sin\left(y - \frac{n+1}{2}\pi\right)\right]. \tag{1.68}$$

In Eq. (1.68) we used asymptotic representation of the Bessel function when its argument is of large value. From Eq. (1.68) it follows that approximate dependence of radius ρ_0 of the first zero of the amplitude $|E_n(\rho,\theta)|$ has the following form ($n \gg 1$):

$$\rho_0 \approx \frac{\lambda f n^2}{4R} \tag{1.69}$$

It is seen from Eq. (1.69) that dependence of the radius of the first zero of the amplitude on SPP number is quadratic. This is in concordance with Figure 1.10, because it is seen that increasing of n leads to increasing of the value of ρ_0, where the first zero of amplitude takes place.

Using Eq. (1.63) we can deduce a recurrent relationship by which the complex amplitudes at different n can be conveniently calculated:

$$\begin{aligned} E_{n+2}(\rho,\theta) = &-\frac{n+2}{n}\exp(i2\theta)E_n(\rho,\theta) \\ &+2(-i)^{n+1}\frac{R}{\rho}\frac{n+1}{n}\exp\left[i(n+2)\theta\right]J_{n+1}\left(\frac{k}{f}R\rho\right). \end{aligned} \tag{1.70}$$

Equation (1.70) holds for all integer n, both even and odd. From Eq. (1.70) one can derive equation for coordinates ρ_m of points at which field amplitude $|E_{n+2}(\rho,\theta)|$ equals zero ($n > 0$):

$$|E_n(\rho_m,\theta)| = 2\frac{R}{\rho_m}\left(\frac{n+1}{n+2}\right)J_{n+1}\left(\frac{kR\rho_m}{f}\right). \tag{1.71}$$

It is seen from Figure 1.11 that maxima of the module of the complex amplitude $|E_n(\rho,\theta)|$ almost coincide with maxima of the function $(2R/\rho)\,(n+1)/(n+2)\;J_{n+1}(kR\rho/f)$, and the minimal coordinate ρ_0, for which the maxima of these two functions coincide exactly, is a first zero of the amplitude $E_{n+2}(\rho,\theta)$. From comparing Figure 1.11a and b it follows that with increasing of n the value of ρ_0 is also increasing.

It is interesting to compare Eq. (1.71) with the equation from Ref. [57], to which coordinates ρ_s of minima and maxima of the field amplitude $E_n(\rho,\theta)$ must satisfy:

$$|E_n(\rho_s,\theta)| = \left(\frac{kR^2}{2f}\right)J_n\left(\frac{kR\rho_s}{f}\right). \tag{1.72}$$

It can be seen from Eq. (1.72) that the following inequality for the intensity of the principal (first) maximum on the diffraction pattern takes place:

$$|E_n(\rho_s,\theta)|^2 \le \left(\frac{kR^2}{2f}\right)^2 J_{0n}, \tag{1.73}$$

where $J_{0n} = \max|J_n(y)|^2$.

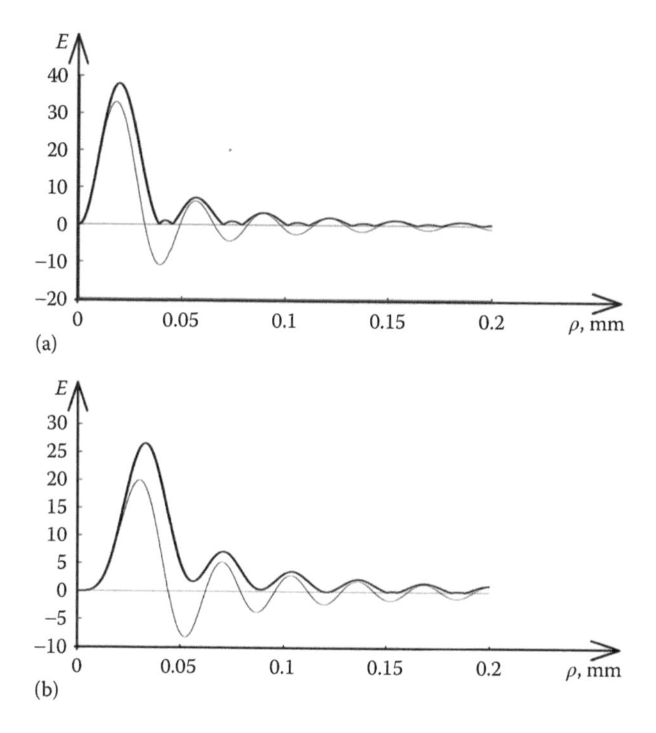

FIGURE 1.11 Plots of the functions from the left and right parts in Eq. (1.71): bold curve is a plot of $|E_n(\rho,\theta)|$ function, thin curve is a plot of the function $(2R/\rho)(n+1)/(n+2)\,J_{n+1}(kR\rho/f)$. Calculation parameters: $R=1$ mm, $f=100$ mm, $\lambda=633$ nm, (a) $n=2$ and (b) $n=4$.

Shown in Figure 1.12 are two plots of the functions from the left and right parts of Eq. (1.72) with the following values of parameters: $n=14$, $R=1$ mm, $f=100$ mm, $\lambda=633$ nm. It is seen that the plots have intersection at points where amplitude $|E_n(\rho_s,\theta)|$ has extreme values.

It is also seen from Figure 1.12 that coordinate ρ_v of the first maximum of the amplitude $|E_n(\rho,\theta)|$ (the vortex radius) is between the first root of the derivative of Bessel function of the nth order and the first root of the Bessel function of the nth order itself:

$$\frac{f\gamma'_{n,1}}{kR} < \rho_v < \frac{f\gamma_{n,1}}{kR}, \tag{1.74}$$

where $\gamma_{n,1}$ and $\gamma'_{n,1}$ are the first roots of the Bessel function of the nth order and its derivative: $J_n(\gamma_{n,1}) = J'_n(\gamma'_{n,1}) = 0$. From inequalities (1.74) it is possible to obtain an approximate expression to determine the radius of the first maximum of the amplitude $|E_n(\rho,\theta)|$:

$$\rho_v \approx \frac{\bar{\gamma}_n f}{kR},\ \bar{\gamma}_n = \frac{\gamma_{n,1}+\gamma'_{n,1}}{2}. \tag{1.75}$$

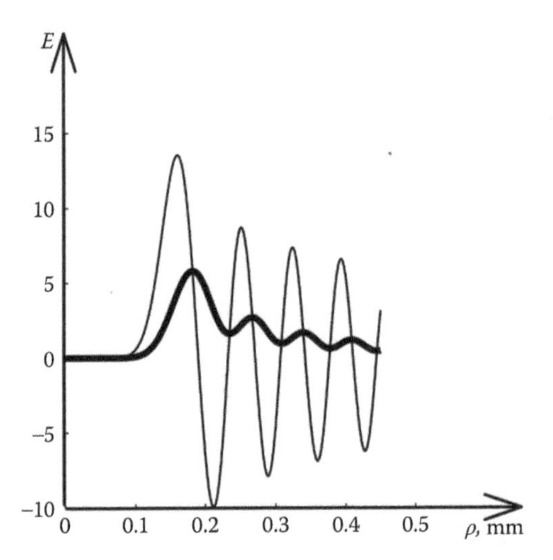

FIGURE 1.12 Plots of the functions from the left and right parts in Eq. (1.72): bold curve is a plot of $|E_n(\rho,\theta)|$ function, thin curve is a plot of the function $(kR^2)/(2f)\,J_n(kR\rho/f)$.

It is interesting to compare Eq. (1.75) with the similar expression for the radius of the vortex, obtained in Ref. [57]:

$$\rho_v = \frac{\gamma_{n-1,1} f}{kR}, \tag{1.76}$$

where $\gamma_{n-1,1}$ is the root of the Bessel function of $(n-1)$th order: $J_{n-1}(\gamma_{n-1,1}) = 0$.

Shown in Table 1.2 are the values of maximal amplitude (vortex radiuses) for n from 1 to 8, calculated by Eq. (1.63) (first line), Eq. (1.75) (second line) and Eq. (1.76) (third line). Radius of the maximal amplitude has been found by Eq. (1.63) and is being considered as more exact, while Eqs. (1.75) and (1.76) give an approximate value of this radius. The table shows that both equations give approximately the same estimation of the vortex radius (when $R = 1$ mm, $f = 100$ mm, $\lambda = 633$ nm) with maximal relative error 4% (for $n > 1$), although Eq. (1.76) is slightly more exact (maximal error 3%).

TABLE 1.2
Approximate Values of the Radius ρ_v of the Maximal Amplitude $|E_n(\rho,\theta)|$

n	1	2	3	4	5	6	7	8
ρ_v (μm), Eq. (1.63)	24.6	39.6	53.0	65.8	78.2	90.4	102.4	114.0
ρ_v (μm), Eq. (1.75)	28.6	41.2	53.3	64.9	76.5	87.8	99.1	110.2
ρ_v (μm), Eq. (1.76)	24.2	38.6	51.7	64.3	76.4	88.4	100.0	111.7

Description of the Fraunhofer Diffraction of the Gaussian Beam by a Helical Axicon Using the Hypergeometric Functions

Let us consider a scalar paraxial diffraction of the collimated Gaussian beam with complex amplitude given by

$$E_0(r)=\exp\left(-\frac{r^2}{w^2}\right) \tag{1.77}$$

by a HA, whose transmittance in the thin transparency approximation is given by Eq. (1.48), where w is the radius of the Gaussian beam waist. Then, the paraxial diffraction of the wave of Eq. (1.77) by the HA of Eq. (1.48) is described by the Fresnel transform:

$$\begin{aligned}F_n(\rho,\theta,z)=&-\frac{ik}{2\pi z}\exp\left(ikz+\frac{ik\rho^2}{2z}\right)\\&\times\int_0^R\int_0^{2\pi}\exp\left[-\frac{r^2}{w^2}+i\alpha r+in\varphi+\frac{ikr^2}{2z}-\frac{ik}{z}\rho r\cos(\varphi-\theta)\right]rdrd\varphi\end{aligned} \tag{1.78}$$

Using the reference integral [38]

$$\begin{aligned}&\int_0^\infty x^{\lambda+1}\exp\left(-px^2\right)J_\nu(cx)dx\\&=\frac{c^\nu p^{-(\nu+\lambda+2)/2}}{2^{\nu+1}\nu!}\Gamma\left(\frac{\nu+\lambda+2}{2}\right){}_1F_1\left[\frac{\nu+\lambda+2}{2},\nu+1,-\left(\frac{c}{2\sqrt{p}}\right)^2\right],\end{aligned} \tag{1.79}$$

we obtain, instead of Eq. (1.78):

$$\begin{aligned}F_n(\rho,\theta,z)=&\frac{(-i)^{n+1}k}{z}\exp\left[in\theta+ikz+\frac{ik\rho^2}{2z}\right]\left(\frac{k\rho}{2z}\right)^n\frac{\gamma^{-(n+2)/2}}{2^{n+1}n!}\\&\times\sum_{m=0}^{\infty}\frac{(i\alpha)^m\gamma^{-m/2}}{m!}\Gamma\left(\frac{m+n+2}{2}\right){}_1F_1\left[\frac{m+n+2}{2},n+1,-\left(\frac{k\rho}{2z\sqrt{\gamma}}\right)^2\right],\end{aligned} \tag{1.80}$$

where:

$\gamma=1/w^2-ik/(2z)$

${}_1F_1(a,b,x)$ is a degenerate or confluent hyper-geometric function:

$$ {}_1F_1(a,b,x) = \sum_{m=0}^{\infty} \frac{(a)_m x^m}{(b)_m m!}. \tag{1.81} $$

From Eq. (1.80) it follows that the diffraction pattern represents a set of concentric rings. At $\rho = 0$ and any $n \neq 0$, there will be zero intensity in the center of the diffraction pattern. Because the complex amplitude in Eq. (1.80) depends on the combination of variables $k\rho/(2z\sqrt{\gamma})$, the radii ρ_l of the diffraction pattern local maxima and minima should satisfy the relation:

$$ \rho_l = \frac{wza_l}{z_0}\left(1 + \frac{z_0^2}{z^2}\right)^{1/4}, \tag{1.82} $$

where:

a_l are constants that depend only on the ring number $l = 1,2, \ldots$ of the diffraction pattern and parameter α

$z_0 = kw^2/2$ is the Rayleigh range

It is seen from Eq. (1.82) that $\rho_l \sim wa_l\sqrt{z/z_0}$ when $z_0 \gg z$, that is, vortex radius increases as $\sqrt{z}$ with increasing of z.

At $\alpha = 0$ (no axicon), from Eq. (1.80) follows a relationship for the complex amplitude of Fresnel diffraction of the Gaussian beam by the SPP:

$$ \begin{aligned} F_n(\rho,\theta,z,\alpha=0) = {} & \frac{(-i)^{n+1} k}{z} \cdot \exp\left[i(n\theta + kz) + \frac{ik\rho^2}{2z}\right]\left(\frac{k\rho}{2z}\right)^n \\ & \times \frac{\gamma^{-(n+2)/2}}{2^{n+1} n!}\Gamma\left(\frac{n+2}{2}\right){}_1F_1\left[\frac{n+2}{2}, n+1, -\left(\frac{k\rho}{2z\sqrt{\gamma}}\right)^2\right]. \end{aligned} \tag{1.83} $$

Taking into account relation between a hypergeometric and Bessel functions:

$$ J_{(n-1)/2}(x) = \frac{\left(\dfrac{x}{2}\right)^{(n-1)/2} \exp(-ix)}{\Gamma\left(\dfrac{n-1}{2}\right)} {}_1F_1\left(\frac{n}{2}, n; 2ix\right) \tag{1.84} $$

and the recurrent relation for the hypergeometric functions

$$ {}_1F_1\left(\frac{n}{2}, n+1; 2ix\right) = \left(i\frac{d}{dx} + 2\right){}_1F_1\left(\frac{n}{2}, n; 2ix\right), \tag{1.85} $$

we can instead use Eq. (1.83) to derive a well-known relation for the Fresnel diffraction of the Gaussian beam by a SPP [16,33]:

$$F_n(\rho,\theta,z,\alpha=0)=\frac{(-i)^{n+1}\sqrt{\pi}}{2}\left(\frac{z_0}{z}\right)^2\left(\frac{\rho}{w}\right)\left[1+\left(\frac{z_0}{z}\right)^2\right]^{-3/4}$$
$$\times\exp\left[i\frac{3}{2}\tan^{-1}\left(\frac{z_0}{z}\right)-i\frac{k\rho^2}{2R_0(z)}+i\frac{k\rho^2}{2z}-\frac{\rho^2}{w^2(z)}+in\theta+ikz\right]$$
$$\times\left\{\begin{array}{l}I_{\frac{n-1}{2}}\left[\rho^2\left(\frac{1}{w^2(z)}+\frac{ik}{2R_0(z)}\right)\right]\\-I_{\frac{n+1}{2}}\left[\rho^2\left(\frac{1}{w^2(z)}+\frac{ik}{2R_0(z)}\right)\right]\end{array}\right\},\tag{1.86}$$

where:

$$w^2(z)=2w^2\left[1+(z/z_0)^2\right]$$

$$R_0(z)=2z\left[1+(z/z_0)^2\right]$$

$I_\nu(x)$ is the Bessel function of the second kind and νth order.

At $z\to\infty$ ($z\gg z_0$), from Eq. (1.80) follows a relationship for the complex amplitude of Fraunhofer diffraction of the Gaussian beam by the HA ($\gamma=1/w^2$):

$$F_n(\rho,\theta,z\to\infty)=\frac{(-i)^{n+1}z_0}{2^n n!z}\exp(in\theta+ikz)\left(\frac{z_0\rho}{zw}\right)^n$$
$$\times\sum_{m=0}^{\infty}\frac{(i\alpha w)^m}{m!}\Gamma\left(\frac{m+n+2}{2}\right){}_1F_1\left[\frac{m+n+2}{2},n+1,-\left(\frac{z_0\rho}{zw}\right)^2\right].\tag{1.87}$$

At $\alpha=0$ (no axicon) and $z\to\infty$ ($z\gg z_0$), from Eq. (1.80) follows a relationship for the complex amplitude of Fraunhofer diffraction of the Gaussian beam by the SPP:

$$F_n(\rho,\theta,z\to\infty,\alpha=0)=\frac{(-i)^{n+1}z_0}{2^n n!z}\exp(in\theta+ikz)\left(\frac{z_0\rho}{zw}\right)^n$$
$$\times\Gamma\left(\frac{n+2}{2}\right){}_1F_1\left[\frac{n+2}{2},n+1,-\left(\frac{z_0\rho}{zw}\right)^2\right].\tag{1.88}$$

It would be interesting to compare the expression in Eq. (1.88) with Eq. (1.54) for the complex amplitude of the Fraunhofer diffraction of a plane, finite-radius wave R and a focal length f by the SPP. From this comparison it follows that when $z = f$ and $R = w$, both complex amplitudes $E_n(x)$ in Eq. (1.54) and Eq. (1.88) depend on the same combination of variables: $x = kR\rho/(2f)$.

Numerical Simulation of the Diffraction of the Gaussian Beam by a Helical Axicon and a Spiral Phase Plate

Shown in Figure 1.13 is the calculated amplitude distribution $|F_n(\rho,\theta)|$ in relative units as a function of the radial variable. These curves present the radial profile of the Fresnel diffraction pattern ($z = 200$ mm) when the Gaussian beam of radius

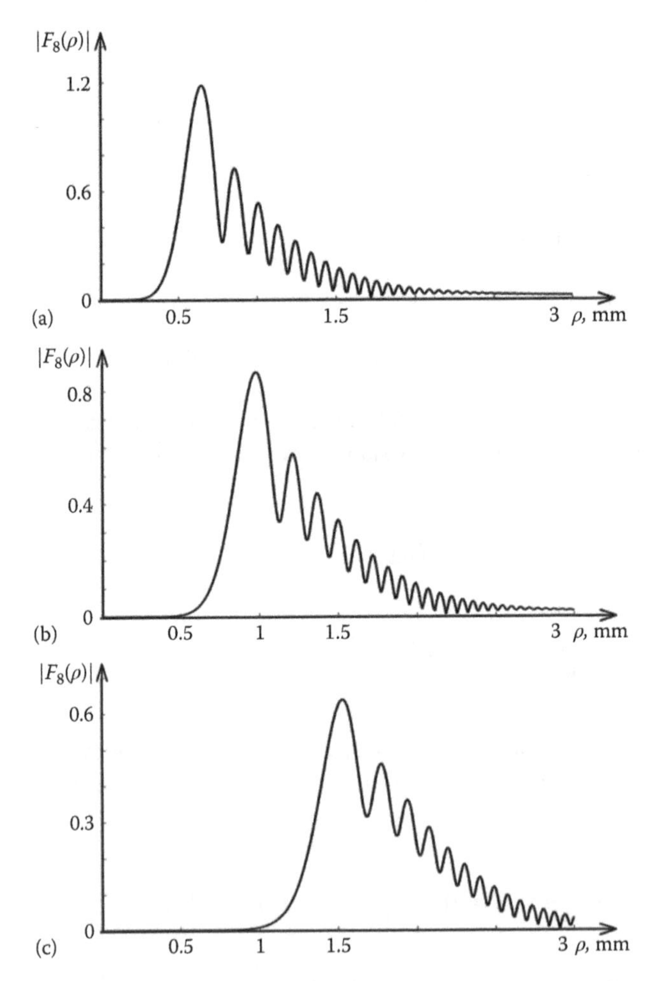

FIGURE 1.13 Radial profile of the Fresnel diffraction pattern (amplitude $|F_n(\rho,\theta)|$ at a distance $z = 200$ mm) for the Gaussian beam ($\lambda = 633$ nm, $w = 1$ mm) diffracted by the HA ($n = 8$): (a) $\alpha = 0$, (b) $\alpha = 20$ mm^{-1}, and (c) $\alpha = 50$ mm^{-1}.

$w = 1$ mm and wavelength $\lambda = 633$ nm is diffracted by an HA ($n = 8$) with the parameter $\alpha = 0\,\text{mm}^{-1}$ (a), $\alpha = 20\,\text{mm}^{-1}$ (b), $\alpha = 50\,\text{mm}^{-1}$ (c).

From Figure 1.13, the radius of a main maximum of the amplitude module plot is seen to increase with increasing α. From comparison of Figures 1.9 through 1.13 it follows that when the Gaussian beam is diffracted by the HA, there are no additional rings inside the ring with maximal intensity.

Figure 1.14 depicts two calculated radial Fresnel diffraction patterns (amplitude $|F_n(\rho,\theta)|$) for Gaussian beam ($w = 1$ mm, $\lambda = 633$ nm) diffracted by an HA ($n = 8$) with the parameter $\alpha = 20$ mm^{-1} with $z = 400$ mm (a) and $z = 500$ mm (b). We can see from Figure 1.14 that as the distance z increases the radius of the diffraction pattern's first bright ring, characterized by the maximal amplitude, also increases. Comparison of Figures 1.13 and 1.14 suggests that the radius of the first ring can be varied either by changing the axicon parameter α, with the distance z remaining unchanged, or by changing the distance z from the axicon to the observation plane. The difference will be in the number of peripheral rings (side petals) of the diffraction pattern. By way of illustration, we can see in Figure 1.13c that 13 peripheral diffraction rings fit in the radius interval from 1.5 to 3 mm. In the meantime, in Figure 1.14a only six side petals

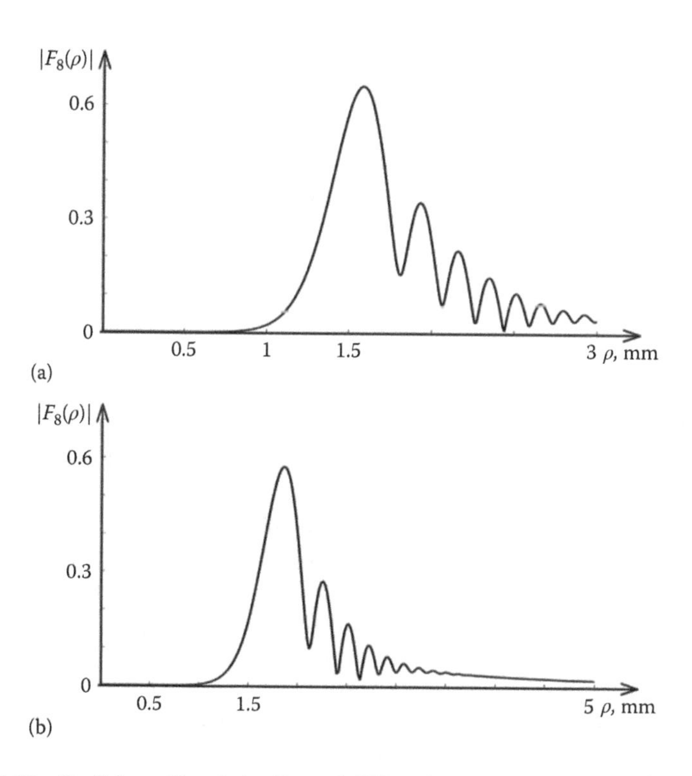

FIGURE 1.14 Radial profile of the Fresnel diffraction pattern (amplitude $|F_n(\rho,\theta)|$) for the Gaussian beam ($\lambda = 633$ nm, $w = 1$ mm) diffracted by the HA ($n = 8$, $\alpha = 20$ mm^{-1}): (a) $z = 400$ mm and (b) $z = 500$ mm.

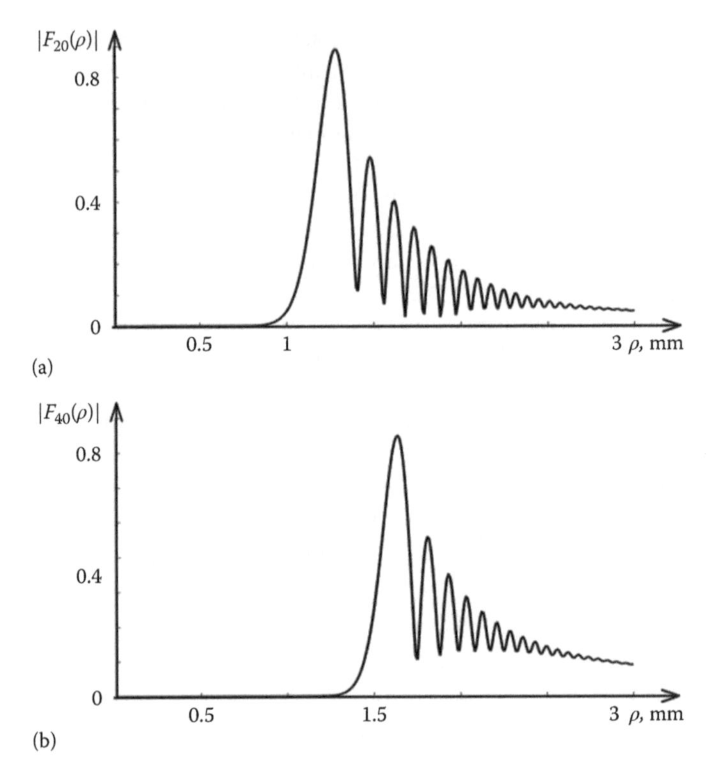

FIGURE 1.15 Radial profile of the Fresnel diffraction pattern (amplitude $|F_n(\rho,\theta)|$ at the distance $z = 200$ mm) for the Gaussian beam ($\lambda = 633$ nm, $w = 1$ mm) diffracted by the HA ($\alpha = 20$ mm^{-1}): (a) $n = 20$ and (b) $n = 40$.

are seen to fit in the same radial variable interval from 1.5 to 3 mm, with the first rings' radii being the same for the both patterns.

Figure 1.15 depicts two calculated radial Fresnel diffraction patterns (amplitude $|F_n(\rho, \theta)|$ at the distance $z = 200$ mm) for Gaussian beam ($w = 1$ mm, $\lambda = 633$ nm) diffracted by an HA ($\alpha = 20$ mm^{-1}) with different number n: 40 (a) and 50 (b). From Figure 1.15 it can be seen that the radius of the diffraction pattern's first ring can be changed through changing both the spiral axicon's number n and the parameter α. Note, however, that in the case of Figure 1.15, in addition to increasing radius of the first ring, an increase in the number n results in a thinner first ring, a greater number of peripheral rings, and increased ring contrast (in comparison with Figures 1.13 and 1.14).

The lithographic technique is known to be easiest to manufacture a phase DOE [70]. However, the phase function of the vortex axicon in (1.48) is not binary. Because of this we shall employ a simple technique for converting the half-tone function into the binary function involving the use of the spatial carrier frequency. In this case, the transmittance of the binary phase axicon is

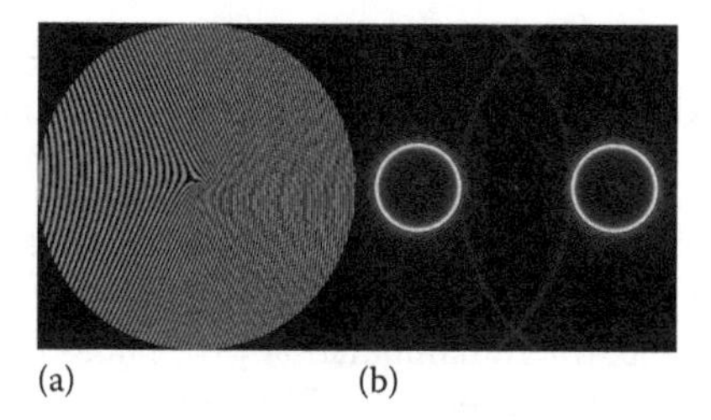

FIGURE 1.16 (a) Phase of a binary vortex axicon of 10th order, (b) calculated diffraction pattern of a plane wave by the DOE of Figure 1.16a.

$$\tau_{n\beta}(r,\varphi)=\text{sign}\left[\cos\left(\alpha r+n\varphi+\beta r\cos\varphi\right)\right]\text{rect}\left(\frac{r}{R}\right), \tag{1.89}$$

where:

$\beta = 2\pi/T$

T is the period of the spatial carrier frequency

The binary DOE is known to mainly generate two identical plus/minus first diffraction orders, each having 41% efficiency [70]. In order for these two diffraction orders (represented by two ring-shaped optical vortices) to be spatially separated the inequality $\beta > \alpha$ must be fulfilled. This is because the ring radius of the Fraunhofer diffraction pattern of the vortex axicon approximately equals $\alpha f/k$. Figure 1.16a depicts the binary phase of a DOE to generate two vortices with identical numbers $n = 10$ but opposite in sign. Shown in Figure 1.16b is the calculated Fraunhofer diffraction pattern of the DOE in Figure 1.16a. The DOE radius is $R = 2$ mm, wavelength is $\lambda = 532$ nm, axicon parameter is $\alpha = 50$ mm^{-1}, spatial carrier frequency is $\beta = 100$ mm^{-1}, and focal length of the spherical Fourier-lens is $f = 420$ mm.

New analytical relationships for the paraxial Fraunhofer diffraction of the plane, finite-radius wave by the HA and SPP and for the paraxial Fraunhofer and Fresnel diffraction of the Gaussian beam by the HA and SPP with an arbitrary integer n have been deduced. For the diffraction of the plane wave by HA, the complex amplitude is described by a series composed either of the hypergeometric functions (Eq. 1.51) or of the Bessel functions (Eq. 1.56). For the SPP, the complex amplitude is given by a finite sum of the Bessel functions (Eq. 1.63). It is noteworthy that the diffraction pattern complex amplitudes produced by the HA are described by different analytical relationships inside and outside a certain radius (Eqs. 1.58 and 1.60). For the SPP, the diffraction pattern is also described by different finite sums for the even and odd numbers n.

Also, new analytical relations to describe the Fresnel diffraction of the Gaussian beam by the HA and SPP have been obtained. For the HA, the complex amplitude is described by a series composed of the hypergeometric functions (Eqs. 1.80 and 1.87). For the SPP, the complex amplitude is given also by a series of the hypergeometric functions (Eqs. 1.83 and 1.88).

1.4 FRAUNHOFER DIFFRACTION OF THE PLANE WAVE BY THE MULTILEVEL (QUANTIZED) SPIRAL PHASE PLATE

Helical beams formed by the SPP have attracted great attention in modern scientific research. The main reasons for such an interest are increased accuracy of fabrication of these diffractive optical elements (DOE), and the possibility to use them in some practical areas, for example, micromanipulation [31]; photolithography with resolution $\lambda/10$, where λ is the light wavelength [71]; spiral interferogram analysis [72], allowing one to distinguish between convex and concave areas in the wavefront; astronomy [73]; and implementation of the radial Hilbert transform [45].

There are several ways to fabricate SPP, for example, fabrication of multisteps on the silica material through multi-etching process [74] or excimer laser ablation of the polyimide substrate [75]. The microrelief of the resulting SPP is multilevel or quantized. The multilevel SPPs were explored in [30,65]. In Ref. [30] the efficiency of the SPP-aided transformation of the Gaussian beam to the Laguerre-Gaussian mode (0,1) was theoretically calculated. Also, an experiment with a 16-level SPP, fabricated through the photolithography, was conducted. In Ref. [65] the minimal numbers of SPP phase levels (for SPP numbers $n < 8$) for which the finite-level SPPs only insignificantly differ from the continuous SPPs were found theoretically. With the help of a finite-level SPP, implemented with use of the liquid crystal cell, helical laser beams with singular numbers up to 6 were formed in [65].

In Refs. [69,76] the achromatic SPP was considered, which can form almost the same helical fields if the wavelength of the irradiance field varies over a rather wide range of 140 nm. In Ref. [77] the multilevel SPP was investigated in detail. It was shown that the orbital angular momentum of the beam generated by the SPP reached its maximum only in special cases. In Refs. [30,65,69,76,77] the SPP was analyzed through the decomposition into a series of angular harmonics. No expressions to describe the diffraction patterns of the multilevel SPP were offered.

In this section a finite-level SPP bounded by a polynomial aperture is considered. The number of the SPP phase quantization levels equals the number of sides of a regular polygon bounding the SPP aperture. In this case it has become possible to obtain an analytical expression in the form of a finite sum of plane waves for the complex amplitude, describing the Fraunhofer diffraction of the plane wave by the multilevel SPP, bounded by the regular polygon.

Let us note that the possibility to form helical fields with the use of non-SPPs was considered earlier [78]. In our case, unlike Ref. [78], when the number of phase quantization levels (or the number of sides of the polygon) increases, the far-zone diffraction pattern tends to that formed by the continuous SPP with a circular aperture.

Complex Amplitude of an Optical Vortex

Let Ω be the polygon with vertexes $A_p(x_p, y_p)$, $p = \overline{0, P-1}$, where P is the number of vertexes. The complex amplitude that describes the Fraunhofer diffraction of the plane wave by such a polynomial aperture is given by [79]:

$$E(\xi,\eta) = -\frac{if}{2\pi k}\sum_{p=1}^{P}\frac{(y_{p+1}-y_p)(x_p-x_{p-1})-(y_p-y_{p-1})(x_{p+1}-x_p)}{\left[\xi(x_{p+1}-x_p)+\eta(y_{p+1}-y_p)\right]\left[\xi(x_p-x_{p-1})+\eta(y_p-y_{p-1})\right]}$$
$$\exp\left[\pm i\frac{k}{f}(\xi x_p+\eta y_p)\right], \tag{1.90}$$

where:
$k = 2\pi/\lambda$ is the wave number
f is the lens focal length
(ξ,η) are the Cartesian coordinates in the Fourier plane

In Eq. (1.89) $x_P = x_0$, $y_P = y_0$, $x_{P+1} = x_1$, $y_{P+1} = y_1$. Let us consider the DOE in the form of a regular polygon $\Omega = A_0A_1...A_{P-1}$, inscribed into a circle of radius R. Then, we have $\Omega = \bigcup_{p=0}^{P-1}\Omega_p$, where Ω_p are triangles OA_pA_{p+1}. Let the DOE transmittance inside each triangle Ω_p have a constant value of $\tau(x, y) = \exp(i\Psi_p)$ (Figure 1.17).

The Fraunhofer diffraction of the plane wave by such a DOE is a sum of diffractions by each of the triangles Ω_p. Diffraction by the triangle Ω_p can be obtained from Eq. (1.90), multiplied by $\tau(x, y)$ (the sum will contain three terms). Therefore, the total diffraction by such a DOE will be a sum of $3P$ terms. So, using Eq. (1.90) for each triangle Ω_p, one can obtain the expression for the complex amplitude in the polar coordinates (ρ,θ), describing the Fraunhofer diffraction of the plane wave by such a DOE (Figure 1.18):

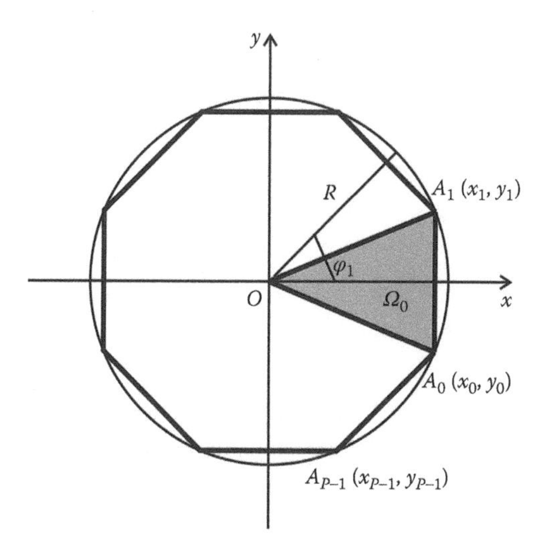

FIGURE 1.17 DOE with aperture in the form of a regular polygon.

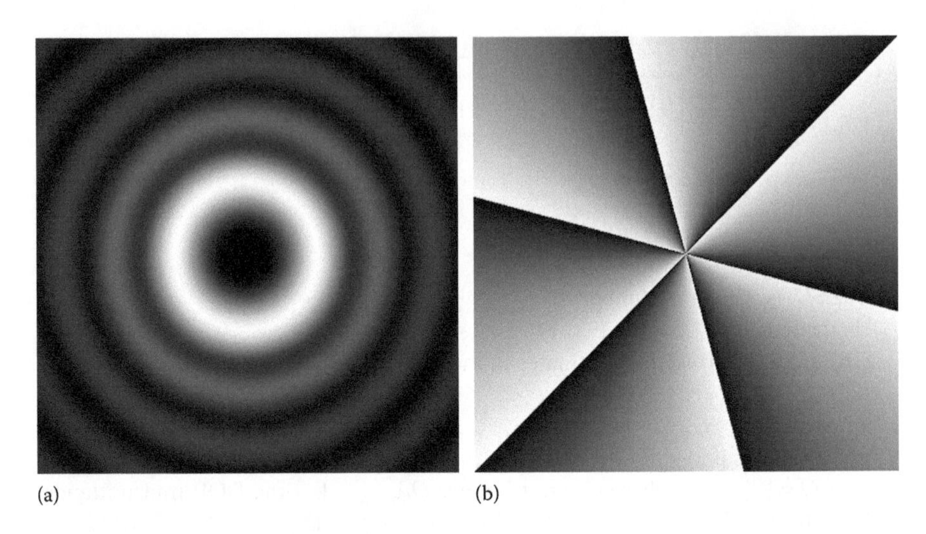

FIGURE 1.18 Fraunhofer diffraction pattern of the plane wave by the continuous limited SPP ($n=6$): (a) amplitude and (b) phase.

$$E(\rho,\theta)=\frac{if\cos\frac{\pi}{P}}{2\pi k\rho^2}\sum_{p=0}^{P-1}\frac{\exp(i\Psi_p)}{\sin(\varphi_p-\theta)\cos\alpha_p\cos\alpha_{p+1}}\times\left[\begin{array}{l}2\sin\frac{\pi}{P}\sin(\varphi_p-\theta)+\cos\alpha_{p+1}\exp\left(-i\frac{kR\rho}{f}\cos\alpha_p\right)\\-\cos\alpha_p\exp\left(-i\frac{kR\rho}{f}\cos\alpha_{p+1}\right)\end{array}\right], \tag{1.91}$$

where:

$\varphi_p=(2\pi/P)p$

$p=\overline{0,P-1}$

$\alpha_p=\varphi_p-\pi/P-\theta,\ P\geq 3$

Each term in the sum (1.91) is a sum of three terms, corresponding to the vertexes of the triangle Ω_p. At $\rho\to 0$, we decompose the exponents in Eq. (1.91) into a Taylor series, retaining only the first four terms. It can be shown that the first two terms will turn to zero, then:

$$E(\rho \to 0,\theta) \approx -\frac{ikR^2 \sin\frac{2\pi}{P}}{4\pi f} \sum_{p=0}^{P-1} \exp(i\Psi_p) - \frac{k^2 R^3 \rho \sin\frac{2\pi}{P} \cos\frac{\pi}{P}}{6\pi f^2} \sum_{p=0}^{P-1} \exp(i\Psi_p) \cos(\varphi_p - \theta). \tag{1.92}$$

From Eq. (1.92) it is seen that if $\Psi_p = n\varphi_p$ (where n is an integer) the azimuth DOE of Eq. (1.91) represents a quantized SPP of integer order. Then, the amplitude on the optical axis is zero, because the first sum in Eq. (1.92) is zero ($n \neq 0$). The first sum in Eq. (1.92) can be equal to zero not only for a SPP. For instance, when $\Psi_{k+P/2} = \Psi_k + \pi$ we will also have zero amplitude on the optical axis. For the SPP, the second term in Eq. (1.92), which is proportional to the variable $\rho \to 0$, will be equal to zero at $n \neq \pm 1$. If, for example, $n = 1$, the complex amplitude in Eq. (1.92) will be proportional to $E(\rho \to 0,\theta) \sim \rho \exp(i\theta)$. This means that a multi-level SPP with the number $n = 1$ and at any $P > 2$ will generate near the optical axis the same far-field diffraction pattern as a continuous SPP. A similar statement can also be made for a multilevel SPP with any number n (see Figure 1.19c and f). From (1.92) it is seen that although there is ρ^2 in the denominator of Eq. (1.91), there is no singularity on the optical axis. When in (1.92) $\theta \to \varphi_q$ or $\theta \to \varphi_q - \pi/P \pm \pi/2$, $q = 0, P-1$, some terms in the sums of Eq. (1.91) will have zero denominators. Nevertheless, one can show that these terms appear in pairs (or two pairs) and compensate for each other. Therefore, there are no irregularities in Eq. (1.91). Equation (1.91) is suited for an arbitrary multilevel azimuthal DOE, that is, a DOE with the phase depending only on the polar angle. In the case of the SPP, we have $\Psi_p = n\varphi_p$.

Numerical Simulation

Shown in Figure 1.18 is the Fraunhofer diffraction pattern of the plane wave by a continuous SPP with a circular aperture, obtained with the finite sums of Bessel functions [80]. The following values were used for the calculation: $\lambda = 633$ nm, $f = 150$ mm, $R = 2$ mm, $n = 6$.

Figure 1.19 shows the Fraunhofer diffraction patterns of the plane wave by a multilevel limited SPP, obtained from Eq. (1.91).

FIGURE 1.19 Fraunhofer diffraction pattern of the plane wave by the multilevel limited SPP ($n = 6$): (a, d, g) DOE phase, (b, e, h) amplitude, and (c, f, i) phase in the Fraunhofer diffraction zone. The number of sectors: (a, b, c) 18, (d, e, f) 30, (g, h, i) 42.

Table 1.3 shows the RMS between the intensities of the Fraunhofer diffraction patterns of the plane wave diffracted by the limited multilevel SPP and the continuous SPP, as a function of the number of sectors of the multilevel SPP.

Table 1.4 shows the minimal number of sectors of the multilevel SPP (for several SPP numbers), for which the RMS between the intensities of the Fraunhofer diffraction patterns of the plane wave diffracted by the limited multilevel SPP and the continuous SPP does not exceed 2%.

TABLE 1.3
The RMS as a Function of the Number of the SPP Sectors ($n = 6$)

Number of SPP Sectors	RMS
18	19.1411
30	1.9003
42	0.1320
54	0.0479

TABLE 1.4
The Minimal Number of Sectors versus the SPP Number

SPP Number	Minimal Number of Sectors
2	19
4	25
6	29
8	35
10	39

In summary, we have obtained an analytical expression describing the paraxial scalar Fraunhofer diffraction of the plane wave by a multilevel SPP, bounded by the aperture in the form of a regular polygon. For several SPP numbers we have numerically obtained the minimal number of SPP sectors for which the RMS between the Fraunhofer diffraction patterns for the multilevel and continuous SPP does not exceed 2%.

2 Elliptic Laguerre-Gauss Beams

The recent years have seen an increased interest in elliptic Gaussian laser beams. A decentered elliptic Gaussian beam propagating in a non-axially-symmetric optical system was discussed in Ref. [81]. Such beams are described using tensors. In Ref. [82] decentered elliptic Gauss-Hermite beams were introduced. An elliptic Gaussian beam of partially coherent light was dealt with in Ref. [83]. Fresnel diffraction of an elliptic (astigmatic) Gaussian beam by a diffraction grating is studied in Ref. [84]. In Ref. [85] the propagation of such a beam in a uniaxial crystal is investigated, whereas Ref. [86] studies second-harmonic generation in a non-linear crystal using an elliptic laser beam.

Gaussian beams of varying ellipticity can be used for beam shaping [87], that is, for producing elliptic laser beams with uniform cross-sectional intensity distributions. A linear combination of elliptic Gauss-Hermite beams can lead to "tubular" laser beams with zero axial intensity [88].

In Refs. [5,89–92] Ince-Gauss beams were introduced and studied both theoretically [5,89–91] and experimentally [92]. Such light fields are particular solutions of the paraxial wave equation (Schroedinger's type) in the elliptic coordinates. In these coordinates, the equation is solved by separation of variables, with the solution derived as a product of the Gaussian function by the Ince polynomials. The Ince polynomials are the solutions to the Whittaker-Hill equation. The Ince-Gauss beams represent an orthogonal basis that generalizes the familiar Hermite-Gauss and Laguerre-Gaussian bases. With the ellipse changing into a circumference (eccentricity $\varepsilon = 1$) the Ince-Gauss modes change to Laguerre-Gaussian modes. For $\varepsilon \to \infty$ (the ellipse reduces to a straight-line segment), the Ince-Gauss modes change to the Hermite-Gaussian modes.

Note that in Ref. [3] similar laser beams were treated and given the name Hermite-Laguerre-Gauss modes, which change into the conventional Hermite-Gaussian and Laguerre-Gaussian modes at a definite value of a certain parameter (the rotation angle of a cylindrical lens about the optical axis).

Elliptic beams can also be produced via an oblique incidence of an axially symmetric beam onto an optical element. In Refs. [93,94] such oblique incidence of a laser beam with a plane wavefront onto a conical axicon and a binary diffractive axicon was studied. The diffraction pattern produced by a collimated He-Ne laser beam falling at oblique incidence (the angle of 8°–16°) onto a conical axicon with an apex angle 0.01 rad and a base diameter 40 mm was simulated and studied experimentally in Ref. [93]. It was demonstrated that, while the axial illumination of the axicon base results in a diffraction-free Bessel beam of zero order, the oblique illumination causes the radial symmetry of the diffraction pattern to be lost, with the beam starting to diverge and changing the cross-sectional intensity distribution.

Similar studies were conducted in Ref. [94] but with a binary diffractive element used instead of the conventional axicon. The incidence angle of light of wavelength $\lambda = 632.8$ nm was up to 10°, the DOE diameter being 16.4 mm, and the axicon parameter being $\alpha = 0.036$. The axicon transmittance is given by $\exp(-ik\alpha r)$, where k is the wavenumber and r is the radial coordinate.

In Ref. [95] the diffraction pattern of a multi-channel binary DOE under oblique illumination by a plane wave was studied experimentally and theoretically. The DOE transmittance was proportional to functions that describe five Bessel beams of amplitudes $J_m(\alpha r)\exp(im\varphi)$, $m = 0, \pm 1, \pm 2$, propagated at different angles with respect to the optical axis. Diverging astigmatic Bessel beams were shown to be produced, with their diffraction patterns rotated by the angle of 45° clockwise at $m > 0$ and anticlockwise at $m < 0$ and remaining unchanged at $m = 0$. Note that the greater the order of the Bessel beam, $|m|$, the more local minima and maxima there are in the diffraction pattern at a fixed distance from the DOE.

In this section, we study the propagation of an elliptic Laguerre-Gaussian (LG) beam, which can no longer be looked upon as a mode. We show that in oblique illumination by a plane wave of a DOE with transmittance proportional to the function of the (m, n)th order GL mode the resulting diffraction pattern is equivalent to the diffraction pattern produced by the elliptic beam. When propagated in free space, the elliptic GL beam produces the diffraction pattern rotated clockwise by 45° at $m > 0$ and anticlockwise at $m < 0$. Note that with increasing distance z, at first, the number of the local maxima increases and, then, decreases. For $z \to \infty$ (far-field diffraction), the diffraction pattern is composed of a set of concentric rings rotated by 90° with respect to the initial diffraction pattern at $z = 0$. The theoretical conclusions are confirmed with a numerical simulation and physical experiments.

2.1 OBLIQUE PARAXIAL LAGUERRE-GAUSSIAN BEAM

Using the reference integral [38]:

$$\int_0^\infty x^{\frac{m}{2}} \exp(-px) J_m\left(b\sqrt{x}\right) L_n^m(cx)\,\mathrm{d}x = \left(\frac{b}{2}\right)^m \frac{(p-c)^n}{p^{m+n+1}} \exp\left(-\frac{b^2}{4p}\right) L_n^m\left(\frac{b^2 c}{4pc - 4p^2}\right), \tag{2.1}$$

where:

$J_m(x)$ is the Bessel function of the mth order and first kind

$L_n^m(x)$ is the attached Laguerre polynomial

n, m are integers, we find an expression for the Fresnel transform of the LG mode

$$\Psi_{mn}(r,\varphi) = \left(\frac{r\sqrt{2}}{w_0}\right)^m \exp\left(-\frac{r^2}{w_0^2}\right) L_n^m\left(\frac{2r^2}{w_0^2}\right) \exp(im\varphi), \tag{2.2}$$

where:

(r,φ) are the polar coordinates

w_0 is the radius of the Gaussian beam waist

The result of using Eqs. (2.1) and (2.2) is an expression for the amplitude of the LG mode at a distance z from the waist:

$$\Psi_{mn}(\rho,\theta,z)$$

$$=\frac{(-i)^{m+1}k}{z}\exp\left(\frac{ik\rho^2}{2z}+im\theta\right)\int_0^\infty\left(\frac{r\sqrt{2}}{w_0}\right)^m\exp\left(-\frac{r^2}{w_0^2}+\frac{ikr^2}{2z}\right)L_n^m\left(\frac{2r^2}{w_0^2}\right)J_m\left(\frac{kr\rho}{z}\right)r\,\mathrm{d}r.$$

$$=\frac{w_0}{w(z)}\left(\frac{\rho\sqrt{2}}{w(z)}\right)^m\exp\left[\frac{-\rho^2}{w^2(z)}+\frac{ik\rho^2}{2R(z)}+im\theta-i(2n+m+1)\operatorname{arctg}\left(\frac{z}{z_0}\right)\right]L_n^m\left(\frac{2\rho^2}{w^2(z)}\right),\tag{2.3}$$

where:

$w(z)=w_0\left[1+z^2/z_0^2\right]^{1/2}$

$R(z)=z\left(1+z_0^2/z^2\right)$

$z_0=kw_0^2/2$

k is the wavenumber

From Eq. (2.3), the structure of the LG mode is seen to be preserved, with the intensity being a radially symmetric function: $I_{mn}(\rho,z)=\left|\Psi_{mn}(\rho,\theta,z)\right|^2$. Using the light field (at $z=0$)

$$\Psi_{mn}^{(1)}(r,\varphi)=\left(\frac{r}{w_0}\right)^{2n+m}\exp\left(-\frac{r^2}{w_0^2}+im\varphi\right),\tag{2.4}$$

it is possible to produce the generalized LG modes that also show radial symmetry and preserve their structure up to a scale upon propagation. We now employ the reference integral [38]

$$\int_0^\infty r^{2n+m}\exp\left(-pr^2\right)J_m(cr)r\,\mathrm{d}r=\frac{n!c^m}{2^{m+1}p^{m+n+1}}\exp\left(-\frac{c^2}{4p}\right)L_n^m\left(\frac{c^2}{4p}\right).\tag{2.5}$$

Then, in view of Eq. (2.5), the Fresnel transform of the initial light field in Eq. (2.4) is

$$\Psi_{mn}^{(1)}(\rho,\theta,z)$$

$$=\frac{(-i)^{m+1}k}{z}\exp\left(\frac{ik\rho^2}{2z}+im\theta\right)\int_0^\infty\left(\frac{r}{w_0}\right)^{2n+m}\exp\left(-\frac{r^2}{w_0^2}+\frac{ikr^2}{2z}\right)J_m\left(\frac{kr\rho}{z}\right)r\,\mathrm{d}r \tag{2.6}$$

$$=\frac{(-i)^{m+1}z_0\,n!}{z}\left(1-\frac{iz_0}{z}\right)^{-n-\frac{m}{2}-1}\exp\left[\frac{ik\rho^2}{2z}+im\theta\right]x^{\frac{m}{2}}\exp(-x)L_n^m(x),$$

where:

$$x=\left[\frac{1}{w^2(z)}+\frac{ik}{2\hat{R}(z)}\right]\rho^2$$

$$\hat{R}(z)=z\left(1+\frac{z^2}{z_0^2}\right).$$

Let an oblique plane wave exp(ikr cosφ sinγ), where γ is the angle of incidence with respect to the axis $x = r\cos(\varphi)$, fall on a planar optical element. The DOE transmittance is proportional to the function of the LG mode $\Psi_{mn}(r,\varphi)$, with its radial part designated as

$$\hat{\Psi}_{mn}(r)=\left(\frac{r\sqrt{2}}{w_0}\right)^m\exp\left(-\frac{r^2}{w_0^2}\right)L_n^m\left(\frac{2r^2}{w_0^2}\right). \tag{2.7}$$

Then, the Fresnel transform of the LG mode in (2.2) in oblique illumination is

$$F_\gamma(\rho,\theta,z)=\frac{(-i)^{m+1}k}{z}\exp\left[\frac{ik\rho^2}{2z}+im\,\mathrm{arctg}\left(\frac{\rho\sin\theta}{z\sin\gamma-\rho\cos\theta}\right)\right]$$
$$\times\int_0^\infty\hat{\Psi}_{mn}(r)\exp\left(\frac{ikr^2}{2z}\right)J_m\left(\frac{kr}{z}\sqrt{z^2\sin^2\gamma+\rho^2-2\rho z\cos\theta\sin\gamma}\right)r\,\mathrm{d}r. \tag{2.8}$$

In polar coordinates with regard to the inclination,

$$\begin{cases}\xi=\rho\cos\theta-z\sin\gamma,\\ \eta=\rho\sin\theta,\end{cases} \tag{2.9}$$

the argument of the Bessel function in the constituent integral in Eq. (2.8) is

$$\xi^2+\eta^2=z^2\sin^2\gamma+\rho^2-2\rho z\cos\theta\sin\gamma. \tag{2.10}$$

Thus, in view of Eqs. (2.9) and (2.10), it follows from Eq. (2.8) that the intensity of the oblique paraxial LG mode will be radially symmetric:

$$I_\gamma(\xi^2+\eta^2,\ z)=\left|F_\gamma(\rho,\theta,z)\right|^2. \tag{2.11}$$

This is the usual property of paraxial beam propagation through a DOE that operates in the paraxial domain. The symmetry is broken, however, if we move into the para-basal domain, that is, consider a beam that is paraxial along its optical axis, which makes a non-paraxial angle with respect to the axis of the element. For the oblique illumination to result in the distorted GL mode we must consider non-paraxial light propagation and use the Kirchhoff transform instead of the Fresnel transform. The Kirchhoff transform of the oblique LG beam is given by

$$F(\xi,\eta,z)=\frac{-ik}{2\pi}\int_{-\infty}^{\infty}\int_{-\infty}^{\infty}\Psi_{mn}(x,y)\frac{\exp(ikR)}{R}\mathrm{d}x\,\mathrm{d}y, \tag{2.12}$$

where $R^2=(\xi-x)^2+(\eta-y)^2+z^2$. In the polar coordinates Eq. (2.12) takes the form

$$F(\rho,\theta,z)=\frac{-ik}{2\pi z}\int_0^{\infty}\int_0^{2\pi}\Psi_{mn}(r,\varphi)\exp\left[ik\sqrt{r^2+\rho^2+z^2-2r\rho\cos(\theta-\varphi)}\right]r\mathrm{d}r\,\mathrm{d}\varphi, \tag{2.13}$$

where we have approximated $R\approx z$ in the denominator of Eq. (2.12).

Then, the non-paraxial propagation of the oblique LG beam is governed by the expression

$$\begin{aligned}F_\gamma(\rho,\theta,z)=\frac{-ik}{2\pi z}\int_0^{\infty}\int_0^{2\pi}\Psi_{mn}(r,\varphi)\exp\Big[&ikr\cos\varphi\sin\gamma\\&+ik\sqrt{r^2+\rho^2+z^2-2r\rho\cos(\theta-\varphi)}\Big]r\mathrm{d}r\mathrm{d}\varphi.\end{aligned} \tag{2.14}$$

Assuming $z\gg r$ and $z\gg\rho$, we decompose in Eq. (2.14) the square root in the exponent into the Taylor series:

$$\begin{aligned}\sqrt{z^2+r^2+\rho^2-2r\rho\cos(\theta-\varphi)}&\approx z+\frac{r^2+\rho^2-2r\rho\cos(\theta-\varphi)}{2z}\\&-\frac{\left(r^4+\rho^4+2r^2\rho^2-(4r^3\rho+4r\rho^3)\cos(\theta-\varphi)+4r^2\rho^2\cos^2(\theta-\varphi)\right)}{8z^3}+\ldots\end{aligned} \tag{2.15}$$

Assuming in Eq. (2.15) $r^4/(8z^3) \ll z$, $4r^3\rho/(8z^3) \ll z$ and $4r^3\rho/(8z^3) \ll z$, we get, instead of Eq. (2.15):

$$\sqrt{z^2+r^2+\rho^2-2r\rho\cos(\theta-\varphi)} \approx \left(z+\frac{\rho^2}{2z}-\frac{\rho^4}{8z^3}\right)+\left(\frac{r^2}{2z}-\frac{r^2\rho^2}{2z^3}\right) - \frac{r\rho}{z}\cos(\theta-\varphi)-\frac{r^2\rho^2}{4z^3}\cos 2(\theta-\varphi). \tag{2.16}$$

In view of Eq. (2.16), Eq. (2.14) is replaced by

$$F_\gamma(\rho,\theta,z)=\frac{-ik}{2\pi z}\exp\left[ik\left(z+\frac{\rho^2}{2z}-\frac{\rho^4}{8z^3}\right)\right]\int_0^\infty \hat{\Psi}_{mn}(r)\exp\left[\frac{ikr^2}{2z}\left(1-\frac{\rho^2}{z^2}\right)\right] \times\left\{\int_0^{2\pi}\exp\left[im\varphi+ikr\cos\varphi\sin\gamma-\frac{ikr\rho}{z}\cos(\theta-\varphi) -\frac{ikr^2\rho^2}{4z^3}\cos 2(\theta-\varphi)\right]\mathrm{d}\varphi\right\}r\mathrm{d}r. \tag{2.17}$$

Let us rewrite in Eq. (2.17) the integral with respect to φ in the braces as a separate formula:

$$I_0=\exp(im\theta)\int_0^{2\pi}\exp\left[im\psi-\frac{ikr\rho_0}{z}\cos(\psi-\nu)-\frac{ikr^2\rho^2}{4z^3}\cos 2\psi\right]\mathrm{d}\psi, \tag{2.18}$$

where:

$$\begin{cases} \psi=\varphi-\theta, \\ \rho_0^2=(\rho-z\sin\gamma\cos\theta)^2+(z\sin\gamma\sin\theta)^2, \\ \nu=\operatorname{arctg}\left(\dfrac{z\sin\gamma\sin\theta}{\rho-z\sin\gamma\cos\theta}\right). \end{cases} \tag{2.19}$$

In Eq. (2.18), we use the designations

$$P=\frac{kr\rho_0}{z},\quad Q=\frac{kr^2\rho^2}{4z^3}. \tag{2.20}$$

Then, the integral in (2.18) takes the form:

$$I_0 = \exp(im\theta)\int_0^{2\pi}\exp\left[im\psi - iP\cos(\psi - v) - iQ\cos 2\psi\right]\mathrm{d}\psi$$
$$= \exp(im\theta)\sum_{p=-\infty}^{\infty} i^p J_p(Q)\int_0^{2\pi}\exp\left[i2p\psi + im\psi - iP\cos(\psi - v)\right]\mathrm{d}\psi \qquad (2.21)$$
$$= 2\pi(-i)^m \exp[im(\theta + v)]\sum_{p=-\infty}^{\infty}(-i)^p J_p(Q) J_{m+2p}(P)\exp(i2pv).$$

Note that in Ref. [95] a similar expression was derived to describe an astigmatic Bessel beam.

In view of Eq. (2.21), the amplitude of the light field (2.17) that describes the astigmatic LG beam (by analogy with the astigmatic Bessel beam of Ref. [95]) is given by:

$$F_\gamma(\rho,\theta,z) = \frac{(-i)^{m+1}k}{z}\exp\left[im(\theta + v) + ik\left(z + \frac{\rho^2}{2z} - \frac{\rho^4}{8z^3}\right)\right]$$
$$\times\sum_{p=-\infty}^{\infty}(-i)^p\exp(i2pv)\int_0^{\infty}\hat{\Psi}_{mn}(r)\exp\left[\frac{ikr^2}{2z}\left(1 - \frac{\rho^2}{z^2}\right)\right]J_p\left(\frac{kr^2\rho^2}{4z^3}\right)J_{m+2p}\left(\frac{kr\rho_0}{z}\right)r\mathrm{d}r. \qquad (2.22)$$

From Eq. (2.22), it can be seen that the astigmatic LG beam is not radially symmetric and does not preserve its structure upon propagation because the azimuth angle θ in Eq. (2.22) enters both ρ_0 and v (Eq. 2.19).

2.2 ELLIPTIC NONPARAXIAL LAGUERRE-GAUSSIAN BEAM

We will demonstrate that by replacing the oblique LG beam by the elliptic one an expression similar to Eq. (2.22) can be derived using the Fresnel transform, rather than Kirchhoff transform. Thus, the paraxial elliptic LG beam will no longer preserve its structure and will lose its ellipticity in the Fresnel diffraction zone.

At $z = 0$, an elliptic LG beam is defined as

$$\Psi_{mn}(x,y;\alpha) = \left(\frac{2x^2 + 2\alpha^2y^2}{w_0^2}\right)^{\frac{m}{2}}\exp\left(-\frac{x^2 + \alpha^2y^2}{w_0^2}\right)L_n^m\left(\frac{2x^2 + 2\alpha^2y^2}{w_0^2}\right).$$
$$\exp\left[im\,\mathrm{arctg}\left(\frac{\alpha y}{x}\right)\right]. \qquad (2.23)$$

In the elliptic coordinates

$$\begin{cases} x = \alpha r \cos\varphi, \\ y = r\sin\varphi, \;\; 0 \le \alpha \le 1, \end{cases} \tag{2.24}$$

we obtain, instead of (2.23),

$$\Psi_{mn}(r,\varphi;\alpha) = \left(\frac{\alpha r\sqrt{2}}{w_0}\right)^m \exp\left(-\frac{\alpha^2 r^2}{w_0^2}\right) L_n^m\left(\frac{2\alpha^2 r^2}{w_0^2}\right)\exp(im\varphi). \tag{2.25}$$

At $\alpha = 1$, Eq. (2.25) is identical to Eq. (2.2). The Fresnel transform of the beam (2.25) in the elliptic coordinates (2.24) takes the form

$$\begin{aligned} F_\alpha(\rho,\theta,z) &= \frac{-ik}{2\pi z}\exp\left[\frac{ik\rho^2}{2z}\left(\cos^2\theta + \alpha^2\sin^2\theta\right)\right] \\ &\times \int_0^\infty \int_0^{2\pi} \hat{\Psi}_{mn}(\alpha r)\exp\left[im\varphi + \frac{ikr^2}{2z}\left(\alpha^2\cos^2\varphi + \sin^2\varphi\right)\right. \\ &\left. - \frac{ik\alpha r\rho}{z}\cos(\theta-\varphi)\right] r\mathrm{d}r\mathrm{d}\varphi, \end{aligned} \tag{2.26}$$

where $\Psi_{mn}(r,\varphi;\alpha) = \hat{\Psi}_{mn}(\alpha r)\exp(im\varphi)$.

In Eq. (2.26), we used the elliptic coordinates in the plane at $z > 0$, rotated by 90° with respect to the coordinates in the plane at $z = 0$:

$$\begin{cases} \xi = \rho\cos\theta, \\ \eta = \alpha\rho\sin\theta. \end{cases} \tag{2.27}$$

From Eq. (2.26), we single out the integral with respect to φ:

$$\begin{aligned} F_\alpha(\rho,\theta,z) &= \frac{-ik}{2\pi z}\exp\left[\frac{ik\rho^2}{2z}\left(\cos^2\theta + \alpha^2\sin^2\theta\right)\right]\int_0^\infty \hat{\Psi}_{mn}(\alpha r)\exp\left[\frac{ikr^2}{4z}(1+\alpha^2)\right] \\ &\times \left\{\int_0^{2\pi}\exp\left[im\varphi - \frac{ikr^2}{4z}(1-\alpha^2)\cos 2\varphi - \frac{ik\alpha r\rho}{z}\cos(\theta-\varphi)\right]\mathrm{d}\varphi\right\} r\mathrm{d}r. \end{aligned} \tag{2.28}$$

From Eq. (2.28), we single out the integral in braces with respect to φ:

$$\hat{I}_0 = \int_0^{2\pi}\exp\left[im\varphi - \frac{ikr^2}{4z}(1-\alpha^2)\cos 2\varphi - \frac{ik\alpha r\rho}{z}\cos(\theta-\varphi)\right]\mathrm{d}\varphi. \tag{2.29}$$

The integral (2.29) is identical to Eq. (2.18) up to the designations and the factor before the integral. Let us introduce the designations:

$$A = \frac{k\alpha r\rho}{z}, \quad B = \frac{kr^2(1-\alpha^2)}{4z}. \tag{2.30}$$

Then we get, instead of Eq. (2.29),

$$\begin{aligned} \hat{I}_0 &= \sum_{p=-\infty}^{\infty} (-i)^p J_p(B) \int_0^{2\pi} \exp\left[i2p\varphi + im\varphi - iA\cos(\varphi-\theta)\right] \mathrm{d}\varphi \\ &= 2\pi \sum_{p=-\infty}^{\infty} (-i)^{p+m} J_p(B) J_{m+2p}(A) \exp[i(m+2p)\theta]. \end{aligned} \tag{2.31}$$

The series in (2.21) and (2.31) are seen to coincide up to designations. Finally, from (2.28) and (2.31) we obtain an expression for the Fresnel diffraction of the elliptic LG beam:

$$\begin{aligned} F_\alpha(\rho,\theta,z) = {} & \frac{(-i)^{m+1}k}{z} \exp\left[\frac{ik\rho^2}{2z}(\cos^2\theta + \alpha^2\sin^2\theta)\right] \times \sum_{p=-\infty}^{\infty} (-i)^p \exp[i(2p+m)\theta] \\ & \times \int_0^{\infty} \hat{\Psi}_{mn}(\alpha r) \exp\left[\frac{ikr^2}{4z}(1+\alpha^2)\right] J_p\left[\frac{kr^2(1-\alpha^2)}{4z}\right] J_{m+2p}\left(\frac{k\alpha r\rho}{z}\right) r\mathrm{d}r. \end{aligned} \tag{2.32}$$

Note that at $\alpha = 1$, Eq. (2.32) is equivalent to Eq. (2.3) for the Fresnel transform of the LG mode. Actually, at $\alpha = 1$, all terms of the series with respect to p, except when $p = 0$, are equal to zero, because $J_p(0) = 0$ at $p \neq 0$ and $J_0(0) = 1$.

Also, note that for $z \to \infty$, so that $r^2/z \ll r\rho/z$, Eq. (2.32) reduces to the Fourier transform of the elliptic LG beam:

$$\begin{aligned} F_\alpha(\rho,\theta,z\to\infty) &\approx \frac{(-i)^{m+1}k}{z} \exp(im\theta) \int_0^{\infty} \hat{\Psi}_{mn}(\alpha r) J_m\left(\frac{kr\alpha\rho}{z}\right) r\mathrm{d}r \\ &= (-i)^m(-1)^n \left(\frac{w_0}{\alpha^2\sigma}\right)\left(\frac{\rho\sqrt{2}}{\sigma}\right)^m \exp\left(-\frac{\rho^2}{\sigma^2}\right) L_n^m\left(\frac{2\rho^2}{\sigma^2}\right), \end{aligned} \tag{2.33}$$

where $\sigma = 2z/(kw_0)$ and $\rho^2 = \xi^2 + (\eta/\alpha)^2$.

From Eq. (2.33), it can be seen that the LG beam, which is elliptic at $z = 0$, while again assuming the elliptic symmetry for the far-field diffraction ($z \to \infty$), turns out to be rotated by 90° with respect to the initial beam at $z = 0$.

At a finite z, the LG beam loses the elliptic symmetry. Then the center of the diffraction pattern ($\rho = 0$) will show a non-zero intensity for the even-numbered elliptic LG beams, $m = 2l \neq 0$. At the same time, at $z = 0$ and $z \to \infty$ the central intensity at $\rho = 0$ will be zero. Actually, at $\rho = 0$, all terms in the series (2.32) are zero, except for the term numbered $p = -m/2$, because $J_{m+2p}(0) = J_0(0) = 1$. Then, from Eq. (2.32), for the even $m = 2l$, we obtain

$$F_\alpha(\rho = 0, \theta, z) \sim \int_0^\infty \hat{\Psi}_{mn}(\alpha r) \exp\left[\frac{ikr^2}{4z}\left(1+\alpha^2\right)\right] J_{\frac{m}{2}}\left[\frac{kr^2}{4z}\left(1-\alpha^2\right)\right] r\mathrm{d}r, \quad (2.34)$$

where ~ symbolizes the proportionality.

When the beam ellipticity is small, $\alpha \to 1$, the Bessel function in Eq. (2.34) can be approximated as

$$J_p(x) \cong \frac{\left(\frac{x}{2}\right)^p}{\Gamma(p+1)}, \quad x \to 0, \quad (2.35)$$

where $\Gamma(p + 1)$ is the gamma function if $p > 0$. If $p < 0$, one should use the equality $J_{-p}(x) = (-1)^p J_p(x)$ if p is an integer. Then, the integral in Eq. (2.34) will be given by

$$F_\alpha(\rho = 0, z) \sim \Gamma^{-1}\left(\frac{m}{2}+1\right)\left[\frac{k\left(\alpha^2-1\right)}{8z}\right]^{\frac{m}{2}}\left(\frac{\alpha\sqrt{2}}{w_0}\right)^m \int_0^\infty r^{2m} \exp\left[-\frac{\alpha^2 r^2}{w_0^2} + \frac{ikr^2\left(1+\alpha^2\right)}{4z}\right] \cdot$$

$$L_n^m\left(\frac{2\alpha^2 r^2}{w_0^2}\right) r\mathrm{d}r. \quad (2.36)$$

Using the reference integral [38]

$$\int_0^\infty x^m \exp(-px) L_n^m(cx)\,\mathrm{d}x = \frac{\Gamma(m+n+1)(p-c)^n}{n!\,p^{m+n+1}} \quad (2.37)$$

the integral in (2.36) reduces to:

$$\int_0^\infty r^{2m} \exp\left[-\frac{\alpha^2 r^2}{w_0^2} + \frac{ikr^2\left(1+\alpha^2\right)}{4z}\right] L_n^m\left(\frac{2\alpha^2 r^2}{w_0^2}\right) r\,\mathrm{d}r$$

$$= \frac{\Gamma(m+n+1)(-1)^n \exp\left[i(2n+m+1)\eta\right]}{2n!\left[\frac{\alpha^4}{w_0^4} + \frac{k^2\left(1+\alpha^2\right)^2}{16z^2}\right]^{\frac{m+1}{2}}}, \tag{2.38}$$

where $\eta = \mathrm{arctg}\left[k\left(1+\alpha^2\right)w_0^2/\left(4z\alpha^2\right)\right]$.

It can be seen that the absolute value of the integral in Eq. (2.38) is always non-zero. It follows from (2.36) that $F_\alpha(\rho = 0, z) = 0$ only for $\alpha = 1$. Thus, we have shown that even a minor ellipticity of the LG mode violates the conditions of the zero central intensity at any z for $\rho = 0$. At the same time, the field of the elliptic beam retains its vortex character. The phase singularity at the center of the order $m = 2l$ appears broken down into p singularities numbered (ordered) m/p, with p points of zero-intensity emerging in the beam center neighborhood. The number p depends on the degree of ellipticity.

In the following, we show that for the weak ellipticity $p = 2$ and at an angle of 45° there appear two intensity zeros of $m/2$th order. For an odd $m = 2l + 1$, the intensity zero is always found at the diffraction pattern center ($\rho = 0$).

Let us consider peculiarities of propagation of a weakly elliptic ($\alpha^2 \approx 1$) Gaussian beam. In Eq. (2.32), we approximate the Bessel function of the pth order under the integration sign by Eq. (2.35). We can do this because the Bessel function's argument tends to zero at weak ellipticity and although the integral in (2.32) also has an infinite upper limit of integration, the Gaussian exponent in the LG mode limits the integration domain by a finite effective mode radius. Then we write, instead of (2.32),

$$F_{\alpha\to 1}(\rho,\theta,z) \approx S(\rho,\theta)\sum_{p=-\infty}^{\infty}(-i)^p\left(\frac{\varepsilon}{2}\right)^{|p|}\Gamma^{-1}\left(|p|+1\right)\exp(i2p\theta)\delta(p)$$

$$\times\int_0^\infty\left(\frac{\alpha r\sqrt{2}}{w_0}\right)^m r^{2|p|}\exp\left[-\frac{\alpha^2 r^2}{w_0^2} + \frac{ikr^2\left(1+\alpha^2\right)}{4z}\right] L_n^m\left(\frac{2\alpha^2 r^2}{w_0^2}\right) J_{m+2p}\left(\frac{k\alpha r\rho}{z}\right) r\,\mathrm{d}r, \tag{2.39}$$

where:

$$\delta(p) = \begin{cases} 1, & p \geq 0, \\ (-1)^{|p|}, & p < 0, \end{cases} \tag{2.40}$$

$$S(\rho,\theta)=\frac{(-i)^{m+1}k}{z}\exp\left[im\theta+\frac{ik\rho^2}{2z}\left(\cos^2\theta+\alpha^2\sin^2\theta\right)\right], \tag{2.41}$$

$\varepsilon=k(1-\alpha^2)/(4z)<<1$ is a small parameter.

Note that in (2.39) the terms with positive and negative numbers p contribute differently to the total sum. If $m > 0$, in Eq. (2.39) the term with $p > 0$ will have a factor $\exp[i\theta(m+2p)]r^{|m|+2|p|}J_{m+2p}(x)$, whereas the term with $p < 0$ will have another factor, $\exp\left[i\theta\left(m-2|p|\right)\right]r^{|m|+2|p|}J_{m-2|p|}(x)$. It is seen that in the first case ($p > 0$) the exponent of the radial variable is equal to the Bessel function's order. Otherwise ($m < 0$), the major contribution to the sum (2.39) will come from the terms with $p < 0$.

The reference integral [38]

$$\int_0^\infty x^{\frac{m+p}{2}}\exp(-cx)L_n^m(cx)J_{m+p}\left(b\sqrt{x}\right)\mathrm{d}x=\frac{(p-c)^n}{p^{m+n+1}}\left(\frac{b}{2}\right)^m\exp\left(-\frac{b^2}{4p}\right)L_n^m\left(\frac{b^2c}{4pc-p^2}\right), \tag{2.42}$$

confirms this indirectly: a modified LG beam can reproduce itself only when the radial variable exponent, $x = r^2$, and the Bessel function order are equal. Note, however, that it would be incorrect to use the integral (2.42) for calculating Eq. (2.39), because the exponent in the integral (2.39) is different from the argument of the attached Laguerre polynomial. Thus, retaining in (2.39) only the terms with $p > 0$ (assuming $m > 0$) and taking account of the weak ellipticity of the LG beam, we retain only the first two terms ($\varepsilon \ll 1$):

$$\begin{aligned}F_{\alpha\to1}(\rho,\theta,z)\approx S(\rho,\theta)\int_0^\infty\left(\frac{\sqrt{2}\alpha r}{w_0}\right)^m L_n^m\left(\frac{2r^2\alpha^2}{w_0^2}\right)\times\exp\left[-\frac{r^2\alpha^2}{w_0^2}+\frac{ikr^2\left(1+\alpha^2\right)}{4z}\right]\\ \times\left\{J_m\left(\frac{k\alpha r\rho}{z}\right)-\frac{i\varepsilon r^2}{2}e^{i2\theta}J_{m+2}\left(\frac{k\alpha r\rho}{z}\right)+O(\varepsilon^2)\right\}r\,\mathrm{d}r,\end{aligned} \tag{2.43}$$

where $O(\varepsilon^2)$ are small terms of order ε^2 and higher.

There is a well-known series [38]

$$\sum_{p=0}^\infty\frac{t^{2p}J_{m+2p}(x)}{(2p)!}=\frac{1}{2}x^{\frac{m}{2}}J_m\left(\sqrt{x^2-2tx}\right)\times\left\{(x-2t)^{-\frac{m}{2}}+(x+2t)^{-\frac{m}{2}}\right\}. \tag{2.44}$$

Assuming $t \ll 1$, we can use the following approximation, instead of (2.44):

$$\sum_{p=0}^\infty\frac{t^{2p}J_{m+2p}(x)}{(2p)!}\approx J_m(x)+\frac{t^2}{2}J_{m+2}(x)+O(t^4). \tag{2.45}$$

In view of (2.44), comparing the expression in the braces in (2.43) with the right-hand side of (2.45) yields, instead of (2.43),

$$F_{\alpha\to 1}(\rho,\theta,z) \approx S'(\rho,\theta)\int_0^\infty \left(\frac{\sqrt{2}\alpha r}{w_0}\right)^m L_n^m\left(\frac{2r^2\alpha^2}{w_0^2}\right)\exp\left[-\frac{r^2\alpha^2}{w_0^2}+\frac{ikr^2\left(1+\alpha^2\right)}{4z}\right]$$

$$\times J_m\left(r\sqrt{\left(\frac{k\alpha\rho}{z}\right)^2-2\sqrt{-i\varepsilon}\,e^{i\theta}\left(\frac{k\alpha\rho}{z}\right)}\right) r\,\mathrm{d}r, \tag{2.46}$$

where:

$$S'(\rho,\theta)=S(\rho,\theta)\frac{1}{2}\left(\frac{k\alpha\rho}{z}\right)^{\frac{m}{2}}\left\{\left(\frac{k\alpha\rho}{z}-2\sqrt{-i\varepsilon}\,e^{i\theta}\right)^{\frac{-m}{2}}+\left(\frac{k\alpha\rho}{z}+2\sqrt{-i\varepsilon}\,e^{i\theta}\right)^{\frac{-m}{2}}\right\}.$$

From Eq. (2.46) it is seen that as $\rho\to\infty$, Eq. (2.46) changes to Eq. (2.3), which is the Fresnel transform of the LG mode:

$$F_{\alpha\to 1}(\rho\to\infty,\theta,z) \approx S(\rho,\theta)\int_0^\infty \left(\frac{\sqrt{2}\alpha r}{w}\right)^m L_n^m\left(\frac{2r^2\alpha^2}{w^2}\right)\times\exp\left[-\frac{r^2\alpha^2}{w^2}+\frac{ikr^2\left(1+\alpha^2\right)}{4z}\right]$$

$$\times J_m\left(\frac{k\alpha\rho r}{z}\right) r\,\mathrm{d}r, \tag{2.47}$$

and is derived using the reference integral (2.1).

Equation (2.47) shows that a weakly elliptic LG beam behaves on the periphery as a conventional mode, but with elliptic symmetry. Thus, the diffraction pattern is represented by a set of concentric ellipses rather than rings. From Eq. (2.46) it follows that for small ρ, the central part of the diffraction pattern of the elliptic LG beam will have isolated intensity zeroes at points where the argument of the Bessel function in Eq. (2.46) turns to zero:

$$\rho=\frac{1}{\alpha}\sqrt{\frac{z}{k}(1-\alpha^2)}\exp\left[i\left(\theta-\frac{\pi}{4}\right)\right]. \tag{2.48}$$

From Eq. (2.48) it follows that two real intensity zeroes of the $m/2$th order are found on the straight line $\theta=\pi/4$ at the distance from the center $\rho=0$ equal to

$$\rho_0=\frac{1}{\alpha}\sqrt{\frac{z(1-\alpha^2)}{k}}. \tag{2.49}$$

From Eq. (2.49), it is seen that the greater the beam's ellipticity ($1 - \alpha^2$) and the distance to the waist, the greater will be ρ_0 and the distance of the intensity zeroes to the center $\rho = 0$.

Note that although from Eq. (2.46) it follows that at $\rho = 0$ there also should be an intensity zero, but this is not the case. The fact is that Eq. (2.46) disregards the terms in Eq. (2.36) with negative $p < 0$, which are small (but not equal to zero) at $\alpha \approx 1$. Earlier, we showed that the major contribution to the intensity of the elliptic LG beam comes from the term given by Eq. (2.34), which is also non-zero at finite z.

To explain why for $z > 0$ at $\rho = 0$ along the line of 45° two intensity zeros arise, let us represent the intensity of the elliptic LG beam as the interference of two fields. From Eq. (2.43) it follows that

$$\left|F_{\alpha\to1}(\rho,\theta,z)\right|^2 \approx \left|F_0(\rho)-\frac{ik(1-\alpha^2)}{8z}e^{2i\theta}F_1(\rho)\right|^2, \tag{2.50}$$

where:

$$F_0(\rho)=\frac{k}{z}\int_0^\infty\left(\frac{\sqrt{2}\alpha r}{w_0}\right)^m L_n^m\left(\frac{2r^2\alpha^2}{w_0^2}\right)\exp\left[-\frac{r^2\alpha^2}{w_0^2}+\frac{ikr^2\left(1+\alpha^2\right)}{4z}\right]J_m\left(\frac{k\alpha\rho r}{z}\right)r\,\mathrm{d}r, \tag{2.51}$$

$$F_1(\rho)=\frac{k}{z}\int_0^\infty\left(\frac{\sqrt{2}\alpha r}{w_0}\right)^m r^2 L_n^m\left(\frac{2r^2\alpha^2}{w_0^2}\right)\exp\left[-\frac{r^2\alpha^2}{w_0^2}+\frac{ikr^2\left(1+\alpha^2\right)}{4z}\right]J_{m+2}\left(\frac{k\alpha\rho r}{z}\right)r\,\mathrm{d}r. \tag{2.52}$$

From Eqs. (2.50) through (2.52), it is seen that the diffraction pattern of the weakly elliptic LG beam is devoid of both radial and elliptic symmetry. The integral (2.51) can be performed using Eq. (2.1). From Eq. (2.50) it is seen that with increasing z the contribution of the second term is decreasing, so that for $z \to \infty$ Eq. (2.50) changes to Eq. (2.33), which describes an elliptic LG beam rotated by 90° with respect to the input LG beam at $z = 0$. From Eq. (2.50), it follows that two light fields, $F_0(\rho)$ and $F_1(\rho)$, will interfere positively along the line $\theta = -\pi/4$. If we considered Eq. (2.39) at $m < 0$ the resulting equation would be similar to Eq. (2.50), but with the light fields $F_0(\rho)$ and $F_1(\rho)$ interfering positively at $\theta = \pi/4$. Thus, the cross-section of the diffraction pattern produced by the elliptic LG beam upon propagation makes it possible to identify the left ($m < 0$) or right ($m > 0$) "phase spin" of the initial LG beam at $z = 0$. Note that the rotation of the diffraction pattern by the angle $\pm\pi/4$ in the Fresnel zone is independent of the absolute value $|m|$. For the far-field diffraction, the elliptic LG beam is rotated by ±90° with respect to the initial beam, so that neither the spin nor the sign of the number $\pm|m|$ can be determined.

Note that as the ellipse parameter α tends to zero, the elliptic LG beam in Eq. (2.23) at $z = 0$ turns into a beam that changes only on the x-axis, with the amplitude remaining unchanged on the y-axis:

$$\Psi_{mn}(x) = \left(\frac{2x^2}{w_0^2}\right)^{\frac{m}{2}} \exp\left(-\frac{x^2}{w_0^2}\right) L_n^m\left(\frac{2x^2}{w_0^2}\right). \tag{2.53}$$

For the far-field diffraction, the LG beam of Eq. (2.53) will change to the one-dimensional Hermite-Gaussian beam with respect to the axis ξ. Actually, using the reference integral [38]

$$\int_0^\infty t^{\left(\frac{m-1}{2}\right)} \exp\left(-\frac{ct}{2}\right) L_n^m(ct) \exp\left(p\sqrt{t}\right) \mathrm{d}t = \frac{2\sqrt{2\pi}}{n!} \cdot \frac{\exp\left(\frac{p^2}{2c} - \frac{i\pi m}{2}\right)}{2^{n+\frac{m}{2}} c^{\frac{m+1}{2}}} H_{n+m}\left(\frac{ip}{\sqrt{2c}}\right) H_n\left(\frac{ip}{\sqrt{2c}}\right), c > 0, \tag{2.54}$$

yields the light field amplitude in a spherical lens focus in the form:

$$\frac{-ik}{2\pi f} \int_{-\infty}^{\infty}\int \Psi_{mn}(x) \exp\left[\frac{-ik}{f}(x\xi + y\eta)\right] \mathrm{d}x\,\mathrm{d}y = (-i)^{m+1} \delta(\eta) \frac{\sqrt{2\pi}}{2^{m+n-1}} \exp\left[-\left(\frac{\xi}{\sigma}\right)^2\right] H_{n+m}\left(\frac{\xi}{\sigma}\right) H_n\left(\frac{\xi}{\sigma}\right), \tag{2.55}$$

where:

$\delta(\eta)$ is the Dirac function

$\sigma = 2f/(kw_0)$ is the Gaussian beam radius at the focal plane of the lens with focus f.

Examples of changing diffraction pattern for the elliptic LG beam are discussed in the subsequent section.

Numerical Simulation of Propagation of the Elliptic Laguerre-Gaussian Beams

The intensity patterns in Figure 2.1 provide proof for the theoretical analysis: as the beam is being formed behind the DOE it acquires the inclination of 45°, with the entire pattern rotated by 90° for the far-field diffraction. As was predicted, with increasing distance z the number of local maxima increases at first and then decreases. In addition, it is seen that at a finite z, there is no intensity zero in the diffraction pattern center, with two intensity zeroes appearing on a line at an angle of 45°.

In general, the light beam in Figure 2.2 behaves similarly to those in Figure 2.1, the only difference being a single local minimum due to the mode index $m = -1$. In addition, in Figure 2.2 we can see the rotation by −45° at the middle-distance.

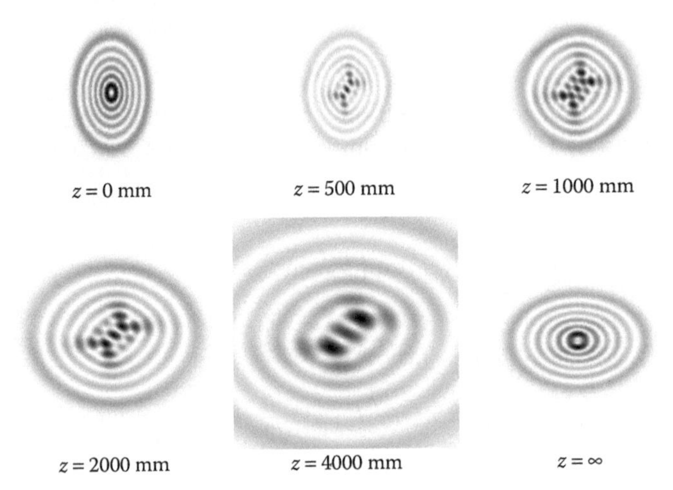

FIGURE 2.1 Propagation of the LG mode (5,2) with the elliptic distortion in free space (negative). The image size is: 5 × 5 mm, 256 × 256 pixels. The characteristic radius of the Gaussian beam is $\sigma = 0.391$, wavelength is $\lambda = 0.63$ μm, the ellipticity coefficient is $\alpha = 0.66$.

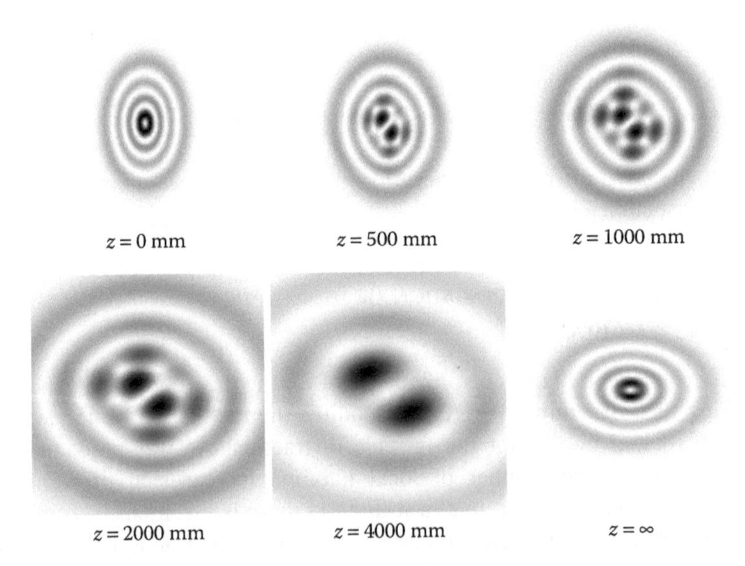

FIGURE 2.2 Propagation of the LG mode (3,–1) with elliptic distortion in free space. The ellipticity coefficient is 0.66.

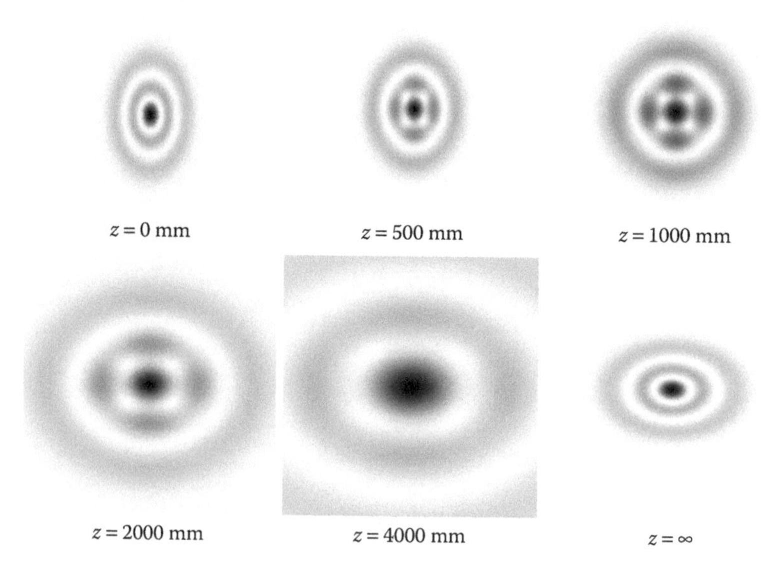

FIGURE 2.3 Propagation of the LG mode (2,0) with elliptic distortion in free space. The ellipticity coefficient is 0.66.

An interesting example is provided by the LG mode (2,0). In this case, no local minima are observed in the central part but there are curious effects in the first mode ring (Figure 2.3). At $m = 0$, the central part of the pattern is transformed to the greatest degree (acquiring a square shape) in the neighborhood of the Fresnel distance $z_0 \approx 1000$ mm. At the same time, the outer rings retain a circular shape instead of becoming ellipses. A similar effect is observed when the magnitude of the elliptic distortion is decreased (Figure 2.4). Nonetheless, the effect of image rotation by 90° in the Fourier plane still takes place.

Figure 2.4 depicts propagation of the LG mode (5,2) with a small ellipticity coefficient, $\alpha = 0.91$. In this case, the number of local maxima/minima is essentially decreased. There remain only two central local minima which allow the mode index, $m = 2$, to be determined. The simulated patterns of propagation of the LG modes show that the number of the near-center local maxima/minima depends on the orders (n, m) of the LG mode and on the ellipticity coefficient. Thus, they vary in number with varying ellipticity coefficient, that is, with the DOE inclination angle for the same LG modes.

We have studied analytically and numerically the propagation of the paraxial elliptic LG beam. It has been shown that, if the initial beam cross-section is in the form of concentric ellipses of zero axial intensity, then upon propagation the central axial intensity for the Fresnel diffraction will be non-zero for the non-zero singularity numbers m of the beam phase.

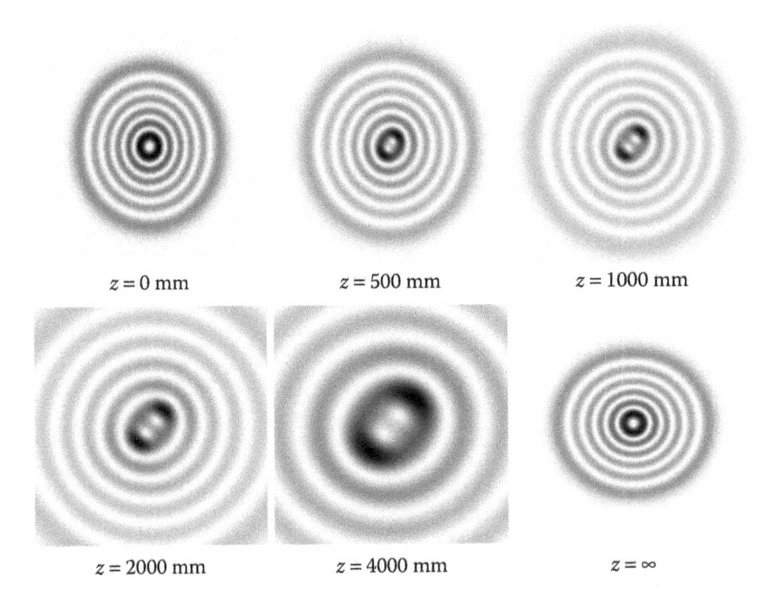

FIGURE 2.4 Propagation of the LG mode (5,2) with elliptic distortion in free space. The ellipticity coefficient is $\alpha = 0.91$.

Note that the axial phase singularity of the mth order "disintegrates" into p singularities of m/pth order, with p intensity zeroes occurring in the neighborhood of the beam cross-section axis. The number of zeroes p depends on the beam ellipticity coefficient, α. By way of illustration, for small ellipticity $(1 - \alpha^2) \ll 0$, the beam complex amplitude can effectively be represented as a superposition of two terms. As a result, there emerge two intensity zeroes ($p = 2$) of the $m/2$th singularity order for even m. These zeroes are found in the beam cross-section on a straight line that makes with the x-axis the angle of 45° for $m > 0$ and −45° for $m < 0$. Upon further axial propagation, the superposition terms with the distance-dependence defined as z^{-q}, $q = 0, \pm 1, \pm 2, \ldots$, will contribute less to the total sum. Thus, for $z \to \infty$ (far-field diffraction) an elliptically symmetric LG beam with mth order singularity and zero axial intensity is formed. The beam's cross-section is identical to the initial beam cross-section (at $z = 0$) but rotated by 90° with respect to it.

3 Hypergeometric Vortices

The Helmholtz equation that describes the propagation of a non-paraxial monochromatic light wave in uniform space admits of solutions with separable variables in 11 different coordinate systems [96]. This implies that there are electromagnetic fields that can propagate without changing their structure. The well-known Bessel modes offer an example [1]. A paraxial analog of the Helmholtz equations is represented by a parabolic equation of Schroedinger type that describes the propagation of paraxial light fields. This equation admits of solutions with separable variables in 17 coordinate systems [96]. The light fields that are described by the previous solutions preserve their structure up to a scale upon propagation. An example is provided by the well-known Hermite-Gauss and Laguerre-Gauss modes [2].

Recent years have seen a dramatic increase in the number of publications dealing with optics-related applications of separable-variable-based solutions of the Helmholtz and Schroedinger equations [4–13]. Novel non-paraxial light beams that preserve their structure upon propagation were discussed in [97,98]. These include parabolic [97] and Mathieu [98] beams. New paraxial light beams that preserve their structure up to scale were presented in Refs. [3,5,33,55,90,92,99,100], including Helmholtz-Gauss [99] and Laplace-Gauss [100] waves, Ince-Gauss modes [5] and elegant Ince-Gauss beams [90], Hermite-Laguerre-Gaussian beams [3], and pure light vortices [33]. Some of the aforementioned beams were realized using laser resonator [3,92], diffractive optical element (DOE) [33], and liquid-crystal microdisplay (LCD) [55].

3.1 HYPERGEOMETRIC MODES

In this section, we deal with yet another family of laser modes that form an orthonormalized basis and are defined by a separable-variable-based solution of the paraxial parabolic equation in the cylindrical coordinate system. In the cylindrical coordinate system, in addition to solutions in the form of Bessel and Laguerre-Gauss modes, the solution of Schroedinger's equation can be found as degenerate hypergeometric (HyG) functions (hence the term HyG modes). The intensity distribution in the cross-section of these beams is similar to that of the Bessel modes, being a set of concentric light rings, but whose amplitude drops faster toward the periphery as $1/r$ with increasing radial variable r. Similar to the Bessel modes, the HyG modes have infinite energy. However, as distinct from the Bessel modes, the radii of the light rings of the HyG modes increase as $z^{1/2}$ with increasing longitudinal coordinate z. An experiment on generating such laser modes with the use of a LCD is also described.

The complex amplitude $E(r, \varphi, z)$ of a paraxial light field satisfies the Schroedinger-type equation in the cylindrical coordinate system (r, φ, z):

$$\left(2ik\frac{\partial}{\partial z}+\frac{\partial^2}{\partial r^2}+\frac{1}{r}\frac{\partial}{\partial r}+\frac{1}{r^2}\frac{\partial^2}{\partial \varphi^2}\right)E(r,\varphi,z)=0, \tag{3.1}$$

where $k = 2\pi/\lambda$ is the wave number of light of wavelength λ. Obeying Eq. (3.1) are the functions that form an orthogonal basis:

$$\begin{aligned}E_{\gamma,n}(r,\varphi,z)&=\frac{1}{2\pi n!}\left(\frac{z_0}{z}\right)^{\frac{1}{2}}\Gamma\left(\frac{n+1+i\gamma}{2}\right)\\&\times\exp\left[-\frac{i\pi}{4}(n-i\gamma+1)-\frac{i\gamma}{2}\ln\frac{z_0}{z}+in\varphi\right]x^{\frac{n}{2}}{}_1F_1\left(\frac{n-1+i\gamma}{2},n+1,ix\right),\end{aligned} \tag{3.2}$$

where $-\infty < \gamma < \infty$, $n=0, \pm1, \pm2, \ldots$ are the continuous and the discrete parameters governing the functions in Eq. (3.2), which will be referred to as mode numbers; $z_0 = kw^2/2$ is an analog of Rayleigh distance; w is the mode parameter analogous to the radius of the Gaussian beam, although in the present context it has a different meaning; $x = kr^2/(2z)$, $\Gamma(\xi)$ is gamma-function; and ${}_1F_1(a, b, y)$ is a degenerate or confluent HyG function [101]:

$${}_1F_1(a,b,y)=\frac{\Gamma(b)}{\Gamma(a)\Gamma(b-a)}\int_0^1 t^{a-1}(1-t)^{b-a-1}\exp(yt)\,\mathrm{d}t, \tag{3.3}$$

where $\mathrm{Re}(b) > \mathrm{Re}(a) > 0$. The solution in Eq. (3.2) in other notations was presented in Ref. [96]. The solution in Eq. (3.2) can also be derived using the reference integral [38] ($\mathrm{Re}(p) > 0$, $\mathrm{Re}(\alpha + \nu) > 0$):

$$\begin{aligned}&\int_0^\infty t^{\alpha-1}\exp(-pt^2)J_\nu(ct)\,\mathrm{d}t\\&=c^\nu p^{-(\alpha+\nu)/2}2^{-\nu-1}\Gamma\left(\frac{\alpha+\nu}{2}\right)\Gamma^{-1}(\nu+1)\,{}_1F_1\left(\frac{\alpha+\nu}{2},\nu+1,-\frac{c^2}{4p}\right).\end{aligned} \tag{3.4}$$

The calculation of the Fresnel transform of the complex amplitude of light field at $z=0$

$$E_{\gamma,n}(r,\varphi,z=0)=\frac{1}{2\pi}\left(\frac{w}{r}\right)\exp\left[i\gamma\ln\left(\frac{r}{w}\right)+in\varphi\right] \tag{3.5}$$

reduces to the integral in Eq. (3.4). The solutions in Eq. (3.2) form a complete orthogonal basis and satisfy the following condition of orthogonality:

$$\int_0^\infty \int_0^{2\pi} E_{\gamma,n}(r,\varphi,z) E^*_{\mu,m}(r,\varphi,z) r dr d\varphi = \delta_{n,m}\delta(\gamma-\mu), \tag{3.6}$$

where $\delta(x)$ is the Dirac delta-function. The condition in Eq. (3.6) can easily be checked for the functions of Eq. (3.5) at $z = 0$, whereas for any z the orthogonality of the modes in Eq. (3.2) holds owing to the validity of the Parseval relation for the Fresnel transform. From (3.3) it can be seen that ${}_1F_1(a,b,y)$ is an integer analytical function. For Eq. (3.2), $\mathrm{Re}(y) = 0$, meaning that Eq. (3.3) is a one-dimensional Fourier-transform of a limited function on the interval (0,1). By Shannon theorem, as $r \to \infty$, the modulation period of the function (3.2) (that is the distance between the neighboring maximums or minimums) in terms of r^2 equals 2π. At large values of the argument, $x \gg 1$, the following asymptotics occur: $x^{n/2}\left|{}_1F_1\left((n+1-i\gamma)/2, n+1, ix\right)\right| \approx 1/x$. Thus, the modulus of the function (3.2) drops as $1/r$ at $r \gg 1$, that is, faster than the modulus of the Bessel function does. In addition, the zeros of the degenerate HyG function ${}_1F_1(a,b,y_{0m}) = 0$ are close to the zeros of the Bessel function $J_{b-1}(y_{b-1,m}) = 0$ [101]: $y_{0,m} \approx \left|y^2_{b-1,m}/(2b-4a)\right|$. The HyG modes in Eq. (3.2) possess infinite energy and are not square-integrable. They have a singular point at $r = 0$ and $z = 0$, having a finite intensity elsewhere. Light modes that satisfy the paraxial equation (3.1) and have infinite energy have been well known and widely used in optics. By way of illustration, there are two familiar types of the paraxial Bessel modes [31,96]:

$$E_{\beta,n}(r,\varphi,z) = J_n(k\beta r)\exp\left[-i\frac{k\beta^2 z}{2} + in\varphi\right], \tag{3.7}$$

$$E_{\rho,n}(r,\varphi,z) = (-i)^{n+1}\left(\frac{\rho}{z}\right) J_n\left(\frac{k\rho r}{z}\right)\exp\left[i\frac{k}{2z}\left(r^2+\rho^2\right) + in\varphi\right], \tag{3.8}$$

where:

β is a dimensionless parameter
ρ is the radius of an infinitely narrow ring at $z = 0$

It is noteworthy that the nonparaxial Bessel mode satisfying the Helmholtz equation is somewhat different from Eq. (3.7): $E_{\beta,n}(r,\varphi,z) = J_n(k\beta r)\exp\left[ikz\sqrt{1-\beta^2} + in\varphi\right]$. The parameters β and ρ specify the scale of the paraxial Bessel modes. In a similar way, the parameter γ specifies the scale of the HyG modes. Despite the fact that the solution in Eq. (3.2) satisfies Eq. (3.1) at any complex γ (Eqs. (3.3) and (3.4) are valid for the complex a), it is only solutions with real γ that form an orthogonal basis (when γ and μ are complex, only the orthogonality with respect to n and m is left in Eq. 3.6). At $z = 0$, the complex amplitude of the HyG mode (3.2) has an exponent that describes the transmittance of a logarithmic axicon $\exp\left(i\gamma\ln(r/w)\right)$. Because of this, similarly to a conventional conic axicon or a cone wave of amplitude $\exp\left(ik\gamma r\right)$, the sign of the γ parameter specifies some difference in the magnitude and scale of the Fresnel diffraction pattern for the HyG mode. For example, n being the same, the phase of light field in Eq. (3.5) has a positive derivative with respect to r at $\gamma > 0$ (a diverging wave), and a negative derivative with respect to r at $\gamma < 0$ (a converging wave). Thus, the diffraction pattern of the field in Eq. (3.5) at distance z will be characterized by larger-radius rings for $\gamma > 0$ than for $\gamma < 0$. At $\gamma = 0$, the HyG function in Eq. (3.2) is proportional to a spherical Bessel function of order $n/2$ [101]. The principal HyG mode is proportional to the Bessel function of zero order, which depends on the radial variable squared:

$$E_{0,0}(r,\varphi,z) = \frac{(1-i)}{2}\left(\frac{z_0}{2\pi z}\right)^{1/2} J_0(x/2)\exp(ix/2)$$

The transverse intensity distribution of the HyG mode (3.2) is represented by a set of concentric rings with their radii obeying the condition $r_m = \left(\alpha_m z\lambda/\pi\right)^{1/2}$, where α_m are constants that depend on the ring's number m and on the mode numbers (γ, n). From this equation it follows that the rings' radii increase as $z^{1/2}$ with increasing z. For comparison, note that the radii of the diffraction pattern rings are increasing linearly with increasing z for the paraxial Bessel modes (3.8), remaining unchanged for the modes in Eq. (3.7). The relation ${}_1F_1\left((n+1+i\gamma)/2, n+1, -ix\right) = \exp(-ix)\,{}_1F_1\left((n+1-i\gamma)/2, n+1, ix\right)$ [101] suggests that the phase of the HyG function is equal to $x/2$: $\arg\left\{{}_1F_1\left((n+1+i\gamma)/2, n+1, -ix\right)\right\} = -x/2$. Thus, the relation for the phase of the HyG mode is given by: $\arg\{E_{\gamma,n}(r,\varphi,z)\} = (\gamma/2)\ln(z/z_0) + n\varphi + kr^2/(4z) - (\pi/4)(n+1) + \arg\Gamma\left((n+i\gamma+1)/2\right)$, where the first term is an analog of Gouy phase, changing its sign at $z > z_0$.

Formation of Hypergeometric Modes

It is possible to generate the HyG modes (3.2) with the use of an optical element of transmittance (3.5). When the optical element (3.5) placed in the plane $z=0$ is illuminated by an unbounded plane wave, the light field of the complex amplitude (3.2) is generated at the distance z. To practically generate the mode (3.2), the optical element (3.5) needs to be used in combination with an annular diaphragm. As a result, the mode (3.2) will be effectively generated at a finite distance: $z_0 < R\left[\tan^{-1}\left(\gamma w/R\right)\right]^{-1}$, where R is the greater radius of the annular diaphragm. In propagation, the light field (3.2) preserves its structure, changing only in scale.

Figure 3.1 illustrates (a) the intensity, (b) the phase, and (c) the intensity radial section derived from the formula (3.2) for an HyG mode with the numbers (γ, n) = (2, 3) at distance $z = 1000$ mm. The wavelength is $\lambda = 633$ nm and $w = 1$ mm. In the experiment, the HyG modes were generated using a LCD "CRL OPTO" of resolution 1316×1024 pixels. A binary DOE 6.5 mm in diameter generated on the microdisplay was illuminated by a linearly polarized plane light beam from a solid-state laser of wavelength 532 nm and power 500 mW.

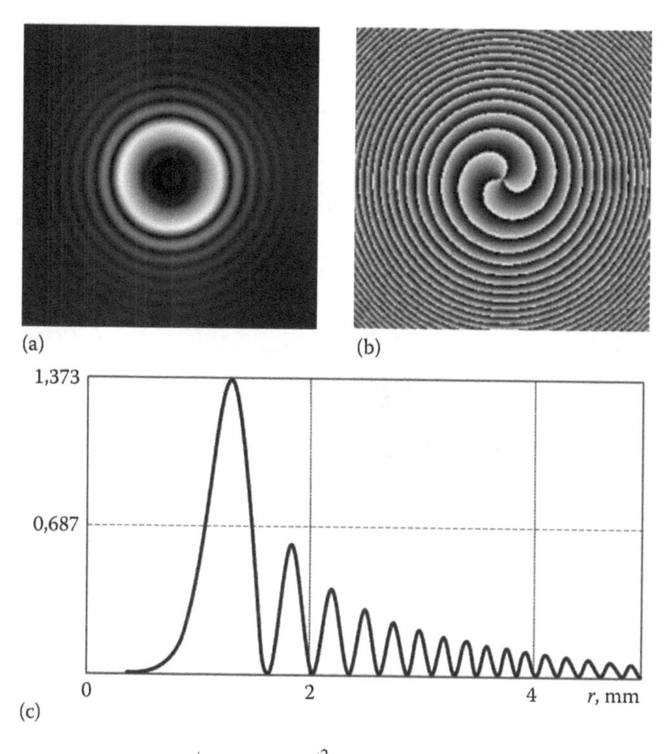

FIGURE 3.1 (a) The intensity $\left|E_{\gamma,n}(r,\phi,z)\right|^2$, (b) the phase, and (c) the intensity radial section derived from the formula (3.2) for a HyG mode with the numbers $(\gamma, n) = (2, 3)$ at distance $z = 1000$ mm. The frame size in (a) and (b) is 8 mm × 8 mm.

Figure 3.2a shows the DOE binary phase $S(r,\varphi)$ that satisfies the equation $S(r,\varphi) = \text{sign}\left\{\cos\left[\gamma \ln(r/w) + n\varphi + kr^2/(2f)\right]\right\}$, where sign($\xi$) is the sign-function and f is the focal length of a spherical lens. Note that the amplitude r^{-1} of the function (3.5) was replaced with a constant value. Figure 3.2b depicts the intensity distribution generated with the LCD with the phase as shown in Figure 3.2a, recorded at distance 700 mm from the display. Figure 3.3 depicts the intensity distributions for the HyG mode at $(\gamma, n) = (3, 3)$ recorded at different distances from the microdisplay. The diffraction pattern structure is seen to be preserved, slightly increasing in scale.

In this section we were the first to discuss theoretically and experimentally a new family of optical vortices, called HyG modes. Satisfying the Schroedinger-type paraxial equation, the HyG modes form an orthogonal basis of functions, preserve their structure up to scale upon propagation, and are close to the familiar Bessel modes. It is noteworthy that at $\gamma = -i$ the HyG modes change to a one-parameter family of pure optical vortices [33].

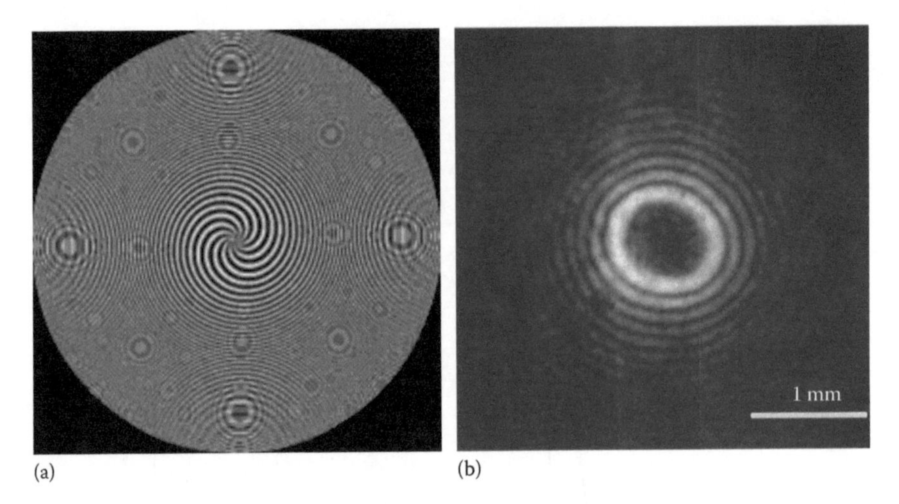

FIGURE 3.2 (a) The binary phase generated with a LCD and (b) the intensity distribution for the HyG-mode $(\gamma, n) = (5, 10)$ recorded with a CCD-camera at a distance of $z = 700$ mm from the display.

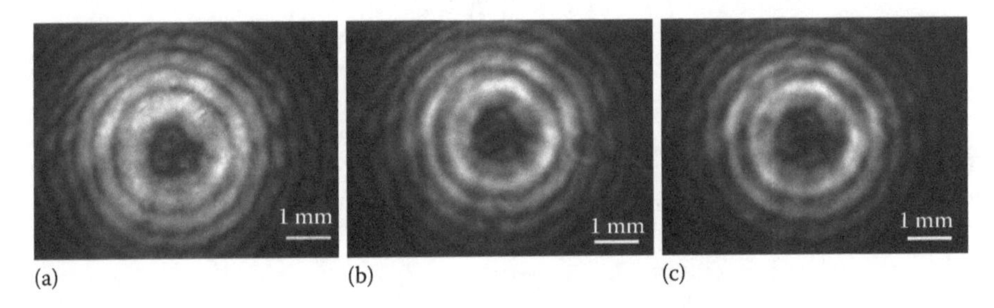

FIGURE 3.3 The intensity distribution for the HyG mode $(\gamma, n) = (3, 3)$ recorded with a CCD-camera at a distance of z: (a) 2 m, (b) 2.1 m, and (c) 2.2 m.

3.2 A FAMILY OF HYPERGEOMETRIC LASER BEAMS

The cutting-edge research has lately shown a significant increase of interest in various types of laser beams. Studies of the well-known Laguerre-Gauss modes have been under way [102]. Non-paraxial Laguerre-Gauss modes were discussed using the Wiegner functions in Ref. [103]. Studies have been continued into the elegant Laguerre-Gauss [9] and Hermite-Gauss beams [10]. These beams are described by polynomials with complex arguments. In Refs. [11,12] elliptic Laguerre-Gauss beams were studied using a number of techniques. A laser beam with elliptic symmetry can, for example, be generated through the oblique incidence of a plane wave onto an axicon. Laser beams generated by the axicon illuminated by the plane wave were discussed in Ref. [104]. A linear combination of laser modes has also attracted interest. By way of illustration, in Ref. [31] it was demonstrated that by fitting an axial linear combination of modes it was possible to obtain the laser beams with rotating cross-section intensity, but devoid of an orbital angular momentum. It is also possible to generate a laser vortex whose diffraction pattern is represented by a solitary bright ring, with the peripheral rings being suppressed [105]. Many of the above-listed beams (e.g., the Laguerre-Gauss modes, the Bessel modes, and the HyG modes) possess phase singularity (phase uncertainty on the optical axis). Such light fields are referred to as vortices. Recent years have seen the publication of a number of articles handling the aberration properties of such vortices [106–109]. In the recent publication by the present authors [6] a new family of paraxial laser beams—HyG modes—were discussed. Being similar to the Bessel modes [1,31], these modes have an infinite energy and can be generated only approximately by use of a DOE. Note that the HyG modes are generated using a logarithmic axicon which was for the first time used as an optical element to focus light into an axial line-segment of uniform intensity [110,111].

In this section, we discuss a generalization of the HyG modes, namely, HyG beams that are generated by a complex amplitude composed of four cofactors: the Gaussian exponent, a logarithmic axicon, a spiral phase plate (SPP), and an amplitude power function with a possible singularity at the origin of coordinates. For such type of beams, the near-field complex amplitude is proportional to the degenerate HyG function, prompting the beams' name—HyG beams. These beams change to the HyG modes when the initial Gaussian beam is replaced with a plane wave or the Gaussian beam's waist radius tends to infinity. For such beams, the cross-section intensity distribution is similar to that of the Bessel modes. The far-field complex amplitude of some types of such beams is shown to be proportional to the Bessel function of integer or semi-integer order. However, the thickness of alternating bright and dark rings decreases with increasing number.

General Form of the Hypergeometric Beams

Let there be a light field with the initial amplitude given by

$$E_{\gamma nm}(r,\varphi)=\frac{1}{2\pi}\left(\frac{r}{w}\right)^m \exp\left(-\frac{r^2}{2\sigma^2}+i\gamma\ln\frac{r}{w}+in\varphi\right), \tag{3.9}$$

where:

- (r,φ) are the polar coordinates in the initial plane $(z=0)$
- w and γ are the real parameters
- σ is the Gaussian beam waist radius
- n is the SPP's integer order
- m is the real value of the amplitude factor's exponent

The complex amplitude in Eq. (3.9) describes the light field of infinite energy and with an amplitude singularity at $r=0$ and $m<0$. Despite this fact, in any other transverse plane at distance z from the initial plane the complex amplitude of the light field generated by the function (3.9) will have no more singularity, being finite. Besides, at $\gamma\neq 0$ the complex amplitude (3.9) has a phase singularity at the center, $r=0$.

With the paraxial propagation of the light field in Eq. (3.9) its complex amplitude at distance z will be determined by the Fresnel transform, which takes the following form in the polar coordinates:

$$E(\rho,\theta,z)=-\frac{ik}{2\pi z}\iint_{R^2} E(r,\varphi,0)\exp\left\{\frac{ik}{2z}\left[\rho^2+r^2-2\rho r\cos(\varphi-\theta)\right]\right\} r\mathrm{d}r\mathrm{d}\varphi, \quad (3.10)$$

where (ρ,θ) are the polar coordinates in a plane perpendicular to the optical axis and offset by distance z from the initial plane. The function (3.9) depends on the azimuth angle φ as a harmonic function: $E(r,\varphi)=A(r)\exp(in\varphi)$. For such a function, instead of Eq. (3.10) we can write

$$E(\rho,\theta,z)=(-i)^{n+1}\frac{k}{z}\exp\left(\frac{ik\rho^2}{2z}+in\theta\right)\int_0^\infty A(r)\exp\left(\frac{ikr^2}{2z}\right)J_n\left(\frac{k\rho r}{z}\right) r\mathrm{d}r, \quad (3.11)$$

where $J_n(x)$ is the Bessel function of the nth order. There is a familiar reference integral [38]:

$$\begin{aligned}&\int_0^\infty x^{\alpha-1}\exp(-px^2)J_\nu(cx)\mathrm{d}x\\&=c^\nu p^{-\frac{\nu+\alpha}{2}}2^{-\nu-1}\Gamma\left(\frac{\nu+\alpha}{2}\right)\Gamma^{-1}(\nu+1)\,{}_1F_1\left(\frac{\nu+\alpha}{2},\nu+1,-\frac{c^2}{4p}\right),\end{aligned} \quad (3.12)$$

where ${}_1F_1(a,b,x)$ is a degenerate HyG function, or Kummer function, where $\Gamma(x)$ is a Γ-function. In view of Eq. (3.12), the Fresnel transform (3.11) for the function (3.9) is given by

$$\begin{aligned}E_{\gamma nm}(\rho,\theta,z)=&\frac{(-i)^{n+1}}{2\pi n!}\left(\frac{z_0}{zq^2}\right)\left(\frac{\sqrt{2}\sigma}{wq}\right)^{m+i\gamma}\left(\frac{k\sigma\rho}{\sqrt{2}qz}\right)^n\exp\left(\frac{ik\rho^2}{2z}+in\theta\right)\\&\times\Gamma\left(\frac{n+m+2+i\gamma}{2}\right){}_1F_1\left(\frac{n+m+2+i\gamma}{2},n+1,-\left(\frac{k\sigma\rho}{\sqrt{2}qz}\right)^2\right),\end{aligned} \quad (3.13)$$

where $z_0=k\sigma^2$, $q=(1-iz_0/z)^{1/2}$.

The modulus of the complex amplitude in Eq. (3.13) is proportional to the Kummer function:

$$\left|E_{\gamma nm}(\rho,\theta,z)\right| \sim x^{\frac{n}{2}}\,{}_1F_1(a,b,-x), \tag{3.14}$$

where x is the complex argument:

$$x = \left(\frac{k\sigma\rho}{\sqrt{2}qz}\right)^2. \tag{3.15}$$

The Kummer function can be expressed as a series:

$${}_1F_1(a,b,-x) = \sum_{l=0}^{\infty} C_l(-1)^l x^l, \tag{3.16}$$

$$C_l = \frac{\Gamma(a+l)}{\Gamma(a)}\frac{\Gamma(b)}{\Gamma(b+l)l!}. \tag{3.17}$$

In view of Eq. (3.13), the Kummer function in Eq. (3.16) can be given by the following series:

$${}_1F_1(a,b,-x) = \sum_{l=0}^{\infty}(-i)^l C_l\left[\frac{k\sigma\rho}{\sqrt{2}z\left(1+z_0^2/z^2\right)^{1/4}}\right]^{2l}\exp\left[-il\tan^{-1}\left(\frac{z}{z_0}\right)\right]. \tag{3.18}$$

From Eq. (3.18) it follows that both amplitude and phase of each term of the function $\left|{}_1F_1\right|$ vary with varying z. This implies that at $l = \text{const}$ each "partial" light field in Eq. (3.18) will propagate in space at its own phase velocity defined by the factor $\exp\left[-il\tan^{-1}(z/z_0)\right]$. Following the axial interference of all terms in Eq. (3.18), the modulus of the function (3.18), and, hence, that of the complex amplitude of the light field in Eq. (3.13), will change its form upon propagation.

Hypergeometric Beams in the Near-Field

At $z \ll z_0 = k\sigma^2$, Eq. (3.13) ceases to be dependent on σ and z_0, because

$$\frac{1}{2\sigma^2} - \frac{ik}{2z} \approx -\frac{ik}{2z},\; q^2 \approx -i\frac{z_0}{z}. \tag{3.19}$$

Then, instead of Eq. (3.13), we can write

$$E_{\gamma nm}(\rho,\theta,z \ll z_0) = \frac{(-i)^{\frac{n-m-i\gamma}{2}}}{2\pi n!}\left(\frac{kw^2}{2z}\right)^{-\frac{m+i\gamma}{2}}\left(\frac{k\rho^2}{2z}\right)^{\frac{n}{2}}\exp\left(\frac{ik\rho^2}{2z}+in\theta\right)$$
$$\times\Gamma\left(\frac{n+m+2+i\gamma}{2}\right){}_1F_1\left(\frac{n+m+2+i\gamma}{2},n+1,-\frac{ik\rho^2}{2z}\right). \quad (3.20)$$

From Eq. (3.20) it follows that at $z \ll z_0$ the modulus of complex amplitude $|E_{\gamma nm}(\rho,\theta,z \ll z_0)|$ will retain its form, changing only in scale. Note that at $\sigma \to \infty$ (the Gaussian beam is replaced by a plane wave), Eq. (3.13) also changes to Eq. (3.20) that describes the paraxial mode beams and generalized HyG modes [6].

Hypergeometric Beams in the Far-Field

If $z \gg z_0 = k\sigma^2$ then $q = (1 - iz_0/z)^{1/2} \approx 1$. Then, Eq. (3.13) can be rewritten as

$$E_{\gamma nm}(\rho,\theta,z \gg z_0) = \frac{(-i)^{n+1}}{2\pi n!}\left(\frac{z_0}{z}\right)\left(\frac{\sqrt{2}\sigma}{w}\right)^{m+i\gamma}\left(\frac{k\sigma\rho}{\sqrt{2}z}\right)^{n}\exp\left(\frac{ik\rho^2}{2z}+in\theta\right)$$
$$\times\Gamma\left(\frac{n+m+2+i\gamma}{2}\right){}_1F_1\left(\frac{n+m+2+i\gamma}{2},n+1,-\left(\frac{k\sigma\rho}{\sqrt{2}z}\right)^2\right). \quad (3.21)$$

The z-dependence of the diffraction pattern is qualitatively changed. Both near- and far-field diffraction patterns get represented by a set of concentric rings with increasing spatial frequency, since the amplitude distribution depends on ρ^2. Note, however, that the diffraction pattern remains unchanged in the near-field zone due to the constant ratio ρ^2/z and in the far-field zone due to the constant ρ/z. That is, for the near-field pattern the diffraction ring radii increase slower with increasing z, compared with the far-field: near-field diffraction rings increase proportionally to $z^{1/2}$, whereas the far-field rings are proportional to z.

Particular Cases of the Hypergeometric Beams

At $m = -1$ and $z = 0$, Eq. (3.1) takes the form:

$$E_{\gamma,n,-1}(r,\varphi) = \frac{1}{2\pi}\left(\frac{w}{r}\right)\exp\left(-\frac{r^2}{2\sigma^2}+i\gamma\ln\frac{r}{w}+in\varphi\right), \quad (3.22)$$

while the complex amplitude of such HyG beams at any z will be given by

$$E_{\gamma,n,-1}(\rho,\theta,z)=\frac{(-i)^{n+1}}{2\pi n!}\left(\frac{z_0}{zq^2}\right)\left(\frac{\sqrt{2}\sigma}{wq}\right)^{-1+i\gamma}\left(\frac{k\sigma\rho}{\sqrt{2}qz}\right)^{n}\exp\left(\frac{ik\rho^2}{2z}+in\theta\right)$$
$$\times\Gamma\left(\frac{n+1+i\gamma}{2}\right){}_1F_1\left(\frac{n+1+i\gamma}{2},\,n+1,\,-\left(\frac{k\sigma\rho}{\sqrt{2}qz}\right)^2\right). \tag{3.23}$$

When the Gaussian beam is replaced by a plane unbounded wave ($\sigma\rightarrow\infty$), Eq. (3.23) turns to

$$E_{\gamma,n,-1}(\rho,\theta,z)=\frac{(-i)^{\frac{n+1-i\gamma}{2}}}{2\pi n!}\left(\frac{kw^2}{2z}\right)^{\frac{1-i\gamma}{2}}\left(\frac{k\rho^2}{2z}\right)^{\frac{n}{2}}\exp\left(\frac{ik\rho^2}{2z}+in\theta\right)$$
$$\times\Gamma\left(\frac{n+1+i\gamma}{2}\right){}_1F_1\left(\frac{n+1+i\gamma}{2},\,n+1,\,-\frac{ik\rho^2}{2z}\right). \tag{3.24}$$

Based on the well-known relation for the Kummer function

$${}_1F_1(a,b,z)=\exp(z)\,{}_1F_1(b-a,b,-z),$$

we obtain an expression for the complex amplitude of the HyG modes, instead of Eq. (3.24) [6,31]:

$$E_{\gamma,n,-1}(\rho,\theta,z)=\frac{w^{-n}\rho^n}{2\pi n!}\left(-\frac{ikw^2}{2z}\right)^{\frac{n+1-i\gamma}{2}}$$
$$\times\exp(in\theta)\Gamma\left(\frac{n+1+i\gamma}{2}\right){}_1F_1\left(\frac{n+1-i\gamma}{2},\,n+1,\,\frac{ik\rho^2}{2z}\right). \tag{3.25}$$

The light field in Eq. (3.25) possesses an infinite energy, similarly to the Bessel modes [1], but unlike them, increases the diameter (of the principal diffraction ring) in direct proportion to $z^{1/2}$ as it propagates. Note that the intensity function $\left|E_{\gamma,n,-1}(\rho,\theta,z)\right|^2$ changes only in scale, with the diffraction pattern remaining structurally the same.

It should be noted that the reference integral (3.12) is defined at $\mathrm{Re}(p)>0$ [38], which is not the case at $\sigma\rightarrow\infty(\mathrm{Re}(p)=0)$. However, when Eq. (3.25) is substituted in the paraxial equation of light propagation

$$\left(2ik\frac{\partial}{\partial z}+\frac{\partial^2}{\partial\rho^2}+\frac{1}{\rho}\frac{\partial}{\partial\rho}+\frac{1}{\rho^2}\frac{\partial}{\partial\theta^2}\right)U(\rho,\theta,z)=0, \tag{3.26}$$

we obtain a true identity, namely, the differential Kummer equation:

$$\left[\chi\frac{\partial^2}{\partial\chi^2}+(b-\chi)\frac{\partial}{\partial\chi}-a\right]{}_1F_1(a,b,\chi)=0, \tag{3.27}$$

$$a=\frac{(n+1-i\gamma)}{2},\, b=n+1,\, \chi=\frac{ik\rho^2}{(2z)}. \tag{3.28}$$

Thus, we have demonstrated that the HyG modes [6] are a particular case of the HyG beams in Eqs. (3.13) and (3.20).

Hypergeometric Modes at $\gamma=0$

Let us consider a one-parameter family of HyG modes of Eq. (3.25) at $\gamma=0$. In this case, the initial amplitude function in Eq. (3.22) at $z=0$ will only be characterized by an amplitude singularity at $\rho=0$, whereas the logarithmic phase singularity at $\rho=0$ will not be there any more due to $\gamma=0$. We shall use a familiar relation for the confluent function [101]

$${}_1F_1\left(\frac{n}{2}+\frac{1}{2},2\frac{n}{2}+1,\frac{ik\rho^2}{2z}\right)=\Gamma\left(1+\frac{n}{2}\right)\exp\left(\frac{ik\rho^2}{4z}\right)\left(\frac{k\rho^2}{8z}\right)^{-\frac{n}{2}}J_{\frac{n}{2}}\left(\frac{k\rho^2}{4z}\right). \tag{3.29}$$

On substituting Eq. (3.29) into Eq. (3.25) we get

$$E_{0,n,-1}(\rho,\theta,z)=\frac{w}{2}\left(\frac{k}{2\pi z}\right)^{\frac{1}{2}}\exp\left[-i\frac{\pi}{4}(n+1)\right]\exp\left(\frac{ik\rho^2}{4z}+in\theta\right)J_{\frac{n}{2}}\left(\frac{k\rho^2}{4z}\right). \tag{3.30}$$

From Eq. (3.30) the complex amplitude of the HyG modes at $\gamma=0$ is seen to be proportional to the Bessel function of the first kind and of the integer and semi-integer orders. It is noteworthy that the above-discussed fields (both the HyG beams (3.13) and modes (3.25)) depend on the radial variable squared. This means that with increasing variable ρ the distance between the adjacent local maxima or minima will be decreased. An unlimited increase in the spatial frequency of the diffraction pattern is due to a (amplitude or phase) singularity of the complex amplitude at the initial plane (or waist) center. In practice, in order for such laser beams to be formed one should employ an annular diaphragm that would "block up" the center-coordinate singularity. In this case, however, the HyG beams can be generated only approximately.

Let us consider the first spiral mode at $n=1$. Then, instead of Eq. (3.30), we get the complex amplitude

$$E_{0,1,-1}(\rho,\theta,z)=\frac{-iw}{\pi\rho}\exp\left(\frac{ik\rho^2}{4z}+i\theta\right)\sin\left(\frac{k\rho^2}{4z}\right). \tag{3.31}$$

The zeros of the complex amplitude (3.31) have the coordinates $\rho_m=(2m\lambda z)^{1/2}$. The amplitude first maximum is derived from the equation $2x=\mathrm{tg}(x)$, being located near the point $\rho_{\max} < (\lambda z)^{1/2}$. Note that the distance between two adjacent intensity zeros of the field in Eq. (3.31) with the numbers m and $m+1$ at $m \gg 1$ is given by $\Delta\rho_m \approx \left[\lambda z/(2m)\right]^{1/2}$. Thus, the spatial frequency of the diffraction pattern tends to infinity, whereas $\Delta\rho_m \to 0$ at $\rho \to \infty$. This can be eliminated by bounding the initial field (3.22) by a diaphragm that blocks the central singularity.

Generalized Hypergeometric Modes

Let us consider more general functions than those given by Eq. (3.17), describing the HyG modes:

$$E(\rho,\theta,0) = \rho^p z^q \exp(in\theta) F\left(s\rho^m z^l\right), \tag{3.32}$$

where F is some function. Substituting the function in Eq. (3.32) into the paraxial equation of light propagation in Eq. (3.26) yields the following relation:

$$\begin{aligned} &s^2 m^2 \rho^{2m-2} z^{2l} F'' + \left[(2p+m) sm\rho^{m-2} z^l + 2iksl\rho^m z^{l-1}\right] F' \\ &+\left[p(p-1)\rho^{-2} + p\rho^{-2} - n^2\rho^{-2} + 2ikqz^{-1}\right] F = 0. \end{aligned} \tag{3.33}$$

At $p = \pm n$, Eq. (3.33) is simplified. For certainty, consider the case when $p = +n$:

$$s^2 m^2 \rho^{2m-2} z^{2l} F'' + \left[sm(2n+m)\rho^{m-2} z^l + 2iksl\rho^m z^{l-1}\right] F' + 2ikqz^{-1} F = 0. \tag{3.34}$$

To transform Eq. (3.34) to the Kummer equation (3.27), the coefficient of the second-order derivative needs to be equal to the argument of the function F. To these ends, dividing both sides of the equation by $sm^2\rho^{m-2}z^l$, we obtain

$$s\rho^m z^l F'' + \left(\frac{2n+m}{m} + \frac{2ikl\rho^2 z^{-1}}{m^2}\right) F' + \frac{2ikq\rho^{2-m} z^{-l-1}}{sm^2} F = 0. \tag{3.35}$$

Next, it is required that the coefficient of the first-order derivative should be equal to the difference between the constant and the argument of the function F. This is only possible when

$$m = 2, l = -1, s = -\frac{2ikl}{m^2} = \frac{ik}{2}. \tag{3.36}$$

Then, we get, instead of Eq. (3.35):

$$\frac{ik}{2}\rho^2 z^{-1}F'' + \left(n+1-\frac{ik}{2}\rho^2 z^{-1}\right)F' + qF = 0. \quad (3.37)$$

Thus, we derive the Kummer equation (3.27), in which

$$a = -q, b = n+1. \quad (3.38)$$

The solution of this equation is given by a confluent function of the form:

$${}_1F_1\left(a, n+1, \frac{ik\rho^2}{2z}\right), \quad (3.39)$$

that is, the solution of the paraxial equation of propagation, Eq. (3.26), is given by the function

$$E_{a,+n}(\rho,\theta,z) = \rho^n z^{-a} \exp(in\theta)\, {}_1F_1\left(a, n+1, \frac{ik\rho^2}{2z}\right). \quad (3.40)$$

At $p = -n$, in a similar way, we derive another equation:

$$E_{a,-n}(\rho,\theta,z) = \rho^{-n} z^{-a} \exp(in\theta)\, {}_1F_1\left(a, 1-n, \frac{ik\rho^2}{2z}\right). \quad (3.41)$$

By uniting Eqs. (3.40) and (3.41), we obtain a family of functions:

$$E_{an}(\rho,\theta,z) = \rho^n z^{-a} \exp(\pm in\theta)\, {}_1F_1\left(a, n+1, \frac{ik\rho^2}{2z}\right). \quad (3.42)$$

The complex amplitude in Eq. (3.42) can be considered as a function describing the generalized paraxial HyG modes, which transfer to the ordinary HyG modes [6] at $a = (n+1+i\gamma)/2$. Note that Eq. (3.20) is another special case of the solution in Eq. (3.42) at $a = (n+m+2+i\gamma)/2$.

Diffraction of the Gaussian Beam by a Spiral Logarithmic Axicon

Let us consider a particular case of the light field (1) at $m = 0$:

$$E_{\gamma n0}(r,\varphi) = \frac{1}{2\pi} \exp\left(-\frac{r^2}{2\sigma^2} + i\gamma \ln\frac{r}{w} + in\varphi\right). \quad (3.43)$$

The complex amplitude in Eq. (3.43) describes the diffraction of the Gaussian beam by a spiral logarithmic axicon. The complex amplitude in the Fresnel diffraction zone is derived from the general relation in Eq. (3.13) at $m=0$:

$$E_{\gamma n0}(\rho,\theta,z)=\frac{(-i)^{n+1}}{2\pi n!}\left(\frac{z_0}{zq^2}\right)\left(\frac{\sqrt{2}\sigma}{wq}\right)^{i\gamma}\left(\frac{k\sigma\rho}{\sqrt{2}qz}\right)^{n}\exp\left(\frac{ik\rho^2}{2z}+in\theta\right)$$
$$\times\Gamma\left(\frac{n+2+i\gamma}{2}\right){}_1F_1\left[\frac{n+2+i\gamma}{2},\ n+1,\ -\left(\frac{k\sigma\rho}{\sqrt{2}qz}\right)^2\right]. \quad (3.44)$$

From Eq. (3.44), we obtain the relation for the near-field complex amplitude similar to Eq. (3.20):

$$E_{\gamma n0}(\rho,\theta,z\ll z_0)=\frac{(-i)^{\frac{n-i\gamma}{2}}}{2\pi n!}\left(\frac{kw^2}{2z}\right)^{\frac{i\gamma}{2}}\left(\frac{k\rho^2}{2z}\right)^{\frac{n}{2}}\exp\left(\frac{ik\rho^2}{2z}+in\theta\right)$$
$$\times\Gamma\left(\frac{n+2+i\gamma}{2}\right){}_1F_1\left(\frac{n+2+i\gamma}{2},n+1,-\frac{ik\rho^2}{2z}\right). \quad (3.45)$$

Equation (3.45) suggests that the near-field intensity pattern will remain unchanged at points where $k\rho^2/(2z)=\text{const}$. That is, with changing z the near-field diffraction pattern remains structurally unchanged up to scale.

Similarly, to Eq. (3.21), for the far-field diffraction pattern we have:

$$E_{\gamma n0}(\rho,\theta,z\gg z_0)=\frac{(-i)^{n+1}}{2\pi n!}\left(\frac{z_0}{z}\right)\left(\frac{\sqrt{2}\sigma}{w}\right)^{i\gamma}y^{\frac{n}{2}}$$
$$\times\exp\left(\frac{ik\rho^2}{2z}+in\theta\right)\Gamma\left(\frac{n+2+i\gamma}{2}\right){}_1F_1\left(\frac{n+2+i\gamma}{2},n+1,-y\right), \quad (3.46)$$

where $y=\left[k\sigma\rho/(\sqrt{2}z)\right]^2$.

Let us show that at $\gamma=0$ (in the absence of the logarithmic axicon) the complex amplitude in Eq. (3.46) reduces to the familiar expression for the Fraunhofer diffraction of the Gaussian beam by the SPP [58]. Actually, putting $\gamma=0$ in Eq. (3.46), we obtain:

$$E_{0n0}(\rho,\theta,z\gg z_0)$$
$$=\frac{(-i)^{n+1}}{2\pi n!}\left(\frac{z_0}{z}\right)\Gamma\left(\frac{n+2}{2}\right)\exp\left(\frac{ik\rho^2}{2z}+in\theta\right)y^{\frac{n}{2}}{}_1F_1\left(\frac{n+2}{2},n+1,-y\right). \quad (3.47)$$

From the recurrent relations for the Kummer function it follows that [101]:

$$ {}_1F_1\left(\frac{n+2}{2}, n+1, -y\right) = \exp(-y)\,{}_1F_1\left(\frac{n}{2}, n+1, y\right), \tag{3.48}$$

$$ {}_1F_1'\left(\frac{n}{2}, n+1, y\right) = \frac{n}{2(n+1)}\,{}_1F_1\left(\frac{n+2}{2}, n+2, y\right), \tag{3.49}$$

where ${}_1F_1'$ is the derivative of the Kummer function. Whence,

$$\begin{aligned} {}_1F_1\left(\frac{n}{2}, n+1, y\right) &= 1 + \frac{n}{2(n+1)} \int_0^y {}_1F_1\left(\frac{n+2}{2}, n+2, y\right) \mathrm{d}y \\ &= 1 + \frac{n}{2(n+1)} \Gamma\left(\frac{n+3}{2}\right) \int_0^y \exp\left(\frac{y}{2}\right)\left(\frac{y}{4}\right)^{-(n+1)/2} I_{(n+1)/2}\left(\frac{y}{2}\right) \mathrm{d}y. \end{aligned} \tag{3.50}$$

Based on a familiar reference integral [38]:

$$\int_0^x x^{-\nu} \exp(x) I_\nu(x) \mathrm{d}x = -\frac{x^{-\nu+1}}{2\nu-1} \exp(x)\left[I_\nu(x) - I_{\nu-1}(x)\right] - \frac{2^{1-\nu}}{(2\nu-1)\Gamma(\nu)}, \tag{3.51}$$

Eq. (3.50) can be rearranged as

$$\begin{aligned} &{}_1F_1\left(\frac{n}{2}, n+1, y\right) \\ &= \Gamma\left(\frac{n+1}{2}\right) 2^{(n-1)/2} \left(\frac{y}{2}\right)^{-(n-1)/2} \exp\left(\frac{y}{2}\right)\left[I_{(n-1)/2}\left(\frac{y}{2}\right) - I_{(n+1)/2}\left(\frac{y}{2}\right)\right]. \end{aligned} \tag{3.52}$$

Then, substituting Eq. (3.52) into (3.48), and Eq. (3.48) into (3.47), the complex amplitude in Eq. (3.47) (at $\gamma = 0, m = 0$) can finally be brought to

$$\begin{aligned} E_{0n0}(\rho, \theta, z \gg z_0) &= \frac{(-i)^{n+1}}{\sqrt{2\pi}} \left(\frac{k\sigma^2}{2z}\right) \sqrt{\xi} \\ &\times \exp\left(\frac{ik\rho^2}{2z} + in\theta\right) \exp(-\xi)\left[I_{(n-1)/2}(\xi) - I_{(n+1)/2}(\xi)\right], \end{aligned} \tag{3.53}$$

where $\xi = \left[k\sigma\rho/(2z)\right]^2$. In deriving Eq. (3.53) we used the property of the Γ-function [38]:

$$\Gamma\left(\frac{n+1}{2}\right)\Gamma\left(\frac{n+2}{2}\right)=\frac{\sqrt{\pi}\,n!}{2^n}. \tag{3.54}$$

The expression (3.53) has the same form as Eq. (3.28) of Ref. [58] derived for the Fraunhofer diffraction of the Gaussian beam by the SPP, if we take into account that the waist radius w in Ref. [58] and the radius σ are related as $w=\sqrt{2}\sigma$.

Laguerre-Gauss Modes of the $(0,n)$th Order

Let us study one more particular case of the HyG beams (3.9), namely, the Laguerre-Gaussian (LG) modes of order $(0,n)$. In Eq. (3.13), putting $\gamma=0$ and $m=n$, we get:

$$\begin{aligned} &E_{0nn}(\rho,\theta,z)\\ &=\frac{(-i)^{n+1}}{2\pi}\left(\frac{z_0}{zq^2}\right)\left(\frac{k\sigma^2\rho}{wq^2z}\right)^n\exp\left(\frac{ik\rho^2}{2z}+in\theta\right){}_1F_1\left(n+1,n+1,-\left(\frac{k\sigma\rho}{\sqrt{2}qz}\right)^2\right) \\ &=\frac{(-i)^{n+1}}{2\pi}\left(\frac{z_0}{zq^2}\right)\left(\frac{k\sigma^2\rho}{wq^2z}\right)^n\exp(in\theta)\exp\left[-\frac{\rho^2}{2\left(\sigma^2+i\,z/k\right)}\right]. \end{aligned} \tag{3.55}$$

Designating $\tau(z)=\sigma\left(1+i\,z/z_0\right)^{1/2}$, we obtain, instead of Eq. (3.55):

$$E_{0nn}(\rho,\theta,z)=-\frac{i}{2\pi}\left(\frac{z_0}{zq^2}\right)\left[\frac{\rho}{w(z)}\right]^n\exp(in\theta)\exp\left[-\frac{\rho^2}{2\tau^2(z)}\right], \tag{3.56}$$

where $w(z)=w\left(1+i\,z/z_0\right)$. From Eq. (3.56), the explicit form of the amplitude and phase components is given by

$$\begin{aligned} E_{0nn}(\rho,\theta,z)&=\frac{1}{2\pi}\left(\frac{\rho}{w}\right)^n\left(1+\frac{z^2}{z_0^2}\right)^{-(n+1)/2} \\ &\times\exp\left[in\theta-i(n+1)\tan^{-1}\left(\frac{z}{z_0}\right)-\frac{\rho^2}{2\sigma^2(z)}+\frac{ik\rho^2}{2R(z)}\right], \end{aligned} \tag{3.57}$$

where $\sigma(z)=\sigma\left(1+z^2/z_0^2\right)^{1/2}$, $R(z)=z\left(1+z_0^2/z^2\right)$. When $w=\sigma$, the complex amplitude of Eq. (3.57) describes the LG mode with zero radial index (up to a normalization constant) [2].

Numerical Simulation

In this section, we discuss the results of numerical simulation derived by directly taking the Fresnel transform (3.11) using the rectangle technique, with 1000 pixels per 10 mm on the ρ-axis.

In the following, we calculate in which way the amplitude $\left|E_{\gamma nm}(\rho)\right|$ depends on the radial variable ρ at $n = 10$, $m = \gamma = 0$. Assume that $\lambda = 514.5$ nm, $w = 0.5$ mm, the plane wave radius is ($\sigma = \infty$) $R = 10$ mm, and the distance from the initial plane is $z = 2000$ mm. In this case, the radial section of the Fresnel diffraction pattern of the bounded plane wave by an SPP is shown in Figure 3.4a.

With increasing $n > 0$, the diffraction pattern in Figure 3.4a is beginning to "shrink," with the left and right edges thereof merging at some n_0. Let us estimate at which value of n_0 this will occur. The effective radius of any paraxial finite-energy beam is increasing with increasing z, similarly to the radius of the Gaussian beam [112]:

$$R(z) = R\sqrt{1 + \left(\frac{z}{z_0}\right)^2}, \quad z_0 = \frac{kR^2}{2}, \tag{3.58}$$

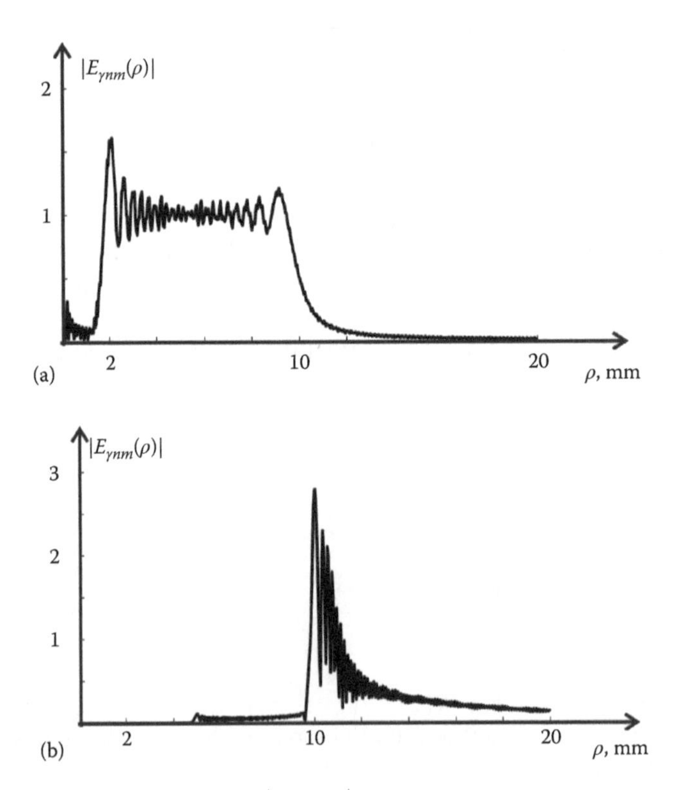

FIGURE 3.4 Amplitude distribution $\left|E_{\gamma nm}(\rho)\right|$ (in relative units) versus the radial variable at distance $z = 2000$ mm for different n: (a) 10, (b) 300.

where R is the beam's initial radius at $z = 0$. With increasing z, the radius of the axial optical vortex in the near-field zone is increasing as [33]:

$$\rho_0(z) = \sqrt{\frac{4\alpha_n z}{k}}, \; \alpha_n \approx 0.54n + 1. \tag{3.59}$$

The left edge of the diffraction pattern in Figure 3.4a is about the same as the right edge when

$$R(z) = \rho_0(z). \tag{3.60}$$

Then, from Eq. (3.60) we obtain an estimate for n_0:

$$n_0 \approx \frac{kR^2}{2z} + \frac{2z}{kR^2} - 2 \approx 250. \tag{3.61}$$

Shown in Figure 3.4b is the radial amplitude distribution $|E_{\gamma nm}(\rho)|$ for $m = \gamma = 0$, $\lambda = 514.5$ nm, $z = 2000$ mm, $R = 10$ mm, $w = 0.5$ mm, and $n = 300$. It is seen from Figure 3.4b that the radius of the principal maximum is $\rho_0 \approx R = 10$ mm and $|E_{\gamma nm}(\rho_0)| \approx 2.75$.

The plots in Figure 3.5 depict the radius ρ_0 (a) and the value $|E_{\gamma nm}(\rho)|$ (b) of the principal maximum against the singularity number n.

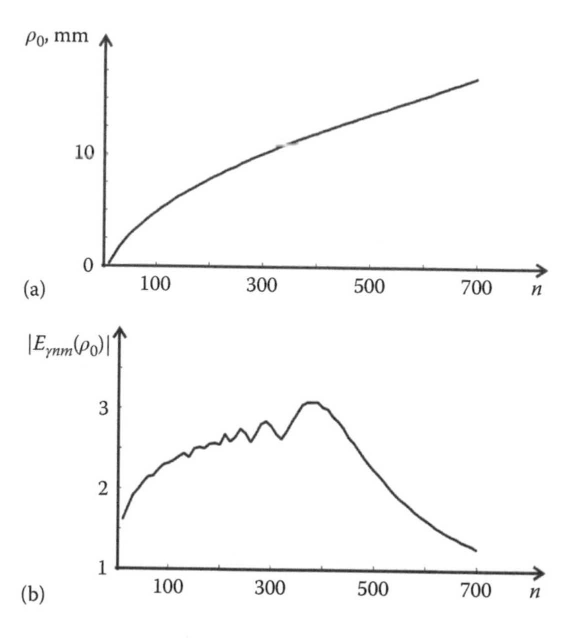

FIGURE 3.5 (a) Radius and (b) magnitude of the diffraction pattern principal maximum versus the singularity number.

It is seen from Figure 3.5 that the amplitude $|E_{\gamma nm}(\rho)|$ reaches on the ring its maximal value of 3 when $n=400$. In this case, the ring radius is $\rho_0=12$ mm. Thus, at designed λ, z, R, and $\gamma=m=0$, by fitting the singularity number n, a near-field narrow bright ring can be generated without use of a lens.

The diffraction pattern shows a similar behavior with increasing parameter γ of the logarithmic axicon at fixed n. Thus, Figure 3.6a depicts the amplitude $|E_{\gamma nm}(\rho)|$ as a function of ρ at $\gamma=150$. The other parameters are $\lambda=514.5$ nm, $m=0$, $n=10$, $R=10$ mm, $w=0.5$ mm, and $z=2000$ mm. From Figure 3.6a, the principal maximum radius, $\rho_0\approx 10$ mm, is seen to coincide with the aperture radius $R=10$ mm. Let us estimate the value of γ at which the left edge of the diffraction pattern (Figure 3.4a) "catches up" with the right one. Equation (3.20) suggests that the local amplitude maxima are found on radii that satisfy the equation

$$\rho_1=\sqrt{\frac{2\beta_n z}{k}}, \tag{3.62}$$

where β_n depends only on n,γ,m. Equating this value of ρ_1 to the aperture radius R, we obtain

$$\beta_n=\frac{kR^2}{2z}. \tag{3.63}$$

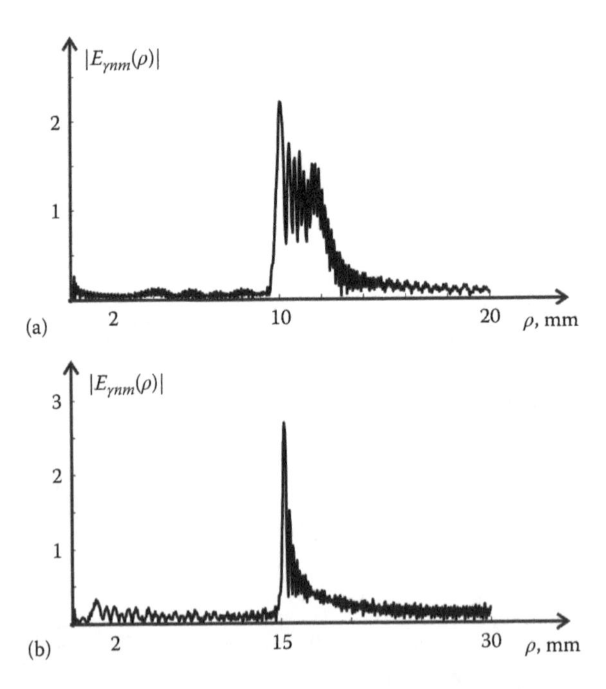

FIGURE 3.6 Radial amplitude distribution at distance $z=2000$ mm at $n=10$, $m=0$, (a) $\gamma=150$, (b) $\gamma=350$.

The right-hand side of Eq. (3.63) coincides with the first term in Eq. (3.61), demonstrating that the order of the magnitude of n and γ at which the right and left edges of the diffraction pattern (Figure 3.4a) coincide is the same.

It can be seen from Figure 3.6a that at $\gamma = 150$ at the left edge of the diffraction pattern the radius reaches the value of $R = 10$ mm. However, because getting closer, the left and right edges of the diffraction pattern "interact" with each other, the "narrowest" diffraction pattern is observed on the radius of $\rho_0 \approx 15$ mm at $\gamma = 350$ (Figure 3.6b).

The rest parameters of the diffraction patterns in Figure 3.6 are the same: $R = 10$ mm ($\sigma = \infty$), $\lambda = 514.5$ nm, and $w = 0.5$ mm.

The plots in Figure 3.7 are analogous to those in Figure 3.5 but depict the γ-dependence, rather than n-dependence. From Figure 3.7 it is seen that the amplitude takes on the ring its maximal value of 2.7, with the ring radius being $\rho_0 = 15$ mm.

Note that for a conventional (not logarithmic) axicon of transmittance $\exp(i\gamma\, r/\omega)$, no "resonance" (that appears as a narrow bright ring) is observed, because at $\gamma > 0$ the conic axicon "displaces" the left and right edges of the Fresnel diffraction pattern as a whole (Figure 3.8).

Figure 3.8 depicts the radial diffraction patterns generated by the conic axicon: $\gamma = -12.5$ (a), $\gamma = 12.5$ (b). It can be seen from Figure 3.8 that the results are identical in absolute value but opposite in sign parameter γ in different diffraction patterns. However, the far-field diffraction pattern (in the spherical lens focus) is the same for both beams. The calculation parameters are $\lambda = 514.5$ nm, $w = 0.5$ mm, $R = 10$ mm,

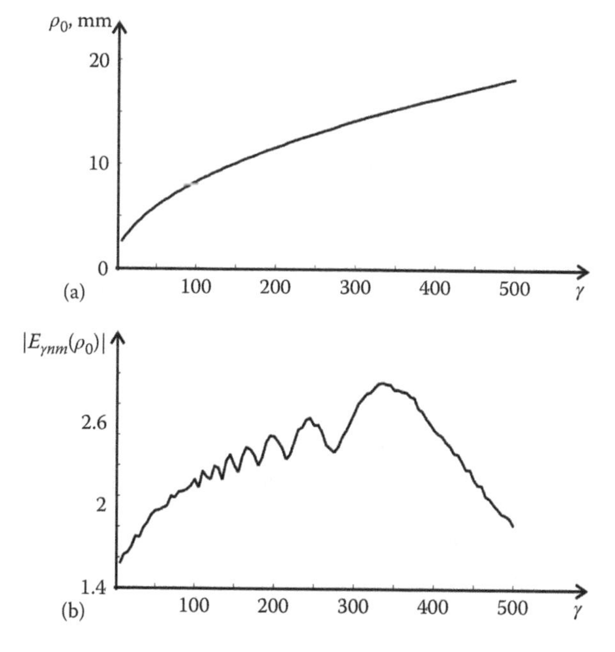

FIGURE 3.7 (a) Radius and (b) principal maximum of on-radius amplitude $|E_{\gamma nm}(\rho_0)|$ for the diffraction pattern at distance $z = 2000$ mm as a function of the logarithmic axicon parameter γ.

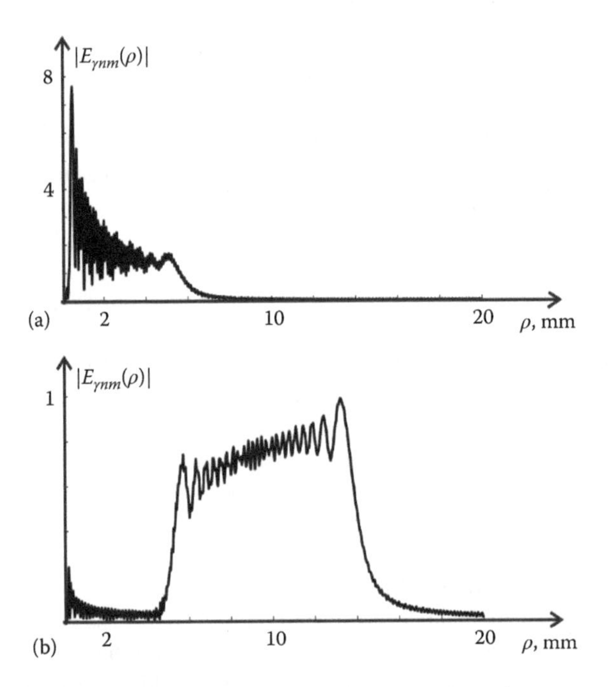

FIGURE 3.8 Radial amplitude distribution for a conic axicon at $z = 2000$ mm, for (a) $\gamma = -12.5$ and (b) $\gamma = 12.5$.

$z = 2000$ mm, $m = 0$, and $n = 10$. When $\gamma = 0$ (Figure 3.4a) the radius of the first maximum can be estimated using Eq. (3.59): $\rho_0 \approx 2$ mm (see Figure 3.4a). Larger $\gamma = 12.5$ will result in a greater radius of the left edge of the diffraction pattern: $\rho_0 \approx 5$ mm (Figure 3.8b).

Propagation of the Hypergeometric Modes

Equation (3.25) describes the complex amplitude of an ideal HyG mode generated using an initial field of Eq. (3.22) at $\sigma = \infty$ (with the Gaussian beam replaced by a plane wave). The light field in Eq. (3.22) has an infinite energy, whereas in practice it should be limited in size. The absolute value of the complex transmittance should also be limited by unity. Figure 3.9a and b show the radial amplitude distributions in the plane $z = 0$, derived from Eq. (3.22): (a) unbounded and (b) bounded by an annular diaphragm with radii $R_1 = 0.2$ mm and $R_2 = 10$ mm. Shown in Figure 3.9c and d are the calculated radial amplitude distributions $|E_{\gamma n,-1}(\rho,\theta,z)|$ in the plane $z = 2000$ mm, derived using (c) Eq. (3.25) and (d) the Fresnel transform. The calculation parameters are: the SPP order is $n = 10$, the logarithmic axicon parameters are $w = 0.5$ mm and $\gamma = 0$, wavelength is $\lambda = 514.5$ nm, and discretization is $\Delta\rho = 0.16$ mm. From Figure 3.9 it is seen that when the initial light field with the central singularity at $\rho = 0$ is bounded by an annular diaphragm, the resulting diffraction pattern of the HyG mode changes inessentially.

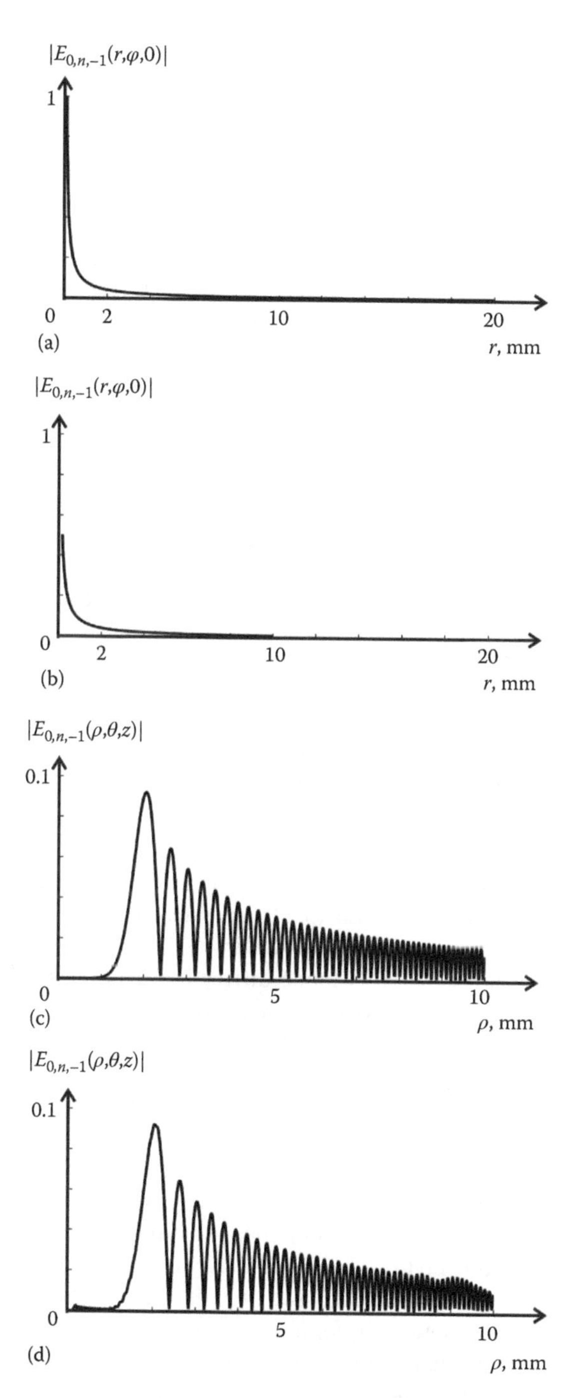

FIGURE 3.9 Radial distribution of the complex amplitude modulus in the planes (a, b) $z = 0$ mm and (c, d) $z = 2000$ mm for the light field in Eq. (3.22) with (a, c) infinite aperture and (b, d) limited by the annular diaphragm of radii $R_1 = 0.2$ mm and $R_2 = 10$ mm.

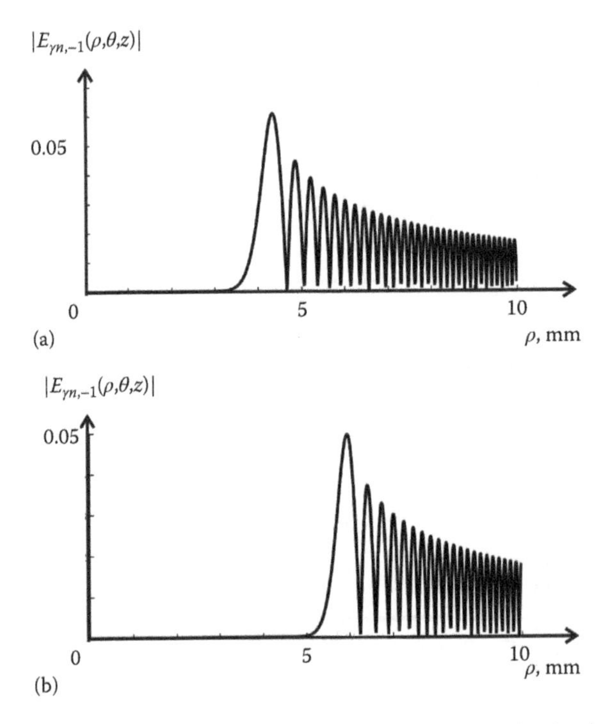

FIGURE 3.10 Radial distribution of the complex amplitude modulus for the initial field in Eq. (3.22) ($n = 10$) in the plane $z = 2000$ mm at (a) $\gamma = 25$ and (b) $\gamma = 50$.

Figure 3.10 shows the calculated radial amplitude distributions at distance $z = 2000$ mm that the initial field of Eq. (3.22) generates after passing the logarithmic axicon with two different values of the parameter $\gamma > 0$. With increasing γ, the radius of the first (major) ring in the HyG mode diffraction pattern is seen to increase. In Figure 3.10, the rest calculation parameters are the same as in Figure 3.9.

We have presented a three-parameter family of paraxial laser beams that are generated by a complex amplitude made up of four cofactors, namely, a Gaussian exponent, a logarithmic axicon, an SPP, and an amplitude power function that can have a singularity at the origin of coordinates. For such beams, the near-field complex amplitude is proportional to a degenerate hypergeometric function and, thus, the beams have been given the name hypergeometric. When the Gaussian beam is replaced with a plane wave the HyG beams change to HyG modes that preserve their structure upon propagation up to scale. The intensity profile of such beams is similar to that of the Bessel beams, being composed of a set of concentric alternating bright and dark rings. We have shown that for a number of such beams the far-field complex amplitude is proportional to a Bessel function of integer or semi-integer order. A specific feature of the HyG beams is that the spatial frequency of the diffraction pattern asymptotically tends to infinity. We have studied particular cases of the HyG modes, which include the absence of the logarithmic axicon or an SPP, and the Laguerre-Gauss modes with the numbers $(0, n)$.

The HyG beams have been simulated numerically. The Fresnel diffraction pattern has been found to present a set of alternating dark and bright rings, predominantly found in another ring constrained by the inner and outer radii. We have derived analytical estimates of the SPP's order and the logarithmic axicon's parameter at which the inner and outer radii of the Fresnel diffraction pattern coincide. Numerical estimates have been derived of the SPP's order and the logarithmic axicon's parameter at which the intensity of the main ring is maximal.

3.3 PROPAGATION OF HYPERGEOMETRIC LASER BEAMS IN A MEDIUM WITH THE PARABOLIC REFRACTIVE INDEX

Paraxial HyG modes were proposed for the first time in 2007 [6]. HyG Gaussian laser beams generated on the basis of the HyG modes followed soon [13]. A large family of HyG beams which includes both particular cases of the HyG modes [6] and the HyG Gaussian beams [13] was discussed in Ref. [113]. The HyG laser beams were experimentally generated using DOEs [114] and computer-synthesized holograms [115]. Recently, analytical expressions that describe the propagation of the HyG modes in a hyperbolic-index medium [116] and a uniaxial crystal [117] have been derived.

On the other hand, a quadratic-index profile is widely used to implement graded-index waveguides, fibers, and lenses [118]. A new technique to fabricate the planar graded-index waveguide in the form of subwavelength binary grating was proposed in [118]. The Gaussian beam [2], the Hermite-Gaussian beam [119], the LG beam [120], the Ince-Gaussian beam [121], and the Airy-Gaussian beam [122,123] are the exact solutions of the paraxial wave equation in a parabolic-index (PI) fiber.

In this section, a relationship that describes the propagation of the HyG beams in a three-dimensional (3D) PI waveguide is proposed. We have shown that the HyG beams are the exact solution of the paraxial wave equation in a PI fiber. The light field of the HyG beams is shown to be periodically varying, with the Fourier spectrum pattern emerging every half-period. According to the geometric optics laws, the beams have been known to be periodically self-reproducing in a PI medium [124]. In this section, we demonstrate that an analogous effect occurs for the paraxial HyG beams. As a particular case, we study the propagation of a circular Gaussian optical vortex in the PI medium. A relationship to describe the maximum intensity on the optical ring is deduced.

Also, a new family of solutions of the nonparaxial Helmholtz equation for the mode propagation in the PI medium is derived. The solutions are proportional to Kummer's functions, becoming divergent at infinity. However, for some values of Kummer's function parameters, the solutions become finite, changing to the familiar nonparaxial LG modes. This finding also shows that the LG modes in a PI medium obey both paraxial and nonparaxial Helmholtz equations of wave propagation.

Recent years have seen a notable increase of interest in the graded-index micro-optical elements. For instance, a graded-index Luneburg microlens was studied in Refs. [125–127]. Experiments with a planar Luneburg lens were conducted in Refs. [125,126] and simulation of a planar photonic-crystal Luneburg lens [127] has shown the lens to generate a focal spot size of full width at half maximum of intensity (FWHM) = 0.44λ for $n_0 = 1.41$ (n_0 is the refractive index at the center of the lens). A planar [128,129] and a 3D Mikaelian's lens with the refractive index varying as a hyperbolic secant (HS) [130] were also reported. In Ref. [130], the graded-index HS lens was shown to form a focal spot of size FWHM = 0.40λ. A planar subwavelength graded-index binary HS lens to generate a focal spot of size FWHM = 159 nm = 0.102λ = 0.35λ/n (refractive index is $n = 3.47$) was discussed in Ref. [131].

In this section, we study the propagation of the Gaussian beam in a PI fiber. A definite length of the fiber can be treated as a PI lens. We show by the numerical simulation that the well-known PI lens can find use for the subwavelength focusing of laser light. In terms of its focusing performance the PI lens is on a par with unconventional lenses such as the Luneburg lens [125–127] and the HS lens [130,131]. The numerical simulation has shown that such a lens produces a focal spot of size FWHM = 0.42λ, for the refractive index on the lens axis of $n_0 = 1.5$. An analytical relationship for the radii of jumps of a binary lens approximating the PI lens has also been derived. The elliptic focal spot's smaller size generated with the PI lens and calculated using the finite-difference time-domain (FDTD) method is FWHM = 0.45λ. It is noteworthy that Refs. [125–131] conducted the analysis of planar-graded index lenses whose diffraction limit was defined by the *sinc*-function, resulting in an FWHM = 0.44λ, whereas in this work we study a 3D PI lens in which the diffraction limit is defined by the Airy disk size of $(2J_1(x)/x)$, producing a larger value of FWHM = 0.51λ.

Paraxial Hypergeometric Beams in a Parabolic-Index Medium

Assume a PI medium with the refractive index in the form:

$$n^2(r) = n_0^2\left[1 - \frac{\left(n_0^2 - n_1^2\right)}{n_0^2}\left(\frac{r}{r_0}\right)^2\right], \tag{3.64}$$

where r is the radial transverse coordinate, n_0 and n_1 are, respectively, the refractive indices on the optical axis ($r = 0$) and at $r = r_0$. Shown in Figure 3.11 is the profile of the refractive index in Eq. (3.64).

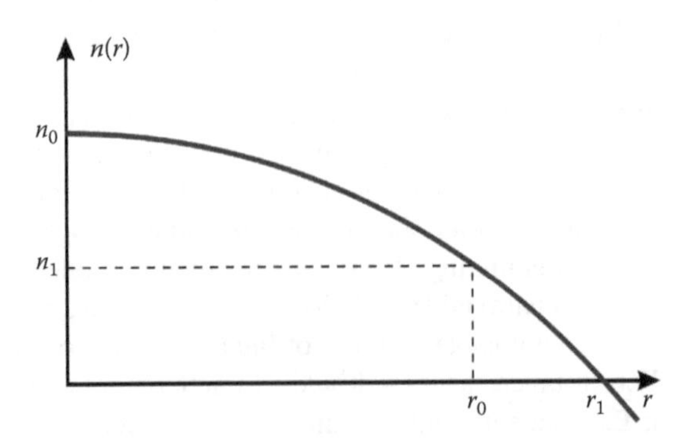

FIGURE 3.11 The parabolic refractive index against the radial coordinate.

The paraxial equation of light propagation in a PI medium in Eq. (3.64) is given by

$$\left[2ik\frac{\partial}{\partial z}+\frac{\partial^2}{\partial x^2}+\frac{\partial^2}{\partial y^2}-(\tau k)^2\left(x^2+y^2\right)\right]E(x,y,z)=0, \tag{3.65}$$

where:

$\tau=(n_0^2-n_1^2)^{1/2}/(r_0 n_0)$

$k=(2\pi/\lambda)n_0$

λ is the free space wavelength

In Ref. [112], the general solution to Eq. (3.65) was shown to take the form:

$$E(x,y,z)=\frac{k\tau}{2\pi i\sin(\tau z)}\exp\left[\frac{i\tau k}{2\tan(\tau z)}\left(x^2+y^2\right)\right]$$
$$\times\int_{-\infty}^{\infty}\int_{-\infty}^{\infty}E_0(\xi,\eta)\exp\left\{\frac{i\tau k}{2\sin(\tau z)}\left[\left(\xi^2+\eta^2\right)\cos(\tau z)-2(x\xi+y\eta)\right]\right\}\mathrm{d}\xi\mathrm{d}\eta. \tag{3.66}$$

Note that the integral transform in (3.66) coincides with the partial Fourier transform within designations [132,133]. The refractive index $n(r)$ becomes imaginary at $r>f_0=r_0n_0/(n_0^2-n_1^2)^{1/2}$ and therefore the medium becomes a perfect absorber. Nevertheless, the solution (3.66) remains valid at $r>r_1$ and in further analysis we will ignore the finiteness of the waveguide. In the cylindrical coordinates and for the initial field $E_0(r,\varphi)=E_0(r)\exp(in\varphi)$, Eq. (3.66) takes the form of an optical vortex with nth topological charge:

$$E(\rho,\theta,z)=(-i)^{n+1}\frac{k}{f_2}\exp\left(\frac{ik\rho^2}{2f_1}+in\theta\right)\int_0^{\infty}E_0(r)\exp\left(\frac{ikr^2}{2f_1}\right)J_n\left(\frac{kr\rho}{f_2}\right)r\mathrm{d}r, \tag{3.67}$$

where:

n is an integer number

$J_n(x)$ is the Bessel function

$$f_1=f_0\tan(z/f_0),\ f_2=f_0\sin(z/f_0). \tag{3.68}$$

Relationships that describe the transformations that the paraxial LG beams undergo in a PI fiber were derived in Ref. [120,134]. Let us consider a Gaussian Beam with power apodization (HyG beam). In the following, we study the propagation of such beams characterized by a complex amplitude in the initial plane in the PI fiber of Eq. (3.64) [113]:

$$E_0(r)=\left(\frac{r}{\delta}\right)^{m+i\gamma}\exp\left(-\frac{r^2}{2\sigma^2}\right), \tag{3.69}$$

where:

m and γ are real numbers

δ and σ are, respectively, a scale factor of the amplitude power component and the Gaussian beam radius

We note that in Eq. (3.69) the exponent ($m + i\gamma$) is complex and different from the number n of the angular harmonic $\exp(in\varphi)$. This is the reason why the HyG beams are essentially different from the well-known LG beams. Real part m means power apodization while imaginary part γ acts as a logarithmical axicon, having the transmittance function $\exp[i\gamma \ln(r/\delta)]$. Substituting Eq. (3.69) into Eq. (3.67) and making use of the reference integral [135] (integral 6.631.1):

$$\int_0^\infty y^{\mu-1} \exp\left(-\beta y^2\right) J_n(cy)dy = \frac{1}{2^{n+1} n!} c^n \beta^{-(n+\mu)/2} \Gamma\left(\frac{n+\mu}{2}\right) {}_1F_1\left(\frac{n+\mu}{2}, n+1, -\frac{c^2}{4\beta}\right), \tag{3.70}$$

where:

$\Gamma(x)$ is the gamma-function

${}_1F_1(a,b,x)$ is Kummer's function [101], Eq. (3.67) is rearranged to

$$E_{m\gamma n}(\rho,\theta,z) = \frac{(-i)^{n+1}}{n!}\left(\frac{k\sigma^2}{f_2}\right)\left(\frac{\sqrt{2}\sigma}{\delta}\right)^{m+i\gamma} \exp\left(\frac{ik\rho^2}{2f_1} + in\theta\right)$$
$$\times\left(1 - \frac{ik\sigma^2}{f_1}\right)^{-(m+i\gamma+2)/2} \Gamma\left(\frac{n+m+i\gamma+2}{2}\right) x^{n/2} {}_1F_1\left(\frac{n+m+i\gamma+2}{2}, n+1, -x\right), \tag{3.71}$$

where:

$$x = \frac{\rho^2}{2\omega^2(z)} + \frac{ik\rho^2}{2R(z)},$$
$$\omega(z) = \sigma\left[\cos^2\left(\frac{z}{f_0}\right) + \left(\frac{f_0}{k\sigma^2}\right)^2 \sin^2\left(\frac{z}{f_0}\right)\right]^{\frac{1}{2}}, \tag{3.72}$$
$$R(z) = \frac{1}{2} f_0 \sin\left(\frac{2z}{f_0}\right)\left[1 + \left(\frac{f_0}{k\sigma^2}\right)^2 \tan^2\left(\frac{z}{f_0}\right)\right].$$

At $z \approx 0$, it follows from Eqs. (3.68) and (3.72) that $f_1 \approx f_2 \approx z$, $w(z) \approx \sigma$, $R(z) \approx z$, $x \approx \rho^2/(2\sigma^2) + ik\rho^2/(2z)$. Hence, the argument x of the HyG function in Eq. (3.71) tends to infinity, although the complex amplitude (3.71) itself is tending to the boundary condition in Eq. (3.69). We note that $R(z)$ does not describe the curvature radius of the HyG beam. At $\gamma = 0$ and $m = n$ function $w(z)$ is the width of the Gaussian envelope. The light field in (3.71) will be self-reproduced in modulus with a period of $L = \pi f_0$. The field amplitudes in planes separated by half-period $L_1 = (\pi f_0)/2$ are related through the Fourier transform:

$$E_{m\gamma n}(\rho,\theta,z=L_1) = \frac{(-i)^{n+1}}{n!}\left(\frac{k\sigma^2}{2f_2}\right)\left(\frac{\sqrt{2}\sigma}{\delta}\right)^{m+i\gamma} \exp(in\theta)$$
$$\times \Gamma\left(\frac{n+m+i\gamma+2}{2}\right)\left(\frac{\rho^2}{2\omega_1^2}\right)^{n/2} {}_1F_1\left(\frac{n+m+i\gamma+2}{2}, n+1, -\frac{\rho^2}{2\omega_1^2}\right), \tag{3.73}$$

where $w_1 = f_0/(k\sigma)$ is the effective radius of the light field in the Fourier plane. The functions $E_{m\gamma n}(\rho, \theta, z)$ of Eq. (3.71) are orthogonal to each other at different values of the topological charge n. Note that at $f_0 = k\sigma^2$ the real part of Kummer's function argument x ceases to be z-dependent: $\omega(z) = \sigma = \text{const}$, whereas the imaginary argument's component remains to be z-dependent: $R(z) = f_0 \tan(z/f_0)$. As a result, the total amplitude of the field of Eq. (3.71) appears to be z-dependent. Because of this, at no values of parameters and numbers m, n, γ can the HyG beams of Eq. (3.71) be represented as modes of a PI medium, except for a particular case of $m = n$ and $\gamma = 0$.

Let us consider in more detail this particular case. From Eq. (3.71) at $\gamma = 0$ and $m = n$, we obtain:

$$E_{n0n}(\rho,\theta,z) = (-i)^{n+1}\left(\frac{k\sigma^2}{f_2}\right)\left(\frac{\sqrt{2}\sigma}{\delta}\right)^{n}\left(1-\frac{ik\sigma^2}{f_1}\right)^{-(n+2)/2}$$
$$\times x^{n/2}\exp\left[-\frac{\rho^2}{2\omega^2(z)}+\frac{ik\rho^2}{2R_1(z)}+in\theta\right], \tag{3.74}$$

where:

$$x = \frac{\rho^2}{2\omega^2(z)} + \frac{ik\rho^2}{2R(z)},$$
$$R_1(z) = 2k\left(\frac{f_0}{k\sigma^2} - \frac{k\sigma^2}{f_0}\right)^{-1}\omega^2(z)\sin^{-1}\left(\frac{2z}{f_0}\right). \tag{3.75}$$

From Eq. (3.75), the wavefront curvature radius $R_1(z)$ is seen to be equal to infinity in the planes $z = \pi f_0 p/2$, $p = 0, 1, 2, \ldots$. In the initial plane ($z = 0$), the terms $(k\sigma^2/f_2)$ and $x^{n/2}$ tend to infinity in the absolute value. However, the term $(1 - ik\sigma^2/f_1)^{-(n+2)/2}$ tends to zero. As a result, these two terms are mutually compensated, because at $z \to 0$, $f_0 \approx f_1 \approx R(z) \approx z$. The beams in Eq. (3.74) can be referred to as Gaussian optical

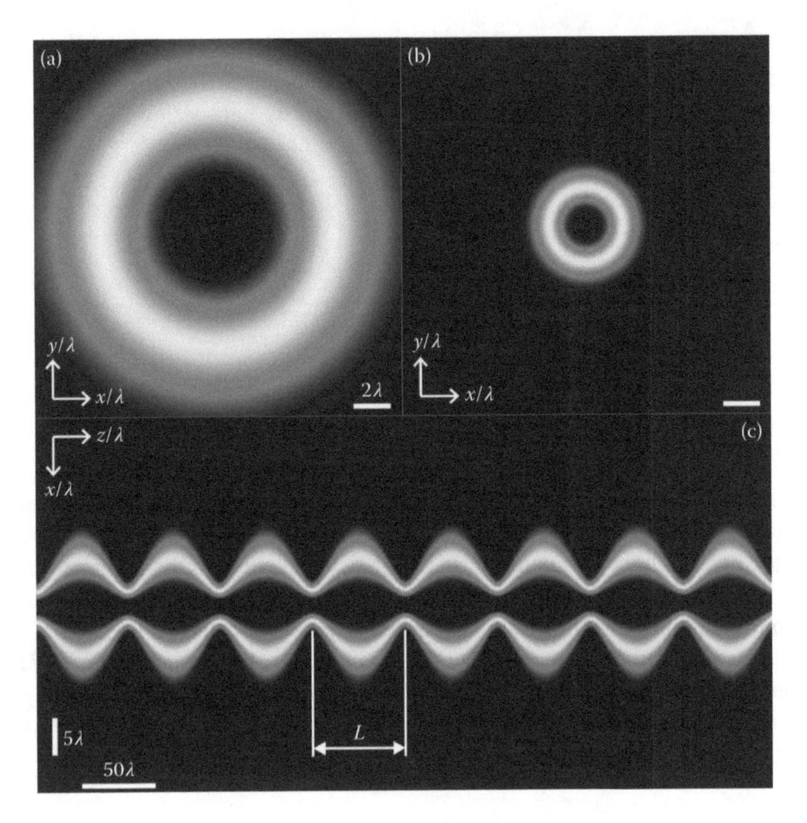

FIGURE 3.12 The intensity pattern from the beam of Eq. (3.74) in the transverse planes (a) $z = L/2$ and (b) $z = L = 20\pi\lambda$, and (c) the longitudinal plane $y = 0$.

vortices that propagate in a PI medium. While propagating in the PI medium, these beams behave in a different way than the LG beams [112]. At $f_0 = k\sigma^2$, the beam of Eq. (3.74) becomes a PI medium mode, with its radius remaining unchanged during propagation. Figure 3.12 depicts the intensity patterns of the beam in Eq. (3.74) in the transverse planes separated by a half-period distance: $z = L/2$ and $z = L$, and in the longitudinal plane $y = 0$. The simulation was conducted for the incident wavelength $\lambda = 532$ nm, the waveguide parameter $\tau = 1/(20\lambda)$, the topological charge $n = 4$, the Gaussian beam waist radius $\sigma = \lambda$, the scale factor $\delta = 2\lambda$, and the dimensionless parameter $\gamma = 0$. The computational domain was $-10\lambda \leq x, y \leq 10\lambda$ in Figure 3.12a, b and $-25\lambda \leq x \leq 25\lambda$, $0 \leq z \leq 500\lambda$ in Figure 3.12c.

From Eq. (3.74) it can be derived that the intensity maximum is found on the ring $\rho = \sqrt{n}\omega(z)$ and equal to

$$I_{\max} = \left(\frac{k\sigma^2}{f_2}\right)^2 \left[1 + \left(\frac{k\sigma^2}{f_1}\right)^2\right]^{-(n+2)/2} \left[\frac{k\sigma^3 n\omega(z)}{\delta^2 f_2}\right]^n \exp(-n). \qquad (3.76).$$

From Eq. (3.76) it is seen that as the beam is alternatively narrowed and widened during propagation, its radius varies from σ to $f_0/(k\sigma)$, with the intensity maximum undergoing a $[f_0/(k\sigma^2)]^2$-fold change. In particular, for the above-specified parameters, the intensity maximum in Figure 3.12b is about 10 times higher than the intensity maximum in Figure 3.12a.

Thus, in this section we have derived a family of exact analytical solutions in Eq. (3.71) of the paraxial equation for PI medium in Eq. (3.65). As a particular case, this solution describes the Gaussian optical vortices of Eq. (3.74), which become the paraxial modes of the PI medium different from the well-known LG modes under the condition $f_0 = k\sigma^2$. In the next section, an attempt is made to derive other non-paraxial modes that would propagate in a PI medium.

Nonparaxial Modes of a Parabolic-Index Medium

Let there be a PI medium of Eq. (3.64): $n^2(r) = n_0^2(1-\tau^2 r^2)$. The solution to Helmholtz' equation for a PI medium

$$\left(\frac{\partial^2}{\partial r^2}+\frac{1}{r}\frac{\partial}{\partial r}+\frac{1}{r^2}\frac{\partial^2}{\partial \varphi^2}+\frac{\partial^2}{\partial z^2}+k_0^2 n^2(r)\right)E(r,\varphi,z)=0 \tag{3.77}$$

with the refractive index of Eq. (3.64) will be sought for as a mode with Gaussian envelope and an optical vortex:

$$E(r,\varphi,z)=r^p\exp\left(-\frac{r^2}{2\omega_2^2}\right)\exp(in\varphi)\exp(i\beta z)F(sr^q), \tag{3.78}$$

where:

F is some function
β is the mode propagation constant
w_2 is the waist radius of the Gaussian envelope
n is the optical vortex topological charge
s, p, and q are some constants to be defined later

Then the Helmholtz' equation (3.77) takes the form:

$$\begin{aligned}&s^2q^2r^{2q-2}F''(sr^q)+\left[(2p+q)sqr^{q-2}-\frac{2sq}{\omega_2^2}r^q\right]F'(sr^q)\\&+\left[\frac{1}{\omega_2^4}r^2-\frac{2(p+1)}{\omega_2^2}+\frac{p^2-n^2}{r^2}+k_0^2n_0^2(1-\tau^2r^2)-\beta^2\right]F(sr^q)=0.\end{aligned} \tag{3.79}$$

Assuming $q = 2$ and $n = \pm p$, and dividing Eq. (3.79) by $4s$ we obtain:

$$\begin{aligned}&sr^2F''(sr^2)+\left(p+1-\frac{r^2}{\omega_2^2}\right)F'(sr^2)\\&+\frac{1}{s}\left[\frac{r^2}{4\omega_2^4}-\frac{p+1}{2\omega_2^2}+\frac{k_0^2n_0^2(1-\tau^2r^2)-\beta^2}{4}\right]F(sr^2)=0.\end{aligned} \tag{3.80}$$

If we suppose that $k_0 n_0 \tau \omega_2^2 = 1$ and $s = 1/\omega_2^2$, then Eq. (3.80) becomes Kummer's differential equation:

$$\xi F''(\xi) + (p+1-\xi)F'(\xi) - \left[\frac{p+1}{2} + \frac{\omega_2^2}{4}\left(\beta^2 - k_0^2 n_0^2\right)\right] F(\xi) = 0. \tag{3.81}$$

where $\xi = r^2/\omega_2^2$. Therefore, solution of Eq. (3.77) can be shown to be given by a family of functions:

$$E(r,\varphi,z) = r^n \exp\left(-\frac{r^2}{2\omega_2^2}\right)\exp(\pm in\varphi)\exp(i\beta z)\,{}_1F_1\left[\frac{n+1}{2} + \frac{\omega_2^2}{4}\left(\beta^2 - k_0^2 n_0^2\right), n+1, \frac{r^2}{\omega_2^2}\right], \tag{3.82}$$

where ${}_1F_1(a,b,x)$ is Kummer's function, as before, $\omega_2 = [1/(k_0 n_0 \tau)]^{1/2}$. Considering that at $\xi \to \infty$, Kummer's function asymptotics are given by [101]

$${}_1F_1(a,b,\xi) = \frac{\Gamma(b)}{\Gamma(a)}\exp(\xi)\xi^{a-b}\left(1 + O(1/\xi)\right), \tag{3.83}$$

where $O(x)$ tends to zero faster than x, the function in (3.82) is diverging at $r \to \infty$, thus the beam has infinite energy. The set of solutions (3.82) also contains non-diverging solutions. Putting Kummer's function first parameter to be a negative integer, Kummer's function becomes equal to the polynomial, whereas the solution (3.82) converges to zero at $r \to \infty$. Thus, under the condition that

$$\frac{n+1}{2} + \frac{\omega_2^2}{4}\left(\beta^2 - k_0^2 n_0^2\right) = -s, \tag{3.84}$$

Eq. (3.82) reduces to the familiar LG modes:

$$\begin{aligned} E(r,\varphi,z) &= r^n \exp\left(-\frac{r^2}{2\omega_2^2}\right)\exp(\pm in\varphi)\exp(i\beta z)\,{}_1F_1\left[-s, n+1, \frac{r^2}{\omega_2^2}\right] \\ &= \frac{s!}{(n+1)^s} r^n \exp\left(-\frac{r^2}{2\omega_2^2}\right)\exp(\pm in\varphi)\exp(i\beta z) L_s^n\left(\frac{r^2}{\omega_2^2}\right), \end{aligned} \tag{3.85}$$

where

$$\beta = k_0 n_0 \sqrt{1 - \frac{2\tau}{k_0 n_0}(2s + n + 1)} \tag{3.86}$$

If in the LG mode of Eq. (3.85) the radius of the Gaussian beam does not meet the condition $w_2 = [1/(k_0 n_0 \tau)]^{1/2}$, the LG beam is no longer a mode. In the unbounded waveguide of Eq. (3.64) only LG modes (3.85) can propagate, but if the waveguide is bounded by radius $r_0 < \infty$, then besides modes (3.85) there would be other modes described by Eq. (3.82).

Thus, in this section we have shown that although there is a wide class of mode solutions of the Helmholtz equation in the cylindrical coordinates for a PI medium, only solutions coincident with the LG modes possess a finite energy, thus being physically realizable. In other words, in this section, we have shown that except for the LG modes, no other nonparaxial modes with cylindrical symmetry propagate in the parabolic medium. This result also proves that the GL modes in the PI medium satisfy both equations: paraxial Eq. (3.65) and nonparaxial Eq. (3.74). Besides, we can infer that both the Gaussian beam of Eq. (3.71) at $f_0 = k\sigma^2$ and the Gaussian beam of Eq. (3.85) at $w_2 = [1/(k_0 n_0 \tau)]^{1/2}$ have the same radius:

$$\omega_2 = \left(k_0 n_0 \tau\right)^{-1/2} = \sigma = \sqrt{f_0/k} = \left(\frac{\lambda r_0}{2\pi\sqrt{n_0^2 - n_1^2}}\right)^{1/2}. \tag{3.87}$$

A Parabolic-Index Microlens

The PI microlens has been well known [124]. However, until now it has not yet been demonstrated that such a microlens can be utilized for the subwavelength focusing of light. In this section, based on the solution in Eq. (3.71) we derive a paraxial estimate of the focal spot size. Considering that its amplitude is periodically self-reproducing during propagation in the PI fiber (3.64), the Gaussian beam enters into the family of solutions (3.71) as a principal beam. Assuming $n = m = \gamma = 0$, from (3.71) we obtain:

$$E_0(\rho,\theta,z) = (\ i)\left(\frac{2k\sigma^2}{f_2}\right)\left(1 - \frac{ik\sigma^2}{f_1}\right)^{-1} \exp\left(-\frac{\rho^2}{2\omega^2(z)} - \frac{ik\rho^2}{2R_1(z)}\right). \tag{3.88}$$

where:

$$R_1^{-1}(z) = \frac{\cos^2(z/f_0) + \left(\frac{f_0}{k\sigma^2}\right)^2 \sin^2(z/f_0) - 1}{f_0 \tan(z/f_0)\left[\cos^2(z/f_0) + \left(\frac{f_0}{k\sigma^2}\right)^2 \sin^2(z/f_0)\right]}. \tag{3.89}$$

From (3.88) and (3.89), it follows that at $f_0 = k\sigma^2$ the Gaussian beam remains unchanged while propagating in the PI fiber, preserving its diameter. If $f_0 \neq k\sigma^2$, the Gaussian beam's radius is changing (Eq. 3.72):

$$\omega(z) = \sigma\left[\cos^2\left(\frac{z}{f_0}\right) + \left(\frac{f_0}{k\sigma^2}\right)^2 \sin^2\left(\frac{z}{f_0}\right)\right]^{\frac{1}{2}}, \tag{3.90}$$

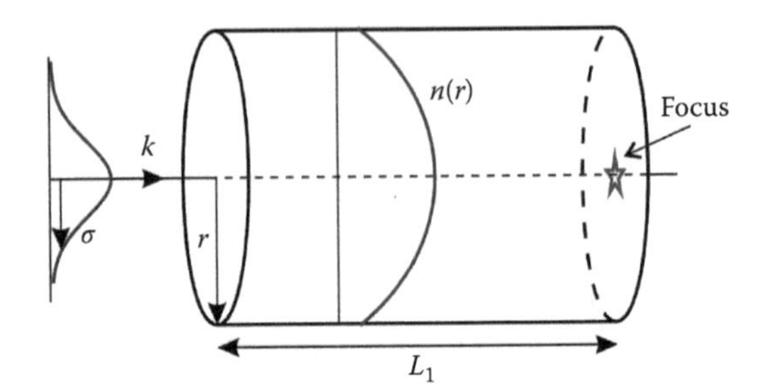

FIGURE 3.13 Focusing of the Gaussian beam with a PI lens.

suggesting that the minimal radius of $w_1 = f_0/(k\sigma)$ (if $f_0 < k\sigma^2$) is attained at distance $L_1 = (\pi f_0)/2$ from the initial point ($z = 0$). The Gaussian beam's diameter at intensity half-maximum equals

$$FWHM = \left(\frac{\sqrt{\ln 2}}{\pi}\right)\frac{\lambda r_0}{\sigma\sqrt{n_0^2 - n_1^2}}. \tag{3.91}$$

Thus, we can consider a PI lens in the form of a section of a PI fiber of radius r_0 and optical-axis length $L_1 = (\pi f_0)/2$. Such a PI lens is able to focus the incident plane Gaussian beam of radius σ into a focal spot of size in Eq. (3.91), which is formed near the lens output surface (Figure 3.13).

From Eq. (3.91), the *NA* of the PI lens ($n_1 = 1$) is given by

$$NA = \frac{\sigma}{r_0}\sqrt{n_0^2 - 1}. \tag{3.92}$$

When $\sigma = r_0$, the *NA* in Eq. (3.92) equals that of a planar HS lens [129]. It should be noted that since Eq. (3.92) has been derived as a paraxial approximation, for it to be used correctly one needs to choose the condition $\sigma \ll r_0$. Note, however, that the well-known relation that describes the diffraction-limited size of the focal spot in a medium with refractive index n has been also deduced in the paraxial approximation (the Airy disk diameter in terms of the intensity FWHM): FWHM = $0.51\lambda/(n \sin\theta)$. Nonetheless, it has been used for the larger-than-unity *NA*s: FWHM = $0.51\lambda/n$.

From Eq. (3.91) it follows that at $n_0 = 1.5$, $n_1 = 1$, $\sigma = r_0$, the PI-lens has the length of $L_1 = 2.1r_0$ and the focal spot size of FWHM = 0.24λ. The latter value is 1.4 times smaller than the diffraction limit in the medium with $n_0 = 1.5$: FWHM = 0.34λ. It should be noted, however, that in deducing Eq. (3.90), the lens was supposed to be of infinite radius, with the refractive index of Eq. (3.64) becoming $n(r) < 1$, thus the estimated focal spot size (FWHM = 0.24λ) cannot be realized for a real lens of limited radius r_0, at which $n(r_0) = 1$. For comparison, we note that the diameter of the intensity FWHM of the minimal nonparaxial LG mode in a PI waveguide can be obtained from Eq. (3.87) at $r_0 = \lambda$:

$$FWHM = 2\sqrt{\ln 2}\left(\frac{\lambda r_0}{2\pi\sqrt{n_0^2 - n_1^2}}\right)^{1/2} = 0.63\lambda. \tag{3.93}$$

Although fabrication techniques for synthesis of continuous-profile graded-index lenses have been widely known, including a chemical vapor deposition method [136] and an ion exchange method [137], they do not enable obtaining an arbitrarily specified refractive index profile. Because of this, in the next section we discuss a binary lens as an approximation of the PI lens. Such a lens can be fabricated by e-beam lithography.

Binary Parabolic-Index Lens

The continuous PI lens can be approximated by a binary PI lens by the rule schematically illustrated in Figure 3.14.

According to Figure 3.14, the PI-lens' radius r_0 is broken down into N equal segments of length Δ: $r_0 = N\,\Delta$, with the segment's origin and end given by the radii $r_p = p\,\Delta$, $p = 0,1,2,\ldots N-1$. The origin of the pth binary ring is at radius r_p, whereas the end $\bar{r}_p$ of the pth binary ring is found from the equation:

$$\pi\left(r_{p+1}^2 - \bar{r}_p^2\right) + n_0\pi\left(\bar{r}_p^2 - r_p^2\right) = 2\pi n_0 \int_{r_p}^{r_{p+1}} \sqrt{1-\tau^2 r^2}\, r dr. \tag{3.94}$$

This equation means that inside each ring $r_p \le r \le r_{p+1}$ average refractive index is the same for both binary and gradient-index lenses. From Eq. (3.94), an explicit relation for the end radius of the pth binary ring is

$$\bar{r}_p^2 = \frac{n_0 r_p^2 - r_{p+1}^2}{n_0 - 1} + \frac{2n_0}{3(n_0-1)\tau^2}\left\{\left[1-\left(\tau r_p\right)^2\right]^{3/2} - \left[1-\left(\tau r_{p+1}\right)^2\right]^{3/2}\right\}, \tag{3.95}$$

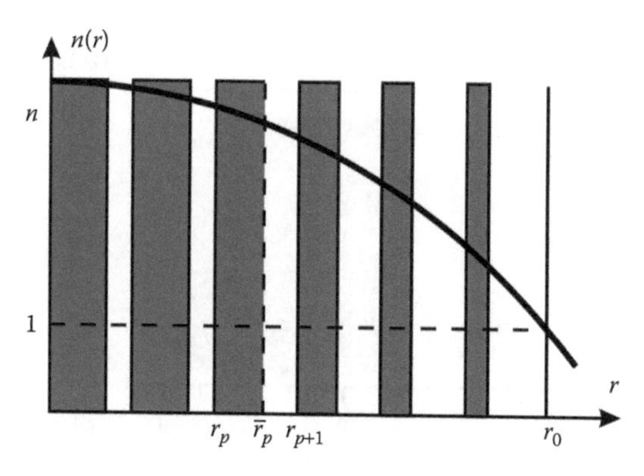

FIGURE 3.14 Schematic representation of the continuous PI lens approximated by a piecewise-constant (binary) lens (dark color: material, white color: air).

where $p = 0,1,2,\ldots N-1$. Note that in Eq. (3.94) it is possible to choose non-equidistant radii r_p of the refractive index jumps, also taking account of the fabrication capabilities with regard to a minimal feasible zone size.

A Planar Parabolic-Index Lens

First, we studied a 2D PI lens. The FDTD simulation of focusing an incident TE-wave was conducted using the commercial program FullWave (Rsoft Design Group). The computational mesh samplings on the optical axis Z and the transverse axis X were $\lambda/130$ (4.1 nm), the time step being $\Delta(cT) = 2.8$ nm.

Figure 3.15 shows a gray-level map of the refractive index in the PI-lens. The lens radius is $r_0 = 1$ µm, $n_0 = 1.5$, $\tau = 0.745$ µm^{-1}, and $\lambda = 0.532$ µm. Considering that $\tau = 1/f_0$, then $f_0 = 1/\tau = 1.342$ µm, the lens length being $L_1 = \pi f_0/2 = \pi/(2\,\tau) =$ 2.1 µm.

Figure 3.16 shows the intensity distribution of the E-field 10 nm behind the lens for a plane incident beam and a Gaussian beam of radius $\sigma = 1$ µm. At this, the Gaussian beam passes partially outside the lens.

The focal spot size is FWHM = 0.388λ = 0.2 µm for the plane wave and FWHM = 0.5λ = 0.27 µm for the Gaussian beam. It can be seen from Figure 3.16 that the intensity side-lobes amount to 30% of the major intensity peak, signifying that the lens length is not optimal.

A more tightly focused spot can be achieved by varying the lens length. Figure 3.17 shows the focal spot size at intensity FWHM as a function of the lens length L_1.

From Figure 3.17, the minimal-size focal spot for the plane wave and the Gaussian beam is seen to be attained at different values of the lens length L_1. The plane wave is most tightly focused at $L_1 = 1.56$ µm, producing FWHM = 0.375λ, whereas the Gaussian beam produces a minimal focal spot size FWHM = 0.394λ at lens length $L_1 = 1.73$ µm. This result reveals a general pattern of laser light focusing: an increase in the side-lobes (Figure 3.16a) is accompanied by a tighter focal spot (and increased focus depth).

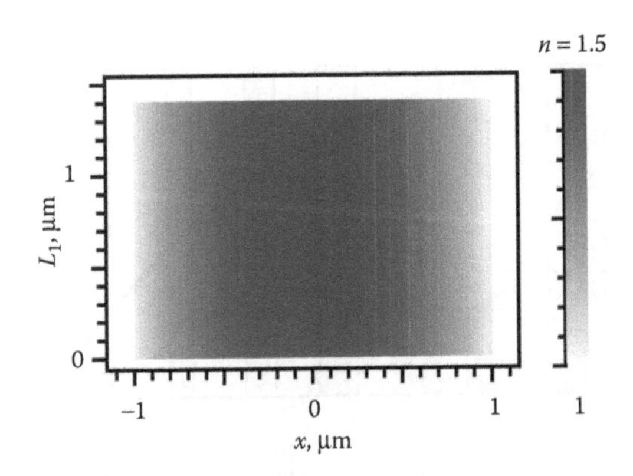

FIGURE 3.15 The refractive index (in gray level) in a planar gradient-index PI lens.

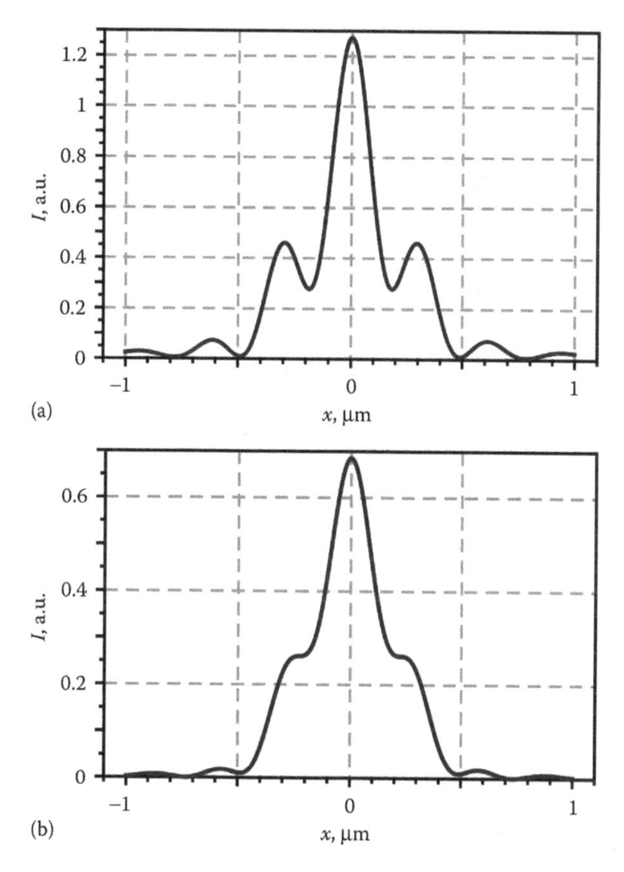

FIGURE 3.16 The intensity distribution ($|E^2|$) in the focal plane found 10 nm behind the lens for (a) a plane incident wave and (b) a Gaussian incident beam.

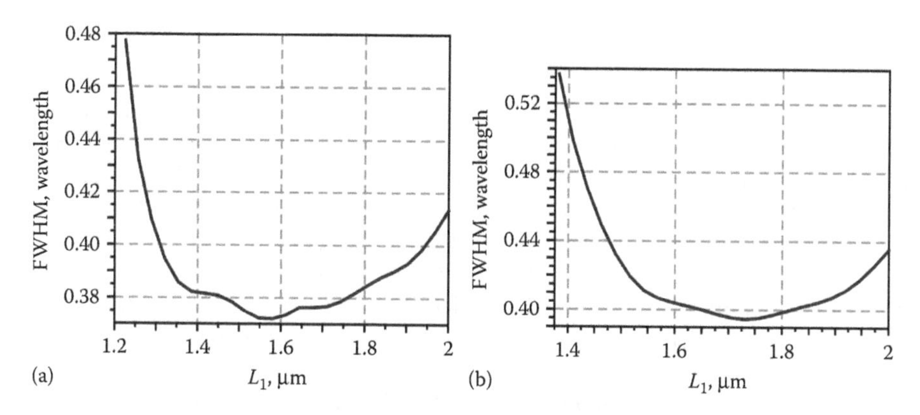

FIGURE 3.17 The focal spot size generated 10 nm away from the lens output as a function of the lens length L_1 for (a) a plane wave and (b) a Gaussian incident beam.

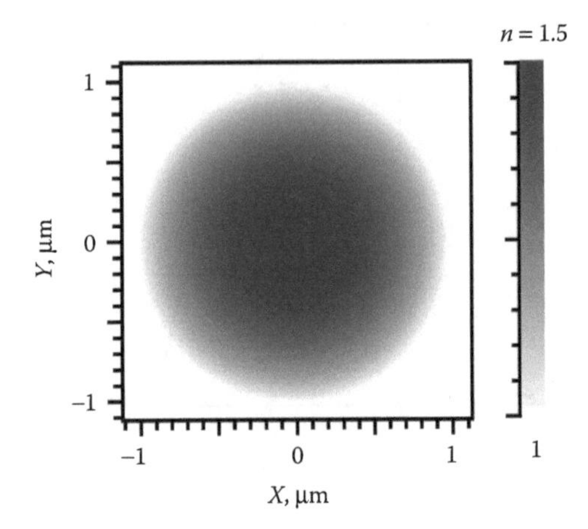

FIGURE 3.18 Lens gray-level graded-index profile in the plane XY.

3D Parabolic-Index Lens

The 3D simulation involved a lens with a graded-index profile in the plane XY, which was dependent on the radius r. The graded refractive index profile of the 3D lens in the plane XY is shown in Figure 3.18.

The 3D simulation was performed on a computational mesh of size $\lambda/40$ (13.3 nm) along all three axes. We simulated the propagation of the linearly polarized light through the lens (for the major incident field component E_y). The incident light was a plane Gaussian beam of radius $\sigma = 1$ μm. The plot in Figure 3.19a shows the focal spot size FWHM on the X-axis as a function of the lens length L_1.

The minimum is seen to be attained at the lens length of $L_1 = 1.6$ μm, with the focal spot size on the X-axis being FWHM = 0.42λ. The intensity profile ($I = |E_x^2| + |E_y^2| + |E_z^2|$) within the lens focal spot for the said parameters is shown in Figure 3.19b. Because the incident light is linearly polarized, the focal spot is widened on the Y-axis, amounting to FWHM = 0.70λ (Figure 3.19c).

3D Binary Parabolic-Index Lens

In practice, it appears difficult to fabricate a lens with a continuous PI profile. Meanwhile, a binary microlens can be fabricated using techniques for fabricating 3D photonic crystal waveguides or Bragg's waveguides [138]. Shown in Figure 3.20 is a binary refractive index distribution in the plane XY for a binary lens of Eq. (3.94), which approximates the continuous-profile PI lens.

The focusing performance of such a lens is somewhat inferior to the continuous-profile PI lens of Eq. (3.64). Figure 3.21a depicts the intensity pattern in the focal plane of the binary PI lens (Figure 3.20) and the distribution cross-sections (Figure 3.21b) drawn through the spot center along the X-axis (curve I) and Y-axis (curve II). The distributions correspond to an optimal lens length of $L_1 \approx 1.9$ μm, whereas the paraxial theory predicts a somewhat larger value of $L_1 = 2.1$ μm. The linearly polarized incident wave has a plane wavefront and a Gaussian amplitude distribution of radius

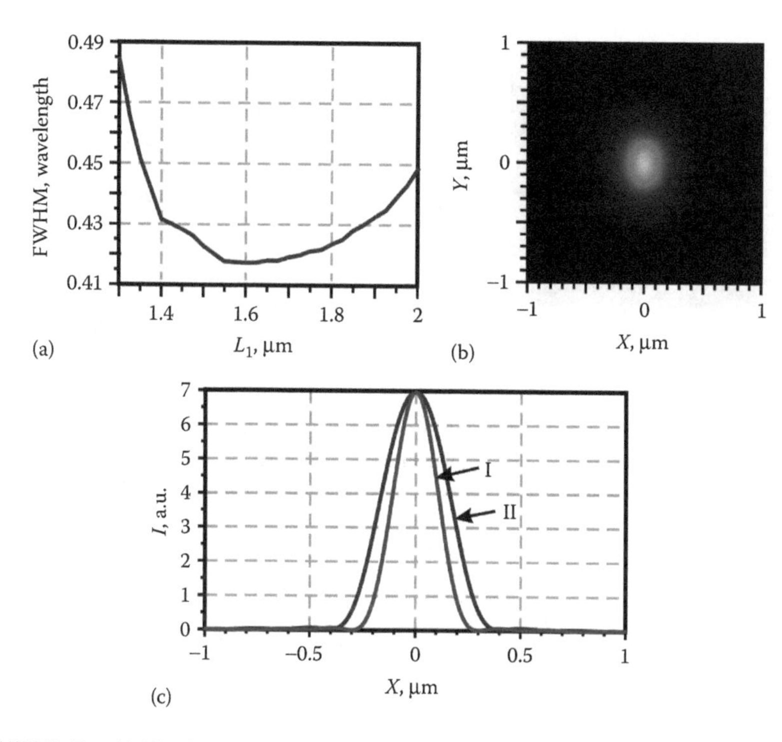

FIGURE 3.19 (a) The focal spot size on the X-axis formed 10 nm away from the lens output plane against the lens length L_1 for the linearly polarized incident Gaussian beam of radius $\sigma = 1$ μm; (b) the intensity pattern in the lens focal spot for an optimal length of $L_1 = 1.6$ μm; (c) the intensity cross-sections passing through the spot center along the X-axis (curve I) and Y-axis (curve II).

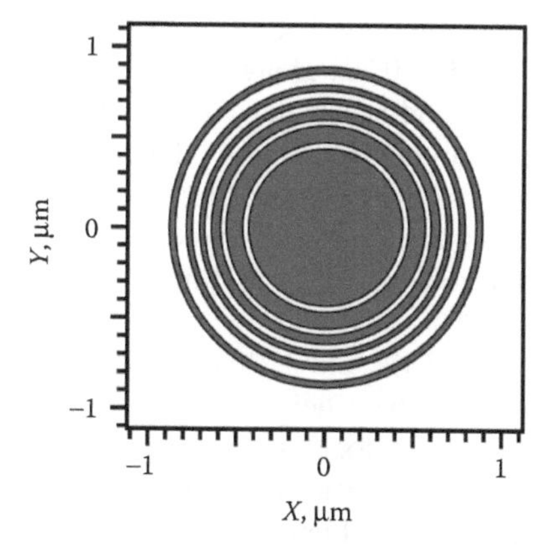

FIGURE 3.20 Binary refractive index distribution in the XY-plane, with dark rings denoting the refractive index $n = 1.5$ and white rings denoting $n = 1$.

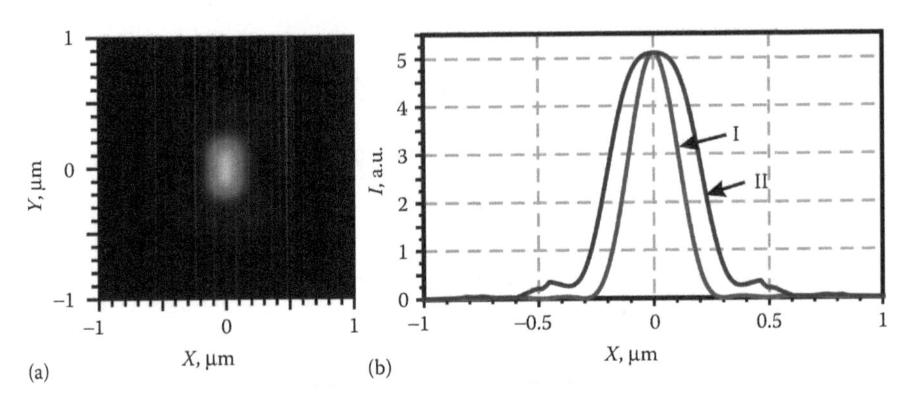

FIGURE 3.21 (a) The intensity distribution in the focal plane (10 nm behind the lens) of the binary PI lens of Figure 3.20 and (b) the intensity cross-sections drawn through the spot center along the X-axis (curve I) and Y-axis (curve II).

$\sigma = 1$ μm. In the binary lens, the minimal zone size (i.e., the difference between the radii of neighboring refractive index jumps) is 35 nm. Note that the implementation of such a lens requires a technology that would enable obtaining a binary microrelief with the aspect ratio of 60 (relief-depth/relief-width ratio). Such fabrication techniques have been utilized for the implementation of X-ray zone plates (ZP). A ZP with the minimal (outermost) zone of width 100 nm and height 1300 nm (aspect ratio = 13) was fabricated using the positive resist PMMA in Ref. [139]. To prevent the collapse of the ZP walls, the resist was dried using a special technique [139].

The collapse of the binary lens in Figure 3.20 can be prevented by replacing the air in the between-wall spacing with a different material. Using a "jelly-roll" technology, a sputtered-slice X-ray ZP composed of two alternating materials with a 160-nm outermost zone and the aspect ratio of greater than 1,000 was fabricated in Ref. [140].

We conducted the simulation on a computational mesh of size $\lambda/70$ (7.6 nm) for all three axes. The initial polarization plane was parallel to the ZY-plane.

For the distributions shown in Figure 3.21, the focal spot size is FWHM = 0.45λ along the X-axis and FWHM = 0.78λ along the Y-axis. The focal spot is seen to be extended along the Y-axis and nearly devoid of side-lobes. The extension of the focal spot along the Y-axis can be explained by the polarization of the field. In the XZ-plane, the field propagates as a TE-wave, that is only E_y component is non-zero. In the YZ-plane, the field propagates as a TM-wave, that is the E_z component is also present, contributing to the total intensity and making the focal spot wider along Y-axis.

The comparison of Figures 3.19c and 3.21b suggests that due to the binary structure of the lens in Figure 3.20 the efficiency of focusing has decreased, with 8% less optical energy coming to the focal spot in Figure 3.21 when compared with that in Figure 3.19. In addition, the focal spot in Figure 3.21b has acquired the side-lobes which the focal spot from the PI lens has nearly been devoid of (Figure 3.19c). However, these drawbacks are an acceptable fee to be paid for the feasibility of fabricating such a lens using the available fabrication techniques [139,140].

The major results of this section are as follows. The relationship for the complex amplitude of a family of paraxial HyG laser beams propagating in a PI fiber has been derived (Eq. 3.71); also, a relationship (Eq. 3.74) that describes a simple particular case of the HyG beams, namely, Gaussian optical vortices in the PI medium (Figure 3.12), has been deduced. Under certain parameters, the said Gaussian optical vortices become modes of the PI medium. A wide class of solutions of the Helmholtz equation in the cylindrical coordinates to describe the propagation of modes in a PI medium has been derived; the solutions are proportional to Kummer's function (Eq. 3.82); it has been shown that only solutions coincident with the nonparaxial LG modes (Eq. 3.85) have a finite energy, thus being physically realizable; there have been no other nonparaxial modes with cylindrical symmetry in the PI waveguide, similar to paraxial modes (3.74). A definite length of the PI fiber has been treated as a parabolic lens, for which relationships for the *NA* and the focal spot size FWHM (Eqs. 3.91 and 3.92) have been derived. The FDTD simulation of focusing a linearly polarized Gaussian beam with a 3D PI lens has shown that an optimal lens length is smaller than that predicted by the scalar theory, with the smaller diameter of the elliptic focal spot being FWHM = 0.42λ (Figure 3.19b). An explicit relation for the ring radii of a binary lens that approximates the PI lens has been derived (Eq. 3.95). It has been estimated that the elliptic focal spot generated with the binary lens has the smaller diameter of FWHM = 0.45λ (Figure 3.21).

3.4 NONPARAXIAL HYPERGEOMETRIC MODES

Recent years have seen an increased interest in the exact solutions of the paraxial Schrödinger-type equation in the cylindrical coordinates. For example, in Ref. [6] the HyG modes were presented. In subsequent publications these beams were generalized, giving rise to hypergeometric-Gauss (HyGG) modes [13], HyG beams [113], and circular beams (CiB) [14]. In Ref. [14] it was indicated that many familiar light beams, for example standard [2] and elegant [141] Laguerre-Gauss modes, quadratic Bessel-Gauss beams [15], and optical Gauss vortices [16,33], could be considered as a particular case of the CiBs.

The Schrödinger equation, however, describes the light propagation in the paraxial approximation, which is not suitable in some cases, for example in tasks involving tight focusing of laser irradiance (tight focusing can be used, e.g., for micromanipulation, for increasing the density of data storage, for surgery applications, and for microlithography).

In this section, we discuss the HyG modes beyond the paraxial approximation. We derive an analytical expression to present the exact solution of the Helmholtz equation in the cylindrical coordinates. This solution is proportional to the product of two Kummer's functions. Then, the solution is represented as two terms that describe the direct nonparaxial hypergeometric (nHyG^+) mode and the inverse nonparaxial hypergeometric (nHyG^-) mode. These beams propagate along the optical axis in the positive and negative directions. At large distances from the initial plane (much larger than the wavelength) the nHyG^+ mode is shown to coincide, up to a constant, with the paraxial HyG mode of Refs. [6,113].

Angular Spectrum of the Plane-Waves for Nonparaxial Hypergeometric Modes

It has been known that any solution of the Helmholtz equation

$$(\Delta + k^2)E(x,y,z) = 0, \tag{3.96}$$

where k is the wavenumber, can be presented as the angular spectrum of the plane-waves

$$E(x,y,z) = \int_{-\pi}^{\pi}\int_{0}^{\pi} f(\theta,\varphi)\exp\left[-ik(x\sin\theta\cos\varphi + y\sin\theta\sin\varphi + z\cos\theta)\right]\sin\theta \mathrm{d}\theta \mathrm{d}\varphi, \tag{3.97}$$

where (θ, φ) are the Euler angles that define a point on the sphere, determining the plane-wave direction. Let us consider the particular angular spectrum

$$f(\theta,\varphi) = \frac{1}{2\pi}\left(\tan\frac{\theta}{2}\right)^{\beta}\sin^{-1}(\theta)\exp(2in\varphi), \tag{3.98}$$

where:

- n is an integer
- β are real numbers

This angular spectrum has been chosen for two main reasons. First, as we will see later, this spectrum allows analytical evaluation of the integral (3.97). Second, when the angle θ becomes small (i.e., paraxial case), we can approximately write $\sin\theta \approx \tan\theta \approx \theta \sim \rho/z$, where ρ is the radial polar coordinate in the input plane and z is the propagation distance. Therefore, the field in the input plane has the two multiples (up to constant), power term $\rho^{\beta-1}$ and angular harmonic $\exp(2in\varphi)$, just as in the paraxial HyG modes [6]. Substituting (3.98) into (3.97) yields

$$E(r,\phi,z) = (-1)^n\exp(i2n\phi)\int_0^{\pi}\exp(-ikz\cos\theta)\left(\tan\frac{\theta}{2}\right)^{\beta}J_{2n}(kr\sin\theta)\mathrm{d}\theta, \tag{3.99}$$

where:

- (r, ϕ) are the polar coordinate in the arbitrary transverse plane $z > 0$ (ϕ must not be confused with φ)
- $J_v(x)$ is the Bessel function

Using a reference integral [38] (integral 2.12.27.3)

$$\begin{aligned}&\int_0^{\pi}\exp(p\cos x)\left(\tan\frac{x}{2}\right)^{2m}J_{2n}(c\sin x)\mathrm{d}x\\ &= \frac{1}{c}\Gamma\begin{bmatrix} m+n+1/2, n-m+1/2 \\ 2n+1, 2n+1\end{bmatrix}M_{m,n}(z_+)M_{m,n}(z_-),\end{aligned} \tag{3.100}$$

where $z_{\pm} = p \pm \left(p^2 - c^2\right)^{1/2}$,

$$\Gamma\begin{bmatrix} a_1,\ldots,a_m \\ b_1,\ldots,b_n \end{bmatrix} = \prod_{i=1}^{m} \Gamma(a_i) \Big/ \prod_{j=1}^{n} \Gamma(b_j),$$

$M_{\chi,\mu}(z)$ is the degenerate (confluent) HyG Whittaker function

$$M_{\chi,\mu}(z) = z^{\mu+1/2} \exp\left(-\frac{z}{2}\right) {}_1F_1\left(\mu - \chi + \frac{1}{2}, 2\mu+1, z\right), \tag{3.101}$$

and ${}_1F_1(a,b,x)$ is the degenerate (confluent) HyG function (Kummer's function) [101, Chapter 13, page 504], we can obtain an explicit analytical expression, instead of (3.99):

$$\begin{aligned} E(r,\phi,z) = {} & \frac{(-1)^n}{[(2n)!]^2} \exp(i2n\phi + ikz) \Gamma\left(\frac{2n+\beta+1}{2}\right) \Gamma\left(\frac{2n-\beta+1}{2}\right) (kr)^{2n} \\ & \times {}_1F_1\left(\frac{2n-\beta+1}{2}, 2n+1, x_+\right) {}_1F_1\left(\frac{2n-\beta+1}{2}, 2n+1, x_-\right), \end{aligned} \tag{3.102}$$

where:

$x_{\pm} = -ikz\left\{1 \pm \left[1 + (r/z)^2\right]^{1/2}\right\}$

(r, φ, z) are cylindrical coordinates

$\Gamma(x)$ is the gamma-function

Equation (3.102) is the exact solution of Eq. (3.96) and describes the sum of two nonparaxial HyG beams:

$$E(r,\varphi,z) = E^+(r,\varphi,z;\beta) + E^-(r,\varphi,z;\beta), \tag{3.103}$$

where E^+ is the direct nHyG$^+$ mode, given by Eq. (3.99) where the θ integral is evaluated from 0 up to $\pi/2$, and E^- is the inverse nHyG$^-$ mode, given by Eq. (3.99) where the θ integral is evaluated from $\pi/2$ up to π.

In the last multiplier of Eq. (3.98) the angular harmonic is $\exp(2in\varphi)$ instead of $\exp(in\varphi)$, because in the reference integral 2.12.27.3 [38] the order of Bessel function is even. Let us note that in book [96] explicit solution of Helmholtz equation is given for parabolic coordinate system. This solution is proportional to the product of two Whittaker functions, like Eq. (3.102). It is supposed in [96] that coefficients of the angular harmonics can be not only even, like in Eqs. (3.98) and (3.99), but odd too. We evaluated this integral numerically and discovered that indeed it can converge for both even and odd orders of the Bessel function. Moreover, we used parameters like $2n = 5$ and $\beta = 0.7 + 1.2i$ (some other values of β also were taken) and found that this integral converged too. It is seen from the view of the integrand (3.99) that it must converge when $\mathrm{Re}\,\beta > 0$ for direct waves and when $\mathrm{Re}\,\beta < 0$ for inverse waves.

It can be shown that $E^-(r,\varphi,z;\beta) = E^+(r,\varphi,-z;-\beta)$. Hence it follows, in particular, that at $z = \beta = 0$ direct and inverse nHyG modes coincide, being defined by the expression

$$E^{-}(r,\varphi,0;0)=E^{+}(r,\varphi,0;0)=0{,}5E(r,\varphi,0)=(-1)^{n}\frac{\pi}{2}\exp(i2n\varphi)J_{n}^{2}(kr/2). \quad (3.104)$$

The last part of Eq. (3.104) has been derived by using a special case of the confluent function when it transforms to the Bessel function [101, Eq. 13.6.1]. This expression can be derived immediately from the following reference integrals [135] (Eqs. 6.519.2 and 6.552.6):

$$\int_{0}^{\pi/2} J_{2\nu}\left(2z\sin x\right)\mathrm{d}x=\frac{\pi}{2}J_{\nu}^{2}\left(z\right),\ \mathrm{Re}\nu>-\frac{1}{2}, \quad (3.105)$$

$$\int_{0}^{1} J_{\nu}\left(xy\right)\frac{\mathrm{d}x}{\left(1-x^{2}\right)^{1/2}}=\frac{\pi}{2}\left[J_{\nu/2}\left(\frac{y}{2}\right)\right]^{2},\ y>0,\ \mathrm{Re}\nu>-1 \quad (3.106)$$

and suggests that the principal nHyG mode at $z=\beta=n=0$ is generated by the squared zero-order Bessel function and has central spot diameter 1.53λ, where λ is the wavelength (by the diameter we mean here a doubled distance from the maximum to the first root of zero-order Bessel function).

Direct and Inverse Nonparaxial Hypergeometric Modes

In general, when $z\neq 0$ or $\beta\neq 0$, expressions describing direct and inverse modes must be derived from Eq. (3.102) in the analytical form. At $z\to\infty$ the following asymptotic expansion for confluent functions takes place [101, Eq. 13.5.1]:

$$\begin{aligned}\frac{{}_1F_1\left(a,b,z\right)}{\Gamma\left(b\right)}&=\frac{\exp\left(\pm i\pi a\right)z^{-a}}{\Gamma\left(b-a\right)}\left[\sum_{n=0}^{R-1}\frac{\left(a\right)_n\left(1+a-b\right)_n}{n!}\left(-z\right)^{-n}+O\left(\left|z\right|^{-R}\right)\right]\\&+\frac{\exp\left(z\right)z^{a-b}}{\Gamma\left(a\right)}\left[\sum_{n=0}^{S-1}\frac{\left(b-a\right)_n\left(1-a\right)_n}{n!}z^{-n}+O\left(\left|z\right|^{-S}\right)\right],\end{aligned} \quad (3.107)$$

where $\left(a\right)_n=\Gamma\left(a+n\right)/\Gamma\left(a\right)$ is the Pochgammer symbol [101] and the upper sign being taken if $-\pi/2<\arg z<3\pi/2$ and the lower sign if $-3\pi/2<\arg z\leq-\pi/2$. Tending R and S in Eq. (3.107) to infinity yields

$$\begin{aligned}\frac{{}_1F_1\left(a,b,z\right)}{\Gamma\left(b\right)}&=\frac{\exp\left(\pm i\pi a\right)z^{-a}}{\Gamma\left(b-a\right)}\,{}_2F_0\left(a,1+a-b,-\frac{1}{z}\right)\\&+\frac{\exp\left(z\right)z^{a-b}}{\Gamma\left(a\right)}\,{}_2F_0\left(b-a,1-a,\frac{1}{z}\right).\end{aligned} \quad (3.108)$$

Substituting (3.108) into (3.102) instead of confluent function with argument x_+ gives

$$E(r,\phi,z)$$

$$=\frac{(-1)^n}{(2n)!}\exp(i2n\phi)\Gamma\left(\frac{2n+\beta+1}{2}\right)\Gamma\left(\frac{2n-\beta+1}{2}\right)(kr)^{2n}{}_1F_1\left(\frac{2n-\beta+1}{2},2n+1,x_-\right)$$

$$\times\left\{\frac{\exp(ikz)}{\Gamma\left(\frac{2n+\beta+1}{2}\right)}\left[+ikz+ik\left(z^2+r^2\right)^{1/2}\right]^{-\frac{2n-\beta+1}{2}}{}_2F_0\left(\frac{2n-\beta+1}{2},\frac{-2n-\beta+1}{2},-\frac{1}{x_+}\right)\right.$$

$$\left.+\frac{\exp\left[-ik\left(z^2+r^2\right)^{1/2}\right]}{\Gamma\left(\frac{2n-\beta+1}{2}\right)}\left[-ikz-ik\left(z^2+r^2\right)^{1/2}\right]^{\frac{-2n-\beta-1}{2}}{}_2F_0\left(\frac{2n+\beta+1}{2},\frac{-2n+\beta+1}{2},\frac{1}{x_+}\right)\right\}.$$

(3.109)

where ${}_2F_0(a,b,x)$ is a HyG function [38]. Applying the Kummer transformation

$${}_1F_1(a,b,z)=\exp(z)\,{}_1F_1(b-a,b,-z), \qquad (3.110)$$

produces

$$E(r,\phi,z)=(-1)^n\exp(i2n\phi)\Gamma\left(\frac{2n+\beta+1}{2}\right)\Gamma\left(\frac{2n-\beta+1}{2}\right)[(2n)!]^{-1}(kr)^{2n}$$

$$\times\left\{\Gamma^{-1}\left(\frac{2n+\beta+1}{2}\right)(-x_+)^{-\frac{2n-\beta+1}{2}}\exp(+ikz)\right.$$

$$\times{}_2F_0\left(\frac{2n-\beta+1}{2},\frac{-2n-\beta+1}{2},-\frac{1}{x_+}\right){}_1F_1\left(\frac{2n-\beta+1}{2},2n+1,x_-\right) \qquad (3.111)$$

$$+\Gamma^{-1}\left(\frac{2n-\beta+1}{2}\right)(+x_+)^{-\frac{2n+\beta+1}{2}}\exp(-ikz)$$

$$\left.\times{}_2F_0\left(\frac{2n+\beta+1}{2},\frac{-2n+\beta+1}{2},+\frac{1}{x_+}\right){}_1F_1\left(\frac{2n+\beta+1}{2},2n+1,-x_-\right)\right\}.$$

Taking into account that $z^{-a}{}_2F_0(a,1+a-b,-1/z)=U(a,b,z)$, where $U(a,b,z)$ is yet another solution of the Kummer equation [101], Eq. (3.16) leads to

$$E(r,\phi,z)=\exp(i2n\phi)\Gamma\left(\frac{2n+\beta+1}{2}\right)\Gamma\left(\frac{2n-\beta+1}{2}\right)[(2n)!]^{-1}(kr)^{2n}$$
$$\times\left\{\frac{\exp(ikz)}{\Gamma\left(\frac{2n+\beta+1}{2}\right)}(-1)^{\frac{\beta-1}{2}}U\left(\frac{2n-\beta+1}{2},2n+1,x_+\right){}_1F_1\left(\frac{2n-\beta+1}{2},2n+1,x_-\right)\right.$$
$$\left.+\frac{\exp(-ikz)}{\Gamma\left(\frac{2n-\beta+1}{2}\right)}(-1)^{-\frac{\beta+1}{2}}U\left(\frac{2n+\beta+1}{2},2n+1,-x_+\right){}_1F_1\left(\frac{2n+\beta+1}{2},2n+1,-x_-\right)\right\}.\tag{3.112}$$

Let us write the expression for the nHyG± modes complex amplitudes:

$$E^{\pm}(r,\phi,z)=\frac{(-1)^{\frac{\pm\beta-1}{2}}}{(2n)!}\Gamma\left(\frac{2n\mp\beta+1}{2}\right)\exp(i2n\phi\pm ikz)(kr)^{2n}$$
$$\times U\left(\frac{2n\mp\beta+1}{2},2n+1,\pm x_+\right){}_1F_1\left(\frac{2n\mp\beta+1}{2},2n+1,\pm x_-\right).\tag{3.113}$$

The expression for $E(r,\phi,z)$ from Eq. (3.112) is the sum of $E^+(r,\phi,z)$ and $E^-(r,\phi,z)$ from Eq. (3.113). The terms $O\left(|z|^{-R}\right)$ and $O\left(|z|^{-S}\right)$ in expansion (3.107) mean that Eqs. (3.11) through (3.113) are not exact for small values of z, that is, near the input plane.

Further, we consider only field $E^+(r,\phi,z)$ propagating from the input plane $z=0$ to plane $z\to\infty$. For large values of z $U(a,b,z)\approx z^{-a}$. Tending z to infinity in Eq. (3.113) for the direct wave $E^+(r,\phi,z)$, we obtain an asymptotic expression

$$E^+(r,\phi,z\gg\lambda)\approx\left[(2n)!\right]^{-1}\exp(i2n\phi+ikz+t)\Gamma\left(\frac{2n-\beta+1}{2}\right)$$
$$\times(2ikz)^{(\beta-1)/2}t^n{}_1F_1\left(\frac{2n+\beta+1}{2},2n+1,-t\right),\tag{3.114}$$

where $t=ikr^2/(2z)$. Electromagnetic field with the same (up to a constant multiplier) slow-varying part of the complex amplitude can be obtained by Fresnel diffraction of the field with the following complex amplitude in the input plane $z=0$ [6,113]:

$$E(r,\phi,z=0)=\frac{1}{2\pi}\left(\frac{r}{w}\right)^m\exp\left(i\gamma\ln\frac{r}{w}+in\phi\right),\tag{3.115}$$

where:

w is the scaling factor

m is the real value of the amplitude factor's exponent

γ is the parameter of a logarithmic axicon (i.e., cone with logarithmic generatrix instead of straight one)

The fields from Refs. [6,113] and Eq. (3.114) coincide if $m = -1$, $i\gamma = \beta$, $w = k^{-1}$ and n is being replaced by $2n$. The disagreement of a constant multiplier is because in paraxial case the angular spectrum (3.98) is only proportional, but not equal to, the initial field (3.115).

In summary, we showed in this section that nonparaxial light field (at $z \gg \lambda$) generated by the angular spectrum (3.98) is similar to the field generated by the field (3.115) in the input plane.

Numerical Simulation

In this section we obtain the intensity distributions in three transverse planes with different propagation distances z: the input plane ($z = 0$) and planes $z = 1$ µm and $z = 1$ mm (Figure 3.22, intensity $I = |E_x|^2$ is shown in relative units). In the input plane we compared two curves—obtained by numeric evaluation of the integral (3.99),

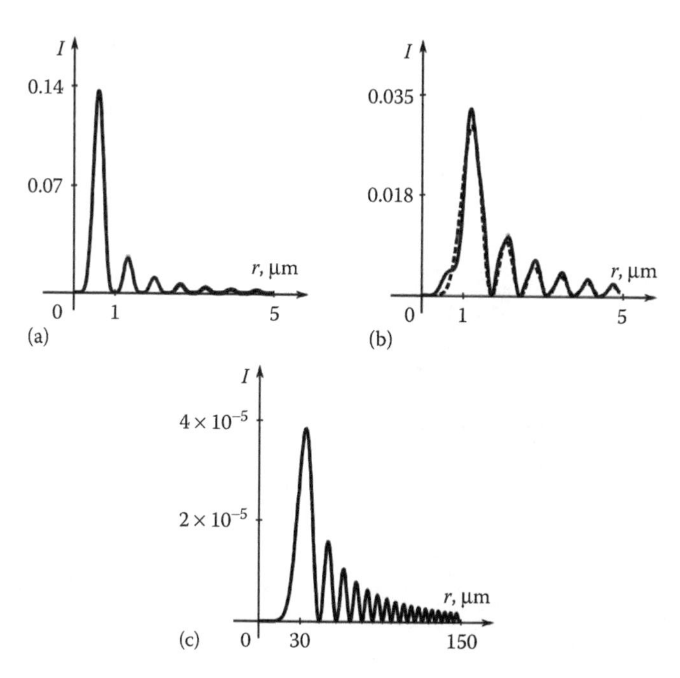

FIGURE 3.22 The intensity distribution in arbitrary units of the nHyG$^+$ mode at $\lambda = 633$ nm, $\beta = 0$, $n = 1$ at the distance z of (a) 0, (b) 1 µm, and (c) 1 mm (solid line—calculated by Eq. (3.98) with upper limit $\pi/2$, dotted line—calculated by Eq. (3.113)).

carried out from 0 to $\pi/2$, and by exact expression (3.104). The r.m.s. error was negligible and therefore there is only one curve in Figure 3.22a. In Figure 3.22b we show the diffraction pattern at $z = 1$ μm. We also compared two curves—obtained by Eq. (3.98) and by asymptotic Eq. (3.113). The dotted line in Figure 3.22b is the intensity distribution, calculated by the Eq. (3.113). In Figure 3.22c we show only analytical curve obtained by Eq. (3.114) because for such large z the integral (3.98) is too hard for numeric evaluation (exponent in the integrand is varying too fast with large z). We used the following simulation parameters: $\lambda = 633$ nm, $\beta = 0$, $n = 1$.

As seen from Figure 3.22, when light propagates near the input plane $z = 0$, changes in the intensity distribution of the nHyG$^+$ beam are mainly found in the side lobes (peripheral rings of the diffraction pattern, Figure 3.22a and b).

When $z \gg \lambda$, the nHyG$^+$ beam coincides with the HyG mode and the intensity changes only in scale with increasing of z, while the diffraction pattern remains unchanged (Figure 3.22c).

From Eq. (3.104) it follows that in the input plane intensity is proportional to $J_0^4(kr/2)$. Since $J_0^4(0.8) \approx 0.51$, we can calculate the size of the focal spot: $FWHM = 0.51\lambda$ and half-maximum area $HMA = 0.204\lambda^2$. It is interesting to note that the same focal spot size has the Airy disk in paraxial optics $2J_1(kr\,NA)/(kr\,NA)$ with maximal numeric aperture $NA = 1$.

For verification of calculations (Figure 3.22b) a simulation was undertaken using of FullWave 6.0 software (RSoft Design, USA, http://www.rsoftdesign.com), aimed for the solution of Maxwell equations by the FDTD method. In the input plane $z = 0$ we generated electric field E_x according to Eq. (3.104), linearly polarized along the x-axis and having the following parameters: $n = 1$, $\lambda = 633$ nm, sampling step $\lambda/20$. In Figure 3.23a we show the diffraction pattern of such a field in a transverse plane at $z = 1$ μm. The pattern has the size of 5×5 μm. In Figure 3.23b we show profile of the diffraction pattern along the horizontal line. It is observed from Figures 3.22b

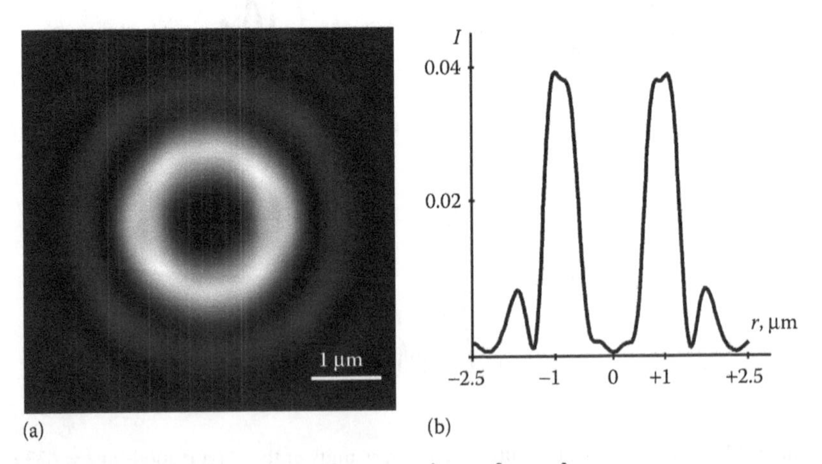

FIGURE 3.23 (a) Diffraction pattern $I = |E_x|^2 + |E_y|^2 + |E_z|^2$ and (b) its profile along the horizontal line (i.e., plane $y = 0$) of nonparaxial HyG beam with electric field E_x as in Eq. (3.104) in the input plane ($z = 0$) at propagation distance $z = 1$ μm.

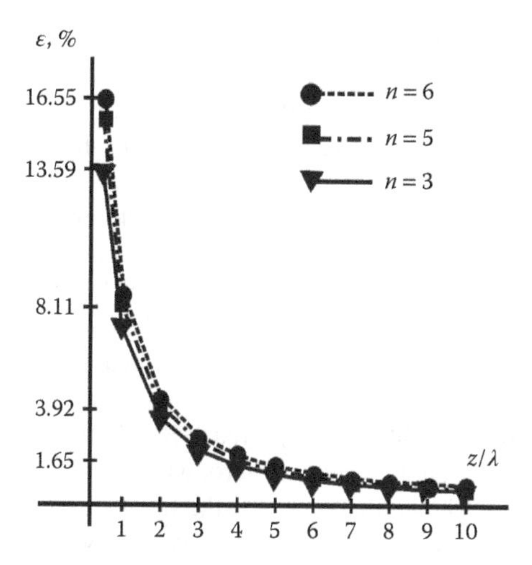

FIGURE 3.24 R.m.s. error between exact intensity evaluated by using integral (3.99) (carried out with limits 0 and $\pi/2$) and evaluated by Eq. (3.113).

(solid line) and 3.23b that they are in good concordance with each other, although in Figure 3.22b the $|E_x|^2$ value is shown, while in Figure 3.23b $I = |E_x|^2 + |E_y|^2 + |E_z|^2$.

In Figure 3.24 we show the r.m.s. error ε between Eq. (3.98), with integration carried out from 0 to $\pi/2$, and (3.113), calculated for several values of topological charge n and propagation distance z (other parameters are $\lambda = 633$ nm, $\beta = 0$). As seen in Figure 3.24, with z starting from about 6λ the error does not exceed 1%. The error is rather large for $z < \lambda$. It is because the argument $1/z$ of confluent function in the expansion (3.107) takes a value exceeding 1.

Summing up, analytical expressions to describe nonparaxial HyG modes propagated along the optical axis in the positive and negative directions have been derived. When the propagation distance is large, the direct nonparaxial HyG mode has been shown to coincide, up to a constant, with the paraxial mode of Refs. [6,113].

3.5 NONPARAXIAL PROPAGATION OF A GAUSSIAN OPTICAL VORTEX WITH INITIAL RADIAL POLARIZATION

Radially polarized beams can be generated using spacevariant dielectric gratings [142], image rotating resonators [143], interferometric techniques [144], or optical fibers [145]. Interest in radially and azimuthally polarized laser beams has been prompted by their potential uses in laser cutting [146] and optical micromanipulation [147], because the focusing of such beams produces a circular subwavelength focal spot. The propagation of radially polarized beams was studied in Refs. [148–159]. Note that the longitudinal field component was not considered in Ref. [148], whereas in Ref. [149] the analysis of the nonparaxial diffraction was conducted for the vortex-free fields. The diffraction of a radially polarized Gaussian beam was studied in Ref. [150] in the paraxial theory without regard for the optical vortices. The propagation of paraxial and nonparaxial vortex beams was investigated in

Refs. [33,57,58,113,151]. The scalar paraxial diffraction of vortex beams was discussed in Refs. [33,57,58,113], and the vector nonparaxial diffraction of the linearly polarized vortex beam was reported in Ref. [151], with Ref. [151] dealing with the scalar nonparaxial diffraction of HyG laser beams. The sharp (nonparaxial) focusing of conventional radially polarized laser beams and elliptically polarized vortex beams was treated in Ref. [152] and Ref. [153], respectively. In Refs. [152,153], the analysis was based on the approximate Debye and Richards-Wolf theories.

In this section, we derive explicit analytic expressions for the radial, azimuthal and axial projections of the electric field vectors that describe the diffraction of nonparaxial Gaussian vortex beams with initial radial polarization. In a particular case of $n = 0$, the azimuthal field component equals zero, with the relationships for the radial and axial components being coincident with those reported in Ref. [150]. At any $n > 1$, the vortex beam intensity on the optical axis is equal to zero, whereas at $n = \pm 1$ an intensity peak is found in the focus. It is shown that at $n = 1$, the focal spot width at half-maximum can be 1.47 times smaller than that predicted by scalar theory. The relationships derived are also verified using the FDTD method. The intensity focus is shown to be closer than the geometric focus. Explicit analytical relationships for the Gaussian vortex beam with initial elliptic polarization are also derived. Expressions that describe the nonparaxial radially polarized Gaussian beam as a linear combination of Gaussian vortex beams with $n = \pm 1$ and left- and right-hand circular polarization are derived.

Integral Transforms to Describe the Propagation of Radially and Azimuthally Polarized Laser Beams in Free Space

The propagation of light in free space along the Oz-axis is described by the well-known Rayleigh-Zommerfeld integrals [154]:

$$\begin{cases} E_x(u,v,z) = -\dfrac{1}{2\pi}\displaystyle\iint_{R^2} E_x(x,y,0)\dfrac{\partial}{\partial z}\left[\dfrac{\exp(ikR)}{R}\right]\mathrm{d}x\mathrm{d}y, \\ E_y(u,v,z) = -\dfrac{1}{2\pi}\displaystyle\iint_{R^2} E_y(x,y,0)\dfrac{\partial}{\partial z}\left[\dfrac{\exp(ikR)}{R}\right]\mathrm{d}x\mathrm{d}y, \\ E_z(u,v,z) = \dfrac{1}{2\pi}\displaystyle\iint_{R^2}\left\{E_x(x,y,0)\dfrac{\partial}{\partial u}\left[\dfrac{\exp(ikR)}{R}\right] + E_y(u,v,0)\dfrac{\partial}{\partial v}\left[\dfrac{\exp(ikR)}{R}\right]\right\}\mathrm{d}x\mathrm{d}y, \end{cases} \tag{3.116}$$

where:

$R^2 = z^2 + (x-u)^2 + (y-v)^2$

(x,y) and (u,v) are the Cartesian coordinates in the input plane and output plane, respectively

$E_{x,y}(x,y,0)$ are the projections of the E-vector of the electromagnetic wave on the x- and y-axes of the plane $z = 0$

$E_{x,y,z}(u,v,z)$ are the projections of the E-vector on the x-, y-, and z-axes in an arbitrary z-plane

k is the wavenumber.

Let us change from the Cartesian components of the E-vector on the x- and y-axis to the radial and azimuthal components:

$$\begin{bmatrix} E_r \\ E_\varphi \end{bmatrix} = \begin{bmatrix} \cos\varphi & \sin\varphi \\ -\sin\varphi & \cos\varphi \end{bmatrix} \cdot \begin{bmatrix} E_x \\ E_y \end{bmatrix}, \tag{3.117}$$

or

$$\begin{bmatrix} E_x \\ E_y \end{bmatrix} = \begin{bmatrix} \cos\varphi & -\sin\varphi \\ \sin\varphi & \cos\varphi \end{bmatrix} \cdot \begin{bmatrix} E_r \\ E_\varphi \end{bmatrix}, \tag{3.118}$$

where:

$$\cos\varphi = \frac{x}{\sqrt{x^2+y^2}}, \quad \sin\varphi = \frac{y}{\sqrt{x^2+y^2}}. \tag{3.119}$$

Then, the radial and azimuthal components (in the cylindrical coordinates) are given by

$$E_r(\rho,\theta,z) = \frac{-1}{2\pi}\iint_{R^2}\left[E_r(r,\varphi,0)\cos(\varphi-\theta) - E_\varphi(r,\varphi,0)\sin(\varphi-\theta)\right]\frac{\partial}{\partial z}\left[\frac{\exp(ikR)}{R}\right] r\mathrm{d}r\mathrm{d}\varphi, \tag{3.120}$$

$$E_\varphi(\rho,\theta,z) = \frac{-1}{2\pi}\iint_{R^2}\left[E_\varphi(r,\varphi,0)\cos(\varphi-\theta) + E_r(r,\varphi,0)\sin(\varphi-\theta)\right]\frac{\partial}{\partial z}\left[\frac{\exp(ikR)}{R}\right] r\mathrm{d}r\mathrm{d}\varphi. \tag{3.121}$$

It can be seen that if the field has a radial symmetry in the original plane, being radially or azimuthally polarized, it will retain radial or azimuthal polarization upon propagation. This can be inferred from the fact that in the polar coordinates $R^2 = z^2 + \rho^2 + r^2 - 2\rho r\cos(\varphi-\theta)$ and integration with respect to the variable φ yields

$$\int_0^{2\pi}\sin(\varphi-\theta)\frac{\partial}{\partial z}\left[\frac{\exp(ikR)}{R}\right]\mathrm{d}\varphi = -\int_0^{2\pi}\frac{\partial}{\partial z}\left[\frac{\exp(ikR)}{R}\right]\mathrm{d}\cos(\varphi-\theta) = -\chi\left[\cos(\varphi-\theta)\right]\Big|_{\varphi=0}^{2\pi} = 0, \tag{3.122}$$

where $\chi(\xi)$ is the antiderivative of the function

$$\frac{\partial}{\partial z}\left[\frac{\exp\left(ik\sqrt{z^2+\rho^2+r^2-2\rho r\xi}\right)}{\sqrt{z^2+\rho^2+r^2-2\rho r\xi}}\right]. \tag{3.123}$$

The common factor in the integrands (3.120) and (3.121) may be approximately simplified to

$$\frac{\partial}{\partial z}\left[\frac{\exp(ikR)}{R}\right]\approx\frac{ikz}{z^2+\rho^2}\exp\left(ik\sqrt{z^2+\rho^2}\right)\exp\left(\frac{ikr^2}{2\sqrt{z^2+\rho^2}}\right)\exp\left[-\frac{ik\rho r\cos(\varphi-\theta)}{\sqrt{z^2+\rho^2}}\right]. \tag{3.124}$$

The expression (3.124) is nonparaxial because the exponent denominator contains both the longitudinal coordinate z and the transverse coordinate ρ in the observation plane. The use of the approximation (3.124) to describe the nonparaxial propagation of light beams is well known [155].

In this case, the expressions for the radial and azimuthal components (3.120), (3.121) may be given in a simpler form:

$$\begin{aligned}E_r(\rho,\theta,z)=&-\frac{ikz}{2\pi\left(z^2+\rho^2\right)}\exp\left(ik\sqrt{z^2+\rho^2}\right)\\&\times\iint_{R^2}\left[E_r(r,\varphi+\theta,0)\cos\varphi-E_\varphi(r,\varphi+\theta,0)\sin\varphi\right]\\&\times\exp\left(\frac{ikr^2}{2\sqrt{z^2+\rho^2}}\right)\exp\left(-\frac{ik\rho r\cos\varphi}{\sqrt{z^2+\rho^2}}\right)rdrd\varphi,\end{aligned} \tag{3.125}$$

$$\begin{aligned}E_\varphi(\rho,\theta,z)=&-\frac{ikz}{2\pi\left(z^2+\rho^2\right)}\exp\left(ik\sqrt{z^2+\rho^2}\right)\\&\times\iint_{R^2}\left[E_\varphi(r,\varphi+\theta,0)\cos\varphi+E_r(r,\varphi+\theta,0)\sin\varphi\right]\\&\times\exp\left(\frac{ikr^2}{2\sqrt{z^2+\rho^2}}\right)\exp\left(-\frac{ik\rho r\cos\varphi}{\sqrt{z^2+\rho^2}}\right)rdrd\varphi.\end{aligned} \tag{3.126}$$

Propagation of Optical Vortex Beams with Initial Radial Polarization

If there is a radially polarized field with a vortex component in the initial plane, that is

$$\begin{cases}E_r(r,\varphi,0)=A_r(r)\exp(in\varphi),\\E_\varphi(r,\varphi,0)=0,\end{cases} \tag{3.127}$$

Eqs. (3.125) and (3.126) suggest that in a general case ($n \neq 0$) the E_r- and E_φ-components will be non-zero. In this section, we conduct a detailed analysis of the relationship for E_r, while the relationship for E_φ is given at the end without derivation. Substituting (3.127) into (3.125) yields:

$$E_r(\rho,\theta,z)=(-i)^n \frac{kz}{2(z^2+\rho^2)} \exp\left(ik\sqrt{z^2+\rho^2}+in\theta\right) \times \int_0^\infty A_r(r)\exp\left(\frac{ikr^2}{2\sqrt{z^2+\rho^2}}\right)\left[J_{n-1}\left(\frac{k\rho r}{\sqrt{z^2+\rho^2}}\right)-J_{n+1}\left(\frac{k\rho r}{\sqrt{z^2+\rho^2}}\right)\right] r\mathrm{d}r, \tag{3.128}$$

where $J_n(x)$ is the Bessel function of the first kind.

The discussion in Ref. [150] was concerned with the propagation of a radially polarized Gaussian beam with the intensity distribution in the initial plane given by

$$\begin{cases} E_x(x,y,0)=E_0\left(\dfrac{\sqrt{2}}{\omega_0}\right)x\exp\left[-\alpha\left(x^2+y^2\right)\right], \\ E_y(x,y,0)=E_0\left(\dfrac{\sqrt{2}}{\omega_0}\right)y\exp\left[-\alpha\left(x^2+y^2\right)\right], \end{cases} \tag{3.129}$$

where:

$\alpha = 1/\omega_0^2 + ik/(2f)$
w_0 is the Gaussian beam waist radius
f is the lens focal length.

Changing from the field in the Cartesian components of Eq. (3.129) to the field with the radial component yields

$$E_r(r,\varphi,0)=E_0\left(\frac{\sqrt{2}}{\omega_0}\right)r\exp\left(-\alpha r^2\right). \tag{3.130}$$

Substitution of Eq. (3.130) for $A_r(r)$ in (3.128) gives

$$E_r(\rho,\theta,z)=\frac{(-i)^n kzE_0}{\sqrt{2}\omega_0\left(z^2+\rho^2\right)}\exp\left(ik\sqrt{z^2+\rho^2}+in\theta\right) \times\int_0^\infty r^2\exp\left(-\beta r^2\right)\left[J_{n-1}(\gamma r)-J_{n+1}(\gamma r)\right]\mathrm{d}r, \tag{3.131}$$

where:

$$\gamma=\frac{k\rho}{\sqrt{z^2+\rho^2}},\quad \beta=\alpha-\frac{ik}{2\sqrt{z^2+\rho^2}}=\frac{\alpha q}{\sqrt{z^2+\rho^2}},\quad q=\sqrt{z^2+\rho^2}-\frac{ik}{2\alpha}. \tag{3.132}$$

The integral in Eq. (3.131) is derived using a table integral (see Ref. [135], Eq. (6.618.1)):

$$\int_0^\infty \exp(-\alpha x^2) J_\nu(\beta x) \mathrm{d}x = \frac{\sqrt{\pi}}{2\sqrt{\alpha}} \exp(-y) I_{\frac{\nu}{2}}(y), \tag{3.133}$$

where:

$y = \beta^2/(8\alpha)$

$I_n(x)$ is the Bessel function of second kind

Differentiating Eq. (3.133) with respect to α we obtain

$$\begin{aligned} &\int_0^\infty x^2 \exp(-\alpha x^2) J_\nu(\beta x) \mathrm{d}x \\ &= -\frac{\sqrt{\pi}}{4\alpha\sqrt{\alpha}} \exp(-y) \left[(\nu - 1 + 2y) I_{\frac{\nu}{2}}(y) - 2y I_{\frac{\nu-2}{2}}(y) \right]. \end{aligned} \tag{3.134}$$

Using Eq. (3.134) and a recurrent relation for the functions $I_\nu(z)$ (Ref. [101], Eq. [9.6.26]), we obtain:

$$\begin{aligned} E_r(\rho, \theta, z) &= \frac{(-i)^n \sqrt{\pi} kz E_0}{4\sqrt{2}\omega_0 \beta \sqrt{\beta} (z^2 + \rho^2)} \\ &\times \left[(n - 4x) I_{\frac{n-1}{2}}(x) + (n + 4x) I_{\frac{n+1}{2}}(x) \right] \exp\left(ik\sqrt{z^2 + \rho^2} - x + in\theta \right), \end{aligned} \tag{3.135}$$

where $x = \gamma^2/(8\beta)$.

It is seen from Eq. (3.135) that at $n = \pm 1$ the following condition will hold: $\left[(\pm 1 - 4x) I_0(x) + (\pm 1 + 4x) I_1(x) \right] \big|_{x=0} = \pm \left[I_0(0) + I_1(0) \right] = \pm I_0(0) \neq 0$. This means that there are no on-axis singularities of the light beam with initial radial polarization and $n = \pm 1$.

In a similar way, we can derive the relationship for the azimuthal component:

$$\begin{aligned} E_\varphi(\rho, \theta, z) &= \frac{(-i)^{n+1} n\sqrt{\pi} kz E_0}{4\sqrt{2}\omega_0 \beta \sqrt{\beta} (z^2 + \rho^2)} \left[I_{\frac{n-1}{2}}(x) + I_{\frac{n+1}{2}}(x) \right] \\ &\times \exp\left(ik\sqrt{z^2 + \rho^2} - x + in\theta \right), \end{aligned} \tag{3.136}$$

From Eq. (3.136) it is also seen that on the optical axis ($x = 0$) at $n = \pm 1$ the sum in the square brackets is not zero ($I_0(0) + I_1(0) \neq 0$), that is, in this case there is no any singularity as well.

Inequalities of complex amplitudes (3.135) and (3.136) to zero at $n = \pm 1$ can be explained. At two points, located symmetrically with respect to the optical axis, complex amplitudes will be different only in sign, but directions of the electric vector will differ only in sign too. Therefore, electric vectors will be the same at these two points.

The same conclusion can be made using Jones vectors formalism, applied earlier to optical vortices in [154] (Figure 3.6b in [156]). Let us note that longitudinal component of the optical vortex is not being considered in [156].

With the topological charge n entering into Eq. (3.136) as a multiplier, at $n = 0$ the azimuthal component becomes equal to zero.

The general relationships (3.135) and (3.136) describe a Gaussian optical vortex of an arbitrary integer order n with the radial initial polarization, Eq. (3.127).

The case of $n = 0$ dealt with in Ref. [150] can also be deduced from Eq. (3.135):

$$E_r(\rho,\theta,z)=\frac{-k^2 zE_0\rho}{2\sqrt{2}\omega_0\alpha^2 q^2\sqrt{z^2+\rho^2}}\exp\left(ik\sqrt{z^2+\rho^2}-\frac{k^2\rho^2}{4\alpha q\sqrt{z^2+\rho^2}}\right). \quad (3.137)$$

Since in the paraxial approximation $q\approx z-ik/(2\alpha)$, $\left(z^2+\rho^2\right)^{1/2}\approx z+\rho^2/(2z)\approx z$, hence

$$E_r(\rho,\theta,z)=\frac{-k^2E_0\rho}{2\sqrt{2}q^2\alpha^2\omega_0}\exp(ikz)\exp\left(\frac{ik\rho^2}{2q}\right). \quad (3.138)$$

A similar relationship was derived in Ref. [150] (see formula (27)), the only difference from Eq. (3.138) being complex conjugation.

Relationships for the Longitudinal Component

The longitudinal component is calculated using the third integral in Eq. (3.116). Changing to the radial and longitudinal components (cylindrical coordinates) yields:

$$\begin{aligned}E_z(\rho,\theta,z)=\frac{1}{2\pi}\iint_{R^2}&\Big\{E_r(r,\varphi,0)\left[r-\rho\cos(\varphi-\theta)\right]\\&+E_\varphi(r,\varphi,0)\rho\sin(\varphi-\theta)\Big\}\frac{1}{R}\frac{\partial}{\partial R}\left[\frac{\exp(ikR)}{R}\right]r\mathrm{d}r\mathrm{d}\varphi.\end{aligned} \quad (3.139)$$

In particular, if the field in the initial plane is radially polarized, using Eq. (3.124) yields

$$\begin{aligned}E_z(\rho,\theta,z)=&\frac{ik}{2\pi\left(z^2+\rho^2\right)}\exp\left(ik\sqrt{z^2+\rho^2}\right)\\&\times\iint_{R^2}E_r(r,\varphi,0)\left[r-\rho\cos(\varphi-\theta)\right]\exp\left(\frac{ikr^2}{2\sqrt{z^2+\rho^2}}\right)\exp\left[-\frac{ik\rho r\cos(\varphi-\theta)}{\sqrt{z^2+\rho^2}}\right]r\mathrm{d}r\mathrm{d}\varphi.\end{aligned} \quad (3.140)$$

If in the initial plane there exists a radially polarized vortex field of Eq. (3.127), by analogy with relations for the transverse components the axial component can be shown to be represented by

$$E_z(\rho,\theta,z)=(-i)^n \frac{ik}{2(z^2+\rho^2)}\exp\left(ik\sqrt{z^2+\rho^2}+in\theta\right)$$
$$\times\left\{2\int_0^\infty r^2 A_r(r)\exp\left(\frac{ikr^2}{2\sqrt{z^2+\rho^2}}\right)J_n\left(\frac{k\rho r}{\sqrt{z^2+\rho^2}}\right)\mathrm{d}r\right. \tag{3.141}$$
$$\left.+i\rho\int_0^\infty rA_r(r)\exp\left(\frac{ikr^2}{2\sqrt{z^2+\rho^2}}\right)\left[J_{n+1}\left(\frac{k\rho r}{\sqrt{z^2+\rho^2}}\right)-J_{n-1}\left(\frac{k\rho r}{\sqrt{z^2+\rho^2}}\right)\right]\mathrm{d}r\right\}.$$

Assume that a radially polarized Gauss beam of Eq. (3.130) propagates in space. Then, substituting (3.130) into (3.141) yields:

$$E_z(\rho,\theta,z)=(-i)^n \frac{ikE_0}{\sqrt{2}\omega_0(z^2+\rho^2)}\exp\left(ik\sqrt{z^2+\rho^2}+in\theta\right)$$
$$\times\left\{2\int_0^\infty r^3\exp(-\beta r^2)J_n(\gamma r)\mathrm{d}r+i\rho\int_0^\infty r^2\exp(-\beta r^2)\left[J_{n+1}(\gamma r)-J_{n-1}(\gamma r)\right]\mathrm{d}r\right\}. \tag{3.142}$$

where:

$$\beta=\alpha-\frac{ik}{2\sqrt{z^2+\rho^2}},\quad \gamma=\frac{k\rho}{\sqrt{z^2+\rho^2}}. \tag{3.143}$$

In the formula earlier, the second integral is similar to Eq. (3.134). Whereas the first integral is calculated by differentiating with respect to α of the following table integral (see Ref. [135], formula 6.631.7):

$$\int_0^\infty x\exp(-\alpha x^2)J_\nu(\beta x)\mathrm{d}x=\frac{\sqrt{\pi}\beta}{8\alpha^{3/2}}\exp\left(-\frac{\beta^2}{8\alpha}\right)\left[I_{\frac{\nu-1}{2}}\left(\frac{\beta^2}{8\alpha}\right)-I_{\frac{\nu+1}{2}}\left(\frac{\beta^2}{8\alpha}\right)\right]. \tag{3.144}$$

Using Eq. (3.144), we obtain:

$$\int_0^\infty x^3 \exp\left(-\beta x^2\right) J_\nu(\gamma x)\mathrm{d}x = \frac{\sqrt{\pi}\gamma}{8\beta^2\sqrt{\beta}}\exp(-t)$$
$$\times\left\{\frac{3}{2}\left[I_{\frac{\nu-1}{2}}(t) - I_{\frac{\nu+1}{2}}(t)\right] + \frac{t}{2}\left[I_{\frac{\nu-3}{2}}(t) - 3I_{\frac{\nu-1}{2}}(t) + 3I_{\frac{\nu+1}{2}}(t) - I_{\frac{\nu+3}{2}}(t)\right]\right\}, \tag{3.145}$$

where $t = \gamma^2/(8\beta)$.

Substituting Eqs. (3.144) and (3.145) into (3.142) yields:

$$E_z(\rho,\theta,z) = (-i)^n \frac{\sqrt{\pi}ikE_0}{4\sqrt{2}\beta^2\sqrt{\beta}\omega_0\left(z^2+\rho^2\right)}\exp\left(ik\sqrt{z^2+\rho^2} - t + in\theta\right)$$
$$\times\left\{\frac{3\gamma}{2}\left[I_{\frac{n-1}{2}}(t) - I_{\frac{n+1}{2}}(t)\right] + \frac{\gamma t}{2}\left[I_{\frac{n-3}{2}}(t) - 3I_{\frac{n-1}{2}}(t)\right.\right. \tag{3.146}$$
$$\left.\left. + 3I_{\frac{n+1}{2}}(t) - I_{\frac{n+3}{2}}(t)\right] - i\beta\rho\left[(n-4t)I_{\frac{n-1}{2}}(t) + (n+4t)I_{\frac{n+1}{2}}(t)\right]\right\}.$$

Using recurrent relationships for the modified Bessel functions and collecting similar terms, we finally obtain:

$$E_z(\rho,\theta,z) = \frac{(-i)^n\sqrt{\pi}ikE_0}{4\sqrt{2}\beta^2\sqrt{\beta}\omega_0\left(z^2+\rho^2\right)}\exp\left(ik\sqrt{z^2+\rho^2} - t + in\theta\right)$$
$$\times\left\{\left[\frac{\gamma(n+2)}{2} - 2\gamma t - i\beta\rho(n-4t)\right]I_{\frac{n-1}{2}}(t)\right. \tag{3.147}$$
$$\left. + \left[\frac{\gamma(n-2)}{2} + 2\gamma t - i\beta\rho(n+4t)\right]I_{\frac{n+1}{2}}(t)\right\}.$$

If the beam is nonvortex in the initial plane (i.e., $n = 0$), then

$$E_z(\rho,\theta,z) = \frac{ikE_0}{\sqrt{2}q^2\alpha^2\omega_0}\left(1 + \frac{ik\rho^2}{2q}\right)\exp\left(ik\sqrt{z^2+\rho^2} - \frac{k^2\rho^2}{4q\alpha\sqrt{z^2+\rho^2}}\right). \tag{3.148}$$

In the paraxial approximation, the following equality is valid:

$$ik\sqrt{z^2+\rho^2} - \frac{k^2\rho^2}{4q\alpha\sqrt{z^2+\rho^2}} \approx ikz + \frac{ik\rho^2}{2z} - \frac{k^2\rho^2}{4q\alpha z} = ikz + \frac{ik\rho^2}{2q}. \tag{3.149}$$

In view of this equality, we get:

$$E_z(\rho,\theta,z)\approx\frac{ikE_0}{\sqrt{2}q^2\alpha^2\omega_0}\left(1+\frac{ik\rho^2}{2q}\right)\exp\left(ikz+\frac{ik\rho^2}{2q}\right). \quad (3.150).$$

This expression is also coincident (up to complex conjugation) with the expression for the longitudinal component of a radially polarized Gaussian beam derived in Ref. [150] (Eq. 3.32).

Propagation of Optical Vortex Beams with Initial Elliptical Polarization

Making use of the diffraction integrals (3.116), the approximation (3.124), the initial field given by

$$\begin{cases} E_x(r,\varphi,0)=B_x\exp\left(-\dfrac{r^2}{\omega^2}\right)\exp(in\varphi), \\ E_y(r,\varphi,0)=B_y\exp\left(-\dfrac{r^2}{\omega^2}\right)\exp(in\varphi), \end{cases} \quad (3.151)$$

and the table integrals (3.134) and (3.144), we can derive analogous relationships for Gaussian vortex beams with the elliptical polarization in the initial plane:

$$\begin{cases} E_{x,y}(\rho,\theta,z)=(-i)^{n+1}\dfrac{B_{x,y}kz\exp\left(in\theta+ik\sqrt{\rho^2+z^2}\right)}{\rho^2+z^2}\dfrac{c\sqrt{\pi}}{8p^{3/2}}\exp(-y)\left[I_{\frac{n-1}{2}}(y)-I_{\frac{n+1}{2}}(y)\right], \\ E_z(\rho,\theta,z)=(-i)^n\dfrac{k}{\rho^2+z^2}\exp\left(ik\sqrt{\rho^2+z^2}+in\theta\right)\dfrac{\sqrt{\pi}}{8p^{3/2}}\exp(-y) \\ \times\left(\dfrac{B_x-iB_y}{2}\exp(i\theta)\left\{(n+3-3y)\left[I_{\frac{n+1}{2}}(y)-I_{\frac{n+3}{2}}(y)\right]+y\left[I_{\frac{n-1}{2}}(y)-I_{\frac{n+5}{2}}(y)\right]\right.\right. \\ \left.-\dfrac{B_x+iB_y}{2}\exp(-i\theta)\left\{(n+1-3y)\left[I_{\frac{n-1}{2}}(y)-I_{\frac{n+1}{2}}(y)\right]+y\left[I_{\frac{n-3}{2}}(y)-I_{\frac{n+3}{2}}(y)\right]\right\}\right. \\ \left.-i(B_x\cos\theta+B_y\sin\theta)c\rho\left[I_{\frac{n-1}{2}}(y)-I_{\frac{n+1}{2}}(y)\right]\right), \end{cases}$$

(3.152)

where:

$$p = \frac{1}{\omega^2} - \frac{ik}{2\sqrt{\rho^2 + z^2}}, \ c = \frac{k\rho}{\sqrt{\rho^2 + z^2}}, \ y = \frac{c^2}{8p}, \tag{3.153}$$

and B_x, B_y are constants.

Cartesian components of the E-vector in Eq. (3.152) describe the nonparaxial diffraction of the elliptically polarized Gaussian beam by a SPP with the topological charge n. Note that the Gaussian beam is circularly polarized at $B_x = \pm iB_y$ and linearly polarized at $B_x \neq 0$, $B_y = 0$.

In particular, as a result of the superposition of two beams (3.152) with the right-hand ($B_x = +iB_y$) and the left-hand ($B_x = -iB_y$) circular polarization and respective $n = 1$ and $n = -1$, Eq. (3.152) changes to a radially polarized beam:

$$\begin{cases} E_r(\rho,\theta,z) = \dfrac{-B_x kzc\sqrt{\pi}}{4(\rho^2 + z^2)p^{3/2}} \exp\left(ik\sqrt{\rho^2 + z^2} - y\right)\left[I_0(y) - I_1(y)\right], \\ E_\varphi(\rho,\theta,z) = 0, \\ E_z(\rho,\theta,z) = \dfrac{iB_x k\sqrt{\pi}}{4(\rho^2 + z^2)p^{3/2}} \exp\left(ik\sqrt{\rho^2 + z^2} - y\right) \\ \qquad \times\left\{(2 - 3y + ic\rho)\left[I_0(y) - I_1(y)\right] + y\left[I_1(y) - I_2(y)\right]\right\}. \end{cases} \tag{3.154}$$

Equation (3.154) describes a radially polarized Gaussian laser beam which was generated interferometrically in Ref. [157]. Note that mathematically speaking the expression for the amplitude E_r in Eq. (3.154) is identical to the expression for the scalar amplitude of the paraxial optical Gaussian vortex at $n = 1$ reported in Ref. [33]. Light beams (3.138) and (3.154) are both Gaussian beams with initial radial polarization. The difference between these beams is in complex amplitude at $z = 0$, respectively:

$$E_1(x, y, z = 0) = \exp\left(-\frac{r^2}{\omega^2}\right) r \begin{bmatrix} \cos\varphi \\ \sin\varphi \end{bmatrix}, \quad E_2(x, y, z = 0) = \exp\left(-\frac{r^2}{\omega^2}\right) \begin{bmatrix} \cos\varphi \\ \sin\varphi \end{bmatrix}, \tag{3.155}$$

where the column vector is a Jones vector for radially polarized light.

In a similar way, the subtraction of the field (3.152) with the right-hand polarization and $n = 1$ from the same field with the left-hand polarization and $n = -1$ produces azimuthally polarized beams:

$$\begin{cases} E_r(\rho,\theta,z) = 0, \\ E_\varphi(\rho,\theta,z) = \dfrac{iB_x kzc\sqrt{\pi}}{4(\rho^2 + z^2)p^{3/2}} \exp\left(ik\sqrt{\rho^2 + z^2} - y\right)\left[I_0(y) - I_1(y)\right], \\ E_z(\rho,\theta,z) = 0. \end{cases} \tag{3.156}$$

The light beams of Eq. (3.156) have been employed in hollow metallic waveguides because they show minimal losses due to wall scattering [158].

Numerical Simulation

When conducting the numerical simulation, the integrals in Eq. (3.116) were taken by the method of rectangles and then compared with values given by the relationships (3.135), (3.136) and (3.147), and relationships derived from (3.138) and (3.150) for the paraxial approximation.

The simulation was conducted for the wavelength of $\lambda = 633$ nm, $E_0 = 1$ (a.u.), the Gaussian beam waist radius of $w_0 = 1$ μm, the focal length of the parabolic lens of $f = 4$ μm, the distance traveled by light of $z = 4$ μm (implying the field in the lens focus), the polar angle at the output plane of $\theta = 0$, and the integration limits of $x, y \in [-4\omega_0, +4\omega_0] \times [-4\omega_0, +4\omega_0]$ when calculating the field using Eq. (3.116), the discretization step for both variables being $8w_0/N$, $N = 300$.

Figure 3.25a shows the amplitude (real part) of the radial component E_r as a function of the radial coordinate ρ. Curve 1 describes the field derived from Eq. (3.116), curve 2 is for the nonparaxial formula (3.135), and curve 3 for the paraxial formula (3.138). Figure 3.25b depicts the same dependencies for the component E_z: curve 1 describes the field derived from (the third integral of) Eq. (3.116), curve 2 is for the

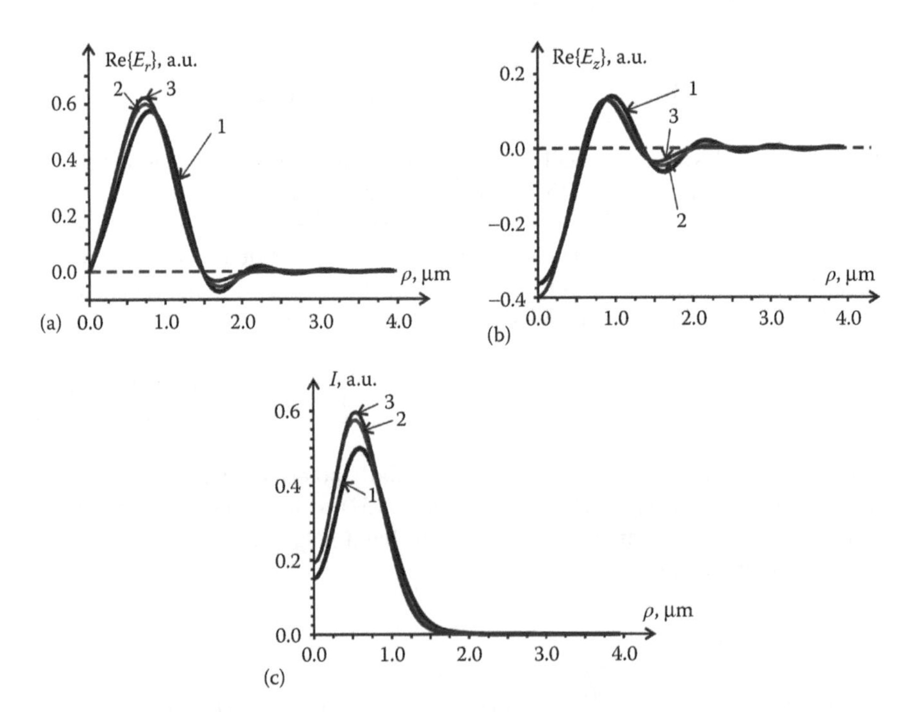

FIGURE 3.25 Amplitude (real part) of a radially polarized Gaussian beam ($n = 0$) in the plane of the lens geometrical focus: (a) radial component E_r, (b) longitudinal component E_z, and (c) intensity I.

nonparaxial Eq. (3.147), and curve 3 for the paraxial formula (3.150). Figure 3.25c depicts the intensity profile $I = |E_r|^2 + |E_z|^2$ derived from Eq. (3.116) (curve 1), curve 2 has been derived from Eqs. (3.135) and (3.147), and curve 3 has been derived from Eqs. (3.138) and (3.150).

Figure 3.25 suggests that in the optical axis vicinity the amplitude distribution derived from the paraxial formulae (3.138) and (3.150) is close to that of Eq. (3.116), but with increasing distance from the axis the nonparaxial formulae become much more accurate.

Similar comparative studies were conducted for the beam of Eq. (3.130) containing an optical vortex with the topological charge $n = 1$. Shown in Figure 3.26 are the amplitude distributions for the E_r-, E_φ- and E_z-components and the intensity distribution I. Curve 1 describes the field derived from Eq. (3.116) and curve 2 is for the nonparaxial formulae (3.135), (3.136) and (3.147). The intensity shown in Figure 3.26d was calculated using Eq. (3.116), $I = |E_r|^2 + |E_\varphi|^2 + |E_z|^2$ (curve 1) and using Eqs. (3.135), (3.136) and (3.147) (curve 2). It is seen from Figure 3.26 that obtained nonparaxial curves almost coincide with exact solution, when $\rho > \lambda$ (Figure 3.26d).

The propagation of the vortex laser beam with the radial polarization in the initial plane, described by Eq. (3.130), was simulated for $n = 0$ and $n = 1$ using the FDTD method (Figures 3.27 and 3.28) in FullWave (RSoft). The simulation parameters for $n = 0$ were the same as in Figure 3.25, and for $n = 1$ the same as in Figure 3.26.

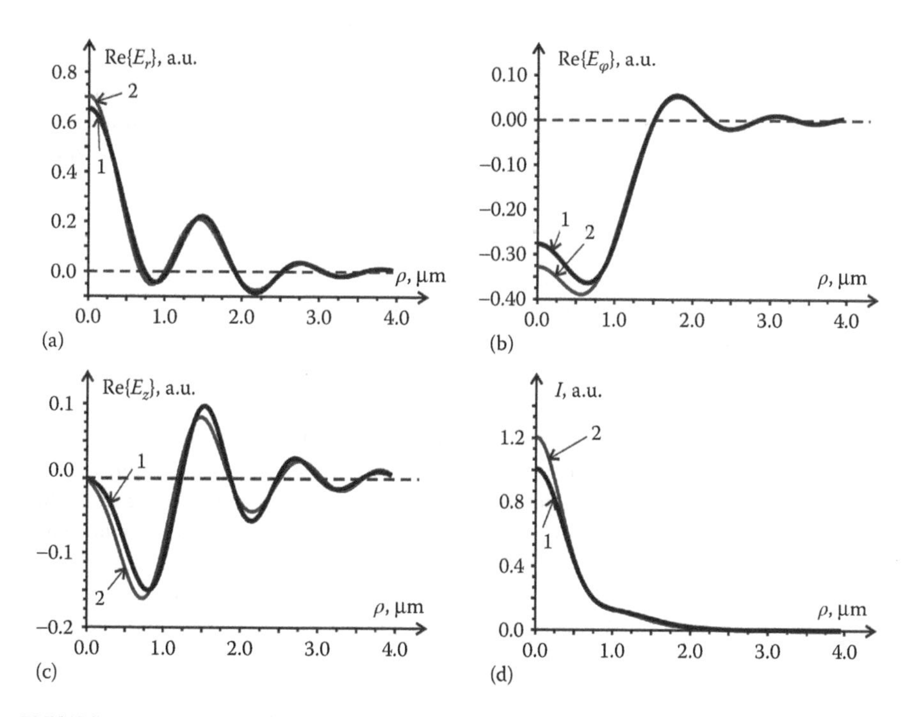

FIGURE 3.26 Amplitude (real part) of a Gaussian vortex beam ($n = 1$) with initial radial polarization in the plane of the lens geometrical focus: (a) radial component E_r, (b) azimuthal component E_φ, (c) longitudinal component E_z, and (d) intensity I.

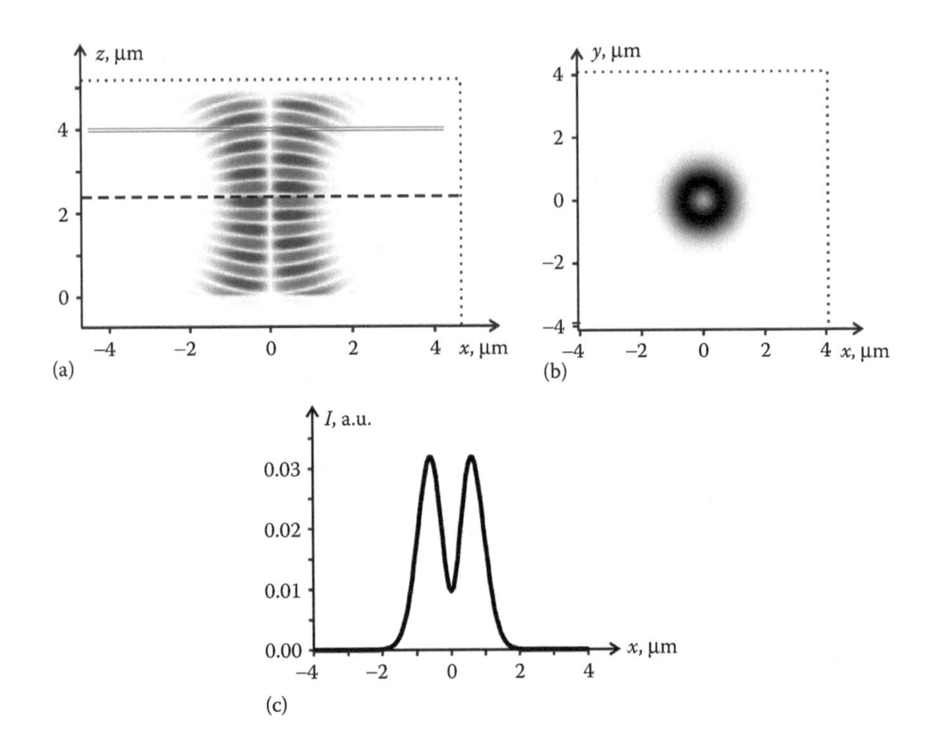

FIGURE 3.27 Modeling a non-vortex Gaussian beam ($n = 0$) with the radial polarization in the initial plane ($z = 0$) using the FDTD method: (a) instantaneous amplitude E_x in the XZ-plane, (b) time-averaged diffraction pattern in the plane $z = 4$ μm, and (c) the corresponding horizontal cross-section (horizontal dashed line in Figure 3.27a is position of focus, dotted lines are borders of simulation area).

Figure 3.27a depicts the instantaneous amplitude E_x in the XZ-plane for a radially polarized nonvortex beam ($n = 0$). Shown in Figure 3.27b is a time-averaged diffraction pattern in the plane $z = 4$ μm, and in Figure 3.27c is its horizontal cross-section, which is seen to be identical (up to a constant) with the intensity distribution in Figure 3.25c.

Figure 3.28a shows the instantaneous amplitude distribution E_x in the XZ-plane for a vortex beam ($n = 1$) with the initial radial polarization. The plot in Figure 3.28b describes a time-averaged diffraction pattern in the plane $z = 4$ μm, and Figure 3.28c shows the corresponding horizontal cross-section, which is structurally similar with the intensity distribution in Figure 3.26c.

It is of interest that the non-vortex Gaussian beam ($n = 0$) gives an annular intensity distribution at the focus, despite the fact that $I \neq 0$ at $\rho = 0$ (it follows from Eq. 3.148). On the other hand, the Gaussian vortex beam ($n = 1$) focuses into a circular focal spot (it follows from Eq. 3.135).

It would be interesting to note that in both cases, whether the beam is vortex ($n = 1$) or non-vortex ($n = 0$), its focus is found closer than geometrical ($z = f$). This is a well-known focal shift discussed in Refs. [159,160]. From Figures 3.27a

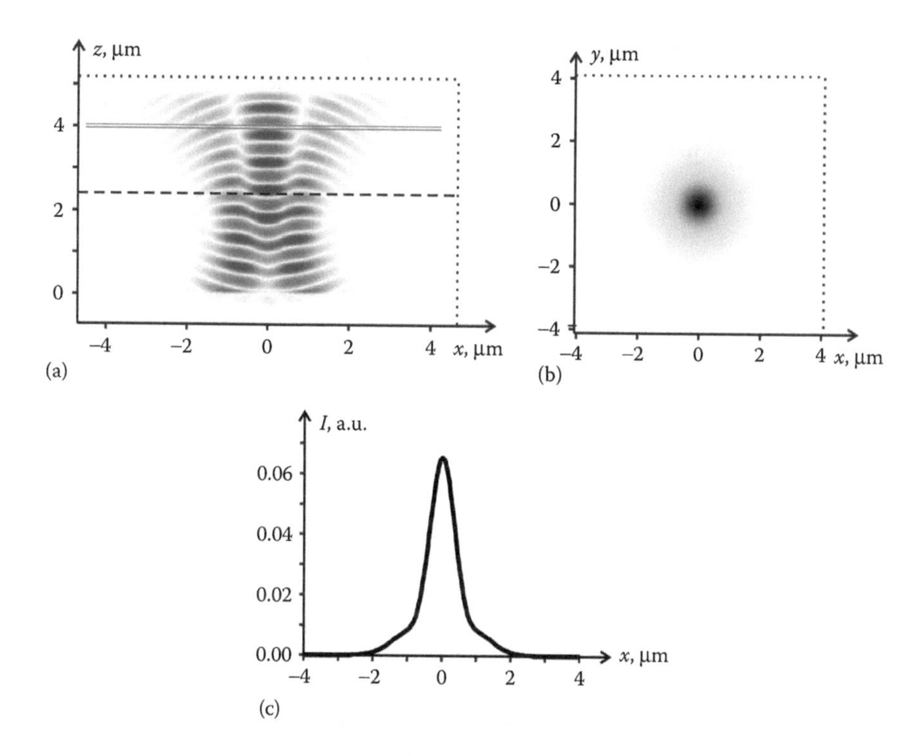

FIGURE 3.28 Modeling a vortex Gaussian beam ($n = 1$) with the radial polarization in the initial plane ($z = 0$) using the FDTD method: (a) instantaneous amplitude E_x in the XZ-plane, (b) time-averaged diffraction pattern in the plane $z = 4$ μm, and (c) the horizontal cross-section (horizontal dashed line in Figure 3.28a is position of focus, dotted lines are borders of simulation area).

and 3.28a, the focal shift from the geometrical focus is seen to be nearly 50%: instead of $z = f = 4$ μm, the focus is found at $z \approx 2$ μm. The axial focal shift for $n = 0$ can be evaluated using Eq. (3.148). From Eq. (3.148) it follows that at $\rho = 0$

$$E_z(\rho,\theta,z) = \frac{ikE_0 \exp(ikz)}{\sqrt{2}\alpha^2 q^2 \omega_0}, \tag{3.157}$$

and the focus will be found at a point on the z-axis when $|E_z|$ is maximal, because $E_r = 0$ at $\rho = 0$ (Eq. 3.137). From Eq. (3.157) it follows that $|E_z|$ will be maximal when $|q| = |z - ik/(2\alpha)|$ is minimal, that is, at

$$z = \frac{f}{1 + \left(\dfrac{f}{z_0}\right)^2}, \tag{3.158}$$

where $z_0 = k\omega_0^2/2$ is the Rayleigh range.

The same equation has been obtained in [150] (Eq. 3.13), although it is based on different physical principles. Instead of maximizing intensity on the optical axis, the condition of plane wavefront has been used for derivation of Eq. (3.13) in [150].

From Eq. (3.158) the focal shift is defined as

$$\frac{z-f}{f}=-\frac{1}{1+\left(\frac{z_0}{f}\right)^2}. \tag{3.159}$$

In our case, at $\lambda=0.633$ μm, $w_0=1$ μm, and $f=4$ μm, the focal shift is $\left|(z-f)/f\right|\approx 0.4$. This value is in good agreement with Figures 3.27a and 3.28a.

We have examined the nonparaxial diffraction of Gaussian vortex beams with the initial radial polarization. Expressions for the radial, azimuthal, and longitudinal components of the E-vector have been deduced. These expressions are more general than obtained earlier in [161,162]. The numerical simulation has shown that in the optical axis vicinity the amplitude distribution derived from the paraxial formulae (3.137) and (3.148) is close to that of Eq. (3.116), but with increasing distance from the axis the nonparaxial formulae derived are much more accurate. For the above-considered case (Figures 3.26d and 3.28c), the diffraction spot at half-maximum is $FWHM\approx 1{,}4\lambda$, although the lens numerical aperture is $NA\approx 1/\sqrt{1+4^2}\approx 0{,}243$. The scalar theory predicts that the focal spot size (at half intensity) should be $FWHM\approx 0{,}5\lambda/NA$. In our case, it should have been $FWHM\approx 2{,}06\lambda$. The modeling has shown also that there is nearly a two-fold difference between the focus z_f and the geometrical focus f.

4 Hankel-Bessel Laser Beams

In optics, there are laser beams whose scalar complex amplitudes are described by the exact solutions of the nonparaxial Helmholtz equation. These include well-known planar and spherical waves [163], as well as more recently proposed Bessel modes [1], Mathieu beams [98], and parabolic laser beams [97]. There also exist paraxial hypergeometric laser beams (HyG beams) [113] with their complex amplitude being described by the Kummer function. The nonparaxial hypergeometric (nHyG) laser beams [151] were derived as a solution of the Helmholtz equation by calculating the integral for the angular spectrum of plane waves with the even numbers of the topological charge of the beam spiral phase. The nHyG beams are described by a complex amplitude in the form of a product of two Kummer's functions ${}_1F_1(a, b, x)$ [101] with different arguments x. They represent a superposition of two identical light waves propagating in the positive and negative directions along the optical axis z. Note that approximate relations for the complex amplitude describing the propagation of the light waves in the positive and negative directions (the HyG laser beams) were also derived in Ref. [151] using the asymptotic decomposition of one of Kummer's functions.

In this section, extending the results reported in Ref. [151], we derive an exact solution of the Helmholtz equation for any integer topological charge to describe a nonparaxial laser beam with the complex amplitude in the form of a product of Kummer's function ${}_1F_1(a, b, x_1)$ and the second solution of Kummer's equation $U(a, b, x_2)$. Considering that under certain parameters the $U(a, b, x_2)$ function is proportional to the Hankel function and Kummer's function—to the Bessel function, such beams have been given the name nonparaxial Hankel-Bessel beams. It is noteworthy that the solution proposed can be found neither in the handbooks of mathematical functions [101,38] nor in the well-known work by W. Miller [96].

The Hankel-Bessel beams ($n = 0$) are generated by a source with infinite energy density located in the initial plane. As they propagate along the positive z-axis, the divergence of the beams is proportional to $\sqrt{z}$. For a non-zero topological charge, $n \neq 0$, the source produces no radiation along the optical axis.

4.1 SOLUTION OF THE HELMHOLTZ EQUATION IN THE PARABOLIC COORDINATES

The Helmholtz equation in the cylindrical coordinates (r, φ, z) takes the form:

$$\left(\frac{\partial^2}{\partial r^2}+\frac{1}{r}\frac{\partial}{\partial r}+\frac{1}{r^2}\frac{\partial^2}{\partial \varphi^2}+\frac{\partial^2}{\partial z^2}+k^2\right)E(r,\varphi,z)=0, \tag{4.1}$$

where k is the wavenumber.

We will seek the solution $E(r, \varphi, z)$ in the following form:

$$E(r,\varphi,z)=E(r,z)r^{p}\exp(in\varphi+ikz), \tag{4.2}$$

where n and p are integer numbers.

Then, for $E(r, z)$, the equation (4.1) is

$$\frac{\partial^2 E}{\partial r^2}+\frac{\partial^2 E}{\partial z^2}+\left(\frac{2p+1}{r}\right)\frac{\partial E}{\partial r}+2ik\frac{\partial E}{\partial z}+\left(\frac{p^2-n^2}{r^2}\right)E=0. \tag{4.3}$$

At $p = \pm n$, the third term in Eq. (4.3) is eliminated. Changing to the parabolic coordinates

$$\begin{cases} x=\sqrt{r^2+z^2}+z, \\ y=\sqrt{r^2+z^2}-z, \end{cases} \tag{4.4}$$

Eq. (4.3) takes the form (at $p = +n$):

$$x\frac{\partial^2 E}{\partial x^2}+y\frac{\partial^2 E}{\partial y^2}+(n+1+ikx)\frac{\partial E}{\partial x}+(n+1-iky)\frac{\partial E}{\partial y}=0. \tag{4.5}$$

Equation (4.5) is solved in separable variables. Assuming in Eq. (4.5) that $E(x, y) = X(x)Y(y)$, we obtain:

$$\left[x\frac{X''}{X}+(n+1+ikx)\frac{X'}{X}\right]=\left[-y\frac{Y''}{Y}-(n+1-iky)\frac{Y'}{Y}\right]=C, \tag{4.6}$$

where C is a constant independent of x and y. Then, Eq. (4.5) is replaced by two equations:

$$\begin{cases} x\dfrac{d^2X}{dx^2}+(n+1+ikx)\dfrac{dX}{dx}-CX=0, \\ y\dfrac{d^2Y}{dy^2}+(n+1-iky)\dfrac{dY}{dy}+CY=0. \end{cases} \tag{4.7}$$

Introducing the designations $\xi = -ikx$, $\eta = iky$ and $C = -ikD$, both equations in (4.7) are transformed into Kummer's equations [101, Eq. 13.1.1]:

$$\begin{cases} \xi\dfrac{d^2X}{d\xi^2}+(n+1-\xi)\dfrac{dX}{d\xi}-DX=0, \\ \eta\dfrac{d^2Y}{d\eta^2}+(n+1-\eta)\dfrac{dY}{d\eta}-DY=0, \end{cases} \tag{4.8}$$

with their solutions described by Kummer's functions ${}_1F_1(a, b, z)$ [101]:

$$\begin{cases} X(\xi) = {}_1F_1(D, n+1, \xi), \\ Y(\eta) = {}_1F_1(D, n+1, \eta). \end{cases} \tag{4.9}$$

Then, $X(x) = {}_1F_1(D, n + 1, -ikx)$ and $Y(y) = {}_1F_1(D, n + 1, iky)$, and the solution of the original Helmholtz equation (4.1) can now be written down as:

$$E(r,\varphi,z) = A_0 (kr)^n \exp(in\varphi + ikz)\, {}_1F_1(D, n+1, -ikx)\, {}_1F_1(D, n+1, iky), \tag{4.10}$$

where A_0 is a constant characterizing the beam power.

The said solution was offered without derivation in Ref. [96]. In particular, when $D = (n + 1)/2$, expressing the Bessel functions through Kummer's functions [101]:

$${}_1F_1\left(\nu + \frac{1}{2}, 2\nu + 1, 2iz\right) = \Gamma(1+\nu)\exp(iz)\left(\frac{z}{2}\right)^{-\nu} J_\nu(z), \tag{4.11}$$

Eq. (4.10) can be rearranged as

$$E(r,\varphi,z) = A_0 \Gamma^2\left(1+\frac{n}{2}\right)(4i)^n \exp(in\varphi) J_{\frac{n}{2}}\left[\frac{k}{2}\left(z+\sqrt{r^2+z^2}\right)\right] J_{\frac{n}{2}}\left[\frac{k}{2}\left(z-\sqrt{r^2+z^2}\right)\right]. \tag{4.12}$$

Note that Eq. (4.12) does not contain multiplier $\exp(ikz)$, so it does not describe the propagation of laser light in a certain direction. What it describes is a standing wave resulting from the interference of two identical waves propagating in the positive and negative directions along the optical axis z.

Actually, for large distances $z \gg r$, using the approximate relations for the Bessel function for small and large values of the argument, the approximate expression of Eq. (4.12) given by

$$E(r,\varphi,z) = A_0 \Gamma\left(1+\frac{n}{2}\right)\exp(in\varphi)\sqrt{\frac{2}{\pi kz}}\left(\frac{2kr^2}{z}\right)^{\frac{n}{2}} \cos\left(kz - \frac{\pi n}{4} - \frac{\pi}{4}\right), \tag{4.13}$$

suggests that the amplitude is proportional to the cosine function of the distance.

Let us separate the light field in Eq. (4.10) into the forward and backward waves. In doing so, we make use the relation between Kummer's function $M(a, b, x) = {}_1F_1(a, b, x)$ and the second solution of Kummer's equation [101]:

$$U(a,b,z) = \frac{\pi}{\sin \pi b}\left[\frac{M(a,b,z)}{\Gamma(1+a-b)\Gamma(b)} - z^{1-b}\frac{M(1+a-b, 2-b, z)}{\Gamma(a)\Gamma(2-b)}\right]. \tag{4.14}$$

Note that the function (4.14) is also defined for the integer values of the parameter b [101].

The inversion of Eq. (4.14) yields the expression of ${}_1F_1(a, b, x)$ through $U(a, b, x)$:

$$ {}_1F_1(a,b,z) = AU(a,b,z) + BU(b-a,b,-z), \tag{4.15} $$

where:

$$ A = \frac{\Gamma(b)}{\Gamma(b-a)} D, $$

$$ B = \frac{\Gamma(b)}{\Gamma(a)} (-1)^b e^z D, \tag{4.16} $$

$$ D = \frac{\sin \pi b}{\sin \pi (b-a) + (-1)^b \sin \pi a}. $$

Making use of Eqs. (4.15) and (4.16) and neglecting the constant factors, the solution of the Helmholtz equation (4.10) takes the form:

$$ \begin{aligned} E(r,\varphi,z) = A_0 (kr)^n \exp(in\varphi) \Bigg[& \frac{\exp(ikz)}{\Gamma(n+1-D)} U(D,n+1,-ikx)\, {}_1F_1(D,n+1,iky) \\ & + (-1)^{n+1} \frac{\exp(-ikz)}{\Gamma(D)} U(n+1-D,n+1,ikx)\, {}_1F_1(n+1-D,n+1,-iky) \Bigg]. \end{aligned} \tag{4.17} $$

Designating $D = (n + \beta + 1)/2$ and multiplying both sides of the equation by $\Gamma(D)\Gamma(n + 1 - D)$, we obtain:

$$ \begin{aligned} E(r,\varphi,z) = {} & i^{2n+1} A_0 (kr)^n \exp(in\varphi) \\ & \times \left[-i\Gamma\left(\frac{n+\beta+1}{2}\right) \exp(ikz) U\left(\frac{n+\beta+1}{2}, n+1, -ikx\right) {}_1F_1\left(\frac{n+\beta+1}{2}, n+1, iky\right) \right. \\ & \left. + i\Gamma\left(\frac{n-\beta+1}{2}\right) \exp(-ikz) U\left(\frac{n-\beta+1}{2}, n+1, ikx\right) {}_1F_1\left(\frac{n-\beta+1}{2}, n+1, -iky\right) \right]. \end{aligned} \tag{4.18} $$

The aforementioned relationship represents a sum of the forward and backward hyper-geometric beams as indicated by the exponent signs, $\exp(\pm ikz)$ in Eq. (4.18). In the following, we discuss the forward beams propagating from the plane $z = 0$ into the semi-space $z > 0$. The positive beam is expressed as

$$ \begin{aligned} E_+(r,\varphi,z) = {} & (-1)^n \Gamma\left(\frac{n+\beta+1}{2}\right) A_0 (kr)^n \exp(in\varphi + ikz) \\ & \times U\left(\frac{n+\beta+1}{2}, n+1, -ikx\right) {}_1F_1\left(\frac{n+\beta+1}{2}, n+1, iky\right). \end{aligned} \tag{4.19} $$

For $n \neq 0$, the field complex amplitude on the optical axis ($r = 0$) equals zero, whereas at $n = 0$ the amplitude is non-zero:

$$E_0\left(r=0,\varphi,z\right)=\Gamma\left(\frac{\beta+1}{2}\right)A_0U\left(\frac{\beta+1}{2},1,-2ikz\right). \quad (4.20)$$

In particular, for $\beta = 0$, Eq. (4.19) that describes the positive beam takes the form:

$$E_+\left(r,\varphi,z\right)=i^{3n+1}\frac{\pi}{2}n!A_0\exp\left(in\varphi\right)H_{\frac{n}{2}}^{(1)}\left[\frac{k}{2}\left(z+\sqrt{r^2+z^2}\right)\right]J_{\frac{n}{2}}\left[\frac{k}{2}\left(\sqrt{r^2+z^2}-z\right)\right]. \quad (4.21)$$

In the initial plane ($z = 0$), the arguments of both the Hankel and Bessel functions equal $kr/2$. The Hankel function is diverging as the argument tends to zero. Using the approximate relations for the Hankel and Bessel functions for small argument values (Ref. [101], the relationships [9.1.7] and [9.1.9]), the field amplitude can be shown to have a finite value at $n \neq 0$ and $r = z = 0$:

$$E_+\left(r\approx 0,\varphi,z=0\right)=\left(-i\right)^n\left(n-1\right)!A_0\exp\left(in\varphi\right). \quad (4.22)$$

From Eq. (4.21), it is seen that if $\beta = 0$, the complex amplitude of the nonparaxial HyG beam is proportional to the product of the Hankel and Bessel functions of integer and half-integer orders. Thus, to distinguish the beams in Eqs. (4.18) and (4.21) from those reported in Ref. [151], the former will be referred to as Hankel-Bessel (HB) beams. From Eq. (4.21), it can be inferred that the HB beam is not a free-space mode because the arguments of the Hankel and Bessel functions depend in a different way on the variables r and z, and, therefore, given the same z, these functions' values remain constant at different r. That the beam in Eq. (4.21) propagates in the positive direction of the z-axis follows from the asymptotic formula of the first-order Hankel function at large z ($kz \gg 1$) [101, relationship 9.2.3]:

$$H_\nu^{(1)}\left(z\right)\sim\sqrt{\frac{2}{\pi z}}\exp\left[i\left(z-\frac{\pi\nu}{2}-\frac{\pi}{4}\right)\right]. \quad (4.23)$$

As distinct from the relationship (4.13), the z-dependence in (4.23) is exponential rather than cosine. When $n = 0$, the intensity on the optical axis ($r = 0$) is

$$I\left(r=0,\varphi,z\right)=\left|i\frac{\pi}{2}A_0H_0^{(1)}\left(kz\right)\right|^2=\frac{\pi^2}{4}A_0^2\left[J_0^2\left(kz\right)+Y_0^2\left(kz\right)\right], \quad (4.24)$$

with the intensity peak attained at a point z that satisfies the equation

$$J_0\left(kz\right)J_1\left(kz\right)+Y_0\left(kz\right)Y_1\left(kz\right)=0\,. \quad (4.25)$$

The simulation has shown that while at $t > 0$ the function $f(t) = J_0(t)J_1(t) + Y_0(t)Y_1(t)$ never attains the zero value, the beam is not focused and the intensity of the diverging

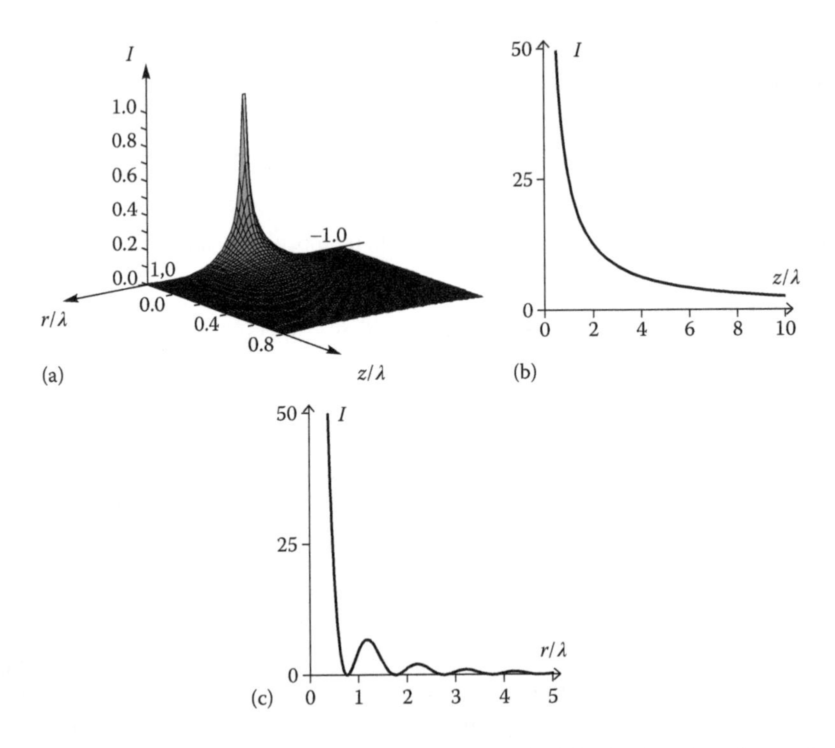

FIGURE 4.1 (a) The squared modulus of the function (4.21) in the Orz-plane for $n = 0$ and $A_0 = 100$, the intensity as a function of (b) the longitudinal axis at $r = 0$ and (c) the radial axis at $z = 0$.

beam is decreased along the optical axis. Moreover, with the function in Eq. (4.24) becoming infinite at $z = 0$, the field in Eq. (4.21) can be assumed to be produced by a source of infinite energy density located at the origin of coordinates. Shown in Figure 4.1 are the squared modulus of the function (4.21) for $z > 0$ in the Orz-plane (Figure 4.1a), the intensity plots on the longitudinal z-axis (Figure 4.1b), and the transverse r-axis (Figure 4.1c). From Eq. (4.21), the HB beams are seen to have infinite energy.

It can be seen from Figure 4.1c that despite the infinite value at $r = z = 0$, the function (4.21) has a strict zero. From Eq. (4.21) follows the zero value at $r_0 \approx 0.76\lambda$. Thus, we can assume that the field (4.21) is generated by a source of infinite energy density and radius 0.76 λ ($n = 0$).

Let us consider another particular case when $\beta = 1$. Then, the positive beam will be expressed as

$$E_+(r,\varphi,z) = (-i)^n (n-1)! \frac{\pi A_0 k^{1-n}}{4} \exp(in\varphi) r \times \left[J_{\frac{n-1}{2}}\left(\frac{ky}{2}\right) + iJ_{\frac{n+1}{2}}\left(\frac{ky}{2}\right) \right] \left[iH^{(1)}_{\frac{n-1}{2}}\left(\frac{kx}{2}\right) + H^{(1)}_{\frac{n+1}{2}}\left(\frac{kx}{2}\right) \right]. \tag{4.26}$$

From Eq. (4.23), it is also seen that the beam in Eq. (4.26) also propagates in the positive direction of the z-axis. When $n = 0$, we obtain a trivial solution of the Helmholtz equation, $E_+(r, \varphi, z) = 0$, because $H_{1/2}^{(1)}(x) + iH_{-1/2}^{(1)}(x) \equiv 0$.

4.2 PECULIARITIES OF THE HANKEL-BESSEL BEAMS

For large z, the HB beam is diverging hyperbolically, as evident from the expression for the field amplitude at $z \gg \lambda$:

$$E_+(r,\phi,z) \approx A_0 \exp(in\phi) H_{n/2}^{(1)}(kz) J_{n/2}\left(\frac{kr^2}{2z}\right). \tag{4.27}$$

Equation (4.27) suggests that the field amplitude remains unchanged if the coordinates r and z are linked by the relation:

$$r = \sqrt{\frac{2\gamma z}{k}}, \tag{4.28}$$

where γ is a constant independent of z. The Hankel function in Eq. (4.21) is diverging when the argument tends to zero, which is only possible when r and z are simultaneously small. In this case, the argument of the Bessel function in (4.21) also tends to zero. Making use of the approximate relations of the Hankel and Bessel functions for small argument values, we can show that

$$E_+(r,\varphi,z) \approx (-i)^n (n-1)! A_0 \exp(in\varphi) \tan^n\left(\frac{\xi}{2}\right), \tag{4.29}$$

where $\tan \xi = r/z$. Thus, the amplitude of a point located near the origin of coordinates depends on the direction from the origin of coordinates toward this point (Figure 4.2).

Figure 4.2a shows the squared modulus of the function (4.21) (at $n = 1$, $A_0 = 100$) in a $2\lambda \times \lambda$ region (black denotes zero, white denotes maximum). Shown in Figure 4.2b–d are the radial intensity profiles in the planes $z = 0.1\lambda$, $z = 0.01\lambda$, and $z = 0.001\lambda$.

Particular Cases of the Hankel-Bessel Beams

For odd n, the Bessel functions of half-integer order become elementary. Let us consider a particular case of $n = 1$. Then,

$$E_1(r,\varphi,z) = \frac{-2i}{kr} A_0 \exp(i\varphi) \sin\left[\frac{k}{2}\left(\sqrt{r^2+z^2} - z\right)\right] \exp\left[\frac{ik}{2}\left(z + \sqrt{r^2+z^2}\right)\right]. \tag{4.30}$$

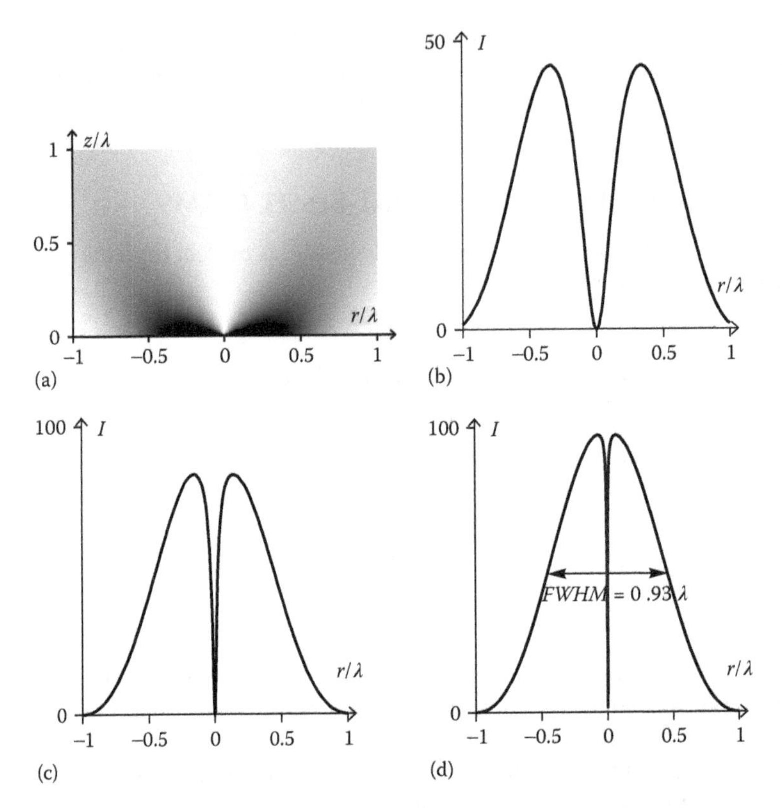

FIGURE 4.2 (a) The intensity in the Orz-plane at $n = 1$, $A_0 = 100$ ($-\lambda \le x \le \lambda$, $0 \le z \le \lambda$; black denotes zero, white denotes maximum intensity. (b) The radial sections of the squared modulus of the function (4.21) in different planes: (b) $z = 0.1\lambda$, (c) $z = 0.01\lambda$, and (d) $z = 0.001\lambda$.

At $r = 0$, there will be zero intensity on the optical axis, despite the fact that r occurs in the denominator of Eq. (4.30) (Figure 4.3). This can be demonstrated, if we assume $r \ll z$:

$$\lim_{r\to 0} E_1(r,\varphi,z) = \frac{-2i}{k} A_0 \exp(i\varphi + ikz) \lim_{r\to 0}\left\{\frac{1}{r}\sin\left(\frac{kr^2}{4z}\right)\right\} = 0. \tag{4.31}$$

The field in Eq. (4.30) is formed from the following amplitude distribution in the initial plane (waist):

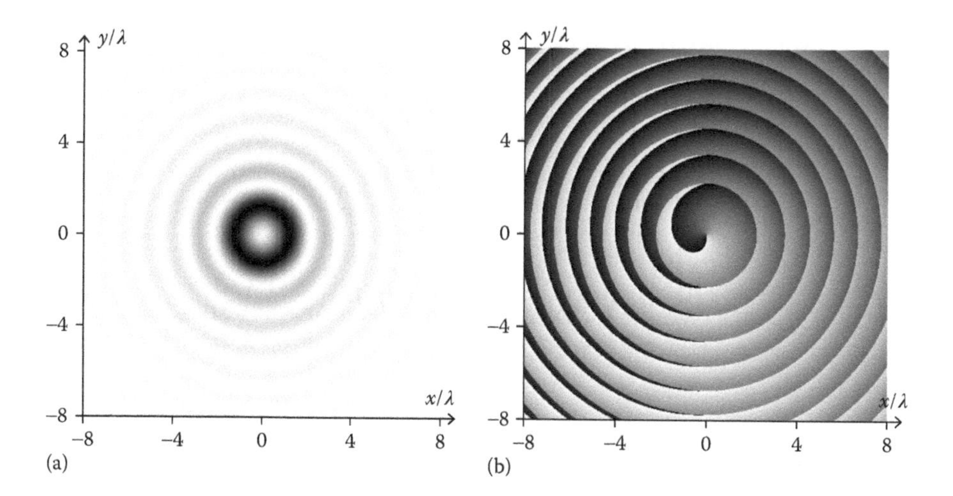

FIGURE 4.3 Simulation results for the beam (4.30) for wavelength $\lambda = 633$ nm: (a) intensity (negative) and (b) phase in the transverse plane $z = 2\lambda$.

$$E_1\left(r,\varphi,z=0\right)=\frac{-2i}{kr}A_0\sin\left(\frac{kr}{2}\right)\exp\left(\frac{ikr}{2}+i\varphi\right). \tag{4.32}$$

The beam waist diameter is $FWHM = 0.93\lambda$ (Figure 4.2d).

Note that being equal to $E_{n=1}(r \approx 0, \varphi, z = 0) = -iE_0 \exp(i\varphi)$ near the original plane center, the amplitude is not defined at the central point itself due to a spiral phase singularity; on the other hand, at $r = z = 0$ the intensity is not equal to zero, $I = |E_1(r = 0, \varphi, z = 0)|^2 = |A_0|^2$.

At large distances $z \gg r$, Eq. (4.30) takes the form:

$$E_1\left(r,\varphi,z\gg r\right)=\frac{-2i}{kr}A_0\sin\left(\frac{kr^2}{4z}\right)\exp\left(i\varphi+ikz\right). \tag{4.33}$$

Simulation Results

Figure 4.4 depicts the radial intensity profiles in the planes $z = 2\lambda$, $z = 4\lambda$, and $z = 6\lambda$. For comparison, the simulation using the finite-difference time-domain (FDTD) method (Figure 4.5) was also conducted. The simulation parameters were as follows: the computation region dimension, $[-8\lambda, 8\lambda] \times [-8\lambda, 8\lambda] \times [0, 8\lambda]$; spatial discretization, $\lambda/16$ (for all coordinates); the simulation time, 20 periods (i.e., 20 λ/c,

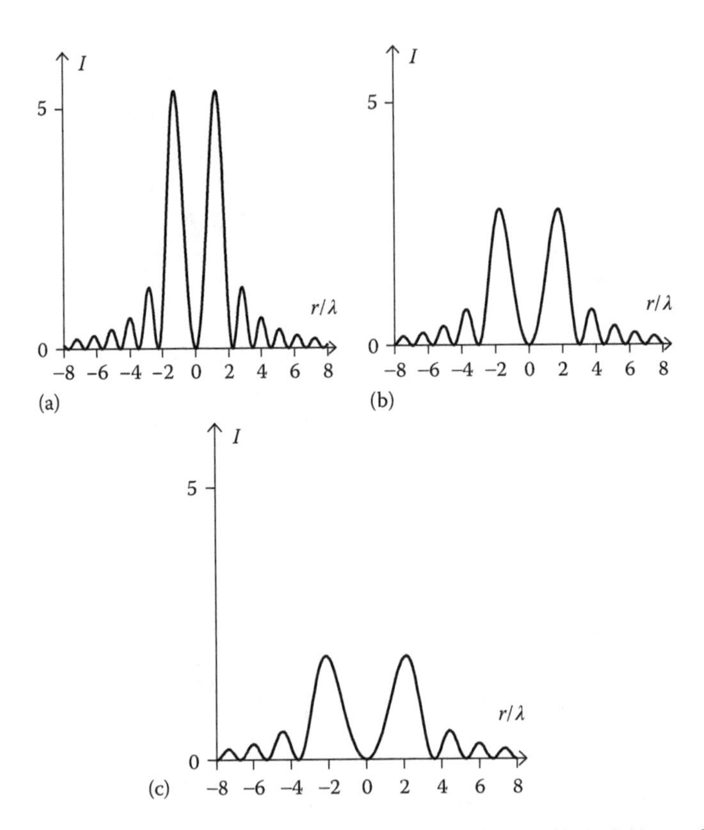

FIGURE 4.4 Intensity profiles in the planes (a) $z = 2\lambda$, (b) $z = 4\lambda$, and (c) $z = 6\lambda$ derived using Eq. (4.21) at $n = 1$, $A_0 = 100$.

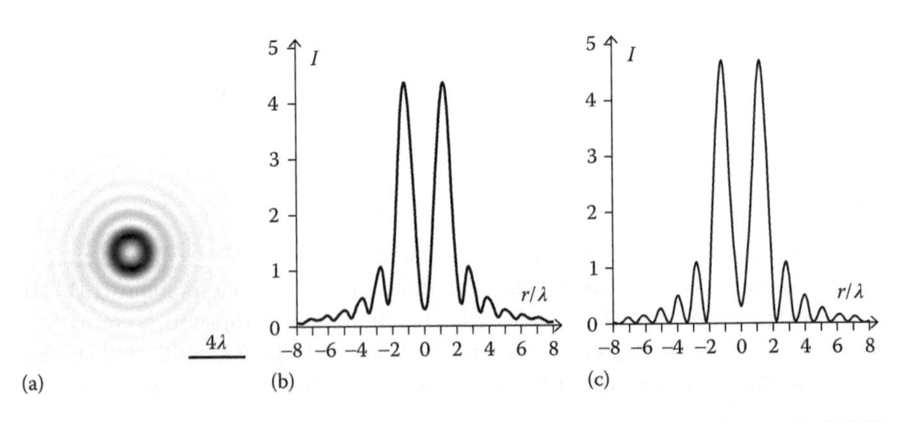

FIGURE 4.5 The simulation of the propagation of the HB beam for $n = 1$ using the FDTD method (TE-polarization, $E_x \neq 0$): (a) time-averaged intensity in the plane $z = 2\lambda$ and its (b) horizontal and (c) vertical profiles.

where c is the speed of light in vacuum); the temporal discretization, $\lambda/(32c)$. The time-averaged intensity in the plane $z = 2\lambda$ is shown in Figure 4.5a and its radial profile is depicted in Figure 4.5b and c.

The central intensity is seen not to fall toward zero, which may be due to the presence of the longitudinal (axial) component E_z (because $n = 1$). It can also be seen that the intensities shown in Figures 4.4a and 4.5c are similar in structure, whereas the intensity in Figure 4.5b is different, not dropping to zero between the bright rings. This may be due to the contribution of the axial component and the linear polarization, which is regarded when simulating in the FDTD method and disregarded in the scalar theory.

Focusing of the Hankel-Bessel Beams

By substitution of variables in Eq. (4.21), $z \to f - z$ (f is the focal length), we obtain:

$$E_+(r,\varphi,z) = i^{3n+1}\frac{\pi}{2}n!A_0\exp(in\varphi) \times H_{\frac{n}{2}}^{(1)}\left\{\frac{k}{2}\left[f - z + \sqrt{r^2+(z-f)^2}\right]\right\} J_{\frac{n}{2}}\left\{\frac{k}{2}\left[\sqrt{r^2+(z-f)^2} - f + z\right]\right\}. \tag{4.34}$$

When $n = 0$:

$$E_0(r,\varphi,z) = \frac{i\pi A_0}{2} H_0^{(1)}\left\{\frac{k}{2}\left[f - z + \sqrt{r^2+(z-f)^2}\right]\right\} J_0\left\{\frac{k}{2}\left[\sqrt{r^2+(z-f)^2} - f + z\right]\right\}. \tag{4.35}$$

When $z > f$ and $r = 0$, the argument of the Hankel function turns to zero and the intensity becomes infinite. In the initial plane $z = 0$, such a field is given by

$$E_0(r,\varphi,z=0) = \frac{i\pi A_0}{2} H_0^{(1)}\left[\frac{k}{2}\left(f + \sqrt{r^2+f^2}\right)\right] J_0\left[\frac{k}{2}\left(\sqrt{r^2+f^2} - f\right)\right]. \tag{4.36}$$

When solved using the FullWave software, this equation describes the propagation of a field with the modulus of the amplitude E_x as shown in Figure 4.6a. Figure 4.6b depicts the intensity in the Oxz-plane. The simulation was conducted for the wavelength $\lambda = 633$ nm and the focal length $f = 4\lambda = 2.53$ μm. The computation region dimension was $[-8\lambda, 8\lambda] \times [-8\lambda, 8\lambda] \times [0, 8\lambda]$. For all coordinates, the sampling was $\lambda/16$. The simulation was conducted over 20 periods. The temporal discretization was $\lambda/(32c)$.

From Figure 4.6c, the beam is seen to be focused into an elliptical spot with the minimal diameter at half-maximum of about FWHM $= 0.65\lambda$. Alongside the focal

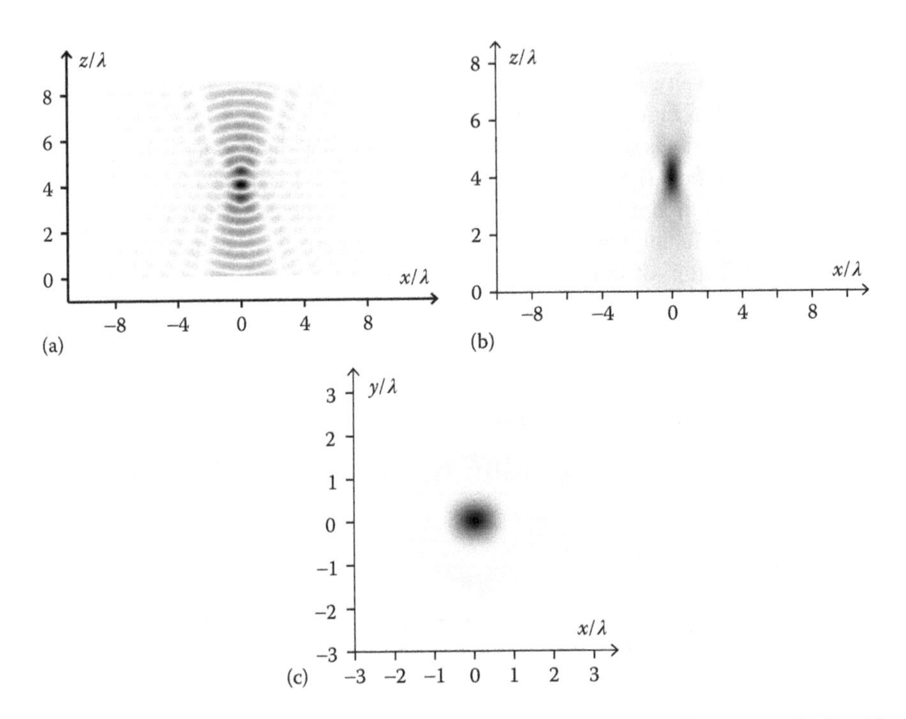

FIGURE 4.6 FDTD-method-based simulation of focusing the HB beam ($n = 0, f = 4\lambda$): (a) the modulus of the amplitude E_x at $t = 20\lambda/c$, (b) intensity $I = |E_x|^2 + |E_y|^2 + |E_z|^2$ in the Oxz-plane, (c) time-averaged intensity in the plane $z = 4\lambda$.

spot, there is a bright ring (side lobe) with the 2.5% peak intensity of that in the spot, which is considerably lower than in the zero-order Bessel beam for which the first-ring intensity is 16% of that in the diffraction pattern center.

The focusing into the axial line $r = 0$ ($z > f$) does not occur, as it might have appeared from Eq. (4.35). This may be because at $z > f$, for the near-axis points ($r \ll z - f$), we can write:

$$E_0\left(r \ll z - f, \varphi, z\right) \approx \frac{i\pi A_0}{2} H_0^{(1)}\left[\frac{kr^2}{4\left(z-f\right)}\right] J_0\left[\frac{k}{2}\left(\sqrt{r^2+\left(z-f\right)^2} - f + z\right)\right]$$
$$\sim -A_0 \ln\left[\frac{kr^2}{4\left(z-f\right)}\right] J_0\left[\frac{k}{2}\left(\sqrt{r^2+\left(z-f\right)^2} - f + z\right)\right]. \tag{4.37}$$

When $r = 0$, the logarithm takes infinite values for any z; however, for any other near-zero r, the logarithm decreases with increasing z. Thus, the simulated intensity was shown to decrease along the optical axis behind the focal point $z = f$.

In a similar way, we simulated the propagation of a field with the vortex $n = 3$. The modulus of the amplitude E_x in the Oxz-plane is shown in Figure 4.7. The simulation parameters are the same. In this case, the axial focus shift also does not occur,

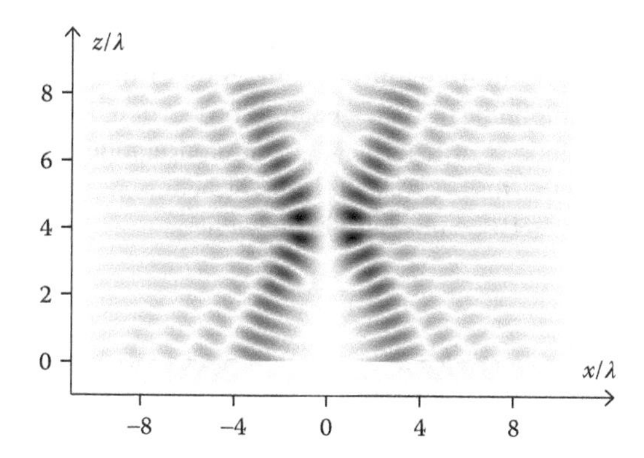

FIGURE 4.7 FDTD-method-based simulation of focusing the HB beam ($n = 3, f = 4\lambda$): the modulus of the amplitude E_x at $t = 20\lambda/c$.

with the beam waist located in the plane $z = f$. The intensity distribution in the transverse plane and the corresponding cross-sections are shown in Figure 4.8.

From Figure 4.2 it might appear that the central intensity minimum can be made infinitesimally small, but this would have required the use of the infinitely wide initial field in Eq. (4.36). With this being actually unfeasible, the FDTD method results in a pattern shown in Figure 4.8. The diffraction ring FWHM is approximately equal to the wavelength of incident light.

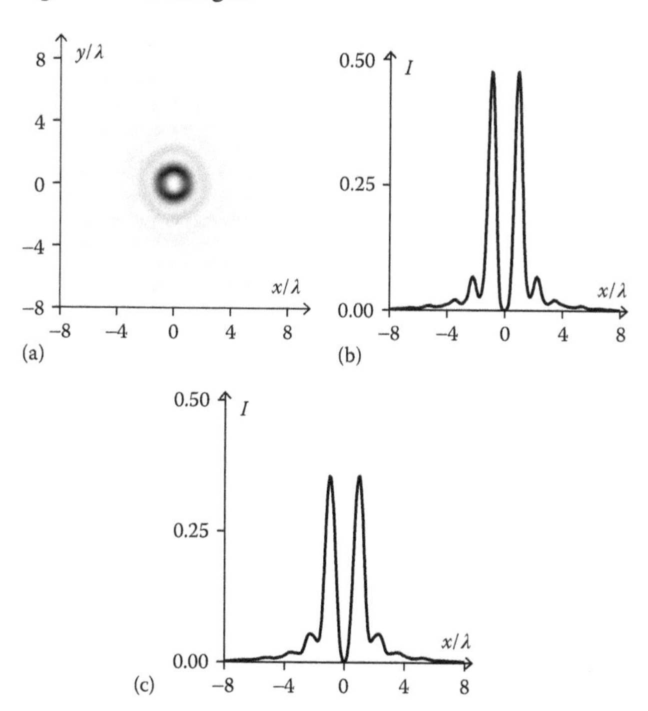

FIGURE 4.8 The intensity distribution in the plane $z = f$: (a) two-dimensional pattern and its (b) vertical and (c) horizontal profiles.

An exact analytical solution of the scalar Helmholtz equation to describe the propagation of the light beam in the positive direction of the optical axis has been derived. The complex amplitude of such a beam is proportional to the product of two linearly independent solutions of Kummer's equation. Relationships for a particular case of such beams in the form of Hankel-Bessel beams have been derived. The focusing properties of the HB beams have been studied.

5 Accelerating Beams

5.1 ACCELERATING AIRY BEAMS

The solution to the paraxial equation of propagation in the form of Airy functions was first offered by Kalnins and Miller in 1974 [164]. Nonspreading wave packets in the context of quantum mechanics were analyzed by Berry and Balazs in 1979 [165]. The solution to the two-dimensional paraxial equation as a product of Airy functions was proposed by Besieris et al. in 1994 [166]. The Airy beams discussed in Refs. [164–166] possess an infinite energy, as the Airy function decays slowly as $x \to \infty$: $\mathrm{Ai}(-x) \approx \pi^{-1/2}x^{-1/4}\sin\left[(2/3)x^{3/2}+(\pi/4)\right]$. In 2007, Siviloglou and Christodoulides [7,167,168] analyzed finite-energy Airy beams in optics. The solution of the equation

$$2i\frac{\partial U}{\partial \xi}+\frac{\partial^2 U}{\partial s^2}=0 \tag{5.1}$$

has been found in the form of a function

$$\begin{aligned} U(s,\xi) &= \mathrm{Ai}\left(s-\xi^2/4+ia\xi\right) \\ &\times\exp\left(is\xi/2+ia^2\xi/2-i\xi^3/12-a\xi^2/2+as\right), \end{aligned} \tag{5.2}$$

which takes the form of the exponentially apodized Airy function at the input, $\xi = 0$:

$$U(s,\xi=0)=\mathrm{Ai}(s)\exp(as), \quad a>0, \tag{5.3}$$

where:

$s = x/x_0$, $\xi = z/(kx_0^2)$ are the dimensionless transverse and longitudinal Cartesian coordinates
$k = 2\pi/\lambda$ is the wavenumber
x_0 is an arbitrary transverse scale
$\mathrm{Ai}(x)$ is the Airy function
a is constant

In Ref. [167] it was also shown that *a* 1D finite-energy Airy beam can be produced by passing a Gaussian beam through a phase system with cubic dispersion on the transverse coordinate, followed by implementing the Fourier transform with a spherical lens. This can be inferred from the Fourier image of the input field (5.3):

$$F(t)=\exp\left(-at^2\right)\exp\left(it^3/3-ia^2t+a^3/3\right). \tag{5.4}$$

The key peculiarity of the Airy beam is that its major lobe propagates along a curved trajectory, which has the form of a parabola. Because of this, the Airy beams are called accelerating or ballistic (a body moving only upon the force of gravity takes a parabolic trajectory referred to as ballistic). However, it has been shown [169] that the "center of gravity" of a finite energy Airy beam in Eq. (5.3) is not displaced upon propagation, with the accelerating effect occurring only at small values of the parameter $a \ll 1$. Analytic expressions for different types of the accelerating beams that also propagate along a parabolic trajectory have been obtained by Bandres [122,170,171]. Propagating beams with their rays forming a desired caustic curve have been examined [172–174]. For instance, a beam with the parabolic caustic has been obtained using a 3/2 phase-only pattern [175]. Accelerating beams described by Bessel functions that propagate on a circular trajectory have also been studied [176,177]. Accelerating laser beams with elliptic trajectories obtained based on the 1D cross-section of a 2D Mathieu beam have been reported [178]. Besides the aforementioned articles, the Airy beams have been dealt with in quite a number of publications.

However, to our knowledge there have been no publications handling the Airy beams accelerating along a hyperbolic trajectory that can be defined by an analytical expression. To fill the gap, we examine the Airy beams of the second kind (Type-II Airy beams) or hyperbolic Airy beams in the following.

Hyperbolic Airy Beams

As a rule, the 1D Airy beams are obtained using the initial field of Eq. (5.4), corresponding to a phase cubic transparency. Then, the Fourier image of the field (5.3) is generated using a spherical lens. As a result, an Airy beam with the complex amplitude (5.2) is generated behind the Fourier plane. Type-II Airy beams are obtained in the Fresnel diffraction zone of the phase transparency (5.4). To prove this, let us find the complex amplitude of the Gaussian beam directly behind the phase transparency:

$$E(x,0)=\exp\left[-\left(\frac{x}{w}\right)^2+i\alpha\left(\frac{x}{x_0}\right)^3+i\beta\left(\frac{x}{x_0}\right)\right], \tag{5.5}$$

where:

w is the Gaussian beam's waist radius
α and β are dimensionless parameters of the phase transparency

Then, the paraxial approximation of the light field amplitude at distance z is given by

$$\begin{aligned}E(x,z)=&\sqrt{\frac{-i2\pi k}{z}}\frac{x_0}{\sqrt[3]{3\alpha}}\exp\left[\frac{1}{3\alpha}\left(\frac{x_0}{w}\right)^2\left(\beta-\frac{kx_0x}{z}\right)+\frac{2}{27\alpha^2}\left(\frac{x_0}{w}\right)^6\left(1-3\frac{z_0^2}{z^2}\right)\right]\\&\times\exp\left[\frac{ikx^2}{2z}-\frac{iz_0}{z}\frac{1}{3\alpha}\left(\frac{x_0}{w}\right)^2\left(\beta-\frac{kx_0x}{z}\right)\right]\exp\left[-\frac{2i}{27\alpha^2}\left(\frac{x_0}{w}\right)^6\left(3\frac{z_0}{z}-\frac{z_0^3}{z^3}\right)+ikz\right]\\&\times\mathrm{Ai}\left\{\frac{1}{(3\alpha)^{1/3}}\left[\beta-\frac{kx_0x}{z}+\frac{1}{3\alpha}\left(\frac{x_0}{w}\right)^4\left(1-\frac{iz_0}{z}\right)^2\right]\right\}.\end{aligned} \tag{5.6}$$

The expression in Eq. (5.6) describes Type-II Airy beams. Eq. (5.6) suggests that unlike the linear phase of the Airy beams (5.2), the Type-II Airy beams have a quadratic phase, thus experiencing the divergence upon propagation. Besides, similarly to Eq. (5.2), the argument of the Airy function in Eq. (5.6) is complex, although the z-dependence is different: the value of the Airy function argument is in direct proportion to z^2 in Eq. (5.2) and in inverse proportion to z in Eq. (5.6). It is possible to obtain the infinite energy Airy beams if the cubic phase transparency is illuminated by a plane wave ($w \to \infty$) rather than a Gaussian beam. Then, we obtain, instead of Eq. (5.6):

$$\begin{aligned} E(x,z) = \sqrt{\frac{2\pi k}{iz}} \frac{x_0}{\sqrt[3]{3\alpha}} \mathrm{Ai}\left[\frac{1}{\sqrt[3]{3\alpha}}\left(\beta - \frac{kx_0 x}{z} - \frac{k^2 x_0^4}{12\alpha z^2}\right)\right] \\ \times \exp\left[\frac{ik}{2z}\left(x^2 + \frac{kx_0^3 x}{3\alpha z} - \frac{\beta x_0^2}{3\alpha} + \frac{k^2 x_0^6}{54\alpha^2 z^2}\right) + ikz\right]. \end{aligned} \tag{5.7}$$

Because the phase relation remains quadratic, the beam in Eq. (5.7) will diverge upon propagation. Putting the Airy function's argument equal to values y_m at which the function has local maxima, we find: $[\beta - kx_0x/z - k^2x_0^4/(12\alpha z^2)]/(3\alpha)^{1/3} = y_m$. This leads to explicit expression of the trajectory of the Type-II Airy beam maximum:

$$x = \frac{\left(\beta - y_m\sqrt[3]{3\alpha}\right)z}{kx_0} - \frac{kx_0^3}{12\alpha z}. \tag{5.8}$$

Unlike parabolic trajectory of the beams (5.2), the Type-II Airy beams propagate along a hyperbolic path (5.8). The beam exhibits acceleration on the path sections at which the first and second derivatives have the same sign. This leads to inequality:

$$\frac{1}{z^2} < \frac{12\alpha}{k^2 x_0^4}\left(y_m\sqrt[3]{3\alpha} - \beta\right), \tag{5.9}$$

which is fulfilled if $\mathrm{sign}(\alpha)\beta < y_m(3|\alpha|)^{1/3}$. In this case, the acceleration will occur at distances

$$z > z_1 = \frac{kx_0^2}{2}\left[3\alpha\left(y_m\sqrt[3]{3\alpha} - \beta\right)\right]^{-1/2}. \tag{5.10}$$

Numerical Results

Unlike the beams (5.2), the acceleration is not uniform and decreases as z^{-3}. Let us analyze the following parameters: $x_0 = \lambda = 532$ nm, $\alpha = -1$, $\beta = 10$, $m = 0$, $y_0 = -1.01879$. In this case, the condition (5.9) is satisfied, so that the trajectory has the acceleration at $z > z_1 \approx 330$ nm. The trajectory for these parameters is plotted in Figure 5.1a, while the field intensity distribution (5.7) is depicted in Figure 5.1b. The size of the computation domain in Figure 5.1b is $-10\,\lambda \le x \le +10\,\lambda$, $0 \le z \le 4\lambda$. Figure 5.2 depicts the intensity profiles in the planes (a) $z = \lambda$, (b) 2λ, and (c) 4λ.

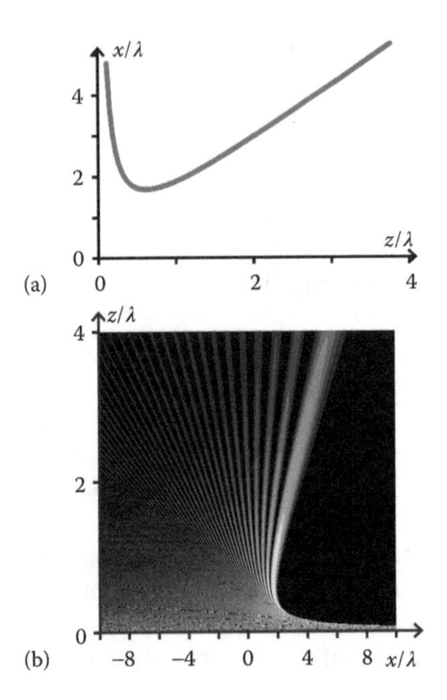

FIGURE 5.1 (a) Trajectory of the accelerating Type-II Airy beam and (b) corresponding intensity profile in the xz-plane.

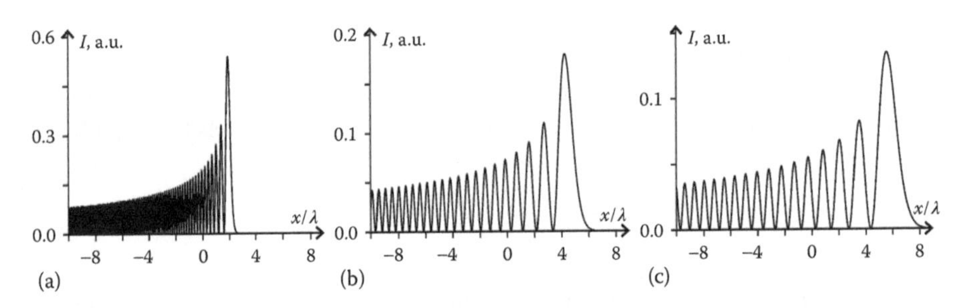

FIGURE 5.2 Intensity profiles for the Type-II Airy beam in the planes (a) $z = \lambda$, (b) 2λ, and (c) 4λ.

For comparison, let us examine the Airy beam of Eq. (5.2) at $a = 0$. Putting the Airy function's argument in Eq. (5.2) equal to values of y_m at which it has local maxima, the trajectory of the Airy beam maximum can be explicitly expressed as $x = x_0 y_m + z^2/(4k^2 x_0^3)$. From the equation, the beam is seen to have a uniform acceleration equal to $1/(2k^2 x_0^3)$, while the Airy beam of Eq. (5.2) is seen to be diffraction–free, because $x_1 - x_2 = x_0(y_m - y_n)$ is independent of z. Meanwhile for the Type-II Airy beams, from Eq. (5.7) it follows that $x_1 - x_2 = (3\alpha)^{1/3} z(y_m - y_n)/(kx_0)$ and the beam shows linear divergence with increasing z (see Figure 5.2). Figure 5.3 shows the intensity profile of the field (5.2) for the following parameters: $\lambda = 532\,\text{nm}$, $x_0 = \lambda/2$. The computation domain in Figure 5.3 remains the same: $-10\lambda \le x \le +10\lambda$, $0 \le z \le 4\lambda$. The acceleration of the beam (5.2) at $a = 0$ equals $1/(\pi^2\lambda)$, whereas

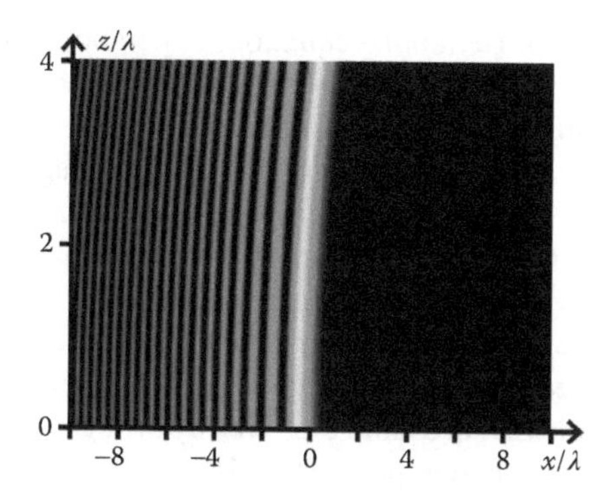

FIGURE 5.3 (Color online) Intensity profile of the (uniformly accelerating) Airy beam in the xz-plane.

for the Type-II Airy beam of Eq. (5.8) shown in Figure 5.1b, the acceleration at $z = z_1 \approx 330$ nm is about $19.87/(\pi^2\lambda)$. This is the reason why the beam's trajectory in Figure 5.1b is more curved.

The "gravity center" for the Type-II Airy beam takes the form:

$$\int_{-\infty}^{\infty} x\left|E(x,z)\right|^2 dx = \sqrt{\frac{\pi}{2}}\left(\frac{3\alpha w^2}{4kx_0^3} + \frac{\beta}{kx_0}\right)wz. \tag{5.11}$$

From Eq. (5.11) it is evident that even in the absence of the linear phase term in Eq. (5.5) ($\beta = 0$), the "gravity center" of the Type-II Airy beam is moving linearly with increasing z, at any w.

Summing up, we have investigated the 1D finite-energy and infinite-energy Airy beams of the second kind (Eqs. 5.6 and 5.7, respectively) that accelerate upon propagation on a hyperbolic trajectory (Eq. 5.8); the Airy beams of the second kind have been shown to accelerate nonuniformly, rapidly decaying with distance by a cubic law, with the beam propagating further along a straight trajectory (Figure 5.1); also, it has been shown that in the near-field (at a distance of several wavelengths behind the phase transparency) the acceleration of the Airy beams of the second kind is by an order of magnitude larger than that of conventional Airy beams propagating along a parabolic path, other conditions being the same; the infinite-energy Airy beams of the second kind have been shown to diverge linearly upon propagation with distance from the input plane and the central lobe of intensity is widened.

5.2 TRANSFORMATION OF DECELERATING LASER BEAMS INTO ACCELERATING ONES

Recently, non-paraxial accelerating Weber beams that retain their shape upon propagation on a parabolic path have been considered [179]. Being similar to the "half-Bessel" beams [176], the Weber beams can be described analytically, which is not the case for the "half-Bessel" beams. The Weber-Hermite beams have also been known to present solutions of the paraxial equation of propagation [180]. A general theory of 3D non-paraxial accelerating beams was developed in Ref. [181] on the basis of familiar solutions of Helmholtz equations in parabolic, oblate, and flattened spheroid coordinates. These beams propagate on a circular arc. The Airy beams that propagate with a non-uniform acceleration on a hyperbolic trajectory were proposed in Ref. [182]. Although such beams are diverging upon propagation, thus not retaining their shape, the final section of their propagation path can be more bent when compared with the conventional Airy beams [168].

In this section, we take a different approach to generating the accelerating beams, which is as follows. There are 2D paraxial light fields in which the complex amplitude function argument is related to variables as x^2/z, where x is the transverse coordinate and z is the longitudinal coordinate. An example is given by the light field generated through diffracting a plane wave by a corner phase step [183] or a well-known solution of the problem of diffraction by the edge of an opaque screen [163]. In this section, other solutions of the paraxial equation of propagation are also discussed. Light fields in which the complex amplitude function has the argument given by x^2/z propagate on a square-root parabola path, $x = z^{1/2}$. Such beams are decelerating because the acceleration, as the second derivative along the trajectory, $x'' = -z^{3/2}$, has the opposite sign to the velocity (the first derivative along the trajectory) $x' = z^{-1/2}$. If, however, the field amplitude at distance z_0 is replaced by the complex-conjugated amplitude, and the axis origin is simultaneously shifted to point z_0, the resulting light field will propagate with acceleration along the path $x = (z_0 - z)^{1/2}$. In this section, we derive analytical relationships for the complex amplitudes of such accelerating beams. These beams will not only be accelerated, but also converge on the segment from the initial plane ($z = 0$) to the focal plane ($z = z_0$).

Also, we analyze paraxial "half-Bessel" beams, which are different from the non-paraxial ones [176].

Accelerating Beams

Assume that at each particular distance z passed by a laser beam the intensity maximum coordinate is given by $x_{max}(z)$. For the beam to experience an acceleration at a certain path section, the first and second derivatives of the intensity maximum coordinate x_{max} with respect to the passed distance z need to have the same sign [182]:

$$\left(\frac{dx_{max}}{dz}\right)\left(\frac{d^2x_{max}}{dz^2}\right) > 0. \tag{5.12}$$

The most widely-known accelerating beams are Airy beams, with their complex amplitude given by [168]:

$$E(x,z) = \text{Ai}\left(\frac{s-\xi^2}{4}\right)\exp\left(\frac{is\xi}{2} - \frac{i\xi^3}{12}\right), \tag{5.13}$$

where:

- (x, z) are the Cartesian coordinates
- $s = x/x_0$, $\zeta = z/(kx_0^2)$, $k = 2\pi/\lambda$ is the wave number
- λ is the wavelength
- x_0 is an arbitrary scaling factor
- $\text{Ai}(x)$ is the Airy function [101, Section 10.4]

The beams' intensity maximums have the coordinates

$$x_{\max} = x_0 y_m + \frac{z^2}{4k^2 x_0^3}, \tag{5.14}$$

where y_m is the coordinate of the mth zero of the function $\mathrm{Ai}'(x)$. Then, $dx_{\max}/dz = z/(2k^2x_0^3)$ and $d^2x_{\max}/dz^2 = 1/(2k^2x_0^3)$. Thus, the condition in Eq. (5.12) is satisfied for any distance $z > 0$, with the acceleration $d^2x_{\max}/dz^2$ having a constant value. In the following, we analyze laser beams characterized by a non-uniform acceleration which is decreasing with the distance propagated.

Airy Beams with a Hyperbolic Path

The Airy beams with a hyperbolic path characterized at $z = 0$ by the complex amplitude

$$E(x,0) = \exp\left[i\alpha\left(\frac{x}{x_0}\right)^3 + i\beta\left(\frac{x}{x_0}\right)\right], \tag{5.15}$$

where:

x_0 is a scaling factor

α, β are dimensionless parameters, were discussed in Refs. [182,184,185]

The beam's trajectory in the Fresnel diffraction region takes the form:

$$x_{\max} = \begin{cases} \dfrac{z}{kx_0}\left(\beta - y_m\sqrt[3]{3\alpha}\right) - \dfrac{kx_0^3}{12\alpha z}, \alpha > 0, \\[2ex] \dfrac{z}{kx_0}\left(\beta + y_m\sqrt[3]{3|\alpha|}\right) + \dfrac{kx_0^3}{12|\alpha|z}, \alpha < 0. \end{cases} \tag{5.16}$$

Based on condition (5.12), the acceleration has been shown [182] to occur at the path section

$$z > z_1 = \begin{cases} \dfrac{kx_0^2}{2\sqrt{3\alpha\left(y_m\sqrt[3]{3\alpha} - \beta\right)}}, \alpha > 0, \\[3ex] \dfrac{kx_0^2}{2\sqrt{3|\alpha|\left(y_m\sqrt[3]{3|\alpha|} + \beta\right)}}, \alpha < 0, \end{cases} \tag{5.17}$$

given that $\mathrm{sign}(\alpha)\beta < y_m(3|\alpha|)^{1/3}$. For such a beam, the acceleration decreases as z^{-3}: $d^2x_{\max}/dz^2 = -kx_0^3/(6\alpha z^3)$.

Hermite-Gauss Beams

It turns out that the well-known Hermite-Gauss beams [186] also possess an acceleration. Actually, at $z = 0$, let the light field be given by the complex amplitude:

$$E(x, z=0) = \exp\left(-\frac{x^2}{w^2}\right) H_n\left(\frac{x}{a}\right), \tag{5.18}$$

where:

(x, z) are the Cartesian coordinates
w is the Gaussian beam waist radius
n and a are, respectively, the order and scale of the Hermite polynomial

Then, by taking the Fresnel transform we can show that at distance z from the initial plane there will be generated a field with the following complex amplitude distribution [187,188]:

$$E(x, z) = \left[\mu(z)\right]^{-(n+1)/2} \left[\nu(z)\right]^{n/2} \exp\left[-\frac{x^2}{w^2 \mu(z)}\right] H_n\left[\frac{x}{a\sqrt{\mu(z)\nu(z)}}\right], \tag{5.19}$$

where:

$z_R = kw^2/2$
$z_a = ka^2/2$
$\mu(z) = 1 + iz/z_R$
$\nu(z) = 1 + iz(1/z_R - 1/z_a)$

For simplicity, let us analyze the case when $n = 1$. Then, the intensity of beam (5.19) in the plane found at distance z from the initial plane will be given by

$$I(x, z) = \left|E(x, z)\right|^2 = \frac{4x^2}{a^2 \left|\mu(z)\right|^3} \exp\left[-\frac{2x^2}{w^2 \left|\mu(z)\right|^2}\right]. \tag{5.20}$$

Differentiating both parts of Eq. (5.20) with respect to x, we obtain the necessary condition for the intensity extremes:

$$2x = \frac{4x^3}{w^2 \left|\mu(z)\right|^2}. \tag{5.21}$$

When $x = 0$, there is a minimum (because the $I(0, z)$ intensity is zero), whereas the maxima coordinates are given by

$$x_{\max} = \pm\sqrt{2}\sigma_0(z), \tag{5.22}$$

where $\sigma_0(z) = (w/2)\left[1 + (z/z_R)^2\right]^{1/2}$ is the square root of the intensity second-order moment for the fundamental mode (Gaussian beam).

It can be easily shown that the maximum intensity curve is defined by a hyperbola. The first- and second-order derivatives of x_{max} with respect to z are:

$$\frac{dx_{max}}{dz} = \pm\frac{w}{\sqrt{2}}\frac{z}{z_R^2\sqrt{1+(z/z_R)^2}},$$
$$\frac{d^2x_{max}}{dz^2} = \pm\frac{w}{\sqrt{2}}\frac{1}{z_R^2\left[1+(z/z_R)^2\right]^{3/2}}. \tag{5.23}$$

From Eq. (5.23), the product $(dx_{max}/dz)(d^2x_{max}/dz^2)$ is seen to be positive at all $z > 0$, which means that in a similar way to the above-discussed Airy beams propagating on a cubic path, the Hermite-Gauss beam is negatively accelerating with distance z by a cubic law. The acceleration of both branches of the Hermite-Gauss beam can be noted at small distances z. Figure 5.4 depicts the intensity of such a beam in the plane Oxz, which was simulated using the beam propagation method (BPM). For the Hermite-Gaussian beam in Figure 5.4, two local maxima are seen to propagate with acceleration on two hyperbolic paths symmetrical to the optical axis.

Note that with increasing number n, the radius (width) of the Hermite-Gauss beam is increasing, with the trajectory of the outermost intensity zeros being described by the relation

$$\left|x_0^n\right| \le \frac{w}{2}\sqrt{n(n-1)}\sqrt{1+\frac{z^2}{z_R^2}}, \tag{5.24}$$

from which, and considering Eq. (5.23), it follows that for large values of n, the acceleration of the Hermite-Gauss beam increases linearly with increasing n.

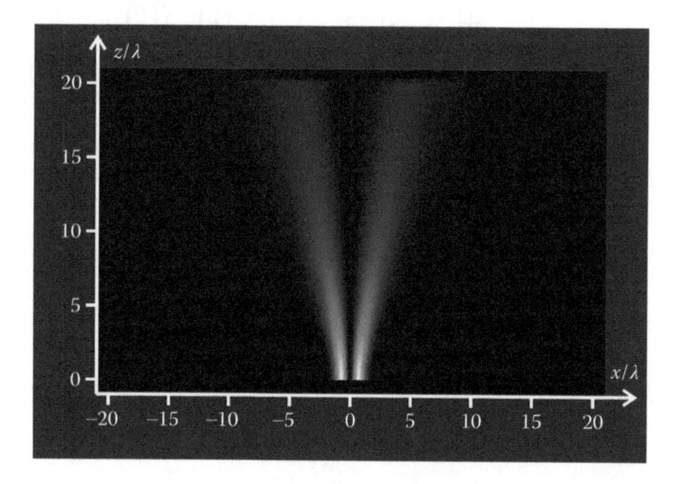

FIGURE 5.4 Accelerating Hermite-Gauss beam ($\lambda = 532$ nm, $w = \lambda$, $a = 2\lambda$, $n = 1$).

No other 2D paraxial accelerating beams that could be defined analytically have been known so far. Because of this, in the next section we show in which way the accelerating beams can be derived from the decelerating ones.

Decelerating Beams

As distinct from the above-discussed accelerating beams of Eqs. (5.13), (5.15), and (5.18), for a beam to be decelerating the first and second derivatives of its maximum intensity coordinate, x_{max}, with respect to the distance passed need to be of opposite sign:

$$\left(\frac{dx_{max}}{dz}\right)\left(\frac{d^2x_{max}}{dz^2}\right) < 0. \tag{5.25}$$

In the following, we consider examples of such beams.

Diffraction of a Plane Wave by an Opaque Semi-Infinite Screen

Assume that a plane wave is propagating along the optical axis z, passing at $z = 0$ through a semi-infinite plane aperture, which is transparent at $x < 0$ (x is the coordinate in the aperture plane). Directly behind it, the complex amplitude is

$$E(x, z=0) = \begin{cases} 1, x < 0, \\ 0, x \geq 0. \end{cases} \tag{5.26}$$

After travelling over a distance z, the light complex amplitude will be given by the Fresnel transform of field (5.26):

$$E(x,z) = \frac{1}{2}\{[1 - C(\xi) - S(\xi)] + i[C(\xi) - S(\xi)]\}, \tag{5.27}$$

where $\xi = [2/(\lambda z)]^{1/2}x$, whereas $C(\xi)$ and $S(\xi)$ are the Fresnel integrals:

$$C(\xi) = \int_0^{\xi} \cos\left(\frac{\pi t^2}{2}\right) dt,$$
$$S(\xi) = \int_0^{\xi} \sin\left(\frac{\pi t^2}{2}\right) dt. \tag{5.28}$$

For this beam, the maximal intensity coordinate x_{max} is

$$x_{max} = \sqrt{\frac{\lambda z}{2}}\xi_m, \tag{5.29}$$

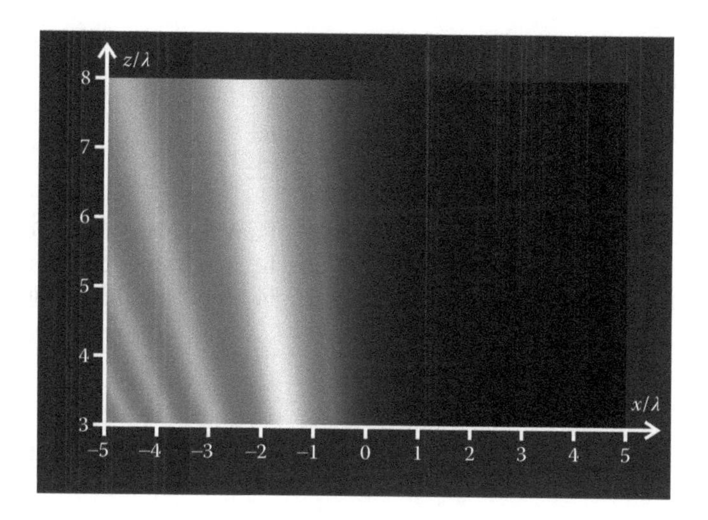

FIGURE 5.5 Diffraction of a plane wave by a semi-infinite aperture: intensity pattern in the Oxz-plane ($\lambda = 532$ nm, $-5\lambda \le x \le 5\lambda$, $3\lambda \le z \le 8\lambda$).

where ξ_m is the coordinate of the mth maximum of the function $I(\xi)=\left[1-C(\xi)-S(\xi)\right]^2+\left[C(\xi)-S(\xi)\right]^2$. Calculating the first and second derivatives of the maximum's coordinate $x_{\max}$ with respect to the passed distance z, we find:

$$\frac{dx_{\max}}{dz} = \frac{1}{2}\sqrt{\frac{\lambda}{2z}}\xi_m,$$
$$\frac{d^2x_{\max}}{dz^2} = -\frac{1}{4z}\sqrt{\frac{\lambda}{2z}}\xi_m. \tag{5.30}$$

From Eq. (5.30), it is seen that condition (5.25) is valid for all $z > 0$. The deceleration of each maximum can be seen in Figure 5.5, which shows the intensity of the beam of Eq. (5.27).

Two-Dimensional Hypergeometric Beams and Bessel Beams

Just as it was done in Ref. [189], the solution to the paraxial equation of propagation

$$2ik\frac{\partial E}{\partial z} + \frac{\partial^2 E}{\partial x^2} = 0, \tag{5.31}$$

will be sought for in the form $E(x, z) = x^p z^q F(sx^m z^n)$, where $F(x)$ is a function and s is a scaling factor. After reducing the resulting second-order differential equation to the

Kummer equation, the relation for the complex amplitude of the 2D analog of the 3D generalized hypergeometric mode [113] takes the form:

$$E(x,z) = z^{-a}\,{}_1F_1\left(a, \frac{1}{2}, \frac{ikx^2}{2z}\right), \tag{5.32}$$

where a is an arbitrary constant. The derivative of any solution of Eq. (5.31) taken with respect to any Cartesian coordinate has also been known to be the solution of Eq. (5.31). Therefore, it is possible to consider a light beam with the amplitude

$$E(x,z) = xz^{-a}\,{}_1F_1\left(a, \frac{3}{2}, \frac{ikx^2}{2z}\right). \tag{5.33}$$

In particular, putting $a = 3/4$, Eq. (5.33) gives the solution of Eq. (5.31) in the form of a fractional-order Bessel beam:

$$E(x,z) = \sqrt{\frac{x}{z+z_0}}\,J_{\frac{1}{4}}\left[\frac{kx^2}{4(z+z_0)}\right]\exp\left[\frac{ikx^2}{4(z+z_0)}\right], \tag{5.34}$$

where z_0 is an arbitrary positive constant (intended to avoid singularity in the plane $z = 0$). In order to derive Eq. (5.34), expression 13.6.1 from [101] has been used. For this beam, the intensity maximum's coordinates are

$$x_{\max} = \sqrt{\frac{4(z+z_0)}{k}\,y_m}, \tag{5.35}$$

where y_m is the mth root of the equation $J_{1/4}(y)\left[J_{1/4}(y) + 4J'_{1/4}(y)\,y\right] = 0$. Relationship (5.35) is similar to (5.29), which implies that $x_{\max}$ is proportional to $z^{1/2}$. Thus, we can infer that beam (5.34) is decelerating, as can be seen from Figure 5.6,

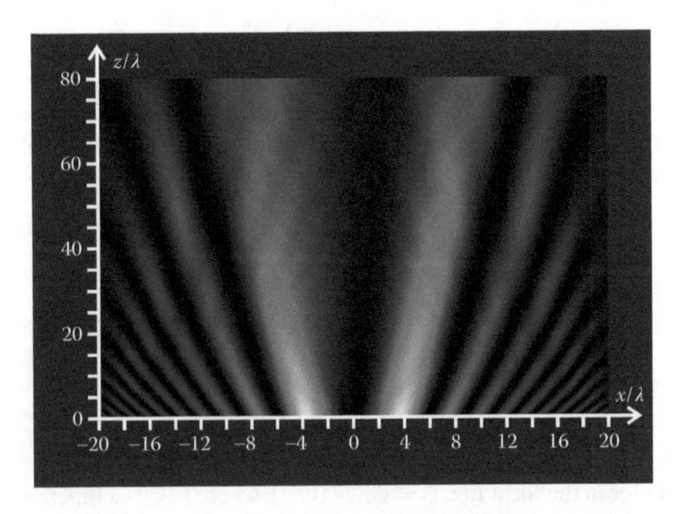

FIGURE 5.6 Intensity pattern in the *Oxz*-plane produced by the light beam of Eq. (5.34).

which depicts the intensity pattern simulated by the BPM (λ = 532 nm, z_0 = 20λ, simulation region: $-20\lambda \le x \le 20\lambda$, $0 \le z \le 80\lambda$).

Transformation of Decelerating Beams into Accelerating Ones

From Eqs. (5.12) and (5.25), a simple change of variables $z \to z_0 - z$ is seen to result in the acceleration turning into deceleration, and vice versa. Actually, let us analyze a light beam with its complex amplitude derived from Eq. (5.27) by use of complex conjugation and substitution of ξ by $\xi = \sqrt{2/\left[\lambda(z_0 - z)\right]}x$:

$$E(x, z < z_0) = \frac{1}{2}\{[1 - C(\xi) - S(\xi)] - i[C(\xi) - S(\xi)]\}. \tag{5.36}$$

The light beam of Eq. (5.37) will be referred to as a Fresnel beam. Considering that

$$\lim_{x\to\infty} C(x) = \frac{1}{2}, \lim_{x\to\infty} S(x) = \frac{1}{2}, \tag{5.37}$$

we find that in the vicinity of the focal plane $z = z_0$, on the assumption that $z < z_0$, the complex amplitude takes the form:

$$E(x, z \to z_0 - 0) = \begin{cases} 0, x > 0, \\ 1, x < 0. \end{cases} \tag{5.38}$$

The intensity pattern produced by beam (5.36) in the plane Oxz is shown in Figure 5.7.

The arguments ξ of Fresnel integrals (5.37) become imaginary immediately behind the plane $z = z_0$, because $2/[\lambda (z_0 - z)] < 0$. Note that since the imaginary part can be both positive and negative, there may be two solutions, with only one of them

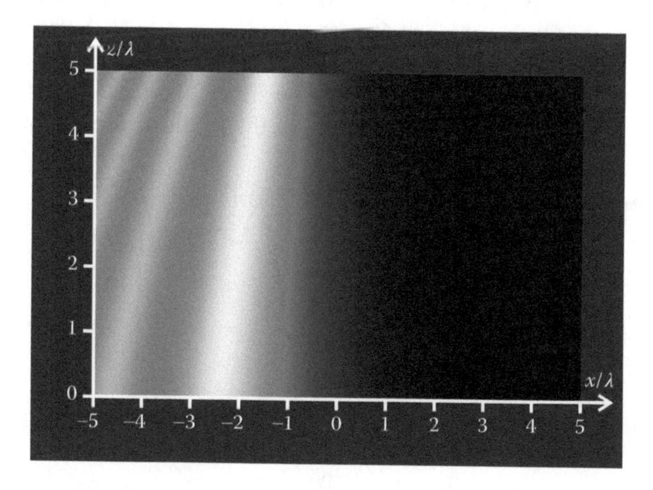

FIGURE 5.7 Intensity pattern produced by light beam (5.36) in the plane Oxz (wavelength $\lambda = 532$ nm, distance from $z = 0$ to the focal plane is $z_0 = 8\lambda$. Simulation domain: $-5\lambda \le x \le 5\lambda$, $0 \le z \le 5\lambda$).

meeting the boundary condition $E(x, z \to z_0 + 0) = E(x, z \to z_0 - 0)$. Making us of the identities for the Fresnel integrals of imaginary variables, $C(iz) = iC(z)$ and $S(iz) = -iS(z)$ ([101], expression 7.3.18), the complex amplitude behind the plane $z = z_0$ can be shown to take the form:

$$E(x, z > z_0) = \frac{1}{2}\{[1 - C(\eta) - S(\eta)] + i[C(\eta) - S(\eta)]\}, \tag{5.39}$$

where $\eta = \sqrt{2/[\lambda(z - z_0)]}x$. The second solution given by

$$E(x, z > z_0) = \frac{1}{2}\{[1 + C(\eta) + S(\eta)] - i[C(\eta) - S(\eta)]\}, \tag{5.40}$$

does not satisfy the boundary condition. From comparison of (5.36) and (5.39), the light beam is seen to accelerate at $z < z_0$, forming a semi-infinite uniform focal spot at $z = z_0$ and reversing to deceleration at $z > z_0$, with the amplitude at $z > z_0$ being a specular reflection of that at $z < z_0$. In Figure 5.8a is depicted the intensity pattern of light beam (5.36), (5.39) in the plane *Oxz*, whereas Figure 5.8b shows the intensity profile in the plane $z = z_0$. The results shown in Figure 5.8 have been derived by FDTD-aided simulation for $\lambda = 532$ nm, $z_0 = 40\lambda$, simulation domain $-20\lambda \le x \le 20\lambda$, $0 \le z \le 60\lambda$. Intensity oscillations in the vicinity of point $x = 0$ appear because the initial field is limited by the simulation domain.

Similarly, substituting in (5.34) $z + z_0$ for $z_0 - z$ and applying complex conjugation, we obtain a light beam described in the initial plane ($z = 0$) by the complex amplitude

$$E(x, z = 0) = \sqrt{\frac{x}{z_0}} J_{\frac{1}{4}}\left(\frac{kx^2}{4z_0}\right) \exp\left(-\frac{ikx^2}{4z_0}\right). \tag{5.41}$$

With the initial field given by Eq. (5.41), the BPM technique was used to calculate the light beam intensity in the plane *Oxz* (at $\lambda = 532$ nm, $z_0 = 40\lambda$), with Figure 5.9a showing that the focused beam is accelerating. Figure 5.9b shows transverse intensity distribution in the focal plane $z_0 = 40\lambda$, the size of the focal spot (full width at half-maximum) is 0.53 μm. The asymmetry of the resulting focal spot is due to a

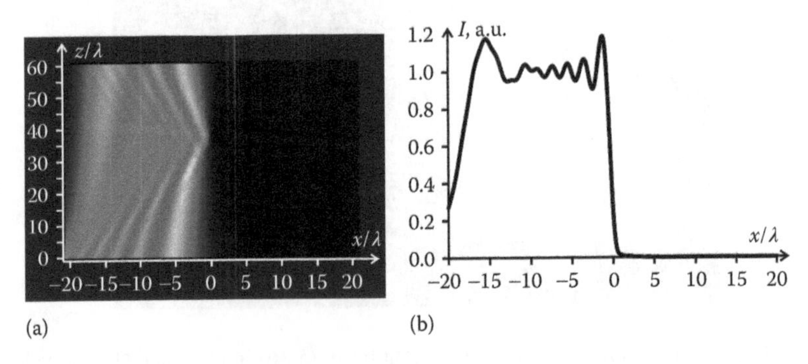

FIGURE 5.8 Intensity profile of beam (5.36), (5.39), calculated by the FDTD-method: (a) intensity pattern in the plane *Oxz* and (b) intensity profile in the plane $z = z_0$.

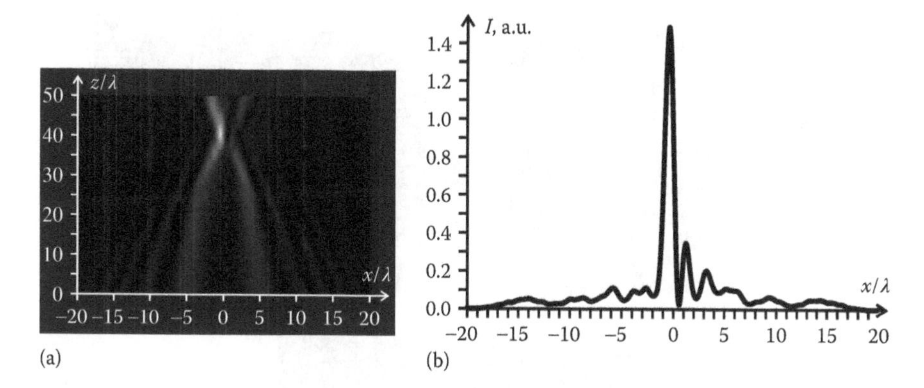

FIGURE 5.9 (a) Intensity pattern in the plane *Oxz* produced by the accelerating light beam with the intensity distribution in the initial plane given by Eq. (5.41), (b) transverse intensity distribution in the focal plane $z_0 = 40\lambda$.

$\pi/2$-phase jump that field (5.41) undergoes at point $x = 0$. This type of focusing of the accelerating beam was earlier reported in [190] for a radially symmetric Airy beam.

Assume that in the initial plane for $x < 0$, the amplitude equals zero:

$$E(x, z=0) = \begin{cases} \sqrt{\dfrac{x}{z_0}} J_{\frac{1}{4}}\left(\dfrac{kx^2}{4z_0}\right)\exp\left(-\dfrac{ikx^2}{4z_0}\right), x \geq 0, \\ 0, x < 0. \end{cases} \quad (5.42)$$

Figure 5.10 shows the intensity pattern for beam (5.42), which was calculated using the BPM technique for the same parameters as in Figure 5.9.

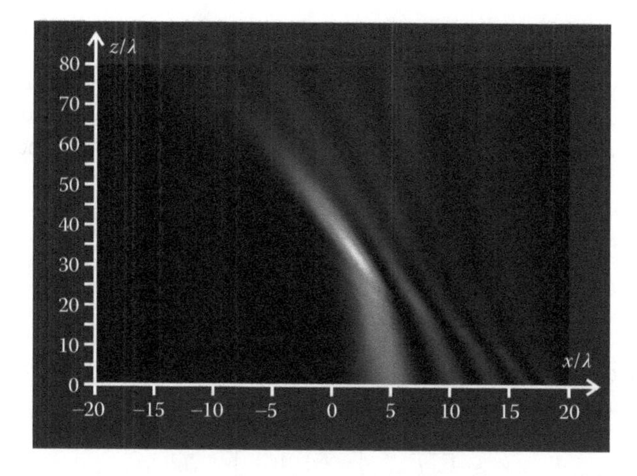

FIGURE 5.10 Intensity pattern in the plane *Oxz* for an accelerating light beam with complex amplitude distribution in the initial plane given by Eq. (5.42).

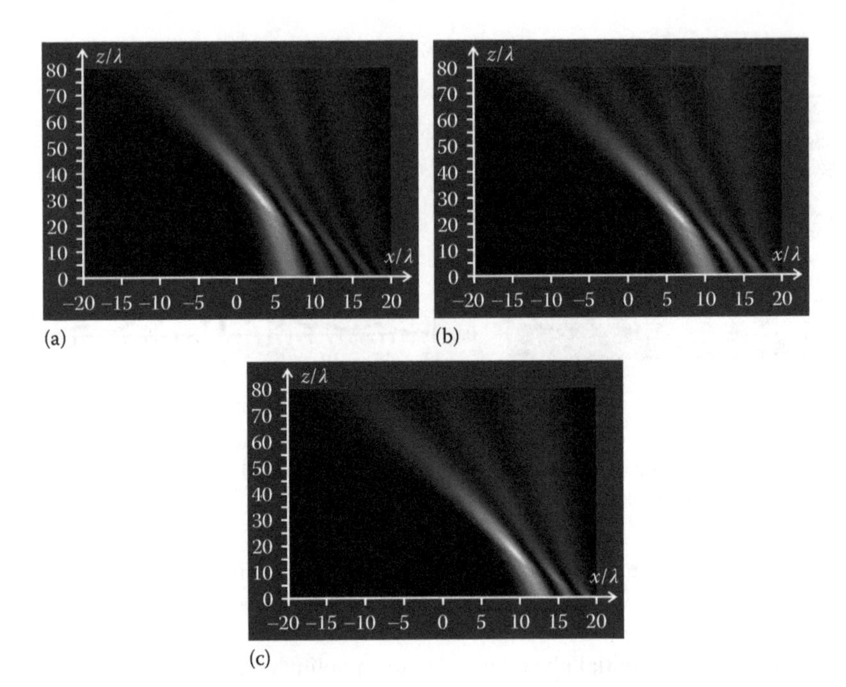

FIGURE 5.11 Intensity pattern for beam (5.42) with different orders of Bessel function: (a) 1, (b) 3, and (c) 5.

Note that for other orders of Bessel function (5.42) a similar intensity pattern is formed (Figure 5.11).

From Figure 5.11, the acceleration is seen to decrease with increasing order of Bessel function. The accelerating paraxial beams in Eq. (5.42) are similar to the non-paraxial "half-Bessel" beams [176], prompting us to call them paraxial "half-Bessel" beams.

Diffraction of a Gaussian Beam by a Semi-Infinite Opaque Screen

In the following, we analyze another example of a decelerating beam described by an analytical function. Let a Gaussian beam of waist radius w pass through a semi-infinite planar aperture. The complex amplitude immediately behind the aperture is

$$E(x,z=0)=\begin{cases}\exp\left(-\dfrac{x^2}{w^2}\right), x<0,\\[2ex] 0, x\geq 0.\end{cases} \tag{5.43}$$

Having traveled over distance z, the beam's complex amplitude will be defined by the Fresnel transform of field (5.43), taking the form [38]:

$$E(x,z)=\sqrt{\frac{-ik}{8pz}}\exp\left[\frac{ikx^2}{2(z-iz_R)}\right]\operatorname{erfc}\left(\frac{-ikx}{2\sqrt{pz}}\right), \tag{5.44}$$

where:

$p = 1/w^2 - ik/(2z)$

$z_R = kw^2/2$ is the Rayleigh range

erfc(x) is a complementary error function:

$$\operatorname{erfc}(z) = 1 - \operatorname{erf}(z) = 1 - \frac{2}{\sqrt{\pi}} \int_0^z \exp(-t^2)\,dt. \tag{5.45}$$

Equation (5.44) can be reduced to the form:

$$E(x,z) = \sqrt{\frac{-ik}{8pz}} \exp\left(\frac{ikx^2}{2z}\right) \exp(-y^2) \operatorname{erfc}(iy), \tag{5.46}$$

where:

$$y = \frac{-kx}{2\sqrt{pz}} = -\left(\frac{x}{w}\right)\left(\frac{z_R}{z}\right)\left(1 + \frac{z_R^2}{z^2}\right)^{-1/4} \exp\left[i\frac{1}{2}\arctan\left(\frac{z_R}{z}\right)\right]. \tag{5.47}$$

When $z \to 0$, the argument of the y variable depends almost not at all on z: $\arg(y) \approx \pi/4$. Because of this, the equation of the beam's path at small distances z takes the form:

$$x_{\max} = -w\eta_m \sqrt{\frac{2z}{z_R}}, \tag{5.48}$$

where η_m is the mth maximum of the function $|\operatorname{erfc}[\eta\,(i-1)]|^2$. Thus, we can infer that similarly to light beam (5.27), beam (5.46) is also decelerating.

By analogy with the generation of a uniform intensity distribution on the semi-plane (using the beam of Eqs. 5.36, 5.39), we shall form the distribution of Eq. (5.43) in the plane $z = z_0$ using an optical beam defined by the complex amplitude distribution:

$$E(x,z) = \sqrt{\frac{ik}{8p(z_0 - z)}} \exp\left[\frac{-ikx^2}{2(z_0 - z + iz_R)}\right] \operatorname{erfc}\left[\frac{ikx}{2\sqrt{p}(z_0 - z)}\right], \tag{5.49}$$

where $p = 1/w^2 + ik/[2(z_0 - z)]$. Contained as a factor in Eq. (5.49) is the probability, or Laplace, integral. Because of this, beams (5.49) will be referred to as Laplace beams. From the path equation (5.48), beam (5.49) is also seen to be accelerating near the plane $z = z_0$. This can be noted from Figure 5.12a, which depicts the intensity pattern for beam (5.49) in the plane Oxz. Shown in Figure 5.12b is the intensity profile in the plane $z = z_0$.

The fields (5.43) and (5.44) automatically follow from the expressions for the fields (5.27) or (5.31) (by A. Torre):

$$E(x,z) = \frac{1}{\sqrt{\mu(z)}} \exp\left[\frac{-x^2}{w^2\mu(z)}\right] \operatorname{erfc}\left[\sqrt{\frac{-ik}{2z}} \frac{x}{\sqrt{\mu(z)}}\right]. \tag{5.50}$$

Thus, we have obtained the following results: (1) the well-known Hermite-Gauss modes and generalized Hermite-Gauss beams, Eq. (5.19), have been shown to be

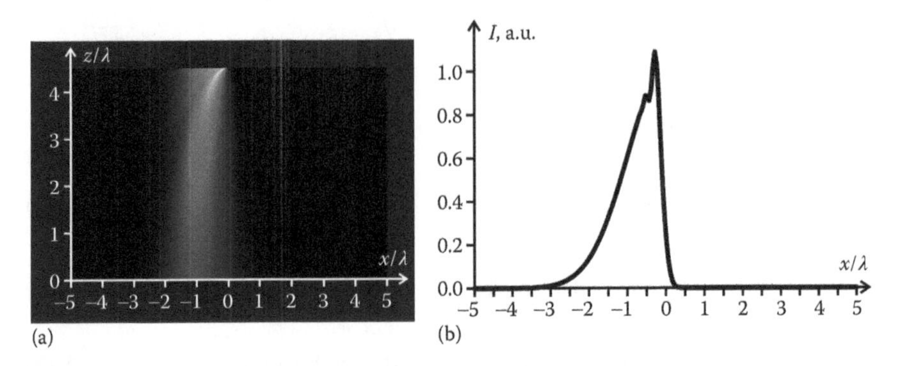

FIGURE 5.12 (a) Intensity pattern of beam (5.49) in the plane Oxz and (b) intensity profile in the plane $z = z_0$.

accelerating, which implies that the two outermost intensity maximums relative to the optical axis propagate on hyperbolic paths with non-uniform acceleration, which decreases as the cube of distance (5.23); (2) a technique to transform 2D decelerating light beams to accelerating ones has been proposed, which consists in complex conjugation and shift along the optical axis of the complex amplitude of the decelerating beam; (3) a Fresnel beam of Eq. (5.36) experiencing the acceleration on a final path segment has been derived by complex conjugation and axis-shift of complex amplitude (5.27) describing the diffraction of a plane wave by an opaque screen; (4) a "half-Bessel" beam of Eq. (5.42) accelerating on a finite segment of a square-root parabola path has been derived by "taking half" of the complex-conjugated and axis-shifted hypergeometric laser beam of Eq. (5.33); (5) in a similar way, a Laplace beam, Eq. (5.49), accelerating on a finite path segment, has been derived from Eq. (5.44), which describes the diffraction of a Gaussian beam by an opaque screen, through complex conjugation and axis-shift of complex amplitude (5.44).

6 Hermite-Gaussian Vortices

6.1 HERMITE-GAUSSIAN MODAL LASER BEAMS WITH ORBITAL ANGULAR MOMENTUM

In 1992 it was shown by Allen [17] that the Laguerre-Gaussian (LG) modes have an orbital angular momentum (OAM). The OAM is also characteristic of all optical vortices or singular laser beams with a phase singularity or wavefront dislocation [191]. The power flux (Poynting vector) of such beams rotates in a spiral about phase singularity points. For the first time, a laser beam with phase singularity was generated in 1979 [192]. The interference of two Hermite-Gaussian (HG) modes, denoted as HG_{01} and HG_{10}, in a chromium ion laser's cavity was shown to produce a mode LG_{01}. In 1989, a term "optical vortex" was proposed [193]. In 1990, an optical vortex was experimentally generated by Soskin [194] using an amplitude diffraction grating with a fork. In 1992, a singular laser beam was generated using a spiral phase plate [24].

In 1991, the conversion of a higher-order HG mode into a LG mode with phase singularity using an astigmatic mode converter was reported by Volostnikov [195]. An interference laser mode $\pi/2$-converter was proposed in Ref. [196]. The laser beams with the OAM have found use in particle micromanipulation, quantum telecommunications, microscopy, interferometry, metrology, and so on. A most recent review of the OAM phenomenon can be found in Ref. [197]. A variety of application areas for the vortex laser beams with OAM were also described in the monograph [198]. The generation of the optical vortices by means of interferometers was dealt with in Refs. [199–201], in which an optical vortex array was generated through the interference of three plane waves. Another interferometric way to generate an optical vortex was considered in a very recent Ref. [202], in which a Mach-Zehnder interferometer has been used to obtain a superposition of two Gaussian beams with a tilt between these beams and with different beam axes. It is worth noting that only vortices with a unitary topological charge can be generated in this way.

In Ref. [195], a relationship enabling a LG mode to be represented as a finite superposition of HG modes was derived. By way of illustration, to obtain a LG mode with topological charge 2, one needs to employ the superposition of at least three HG modes with definite complex coefficients. In this section, we demonstrate that it is possible to obtain a light field with an arbitrary integer OAM through the interference of just two HG modes with definite indices.

It is noteworthy that recently an optical laser vortex with the largest OAM and the largest topological charge of 5050 was generated with the aid of an aluminum reflection diffractive optical element [203]. There are different methods for generating laser beams with fractional OAM [3,204,205]. In [204], laser beams with half-integer OAM were formed as a linear combination of the LG modes. In [3], Hermite-Laguerre-Gauss beams with a fractional OAM were generated using an astigmatic mode converter. In [205], a Bessel beam with a smaller-than-unity OAM was generated based on the conical diffraction of a circularly polarized Gaussian beam.

In this section, we calculate the OAM for a linear combination of two HG beams whose double indices are composed of adjacent integer numbers taken in direct and inverse order with a phase shift of $\pi/2$ between the beams. We analyze generalized HG beams, which change to the HG modes and elegant HG beams under certain parameters. It is shown that the modulus of two corresponding HG modes is an integer number, whereas the modulus of the OAM for two elegant HG beams is always equal to unity. In the superposition of two corresponding hybrid HG beams, the modulus of the OAM is always a fractional number. A trivial case of the superposition of two generalized HG modes (0,1) and (1,0) is an exception. In this case, the modulus of the OAM equals unity, similar to the LG modes.

Generalized Hermite-Gaussian Laser Beams

The HG modes have been known in optics for quite a long time [186]. The elegant HG beams, which are described by functions of complex argument, were for the first time proposed in 1973 by Siegman [112]. However, there are HG beams that also represent a solution of the paraxial equation of propagation and can be expressed in an explicit analytical form. Under definite parameters, such beams change to the HG modes [186] and elegant HG beams [112].

Assume that the complex amplitude of light in the initial plane $z=0$ is given by $E_{nm}(x, y) = E_1(x)E_2(y)$, where $E_1(x) = \exp[-(x/a)^2]\ H_n(x/b)$ and $E_2(y) = \exp[-(y/c)^2]\ H_m(y/d)$, where a, b, c, d are real numbers. Considering that the complex amplitude is a product of two functions dependent on the different Cartesian coordinates, the propagation of the entire 2D beam can be considered as the propagation of a 1D beam along any of the transverse coordinates. For such a 1D light field, the complex amplitude at distance z is calculated in the paraxial approximation using a Fresnel transform, being given by

$$E_1(x,z) = \left(\frac{-iz_0}{z}\right)^{1/2} i^n \left(1-\frac{iz_0}{z}\right)^{-(n+1)/2} \left[\left(\frac{a}{b}\right)^2 - 1 + \frac{iz_0}{z}\right]^{n/2} \times \exp\left[-\left(\frac{x}{a(z)}\right)^2 + \frac{ikx^2}{2R(z)}\right] H_n\left(\frac{x}{b(z)}\right), \tag{6.1}$$

where:

$$z_0 = \frac{ka^2}{2}, \quad a(z) = a\left[1+\left(\frac{z}{z_0}\right)^2\right]^{1/2},$$
$$R(z) = z\left[1+\left(\frac{z_0}{z}\right)^2\right], \quad b(z) = b\left(\frac{z}{z_0}\right)\left(1-\frac{iz_0}{z}\right)^{1/2}\left[\left(\frac{a}{b}\right)^2-1+\frac{iz_0}{z}\right]^{1/2}. \tag{6.2}$$

For the complex amplitude $E_2(y, z)$, relations similar to Eqs. (6.1) and (6.2) can be derived by replacing x, n, a, b with y, m, c, d, respectively. From (6.1), (6.2) at $a/b = c/d = \sqrt{2}$ follows a well-known relationship for the HG modes:

$$E_n(x,z) = i^n\left[\frac{a}{a(z)}\right]H_n\left[\frac{\sqrt{2}x}{a(z)}\right]$$
$$\times\exp\left[-\frac{x^2}{a^2(z)}+\frac{ikx^2}{2R(z)}-i(n+1/2)\text{arctg}\left(\frac{z}{z_0}\right)\right], \tag{6.3}$$

In the 2D case, the 1D modes of Eq. (6.3) are multiplied, producing a 2D HG mode $E_{nm}(x, y, z) = E_n(x,z)E_m(y,z)$. If the condition $a/b = c/d = 1$ is observed, Eqs. (6.3) and (6.2) are reduced to an expression for the elegant HG beams [206]:

$$E_e(x,z) = (q(z))^{-(n+1)/2}\exp\left[-\left(\frac{x}{aq(z)}\right)^2\right]H_n\left(\frac{x}{aq(z)}\right), \tag{6.4}$$

where $q(z) = (1 + iz/z_0)^{1/2}$. Note that the generalized HG beams of Eq. (6.1) and the elegant HG beam of Eq. (6.4) are not free-space modes, with the structure of their transverse intensity distribution being changed upon propagation. It stands to reason that the 2D generalized HG beams are generated by multiplying the corresponding functions of Eqs. (6.1), (6.3), and (6.4). It is possible to generate a hybrid HG beam that is described by a HG mode on one coordinate and by an elegant HG beam on the other coordinate:

$$E_h(x,y,z=0) = \exp\left[-\left(\frac{x}{a}\right)^2-\left(\frac{y}{c}\right)^2\right]H_m\left(\frac{\sqrt{2}x}{a}\right)H_n\left(\frac{y}{c}\right). \tag{6.5}$$

The HG beams of Eq. (6.1) are devoid of the OAM. A linear combination of the HG beams with real coefficients also has a zero-valued OAM. Only the linear combination of the HG beam with complex coefficients can have a non-zero OAM. In the subsequent sections, we derive the OAM for the superposition of two generalized HG beams characterized by a phase delay of $\pi/2$.

Orbital Angular Momentum of a Linear Combination of Two Hermite-Gaussian Modes

Assume that the complex amplitude of light in the initial plane is given by

$$E(x,y,0)=\exp\left[-\frac{w^2}{2}\left(x^2+y^2\right)\right]\left[H_{2p}(cx)H_{2s+1}(cy)+i\gamma H_{2s+1}(cx)H_{2p}(cy)\right], \quad (6.6)$$

where w, c, γ are real numbers. The OAM can be derived from [112]:

$$J_z=\operatorname{Im}\left\{\iint_{\mathbb{R}^2}E^*\left(x\frac{\partial E}{\partial y}-y\frac{\partial E}{\partial x}\right)dxdy\right\}. \quad (6.7)$$

Strictly speaking, Eq. (6.7) defines not the entire OAM but its optical-axis projection determined up to a constant and averaged over the transverse plane. In addition, if Eq. (7) were written in the SI system of units, it would have contained the ratio $\varepsilon_0/(2\omega)$ [207], where ε_0 is vacuum permittivity and ω is a cyclic frequency of monochromatic light. This ratio we have put to be equal to unity. Because the OAM of Eq. (6.7) is preserved upon the beam propagation [112], it can be calculated at an arbitrary plane, for example, at $z=0$. Substituting (6.6) into (6.7) yields:

$$\begin{aligned}J_z&=4\gamma c\int_{-\infty}^{+\infty}x\exp\left(-w^2x^2\right)H_{2p}(cx)H_{2s+1}(cx)dx\\&\times\left[2p\int_{-\infty}^{+\infty}\exp\left(-w^2y^2\right)H_{2s+1}(cy)H_{2p-1}(cy)dy\right.\\&\left.-(2s+1)\int_{-\infty}^{+\infty}\exp\left(-w^2y^2\right)H_{2p}(cy)H_{2s}(cy)dy\right].\end{aligned} \quad (6.8)$$

Considering that in Eq. (6.8) the integrands are in the form of polynomials, the integrals can be calculated and represented as finite sums:

$$\begin{aligned}J_z&=\frac{4\pi\gamma\left[(2p)!(2s+1)!\right]^2}{w^{4(p+s+1)}}\sum_{k=0}^{\min(p,s+1)}\frac{\left[(s+1)c^2-kw^2\right]\left(c^2-w^2\right)^{p+s-2k}\left(2c^2\right)^{2k}}{(p-k)!(s+1-k)!(2k)!}\\&\times\left[\sum_{k=0}^{\min(s,p-1)}\frac{\left(2c^2\right)^{2k+1}\left(c^2-w^2\right)^{p+s-2k-1}}{(p-1-k)!(s-k)!(2k+1)!}-\sum_{k=0}^{\min(p,s)}\frac{\left(2c^2\right)^{2k}\left(c^2-w^2\right)^{p+s-2k}}{(p-k)!(s-k)!(2k)!}\right].\end{aligned} \quad (6.9)$$

The relationship (6.9) is cumbersome, making it hard to conclude when, specifically, the OAM will take integer, fractional, or zero values. What it only allows is making some conclusions under certain conditions. For instance, if $p > s$, $c = w + \delta$, $\delta \ll w$, then at $(p - c)(2s + 1)^{-1}w > \delta$ in Eq. (6.9) $\mathrm{Im}(J_z) > 0$. If in Eq. (6.9) $c = w$, all sums get cancelled, with only terms with maximal number k being retained. From Eq. (6.6) it is seen that at $c = w$, two HG modes with permuted indices are superimposed. Considering that the HG modes are orthogonal, one can infer that the non-zero OAM can be obtained only for a linear combination of modes with two consecutive indices, that is, when $p = s$. In this case, Eq. (6.9) is replaced with

$$J_z = -\frac{2^{4p+2}\pi\gamma\left[(2p+1)!\right]^2}{w^2}. \tag{6.10}$$

For the OAM to be independent of the laser beam power let us analyze the OAM normalized with respect to the intensity. The power of the beam in Eq. (6.6) is described by the relation ($c = w$, $p = s$):

$$I = \iint_{\mathbb{R}^2} E^* E dx dy = \frac{\pi\left(1+\gamma^2\right)}{w^2} 2^{1+4p}(2p)!(2p+1)!. \tag{6.11}$$

Thus, the normalized OAM (the OAM divided by the beam's power) for the linear combination of two HG modes with permuted adjacent indices is

$$\frac{J_z}{I} = -\left(\frac{2\gamma}{1+\gamma^2}\right)(2p+1). \tag{6.12}$$

From Eq. (6.12) it follows that at $\gamma = 1$, the modulus of the OAM of the linear combination of two HG modes

$$E_m(x,y) = \exp\left[-\frac{w^2}{2}\left(x^2+y^2\right)\right]\left[H_{2p}(wx)H_{2p+1}(wy) + iH_{2p+1}(wx)H_{2p}(wy)\right] \tag{6.13}$$

is an integer number

$$\frac{J_z}{I} = -(2p+1). \tag{6.14}$$

Note that at some values of p, the OAM will also be an integer number even when $\gamma \neq 1$. For instance, at $\gamma = 1/2$, the OAM in Eq. (6.12) will be integer at $p = 2$: $J_z/I = -4$, at $p = 7$: $J_z/I = -12$, and so on.

For the linear combination of the HG modes given by

$$E_m(x,y,0) = \exp\left[-\frac{w^2}{2}\left(x^2+y^2\right)\right]\left[H_n(wx)H_{n+1}(wy) + i\gamma H_{n+1}(wx)H_n(wy)\right] \tag{6.15}$$

the normalized OAM for any integer n can be derived in a similar way to Eq. (6.14):

$$\frac{J_z}{I} = -\frac{2\gamma(n+1)}{1+\gamma^2}. \tag{6.16}$$

Note that because for the two modes in Eq. (6.15) the sum of the indices is the same (with the two modes also having the same Gouy phase defined as $(m + n + 1)$ $\arctan(z/z_0)$), the linear combination in Eq. (6.15) will also form a mode, which will preserve its form upon propagation, changing only in scale. Thus, a remarkable result is arrived at: the mode in Eq. (6.15) has an OAM, meaning that the Poynting vector is locally describing a spiral about the optical axis in space; in the meantime, the beam is not rotating, preserving its structure upon propagation.

The beam of Eq. (6.6) or Eq. (6.15) at $c = w$ can be experimentally generated using a Mach-Zehnder interferometer. The HG mode $E_{nm}(x,y)$ generated at the laser output is divided by a 50%-mirror into two identical beams that are coupled into different interferometers' arms. In one interferometer's arm, the HG mode is rotated by 90° using a Duve prism, forming the mode $E_{mn}(x, y)$. At the interferometer's output, the modes are superimposed into a single beam with a relative phase delay of $\pi/2$.

Orbital Angular Momentum of a Linear Combination of Two Elegant Hermite-Gaussian Beams

Let us calculate the OAM for a linear combination of two elegant HG beams. To these ends, we put in Eq. (6.6) $c = w/\sqrt{2}$ and $p = s$, obtaining:

$$E_e(x, y) = \exp\left[-\frac{w^2}{2}(x^2 + y^2)\right] \times\left[H_{2p}\left(\frac{wx}{\sqrt{2}}\right)H_{2p+1}\left(\frac{wy}{\sqrt{2}}\right) + i\gamma H_{2p+1}\left(\frac{wx}{\sqrt{2}}\right)H_{2p}\left(\frac{wy}{\sqrt{2}}\right)\right]. \tag{6.17}$$

Then, the normalized OAM similar to that in Eq. (6.12) takes the form:

$$\frac{J_z}{I} = -\frac{2\gamma}{1+\gamma^2}. \tag{6.18}$$

Equation (6.18) suggests that the linear combination of two elegant HG beams (6.17) will always have an OAM equal to unity in modulus (at $\gamma = 1$) at all possible values of the number p.

This is an extremely unexpected result. It turns out that the OAM in Eq. (6.14) for two HG modes is determined by the maximal number of the constituent mode of the linear combination. Consequently, the larger the number of the HG mode, the larger the laser beam's OAM in Eq. (6.13). For the elegant beams in Eq. (6.17), it follows from Eq. (6.18) that the OAM is determined by the difference of two indices of the constituent beams of the linear combination. Thus, because the difference of the beam indices in Eq. (6.17) equals unity, the OAM also equals unity in modulus (at $\gamma = 1$).

For the linear combination of elegant modes with different numbers (let $k = 2l + 1$ be odd)

$$E_e(x, y) = \exp\left[-\frac{w^2}{2}(x^2 + y^2)\right] \times \left[H_{2p}\left(\frac{wx}{\sqrt{2}}\right)H_{2p+k}\left(\frac{wy}{\sqrt{2}}\right) + i\gamma H_{2p+k}\left(\frac{wx}{\sqrt{2}}\right)H_{2p}\left(\frac{wy}{\sqrt{2}}\right)\right], \tag{6.19}$$

the OAM is obtained in the form:

$$\frac{J_z}{I} = \left(-\frac{\gamma}{1+\gamma^2}\right)\frac{k(4p+k)\Gamma^2(2p+k/2)}{\Gamma(2p+1/2)\Gamma(2p+k+1/2)}, \tag{6.20}$$

where $\Gamma(x)$ is the gamma-function. Equation (6.20) is identical to Eq. (6.18) at $k = 1$. For the subsequent number $k = 3$, from Eq. (6.20) it follows:

$$\frac{J_z}{I} = \left(-\frac{2\gamma}{1+\gamma^2}\right)\frac{3(4p+1)}{(4p+5)}. \tag{6.21}$$

At $\gamma = 1$ and large p, the modulus of the OAM (6.21) is close to 3, that is, close to the difference of numbers of the elegant HG beam of Eq. (6.19).

Orbital Angular Momentum of a Linear Combination of Two Hybrid Hermite-Gaussian Beams

In the following, by the hybrid HG beam is meant a beam which is described by a HG mode on one axis and by an elegant HG mode on the other axis. In this case, there are two variants of the linear combination of the hybrid beams.

Let us analyze the sum of two hybrid HG beams in which the HG mode has a larger number than the elegant HG beam:

$$E_{h1}(x, y) = \exp\left[-\frac{w^2}{2}(x^2 + y^2)\right] \times \left[H_{2p}\left(\frac{wx}{\sqrt{2}}\right)H_{2p+1}(wy) + i\gamma H_{2p+1}(wx)H_{2p}\left(\frac{wy}{\sqrt{2}}\right)\right]. \tag{6.22}$$

Then, the normalized OAM of the beam in Eq. (6.22) is given by

$$\frac{J_z}{I} = -\left(\frac{2\gamma}{1+\gamma^2}\right)\frac{(2p+1)!}{(4p-1)!!}. \tag{6.23}$$

Because the numerator of Eq. (6.23) will always contain both even and odd terms (due to $(2p + 1)!$), whereas the denominator will only contain odd terms (due to $(4p - 1)!!$), the OAM of Eq. (6.23) will never be an integer at $\gamma = 1$, except for a trivial case of $\gamma = 1$ and $p = 0$. For instance, at $p = 2$ and $\gamma = 1$, from Eq. (6.23) follows $J_z/I = -8/7$. This OAM is slightly larger than unity in modulus.

In the other case, when the HG mode has a smaller number than the elegant HG beam in the superposition of two hybrid HG beams

$$E_{h2}(x,y) = \exp\left[-\frac{w^2}{2}\left(x^2+y^2\right)\right] \times\left[H_{2p}(wx)H_{2p+1}\left(\frac{wy}{\sqrt{2}}\right)+i\gamma H_{2p+1}\left(\frac{wx}{\sqrt{2}}\right)H_{2p}(wy)\right], \tag{6.24}$$

Eq. (6.23) for the OAM is replaced with

$$\frac{J_z}{I} = \left(\frac{2\gamma}{1+\gamma^2}\right)\frac{(2p+1)!(2p+1)(p^2-1)}{(4p+1)!!}. \tag{6.25}$$

Note that the OAM of Eq. (6.25) has the opposite sign with respect to all above-calculated OAMs. From Eq. (6.25) it also follows that the OAM for the beam in Eq. (6.24) will always be a fractional number, except for a trivial case of $\gamma = 1$ and $p = 0$. For example, at $p = 2$ and $\gamma = 1$, Eq. (6.25) yields $J_z/I = 40/21$. This OAM is slightly smaller than two in modulus. It should be noted that in the above-considered cases the change of sign of the parameter γ results in the change of sign of the OAM. Note also that the common factor in Eqs. (6.12), (6.18), (6.21), (6.23), and (6.24), $2\gamma/(1+\gamma^2)$ is always smaller than or equal to unity.

Numerical Simulation

For simulation, a linear combination of two generalized HG beams will be considered. Assume that in the initial plane ($z = 0$) the light field has the complex amplitude

$$E(x,y,0) = \exp\left(-\frac{x^2+y^2}{w^2}\right)\left[H_{2p}(bx)H_{2s+1}(cy)+i\gamma H_{2s+1}(cx)H_{2p}(by)\right]. \tag{6.26}$$

The total intensity of the beam (6.26) for wavelength λ and $p = s = 2$ (i.e., for the Hermite polynomials of fourth and fifth degree) in the initial plane (at $-7\lambda \le x \le 7\lambda$, $-7\lambda \le x \le 7\lambda$) at some values of the scaling coefficients b and c is shown in Figures 6.1 through 6.4.

Figures 6.1b through 6.4b depict the intensity of the beams of Eq. (6.26) coherently superimposed with an oblique plane wave

$$I(x,y,z=0) = \left|E(x,y,0)+C\exp(i\alpha x)\right|^2, \tag{6.27}$$

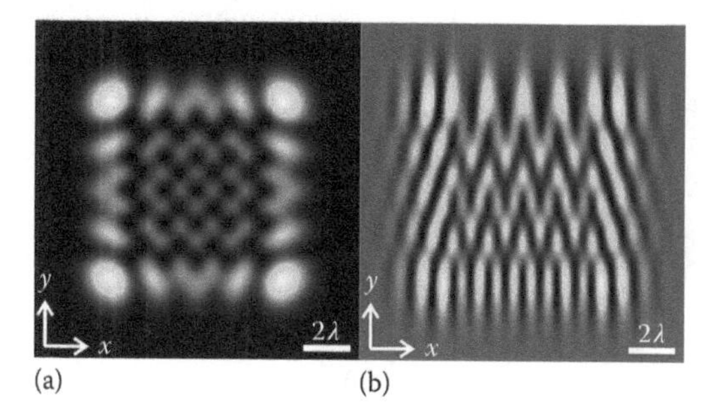

FIGURE 6.1 Intensity of the beam (6.26) (a) without and (b) with the carrier frequency at $w = 2\lambda$, $b = c = \sqrt{2}/w$ (two HG modes).

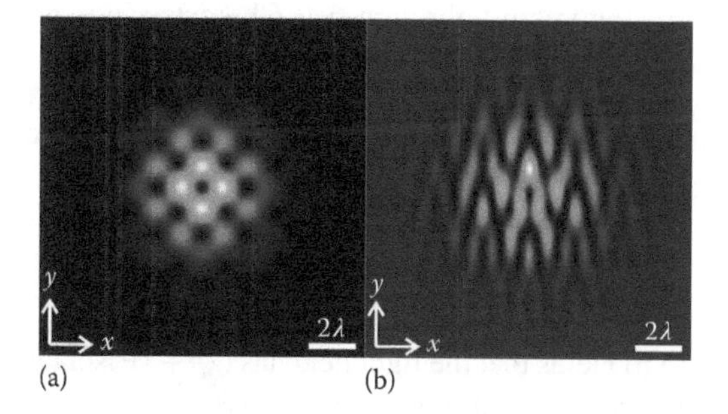

FIGURE 6.2 Intensity of the beam (6.26) (a) without and (b) with the carrier frequency at $w = 2\lambda$, $b = c = 1/w$ (two elegant HG beams).

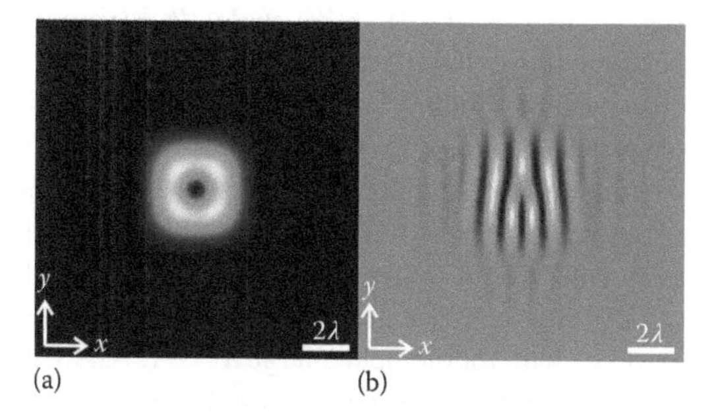

FIGURE 6.3 Intensity of the beam (6.26) (a) without and (b) with the carrier frequency at $w = 2\lambda$, $b = 1/(7\lambda)$, $c = 1/(3\lambda)$ (two generalized HG beams with different width on the axes).

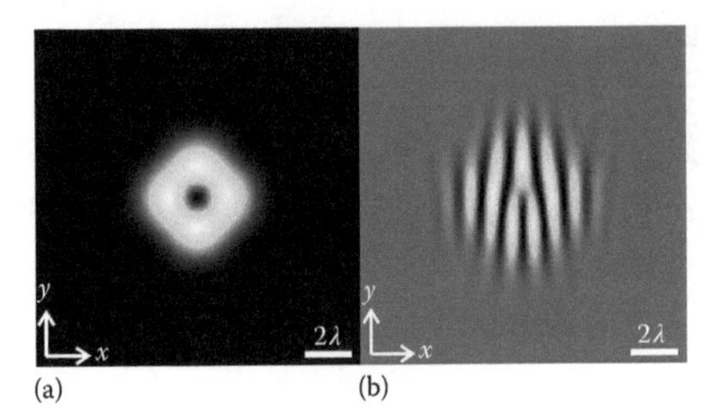

FIGURE 6.4 Intensity of the beam (6.26) (a) without and (b) with the carrier frequency at $w = 2\lambda$, $b = 1/(5\lambda)$, $c = 1/(5\lambda)$ (two generalized HG beams with equal width on the axes).

where the amplitude C and the spatial frequency α were chosen so as to ensure a better visualization of Figures 6.1 through 6.4. Characteristic "forks" that can be observed in Figures 6.1b through 6.4b in the interference fringes show the location of isolated intensity zeroes and phase singularities. The modules of the OAM for the beams in question are equal to 5 (Figure 6.1a), 1 (Figure 6.2a), 0.95 (Figure 6.3a), and 0.92 (Figure 6.4a).

Discussion

Putting $p = s$ in Eq. (6.6) yields that the light field has $(2p+1)^2$ isolated intensity zeros with topological charge $n = 1$ found at the intersections of the horizontal lines formed by zeros of the polynomial $H_{2p+1}(y)$ and the vertical lines of zeros of the polynomial $H_{2p+1}(x)$, also having $(2p)^2$ isolated zeros with topological charge $n = -1$ found at the intersections of the horizontal lines of zeros of the polynomial $H_{2p}(y)$ and the vertical lines of zeros of the polynomial $H_{2p}(x)$. Thus, if all the said zeros were grouped very near to the optical axis, the maximum OAM of the field (6.6) would have been equal to the difference $(2p + 1)^2 - (2p)^2 = 4p + 1$. However, in practice the OAM of the field of Eq. (6.6) is defined by Eqs. (6.12), (6.18), (6.21), (6.23), and (6.25).

A visual picture of how the isolated zeros of the field (6.6) are arranged can be illustrated by a specific example. Figure 6.5 depicts (a) an interference pattern similar to that shown in Figure 6.1b, except for the 2.5 times increased carrier frequency, and (b) the phase distribution for the intensity pattern in Figure 6.5a.

Figure 6.5 suggests that for $p = s = 2$ the modal beam in Eq. (6.26) has five horizontal rows of isolated zeros, or optical vortices with topological charge $n = +1$ (with each row having five zeros), and four additional rows located in-between, each having four optical vortices with topological charge $n = -1$. All in all, in the patterns shown in Figure 6.1 or 6.5, 25 optical vortices with $n = +1$ and 16 optical vortices with $n = -1$ are seen to be formed. The maximal OAM that an optical beam with such a number of zeros can have equals the difference of the "right" and "left" isolated intensity zeros, thus being equal to $25 - 16 = 9$. However, from Eq. (6.12), the OAM is seen to be equal to 5 (Figures 6.1 and 6.5). The fact is that the contribution of an individual isolated zero into the OAM decreases if the zero in question

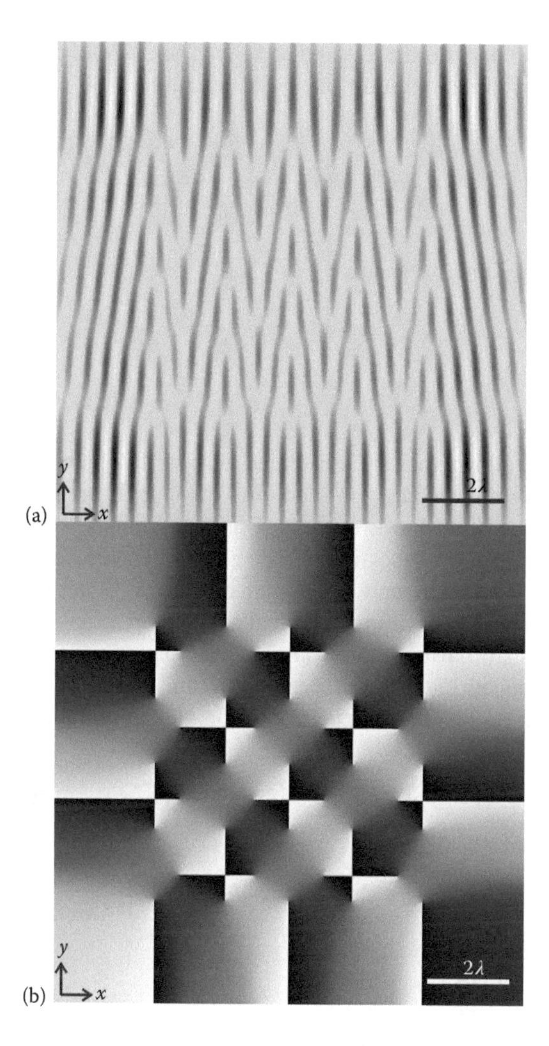

FIGURE 6.5 Patterns of (a) intensity and (b) phase of the beam (6.26) with a carrier frequency for the parameters $w = 2\lambda$, $b = c = \sqrt{2}/w$ (mode).

is moving away from the optical axis toward the Gaussian beam's periphery. To demonstrate this, we shall analyze a simple case of a Gaussian beam that contains a single isolated zero with topological charge $n = +1$, found on the x-axis at distance $r_0 > 0$ from the center. Then, the complex amplitude of the field is given by

$$\begin{aligned} E(x, y) &= \exp\left(-\frac{x^2 + y^2}{2\sigma^2}\right)\left[(x - r_0) + iy\right] \\ &= \exp\left(-\frac{r^2}{2\sigma^2}\right)\left[r\exp(i\phi) - r_0\right], \end{aligned} \tag{6.28}$$

where (r, φ) are the polar coordinates and $\sigma = 1/w$. The OAM of the field in Eq. (6.28) can be derived from a relation in the polar coordinates:

$$\frac{J_z}{I} = \text{Im}\left\{\int_0^{2\pi} d\phi \int_0^{\infty} r dr E^* \frac{\partial E}{\partial \phi}\right\}\left[\int_0^{2\pi} d\phi \int_0^{\infty} r dr |E|^2\right]^{-1}. \tag{6.29}$$

Substituting (6.28) into (6.29) yields:

$$\frac{J_z}{I} = \frac{1}{1 + \left(\frac{r_0}{\sigma}\right)^2}. \tag{6.30}$$

From Eq. (6.30) it follows that if the isolated zero with topological charge $n = +1$ is found on the optical axis at $r_0 = 0$, the OAM equals 1, but if the optical vortex is displaced from the optical axis onto the Gaussian beam's periphery ($r_0 > 0$), the OAM decreases in inverse proportion to the square of the distance from the optical axis. The OAM is reduced by half if the intensity zero is offset by a distance of the Gaussian beam's waist $r_0 = \sigma$ from the optical axis.

Let us analyze a similar but more complex example: two isolated identical zeros with $n = +1$ introduced into the Gaussian beam and found on the horizontal axis symmetrically with respect to the center. The complex amplitude of the field is

$$\begin{aligned} E(x, y) &= \exp\left(-\frac{x^2 + y^2}{2\sigma^2}\right)\left[(x - r_0) + iy\right]\left[(x + r_0) + iy\right] \\ &= \exp\left(-\frac{r^2}{2\sigma^2}\right)\left[r^2 \exp(2i\phi) - r_0^2\right], \end{aligned} \tag{6.31}$$

Substituting (6.31) into (6.29) yields:

$$\frac{J_z}{I} = \frac{4}{2 + \left(\frac{r_0}{\sigma}\right)^4}. \tag{6.32}$$

From Eq. (6.32) it follows that when both zeros with $n = +1$ are simultaneously found on the optical axis, the OAM equals 2. If both zeros are symmetrically moving away from the optical axis, $r_0 > 0$, the OAM (6.32) decreases as the fourth degree of the distance from the optical axis. The OAM will be equal to 1 at $r_0/\sigma = 2^{1/4} \approx 1.19$. Thus, the OAM is reduced by half if both zeros are offset from the optical axis by a distance somewhat larger than the Gaussian beam's waist radius.

Therefore, we can infer that the contribution of the optical vortices in Figures 6.1 and 6.5 will be smaller than $25 - 16 = 9$, because all of them, except for the central zero, are found at different distances from the Gaussian beam's center.

Equations (6.30) and (6.32) also explain why the linear combination of the elegant HG beams has a smaller OAM. From comparison of (6.6) with (6.17) and Figure 6.1

with Figure 6.2, the Gaussian beam's waist of the elegant HG beams is seen to be half as large as that of the HG modes. This is the reason why the isolated zeros (optical vortices) in Figure 6.2, while found at the same distance from the optical axis as those in Figure 6.5, turn out to be located on the periphery of the Gaussian beam with the two-times smaller waist radius. Because of this, the OAM of the beam in Figure 6.2 equals just 1 rather than 5, as is the case in Figure 6.1.

What remains to find out is why the beam of Eq. (6.13) or Eq. (6.15), which has an OAM and has no radial symmetry, does not rotate upon propagation?

An insight into this question can be gained using simple examples. Let us return to the simplest example of Eq. (6.28), when an optical vortex (i.e., an isolated intensity zero) is offset from the optical axis and introduced into the Gaussian beam. Such a field in Eq. (6.28) has an OAM and no radial symmetry. By calculating the Fresnel transform of the complex amplitude (6.28), the intensity zero can be shown to rotate and move away from the optical axis upon propagation in accordance with the relation:

$$\begin{cases} \tan\theta = \dfrac{z}{z_0}, \\ \rho = r_0\sqrt{1+\left(\dfrac{z}{z_0}\right)^2}, \end{cases} \tag{6.33}$$

where:

$z_0 = k\sigma^2$ is the Rayleigh range

(ρ, θ) are the polar coordinates in a plane perpendicular to the optical axis at distance z from the waist

From Eq. (6.33), the isolated intensity zero (6.28) is seen to rotate during propagation. At $z = z_0$, the zero will rotate anti-clockwise by the angle $\theta = \pi/4$, rotating by the angle $\theta = \pi/2$ at $z \to \infty$. The second equation in (6.33) shows that as the beam propagates, the intensity zero will be moving away from the optical axis as fast as the Gaussian beam is diverging.

Let us analyze a more complex example in Eq. (6.31), when two isolated intensity zeros with the same topological charge are introduced into the Gaussian beam. The light field in Eq. (6.31) also has the OAM (6.32), being devoid of radial symmetry. Similarly to Eq. (6.33), the formula to describe the rotation and moving away from the optical axis of two intensity zeros in Eq. (6.31) during propagation is given by

$$\begin{cases} \tan(2\theta) = \left(\dfrac{2z}{z_0}\right)\left[1-\left(\dfrac{z}{z_0}\right)^2\right]^{-1/2}, \\ \rho = r_0\sqrt{1+\left(\dfrac{z}{z_0}\right)^2}. \end{cases} \tag{6.34}$$

From Eq. (6.34), both intensity zeros are seen to rotate anti-clockwise as a whole during propagation. As previously, at $z = z_0$, the zeros will rotate by the angle $\theta = \pi/4$, rotating by the angle $\theta = \pi/2$ at infinity $z \to \infty$.

If two intensity zeros have the topological charges of opposite sign, they show the entirely different behavior during propagation. In this case, it is impossible to derive a relationship similar to Eq. (6.33) or (6.34). The reason is that depending on where the intensity zeros are located in the Gaussian beam, the OAM can be positive, negative, or zero. For instance, assume that there are two isolated intensity zeros with the topological charges of $n = 1$ and $n = -1$ introduced into the Gaussian beam's waist and located on the horizontal axis at points r_0 and $-r_1$. At $z = 0$, the complex amplitude of such a field is

$$\begin{aligned} E(x, y) &= \exp\left(-\frac{x^2 + y^2}{2\sigma^2}\right)\left[(x - r_0) + iy\right]\left[(x + r_1) - iy\right] \\ &= \exp\left(-\frac{r^2}{2\sigma^2}\right)\left[r^2 - r_0 r_1 + r r_1 \exp(i\phi) - r r_0 \exp(-i\phi)\right]. \end{aligned} \tag{6.35}$$

The OAM of the field (6.35) takes the form:

$$\frac{J_z}{I} = \frac{\left(r_1^2 - r_0^2\right)}{\left(r_1 - r_0\right)^2 + 2\sigma^2 + \left(\frac{r_0 r_1}{\sigma}\right)^2}. \tag{6.36}$$

From Eq. (6.36) it follows that the OAM of the field (6.35) is equal to zero at $r_1 = r_0$, to a positive number at $r_1 > r_0$, and to a negative number at $r_1 < r_0$. At $r_1 \to \infty$, the OAM (6.36) is coincident with the OAM of a single isolated intensity zero in Eq. (6.30).

From the above, one may infer that two intensity zeros with the topological charges of opposite sign can rotate clockwise, anti-clockwise, or not at all during propagation. With a larger number of optical vortices, as in Figure 6.5, the number of differently arranged combinations of the intensity zeros will essentially increase, consequently increasing the number of feasible behavioral patterns of the zeros during propagation. Therefore, the question of why the light beam of Eq. (6.13) or (6.15), while having the OAM and being devoid of radial symmetry, does not rotate during propagation can be addressed only in general.

A feasible explanation can be found in Ref. [208], where using the expansion of the light field in terms of the LG modes, the rotation condition of the laser beam intensity cross-section during propagation (Eq. (6.3) in [208]) was shown to be essentially different from the non-zero OAM condition of the beam (Eq. (6.4) in [208]). One can always find laser beams (as a combination of a finite number of LG modes) that rotate during propagation, while being devoid of the OAM and, vice versa, there are laser beams that have the OAM but do not rotate during propagation. The light fields in Eqs. (6.13) and (6.15) are the illustration of such beams. It should be noted that in Ref. [3] the Hermite-Laguerre-Gaussian beams have the same properties as the beams (6.13) and (6.15), that is, these beams are spatial modes that do not rotate upon propagation but have the OAM.

Let us also note that if instead of Eq. (6.26) at $s = p$, we add the HG beams with permuted indices:

$$\tilde{E}(x,y,0) = \exp\left[-\frac{w^2}{2}(x^2+y^2)\right] \times\left[H_{2p+1}(bx)H_{2p}(cy)+i\gamma H_{2p}(cx)H_{2p+1}(by)\right], \tag{6.37}$$

then for the superposition of modes ($b = c = w$), the normalized OAM (Eq. (6.12)) only changes sign:

$$\frac{J_z}{I} = \left(\frac{2\gamma}{1+\gamma^2}\right)(2p+1), \tag{6.38}$$

for the superposition of elegant beams ($b = c = w/\sqrt{2}$), the normalized OAM (Eq. (6.18)) only changes sign as well:

$$\frac{J_z}{I} = \frac{2\gamma}{1+\gamma^2}, \tag{6.39}$$

while for two hybrid HG beams the normalized OAMs not only change sign but also swap. For $b = w/\sqrt{2}$, $c = w$ instead of Eq. (6.23) we have

$$\frac{J_z}{I} = -\left(\frac{2\gamma}{1+\gamma^2}\right)\frac{(2p+1)!(2p+1)(p^2-1)}{(4p+1)!!}, \tag{6.40}$$

and for $b = w$, $c = w/\sqrt{2}$ instead of Eq. (6.25) we obtain

$$\frac{J_z}{I} = \left(\frac{2\gamma}{1+\gamma^2}\right)\frac{(2p+1)!}{(4p-1)!!}. \tag{6.41}$$

Thus, the following results have been obtained in this section. An expression for the complex amplitude of the generalized paraxial Hermite-Gauss beams of Eq. (6.1) has been derived. These beams have been shown to change to the familiar HG modes and elegant HG beams under certain parameters. An OAM of a linear combination of two generalized HG beams with double indices composed of two adjacent integer numbers taken in direct and inverse order and a phase delay of $\pi/2$ has been calculated. The modulus of the OAM has been shown to be an integer number for the HG mode for the superposition of two HG modes, always equal to unity for the superposition of two elegant HG beams, and a fractional number for two hybrid HG beams.

6.2 VORTEX HERMITE-GAUSSIAN LASER BEAMS

The laser beams with the OAM have found use in particle micromanipulation, quantum telecommunications, microscopy, interferometry, and metrology. A most recent review of the OAM phenomenon can be found in Ref. [197]. In 1992 it was shown by Allen [17] that the LG modes have an OAM, while the HG modes have zero OAM. In this work, we show that a definite superposition of HG beams has the OAM. HG modes have been around in optics since 1966 [186]. The elegant HG beams that were first introduced by A. E. Siegman in 1973 [206] are described by functions with a complex argument. Generalized HG beams, which also represent the solution of a paraxial equation of propagation in an explicit analytic form, have also been proposed [187,188]. At certain parameters, the said beams change to the HG modes [186] and elegant HG beams [206]. If the radii of the Gaussian beams are different in Cartesian coordinates, then the generalized HG beams become elliptical HG beams [209]. Using an astigmatic mode converter, a higher-order HG mode was reported to be transformed into a LG mode with phase singularity [195]. The use of astigmatic laser mode $\pi/2$-converters has been discussed [196]. A relation that allows the LG mode to be expressed as a finite sum of HG modes has also been derived [195]. Using an astigmatic mode converter [3], Hermite-Laguerre-Gaussian beams with fractional OAM have been generated.

In this section, we study vortex Hermite-Gaussian (vHG) beams, which are formed as a superposition of $(n + 1)$ generalized HG beams. In the vHG beams, the complex amplitude is proportional to the nth order Hermite polynomial with its argument being a function of a real parameter a. For $|a| < 1$, on the vertical axis of the beam cross-section there are n isolated optical nulls that produce optical vortices with topological charge +1 ($a < 0$) or −1 ($a > 0$). For $|a| > 1$, similar isolated optical nulls of the vHG beams are found on the horizontal axis. When $|a| = 1$, all n isolated optical nulls are collected on the optical axis at the beam center, producing an nth order optical vortex and making the vHG mode similar to the LG mode of the order $(0, n)$; whereas at $a = 0$, the vHG mode is coincident with the HG mode of the order $(0, n)$. In the following, we derive the OAM of the vHG beams, which is found to be a function of the parameter a and vary from 0 (at $a = 0$ and $a \to \infty$) to n ($a = 1$). We show that two vHG beams with different numbers n and m are orthogonal, whereas two same-number modes with different parameter a are unorthogonal.

Hermite-Gaussian Beam with Complex Argument

In Refs. [187,188] generalized HG beams were proposed. In the one-dimensional case, their complex amplitude reads as

$$E_1(x,z) = \left(\frac{-iz_0}{z}\right)^{1/2} i^n \left(1 - \frac{iz_0}{z}\right)^{-(n+1)/2} \left[\left(\frac{p}{c}\right)^2 - 1 + \frac{iz_0}{z}\right]^{n/2} \times \exp\left[-\left(\frac{x}{p(z)}\right)^2 + \frac{ikx^2}{2R(z)}\right] H_n\left(\frac{x}{c(z)}\right), \tag{6.42}$$

where:

$$z_0 = \frac{kp^2}{2},\ p(z) = p\left[1+\left(\frac{z}{z_0}\right)^2\right]^{1/2},\ R(z) = z\left[1+\left(\frac{z_0}{z}\right)^2\right],$$
$$c(z) = c\left(\frac{z}{z_0}\right)\left(1-\frac{iz_0}{z}\right)^{1/2}\left[\left(\frac{p}{c}\right)^2 - 1 + \frac{iz_0}{z}\right]^{1/2}, \tag{6.43}$$

p is the Gaussian beam waist radius, $R(z)$ is the Gaussian beam's wavefront curvature radius, z_0 is the Rayleigh range, k is the wave number of light, $H_n(x)$ is the Hermite polynomial, and c is the scaling factor. In the 2D case, the 1D beams of Eq. (6.42) are multiplied, producing a 2D HG beams $U_{nm}(x, y, z) = U_n(x, z)U_m(y, z)$. The HG beams in Eq. (6.42) are devoid of OAM. A linear combination of the HG beams with real coefficients also has a zero-valued OAM. The OAM can only take non-zero values for a linear combination of the HG beams with complex coefficients [188,195]. In the following, we analyze a linear combination of the HG beams of Eq. (6.42) at $z=0$ for definite coefficients:

$$U_n(x,y,z=0) = i^n \exp\left(-\frac{x^2}{p^2}-\frac{y^2}{q^2}\right)\left(1+a^2\right)^{-n/2}$$
$$\times \sum_{p=0}^{n} \frac{n!(ia)^p}{p!(n-p)!} H_p\left(\frac{x}{c}\right) H_{n-p}\left(\frac{y}{d}\right), \tag{6.44}$$

where:

- a is a real parameter
- q is the Gaussian beam waist radius along the y-axis
- c, d are the scaling factors

Making use of the reference relationship [11]

$$\sum_{p=0}^{n} \frac{n!t^p}{p!(n-p)!} H_p(\xi)H_{n-p}(\eta) = \left(1+t^2\right)^{n/2} H_n\left(\frac{t\xi+\eta}{\sqrt{1+t^2}}\right), \tag{6.45}$$

we reduce Eq. (6.44) to the relation for the complex amplitude of the vHG beam:

$$U_n(x,y,z=0) = i^n \exp\left(-\frac{x^2}{p^2}-\frac{y^2}{q^2}\right)\left(\frac{1-a^2}{1+a^2}\right)^{n/2} H_n\left[\frac{iadx+cy}{cd\sqrt{1-a^2}}\right]. \tag{6.46}$$

At $p=q=w$ and $c=d=w/\sqrt{2}$ the beams (6.46) are vHG modes, which preserve their shape upon propagation. These vortex modal beams are a special case of the

Hermite-Laguerre-Gaussian modes in [3]. At $p = c\sqrt{2} = w_x$ and $q = d\sqrt{2} = w_y$ the beams (6.46) are elliptical vHG beams and instead of Eq. (6.46) their complex amplitude reads as

$$U_n(x, y, z = 0) = i^n \exp\left(-\frac{x^2}{w_x^2} - \frac{y^2}{w_y^2}\right)\left(\frac{1-a^2}{1+a^2}\right)^{n/2} H_n\left[\sqrt{2}\frac{iaw_y x + w_x y}{w_x w_y \sqrt{1-a^2}}\right]. \quad (6.47)$$

At $p = c = q = d = w$, the beams (6.46) are elegant vHG beams.

Orbital Angular Moment of Vortex Hermite-Gaussian Beam

In the following, we derive a relationship for the OAM of the light field in Eq. (6.47), making use of the formula [17,112,197] (up to a constant multiplier):

$$J_z = -i\iint_{\mathbb{R}^2} U^*\left(x\frac{\partial U}{\partial y} - y\frac{\partial U}{\partial x}\right)dxdy. \quad (6.48)$$

Strictly speaking, Eq. (6.48) defines not the entire OAM but its projection onto the optical axis. As the beam propagates, OAM in Eq. (6.48) is preserved [112], that is, it can be calculated in an arbitrary plane, for example at $z = 0$. Substituting (6.47) into (6.48) yields:

$$J_z = -\pi 2^{n-1} n! na\left(1 + a^2\right)^{-1}\left(w_x^2 + w_y^2\right). \quad (6.49)$$

To obtain the OAM independent of the laser beam power let us consider the OAM normalized to intensity. The power of the beam in Eq. (6.46) is given by

$$I = \iint_{\mathbb{R}^2} E^* E dxdy = \pi 2^{n-1} n! w_x w_y. \quad (6.50)$$

Thus, the normalized OAM of the vHG beam is obtained by dividing by the beam power

$$\frac{J_z}{I} = \left(\frac{-na}{1+a^2}\right)\left(\frac{w_x^2 + w_y^2}{w_x w_y}\right). \quad (6.51)$$

Further, we introduce a parameter of a Gaussian beam ellipticity ε: $w_x = w$, $w_y = \varepsilon w$. So, Eq. (6.51) now reads as follows

$$\frac{J_z}{I} = -n\left(\frac{a}{1+a^2}\right)\left(\frac{1+\varepsilon^2}{\varepsilon}\right). \quad (6.52)$$

It is seen in Eq. (6.52) that the OAM tends to infinity when $\varepsilon \to \infty$ or $\varepsilon \to 0$. Eq. (6.52) can be used for obtaining laser beams with a different transverse intensity distribution, but with the same OAM. For equal radii of the Gaussian beam ($w_x = w_y = w$), Eq. (6.52) becomes simpler:

$$\frac{J_z}{I} = \frac{-2na}{1+a^2} = \frac{-2nb}{1+b^2}, \quad b = a^{-1}. \tag{6.53}$$

It is seen from Eq. (6.53) that the sign of the OAM depends on the sign of the product na. It is interesting to note that the OAM of Eq. (6.53) is coincident with that of a beam built as a linear combination of just two HG modes, which have recently been considered by the present authors [188]. From Eq. (6.53) it follows that at $a = 1$ the OAM of the vHG modes is defined by an integer $J_z/I = -n$. From Eq. (6.53), one can infer that because at $a = 1$ the vHG mode is similar to a standard LG mode with the number $(0, n)$, its OAM equals in modulus the topological charge of an optical vortex $\exp(in\varphi)$.

Computer Simulation

Using Eq. (6.47), we calculated the intensity and phase of the vHG modes in the initial plane ($z = 0$) at different values of the topological charge n and parameter a. The simulation results shown in Figure 6.6 were obtained for the monochromatic light wavelength $\lambda = 532$ nm, topological charge (or Hermite polynomial's order) $n = 3$ (a–l) and $n = 10$ (m–r), Gaussian beam waist radius $w_x = w_y = w = 10\lambda$ (at $z = 0$), and the computation domain $-25\lambda \le x, y \le +25\lambda$ (a–l) and $-40\lambda \le x, y \le +40\lambda$. Figure 6.6 depicts (a–c, g–i, m–o) the intensity and (d–f, j–l, p–r) phase patterns of the vHG mode in the initial plane ($z = 0$) at different parameters a: (a, d) 0.1, (b, e) 0.25, (c, f) 0.5, (g, j) 0.75, (h, k) 1.0, (i, l) 1.25, (m, p) 0.1, (n, q) 0.25, and (o, r) 0.5. For the modes presented in the illustrations, the OAM was derived using Eq. (6.53), being equal to $J_z/I = -2.4$ (Figure 6.6a and d), $J_z/I = -2.77$ (Figure 6.6b and e), and $J_z/I = -3$ (Figure 6.6c and f). The plot of OAM as a function of the a parameter is shown in Figure 6.7.

From Figure 6.7, at a given n, the OAM is seen to take the maximal value at $a = 1$.

Isolated Optical Nulls of Vortex Hermite-Gaussian Beam

Equation (6.47) enables deriving the coordinates of the vHG beam optical nulls in the initial plane. By equating the Hermite polynomial's argument to the value of the polynomial's root γ_{nk}, we obtain $H_n(\gamma_{nk}) = 0$.

Then, the isolated optical nulls will be found at the coordinates:

$$x_{nk} = 0, \quad y_{nk} = \gamma_{nk}\left(\frac{w_y}{\sqrt{2}}\right)\sqrt{1-a^2}\,, |a| < 1. \tag{6.54}$$

From Eq. (6.54), at $|a| < 1$ the nth order vHG beam in Eq. (6.47) is seen to generate n isolated optical nulls on the y-axis with their coordinates proportional to the

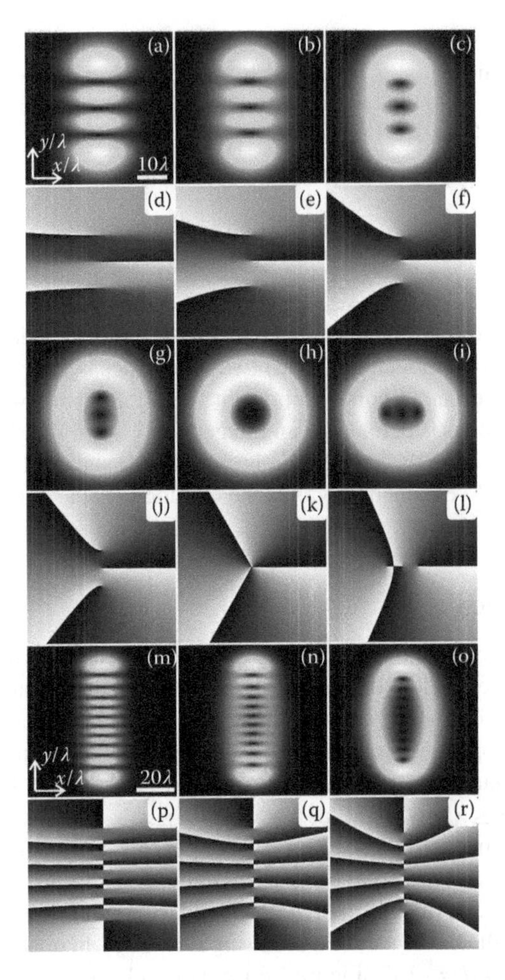

FIGURE 6.6 Intensity (a–c, g–i, m–o) and phase (d–f, j–l, p–r) patterns of the vHG mode in the initial plane ($z=0$) at different values of the topological charge n ((a–l) $n=3$, (m–r) $n=10$) and at different values of the parameter a: (a, d) 0.1; (b, e) 0.25; (c, f) 0.5; (g, j) 0.75; (h, k) 1.0; (i, l) 1.25; (m, p) 0.1; (n, q) 0.25; and (o, r) 0.5. All figures (a–r) were obtained for the wavelength $\lambda = 532$ nm and Gaussian beam waist radius $w_x = w_y = w = 10\lambda$.

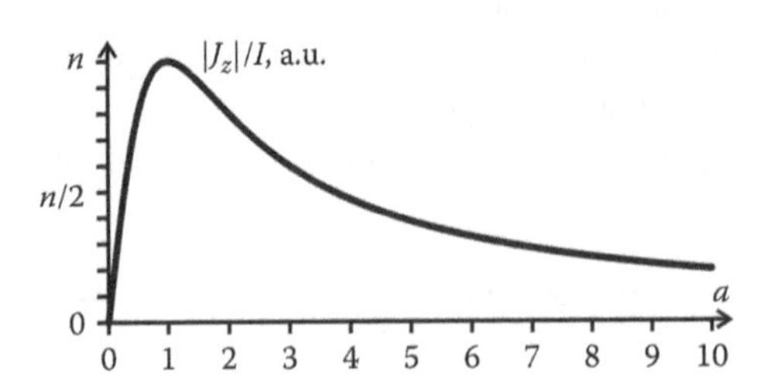

FIGURE 6.7 OAM (6.53) as a function of the parameter a ($\varepsilon = 1$).

Hermite polynomial's roots γ_{nk} but being $[w_y(1-a^2)]^{-1/2}$-times decreased. Note that at $a=0$, when the vHG beam becomes similar to the standard HG mode with the number (0, n), the optical nulls have coordinates defined by the roots γ_{nk}, with the optical nulls ceasing to be isolated but rather belonging to zero-intensity lines parallel to the x-axis. With the a parameter increasing from 0 to 1, the isolated optical nulls of Eq. (6.54) are shifted along the y-axis toward the origin ($x=y=0$), whereas at $a=1$ all optical nulls "merge" into a single isolated (n-times degenerate) optical null at the origin. The isolated optical nulls of the vHG beams behave in a similar way along the x-axis at $|a|>1$. With increasing a, the optical nulls "move away" from the origin, so that in the limit $a \to \infty$ the coordinates of all optical nulls are equal to the roots γ_{nk}, while the optical nulls themselves belong to null lines located in parallel to the y-axis. As suggested by Eq. (6.47), each isolated optical null of the vHG mode is associated with an optical vortex of topological charge -1 ($a>0$) or $+1$ ($a<0$). To describe the propagation of the vHG beam of Eq. (6.47), the Gaussian beam waist radius w_y in the Hermite polynomial's argument should be replaced with the relationship:

$$w_y(z)=w_y\left(1+\frac{z^2}{z_{0y}^2}\right)^{1/2},\quad z_{0y}=\frac{kw_y^2}{2}. \tag{6.55}$$

Substituting (6.55) in (6.54) yields the isolated nulls' coordinates:

$$x_{nk}=0,\quad y_{nk}=\gamma_{nk}\left(\frac{w_y\sqrt{1+z^2/z_{0y}^2}}{\sqrt{2}}\right)\sqrt{1-a^2},\ |a|<1. \tag{6.56}$$

From Eq. (6.56) it can be inferred that the isolated optical nulls (and the entire intensity pattern) from the vHG mode change only in scale upon propagation: while remaining on the y-axis, the optical nulls move away from $y=0$ with increasing z. From Eq. (6.54) follows a remarkable property of the isolated optical nulls: not being limited by the diffraction limit or wavelength, their minimal separation can take an arbitrary value.

Orthogonality of Vortex Hermite-Gaussian Beams

Let us consider a scalar product of two elliptical vHG beams (6.47) with the same radii w_x and w_y, but with different topological charges n and m, and different asymmetry parameters a_1 and a_2:

$$\langle U_{na1},U_{ma2}\rangle=\left(1+a_1^2\right)^{-n/2}\left(1+a_2^2\right)^{-m/2}$$
$$\times\sum_{s=0}^{n}\sum_{t=0}^{m}\frac{n!(-ia_1)^s}{s!(n-s)!}\frac{m!(ia_2)^t}{t!(m-t)!}\left\{2^s s!\sqrt{\frac{\pi}{2}}w_x\delta_{st}\right\}\times\left\{2^{n-s}(n-s)!\sqrt{\frac{\pi}{2}}w_y\delta_{n-s,m-t}\right\}, \tag{6.57}$$

where δ_{nm} is the Kronecker symbol. Eq. (6.57) can be obtained by using Eq. (6.45) and the reference integral [135] (formula 7.374.1).

$$\int_{-\infty}^{+\infty} \exp(-x^2) H_n(x) H_m(x) dx = 2^n n! \sqrt{\pi} \delta_{nm}. \quad (6.58)$$

Because of the property of Kronecker delta, only one term is non-zero in the sums of Eq. (6.57), if $n = m$:

$$\langle U_{na1}, U_{ma2} \rangle = 2^{n-1} n! \pi w_x w_y \frac{(1 + a_1 a_2)^n}{(1 + a_1^2)^{n/2} (1 + a_2^2)^{n/2}} \delta_{nm}. \quad (6.59)$$

When $n = m$ and $a_1 = a_2$, an equation for the beam power can be obtained from Eq. (6.59), which coincides with Eq. (6.50). From Eq. (6.59) the vHG beams are seen to be orthogonal with respect to the number n and unorthogonal with respect to the parameter a. Only when $a_1 a_2 = -1$, the vHG beams become orthogonal.

Experiment

To generate the vHG beams we utilized a spatial light modulator (SLM) PLUTO-VIS (1920 × 1080 pixels resolution, 8-μm pixel size). The output beam from a solid-state laser ($\lambda = 532$ nm) was attenuated using a neutral density filter. A system composed of a microobjective (40x, NA = 0.6), a lens (f = 350 mm), and a pinhole (40-μm aperture) was utilized to generate a homogeneous Gaussian intensity profile of the laser beam incident on the SLM. Then, a lens ($f_2 = 150$ mm) formed an image on the matrix of a CMOS-camera MDCE-5A (1/2″, 1280 × 1024-pixel resolution). The system's focus was found 550 mm apart from the plane $z = 0$, which was conjugate to the SLM display. The zero and first diffraction orders were spatially separated by the superposition of the original phase function and a linear phase mask. With the phase pattern of 1024 × 1024 pixels used in the experiment, the size of the phase element at the SLM output was approximately equal to 8.2 × 8.2 mm. In the experiment, the illuminating beam diameter was about 3 mm. The encoded phase ($n = 10$) obtained with regard for the vHG beam's amplitude at $z = 0$ is shown in Figure 6.8a. In this case, the experimental intensity patterns (Figure 6.8c–e) observed at different distances from the SLM are seen to, first, also preserve the structure upon propagation and, second, accurately reproduce the intensity pattern characteristic of the ideal vHG beam of Figure 6.8b.

Thus, the results obtained are as follows. We have analysed elliptical vHG beams whose complex amplitude is proportional to the nth-order Hermite polynomial with its argument being a function of the parameter a. At $|a| < 1$ on the vertical axis of the beam cross-section there are n isolated optical nulls, which generate optical vortices with topological charge +1 ($a < 0$) or −1($a > 0$). At $|a| > 1$, similar isolated optical nulls of the vHG mode are found on a horizontal axis. At $|a| = 1$ and $w_x = w_y = w$,

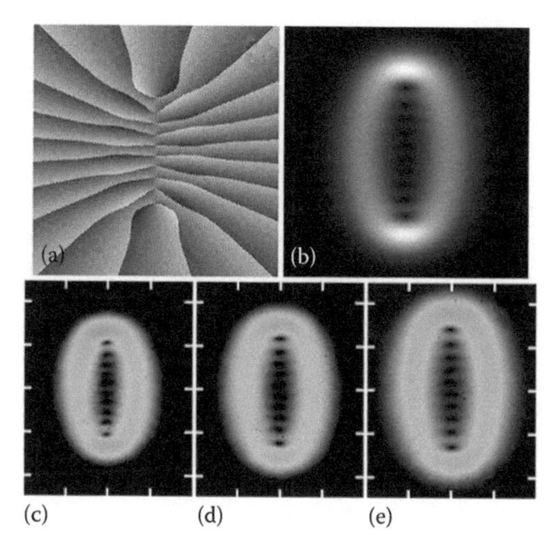

FIGURE 6.8 Encoded phase ($n=10$) to generate (a) a vHG beam, (b) a numerically simulated intensity pattern, and experimental intensity patterns generated at different distances z: (c) 100 mm, (d) 150 mm, and (e) 200 mm. Mesh step is 0.5 mm.

all n isolated optical nulls merge at the beam center, giving rise to an nth order optical vortex with the vHG mode becoming identical to the LG mode of order $(0, n)$; whereas at $a=0$, the vHG mode becomes similar to the HG mode of order $(0, n)$. We have derived the OAM of the vHG modes, which has been found to depend on the parameter a, changing in the range from 0 (at $a=0$ and $a \to \infty$) to n ($a=1$). The derived equation allows the transverse intensity of the vHG-beam to be changed without changing its OAM. Two vHG beams have been shown to be orthogonal to each other if they have different numbers n, and not orthogonal if they have different parameters a and the same number n. In the latter case the beams are orthogonal when the product of two different parameters a is equal to minus one.

7 Asymmetric Vortices Bessel Beams

7.1 ASYMMETRIC BESSEL MODES

The Helmholtz equation [96]

$$(\nabla^2 + k^2)E(x, y, z) = 0, \tag{7.1}$$

where $k = 2\pi/\lambda$ is the wavenumber of monochromatic light of wavelength λ, has a solution in the form of a Bessel function in the cylindrical coordinate system (r, φ, z):

$$E_n(r, \phi, z) = \exp\left(ikz\cos\theta_0 + in\varphi\right)J_n(kr\sin\theta_0), \tag{7.2}$$

where:

θ_0 is the angle of a conical wave that generates the Bessel mode
$J_n(x)$ is the nth order Bessel function of the first kind

The θ_0 angle defines the amplitude of the plane-wave spectrum of the Bessel mode (7.2) on a unit sphere $F_n(\theta,\phi) = (-i)^n \exp(in\varphi)\delta(\theta - \theta_0)$, where $\delta(x)$ is the Dirac delta-function. Possessing an infinite energy, the Bessel modes in Eq. (7.2) preserve their intensity during propagation in free space (squared modulus of amplitude [7.2]) and, thus, are referred to as diffraction-free Bessel beams [210]. A linear combination of solutions (7.2) to Eq. (7.1) with arbitrary coefficients is also a solution to Eq. (7.1). An algorithm for designing a phase optical element to generate diffraction-free Bessel beams with a preset mode composition

$$E(r, \varphi, z = 0) = \sum_{n=0}^{\infty} C_n \exp(in\varphi)J_n(kr\sin\theta) \tag{7.3}$$

was proposed in Ref. [211]. In Ref. [98], a Mathieu beam

$$E_n(\xi, \eta, z) = Ce_n(\xi, q)ce_n(\eta, q)\exp\left(ikz\cos\theta_0\right), \tag{7.4}$$

was proposed as an alternative to the Bessel beams, where $Ce_n(\xi,q)$, $ce_n(\eta,q)$ are nonperiodic and periodic even Mathieu functions. The relationship (7.4) is the solution of Eq. (7.1) in an elliptical coordinates system [1]: $x = d\,\mathrm{ch}\,\xi\cos\eta$, $y = d\,\mathrm{sh}\,\xi\sin\eta$, $z = z$, $q = (kd\sin\theta_0)^2/4$. In Ref. [212] it was shown that although the Mathieu beams were devoid of the orbital angular momentum (OAM), a linear combination of the even and odd Mathieu beams with complex coefficients had an OAM. It is interesting [188] that a linear combination of two zero-OAM Hermite-Gauss modes with complex coefficients has a non-zero OAM. The periodic Mathieu functions can be expanded into a Fourier series [96], whereas, for example, an even Mathieu beam in Eq. (7.4) can be represented as a linear combination of Bessel modes:

$$E_{2n}(\xi,\eta,z) = \exp(ikz\cos\theta_0)\times\sum_{m=0}^{\infty}(-1)^m A_{2m}^{2n}\cos(2m\varphi)J_{2m}(kr\sin\theta_0), \quad (7.5)$$

where (ξ, η, z) and (r, φ, z) are the elliptic and cylindrical coordinates. Diffraction-free beams defined by a linear combination of the Bessel modes in Eq. (7.5) were described in Ref. [213]. In fact, the beams in Eq. (7.5) may be interpreted as the Mathieu beams of Eq. (7.4) in cylindrical coordinates.

Superposition of Bessel Modes

In this section, we propose a linear combination of the Bessel modes in Eq. (7.2) with the coefficients such that the series in Eq. (7.3) can be calculated and equals a Bessel function with complex argument. Such a diffraction-free asymmetric elegant Bessel mode (aB-mode) is shown to have an enumerable number of isolated intensity zeros located on the x-axis. All the zeros (except for the axial one) generate optical vortices with unit topological charge and opposite sign on opposite sides of 0. The intensity zero on the optical axis generates an optical vortex with topological charge n. For such a beam, the OAM can be exactly calculated. Let us analyze the following superposition of Bessel modes:

$$E_n(r,\varphi,z=0;c) = \sum_{p=0}^{\infty}\frac{c^p\exp(in\varphi+ip\varphi)}{p!}J_{n+p}(\alpha r). \quad (7.6)$$

The field in Eq. (7.6) forms a diffraction-free Bessel mode at any integer n and any complex constant c. For simplicity, we assume that the constant is real and positive, $c \geq 0$. It is worth noting that when $c = 0$, just a single term remains to be non-zero in Eq. (7.6) at $p = 0$, with the aB-mode becoming a conventional Bessel mode $E_n(r,\varphi,z=0;c=0) = \exp(in\varphi)J_n(\alpha r)$. At distance z, the complex amplitude (7.6) should be multiplied by a phase term $\exp[i(k^2 - \alpha^2)^{1/2}z]$. There is a reference series (Eq. (5.7.6.1) in Ref. [38]:

$$\sum_{k=0}^{\infty}\frac{t^k}{k!}J_{k+\nu}(x) = x^{\nu/2}(x-2t)^{-\nu/2}J_\nu\left(\sqrt{x^2-2tx}\right). \quad (7.7)$$

Using Eq. (7.7), the series in Eq. (7.6) is reduced to

$$E_n(r,\varphi,z=0;c)=\left[\frac{\alpha r}{\alpha r-2c\exp(i\varphi)}\right]^{n/2}$$
$$\times J_n\left\{\sqrt{\alpha r\left[\alpha r-2c\exp(i\varphi)\right]}\right\}\exp(in\varphi). \tag{7.8}$$

Equation (7.8) describes a three-parameter family of nonparaxial scalar modes (diffraction-free beams) of free space, which will be referred to as aB-modes. In the aB-modes, two real parameters define the scale (α) and asymmetry degree (c). In combination, the integer n and continuous c define the OAM magnitude and the major intensity lobe's radius (or the maximum's position) in the aB-mode's cross-section: $r_n = (n + c)/\alpha$. The Bessel function's argument in (7.8) is complex but it can become purely imaginary only when $0 < r < 2c/\alpha$ and $\varphi = 0$. Thus, the Bessel function in Eq. (7.8) is not diverging at infinity. When the denominator in (7.8) is zero, the Bessel function's argument is also zero. The 0/0 indeterminacy can be resolved. The zeros of the aB-mode (7.8) are found on the x-axis (at $\varphi = \pi m$, where $m = 0, 1, 2,\ldots$) at points defined by

$$x_{+p}=\frac{c+\sqrt{c^2+\gamma_p^2}}{\alpha},\quad x_{-p}=\frac{c-\sqrt{c^2+\gamma_p^2}}{\alpha}. \tag{7.9}$$

In Eq. (7.9), γ_p are the roots of the Bessel function: $J_n(\gamma_p) = 0$. From Eq. (7.9) it follows that the aB-mode's intensity profile is asymmetrical relative to the origin of coordinates, since $x_{+p} > -x_{-p}$. Note that with increasing $c > 0$, the aB-mode becomes more asymmetric, with the isolated zeros of the aB-mode disappearing at $c = 0$ and changing to zero-intensity rings of the radially symmetric Bessel mode. All the aB-mode's zeros (except for the axial one at $r = 0$) generate optical vortices with unit topological charge and opposite sign on opposite sides of 0. At $x < 0$, the optical vortex has a topological charge of +1, whereas at $x > 0$, the topological charge equals −1. To alter the optical vortices' signs, Eq. (7.8) should be replaced with a complex conjugated one. The intensity zero on the optical axis in Eq. (7.8) generates an optical vortex with topological charge n. These conclusions can be verified from the analysis of the phase patterns in Figures 7.1 and 7.2.

Numerical Simulation

Shown in Figures 7.1 and 7.2 are the intensities of the light beams in Eq. (7.8) for different values of the parameters in the initial plane and intensity profiles at $z = y = 0$ and $z = x = 0$. The simulation parameters were as follows: wavelength, $\lambda = 532$ nm, $c = 1$, $\alpha = 0.2$ (μm)$^{-1}$, calculation domain boundaries, $-150\lambda \leq x, y \leq 150\lambda$, and the number of pixel on each axis, 200. In Figures 7.1 and 7.2 the beam's order was $n = 0, 3$.

It is seen from Figure 7.1c that the maximum intensity of the zero-order aB-mode $E_0(r,\phi) = J_0(\{\alpha r[\alpha r - 2c\exp(i\phi)]\}^{1/2})$ is larger than unity: $I_{\max} = 1.6$ (at $c = 1$).

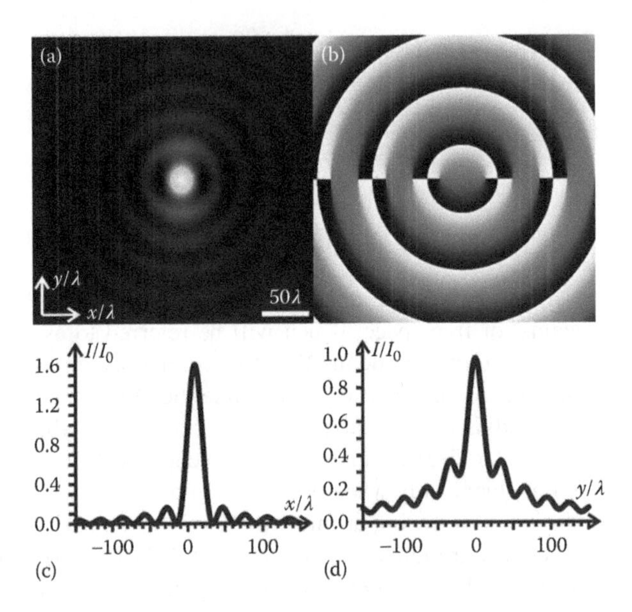

FIGURE 7.1 (a) Intensity and (b) phase patterns of the zero-order light beam in Eq. (7.8) ($n = 0$) in the initial plane and intensity profiles at (c) $z = y = 0$ and (d) $z = x = 0$.

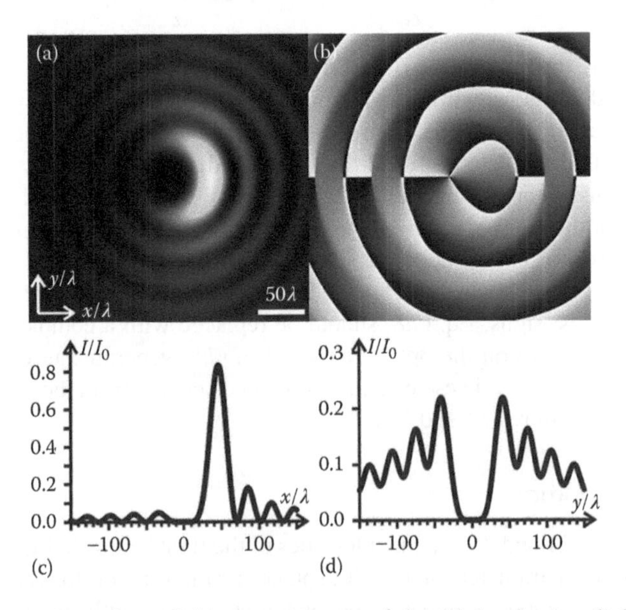

FIGURE 7.2 (a) Intensity and (b) phase patterns of the third-order ($n = 3$) light beam of Eq. (7.8) in the initial plane and intensity profiles at (c) $z = y = 0$ and (d) $z = x = 0$.

This can be explained. Considering that at $\alpha r < 2$ and $\varphi = 0$, the argument in (7.8) is purely imaginary, one can represent the aB-mode on the above-said interval of the x-axis as $E_0(r,\varphi) = I_0\{[\alpha x(2-\alpha x)]^{1/2}\}$, where $I_0(x)$ is a modified Bessel function of zero order. At the ends of the interval [0, $2/\alpha$], the Bessel function's argument takes a zero value, with the Bessel function being equal to $J_0(0) = I_0(0) = 1$. On the interval [0, $2/\alpha$], the $I_0(x)$ function achieves a maximum value at the point of the maximum argument, which is found at the interval center, $x = 1/\alpha$. Thus, the maximal intensity of the aB-mode equals $I_{\max} = \left|E_0(r = 1/\alpha, \varphi = 0)\right|^2 = I_0^2(1) \cong 1{,}60$ From this, in particular, it follows that the maximal intensity of the side-lobes of the zero-order aB-mode (Figure 7.1), which equals that of the conventional Bessel mode $J_0(x)$, accounts for a smaller proportion of the major peak's maximal intensity for the aB-mode (10%) when compared with the conventional Bessel mode (16%). Note that with increasing $c > 1$, the maximal intensity will be increasing as $I_{\max} = \left|E_0(r = 1/\alpha, \varphi = 0; c)\right|^2 = I_0^2(c)$, whereas the relative intensity of the side-lobes will be decreasing. From Figure 7.2, the major lobe of the higher-order aB-mode is seen to be in the form of a right-side crescent.

Asymmetric Modes with a Left-Side Optical Crescent

It is also possible to derive an expression for an aB-mode generating a left-side optical crescent. An aB-mode with the intensity distribution bilaterally symmetric to the beam of Eq. (7.8) with respect to the axis $x = 0$ is described by the amplitude

$$\begin{aligned} E_n(r,\varphi) &= \sum_{p=0}^{\infty} \frac{(-c)^p \exp\left[i(n+p)\varphi\right]}{p!} J_{n+p}(\alpha r) \\ &= \left[\frac{\alpha r}{\alpha r + 2c\exp(i\varphi)}\right]^{n/2} J_n\left\{\sqrt{\alpha r\left[\alpha r + 2c\exp(i\varphi)\right]}\right\} e^{in\phi}. \end{aligned} \tag{7.10}$$

Let us find the Fourier spectrum of the aB-mode. Substituting the Bessel beam's angular spectrum $F_n(\theta,\varphi) = (-i)^n \exp(in\varphi)\delta(\theta - \theta_0)$ into each term of the sum on the right-hand side of Eq. (7.6) yields the Fourier-spectrum of the aB-mode:

$$\begin{aligned} A(\theta,\varphi) &= \sum_{p=0}^{\infty} \frac{(-ic)^p \exp\left[i(n+p)\varphi\right]}{p!} \delta(\theta - \theta_0) \\ &= \frac{(-i)^n}{2\pi \sin\theta_0} \exp\left[in\varphi - ic\exp(i\varphi)\right] \delta(\theta - \theta_0). \end{aligned} \tag{7.11}$$

Substituting Eq. (7.11) into the expansion in terms of plane waves

$$E_n(r,\phi,z) = \int_0^{2\pi} \int_{-\pi}^{\pi} A(\theta,\varphi) \exp\left[ikr\cos(\phi - \varphi)\sin\theta + ikz\cos\theta\right] \sin\theta \, d\theta \, d\varphi \tag{7.12}$$

yields Eq. (7.8). Based on Eq. (7.11), we can predict what kind of symmetry is characteristic of the aB-modes. Actually, the azimuthal component of the spectrum $A(\theta,\varphi) = A_1(\theta)A_2(\varphi)$ is defined by the relation $A_2(\varphi) = \exp(in\varphi - ic\cos\varphi)\exp(c\sin\varphi)$, suggesting that at $c > 0$, not only the phase but also the amplitude $|A_2(\varphi)| = \exp(c\sin\varphi)$ undergo a change along the ring $\delta(\theta-\theta_0)$. On the upper semicircle of the ring $\delta(\theta-\theta_0)$, when $0 < \varphi < \pi$, the amplitude achieves the maximum of $|A_2(\varphi = \pi/2)| = \exp(c)$, whereas on the lower semicircle, when $\pi < \varphi < 2\pi$, the amplitudes achieves the minimum of $|A_2(\varphi = 3\pi/2)| = 1/\exp(c)$. Note that with increasing $c > 0$, the maximum/minimum ratio is increasing, implying that asymmetry of the angular spectrum in Eq. (7.11) is also increasing. Because of this, the transverse intensity profile of the aB-mode in Eq. (7.8) has an asymmetry similar to that of the spectrum (7.11) but rotated clockwise by 90°. This can be seen from Figure 7.3.

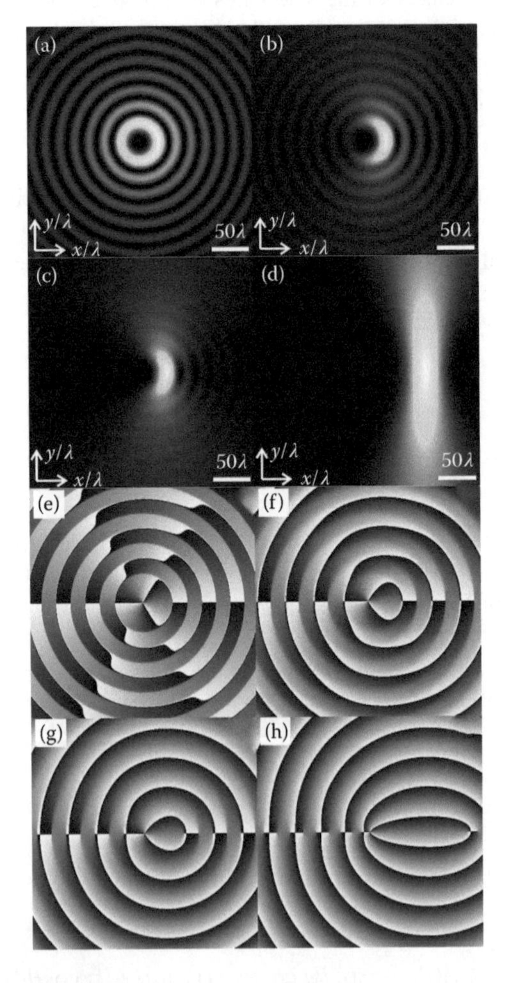

FIGURE 7.3 (a–d) Intensity and (e–h) phase of the third-order aB-mode in Eq. (7.8) ($n = 3$) for different values of c: (a, e) 0.1, (b, f) 1, (c, g) 2, (d, h) 10.

Figure 7.3 shows in which way the parameter c affects the intensity pattern's asymmetry. The other parameters in Figure 7.3 are $\lambda = 532$ nm, $n = 3$, $\alpha = 1/(10\lambda)$, the calculation domain boundaries, $-250\lambda \le x, y \le 250\lambda$, and the number of pixels on each axis, 600.

Orbital Angular Momentum of Asymmetric Bessel Modes

The OAM's projection J_z onto the optical axis and the beam's total intensity I in a plane perpendicular to the optical axis are given by [188]

$$J_z = \mathrm{Im}\left\{ \lim_{R\to\infty} \int_0^R \int_0^{2\pi} E^* \frac{\partial E}{\partial \varphi} r dr d\varphi \right\}, \tag{7.13}$$

$$I = \lim_{R\to\infty} \int_0^R \int_0^{2\pi} E^* E r dr d\varphi. \tag{7.14}$$

Substituting the complex amplitude of the right-hand side of Eq. (7.6) into Eqs. (7.13) and (7.14), we obtain the OAM and the total intensity of the aB-mode:

$$J_z = 2\pi \lim_{R\to\infty} \sum_{p=0}^{\infty} \frac{c^{2p}(n+p)}{(p!)^2} \int_0^R J_{n+p}^2(\alpha r) r dr, \tag{7.15}$$

$$I = 2\pi \lim_{R\to\infty} \sum_{p=0}^{\infty} \frac{c^{2p}}{(p!)^2} \int_0^R J_{n+p}^2(\alpha r) r dr. \tag{7.16}$$

The integrals in Eqs. (7.15) and (7.16) are reference integrals described in [135] (Eq. 5.54.2):

$$\int J_p^2(\alpha r) r dr = \frac{r^2}{2}\left[J_p^2(\alpha r) - J_{p-1}(\alpha r) J_{p+1}(\alpha r) \right]. \tag{7.17}$$

Based on the asymptotics of the Bessel function at large values of the argument (Eq. 9.2.1 in [101]), we find that all the integrals in Eqs. (7.15) and (7.16) are independent of the Bessel function order, being equal to $R/(\pi\alpha)$. Then, using the number series 0.246.1 and 0.246.2 from [135] and dividing (7.15) by (7.16), we obtain a relationship for the OAM normalized with respect to intensity:

$$\frac{J_z}{I} = n + \sum_{p=0}^{\infty} \frac{c^{2p} p}{(p!)^2} \left[\sum_{p=0}^{\infty} \frac{c^{2p}}{(p!)^2} \right]^{-1} = n + \frac{c I_1(2c)}{I_0(2c)}, \tag{7.18}$$

where $I_n(z)$ is a modified Bessel function. From Eq. (7.18), it follows that the aB-mode's OAM can take both integer and fractional values, linearly increasing with increasing order n and increasing constant $c > 0$. From Eq. (7.18) it also follows that the zero-order aB-mode ($n = 0$) can have any OAM, which near-linearly increases with increasing parameter c and equals $J_z/I = cI_1(2c)/I_0(2c)$. It should be noted that the approach adopted for calculating the OAM in the present context is also suitable for calculating the OAM of any optical field represented as a superposition of Bessel modes.

Orthogonality of Asymmetric Bessel Modes

In a similar way to calculating the OAM, it is possible to derive a scalar product of two aB-modes, namely, of the nth order beam with the parameters α and c and the mth order beam with the parameters β and d. We obtain:

$$\left(E_{n\alpha c}, E_{m\beta d}\right) = 2\pi \frac{\delta(\alpha-\beta)}{\alpha}\left(\frac{d}{c}\right)^{\frac{n-m}{2}} I_{n-m}\left(2\sqrt{cd}\right), \tag{7.19}$$

where $I_{n-m}(x)$ is a modified Bessel function. From Eq. (7.19) it is seen that as distinct from the Bessel modes, the aB-modes are orthogonal only in terms of the scale factor, being non-orthogonal in terms of the Bessel function order and the asymmetry parameter.

Thus, we have analyzed a three-parameter family of diffraction-free nonparaxial asymmetric elegant Bessel modes, which is described by the integer-order Bessel functions of the first kind with complex argument. For $c = 0$, the aB-modes are identical to the conventional Bessel modes. The aB-modes have been found to have the enumerable number of isolated intensity zeros located on the x-axis, which are characterized by a unit topological charge (except for the axial intensity zero) and opposite sign on the opposite optical axis sides. The topological charge of the axial intensity zero is equal to the Bessel function's order. The aB-modes can have both integer and fractional OAM, which grows linearly with increasing mode number n and parameter c. For the zero-order aB-mode: (1) the intensity maximum is off-set from the optical axis by c/α, (2) isolated intensity zeros are on the horizontal axis of the beam's cross-section, and (3) the OAM per photon can take any value equal to $\hbar c I_1(2c)/I_0(2c)$. The aB-modes are orthogonal in terms of the continuous scale parameter α and non-orthogonal in terms of the integer topological parameter n and continuous asymmetry parameter c.

7.2 ASYMMETRIC BESSEL-GAUSS BEAMS

Bessel-Gauss (BG) beams were first proposed by F. Gori in 1987 [214]. The complex amplitude of such beams is described by the product of a Gaussian beam by the nth order Bessel beam of the first order and a phase function describing the angular harmonic. The complex amplitude of the BG beam satisfies a paraxial equation of propagation. The BG beams have a radially symmetric intensity distribution and carry OAM. They preserve radial symmetry during propagation. Note, however, that the BG beams do not present free-space modes because in addition to changing in scale during propagation, they also experience the energy redistribution between different diffraction rings in the beam's transverse plane. The generalization of the BG beams was discussed in a number of articles [99,215,216].

For example, the BG beams [214] are a particular case of more general Helmholtz-Gauss beams [99]. If the argument of the Bessel function is complex we obtain generalized BG (gBG) beam [217] or dual BG (dBG) beam [218]. The BG beams have finite energy, but infinite-energy Bessel modes have also been known [210]. The Bessel modes offer the solution of the Helmholtz equation [96] and, while preserving their intensity during propagation in free space, they are also referred to as non-diffracting Bessel beams [210]. A linear combination of Bessel modes with arbitrary coefficients also gives the solution of the Helmholtz equation. An algorithm for designing a phase optical element to generate the non-diffracting Bessel beams with a designed mode content has been proposed [211]. The Mathieu beam has been proposed as an alternative to the Bessel beam [98]. A linear combination of an even and an odd Mathieu beam with complex coefficients has been shown [212] to carry OAM. Note, however, that the Mathieu beams are devoid of OAM. Interestingly, a linear combination of two Hermite-Gauss modes with complex coefficients, individually having zero OAM, carry non-zero OAM [188]. The periodic Mathieu functions can be decomposed into a Fourier series [96]. For instance, the even Mathieu functions are decomposed in terms of cosines of polar angle in a cylindrical coordinate system, while the odd ones are decomposed in terms of the sine function. Therefore, a non-diffracting Mathieu beam can be represented as a linear combination of Bessel modes, which have recently been discussed [213].

It is of interest to find linear combinations of BG beams which would be described by simple analytical functions, enabling certain properties of the beams to be calculated. In this section, we study a linear combination of BG beams that is described by a Bessel function with complex argument. The resultant asymmetric BG (aBG) beam is shown to have in the initial plane a countable number of optical nulls located on the x-axis. All those nulls (except for the one at the origin) generate optical vortices with unit topological charge and opposite signs on the different sides of the origin. Arising from the optical null at the origin is an optical vortex with topological charge n. During propagation in free space, the aBG beams rotate around the optical axis. We show that the aBG beams carry OAM, which increases with increasing number n and increasing beam's asymmetry parameter c. It is noteworthy that the aBG beams can have both integer and fractional OAM. Note that with the Gaussian beam's radius tending to infinity, the aBG beams change to the non-diffracting asymmetric Bessel (aB) modes [219].

Linear Combination of Bessel-Gaussian Beams

In the initial plane at $z=0$, the complex amplitude of the BG beam [214] is given by

$$E_n(r,\varphi,z=0)=\exp\left(-\frac{r^2}{\omega_0^2}+in\varphi\right)J_n(\alpha r), \tag{7.20}$$

where:

$\alpha=k\sin\theta_0=(2\pi/\lambda)\sin\theta_0$ is the scale factor

$k=2\pi/\lambda$ is the wavenumber of the incident wave of wavelength λ

θ_0 is the angle of the conical wave that forms the Bessel beam

At any other z, the complex amplitude of Eq. (7.20) takes the form:

$$E_n(r,\varphi,z) = q^{-1}(z)\exp\left(ikz - \frac{i\alpha^2 z}{2kq(z)}\right)\exp\left(-\frac{r^2}{\omega_0^2 q(z)} + in\varphi\right)J_n\left[\frac{\alpha r}{q(z)}\right], \quad (7.21)$$

where:

$q(z) = 1 + iz/z_0$, $z_0 = k\omega_0^2/2$ is the Rayleigh range
ω_0 is the Gaussian beam's waist radius
$J_n(x)$ is the nth order Bessel function of the first order

Considering that the Bessel function has complex argument, the beams of Eq. (7.21) do not present paraxial modes of free space. Let us consider the following superposition of the BG beams $q = q(z)$:

$$E_n(r,\varphi,z;c) = q^{-1}\exp\left(ikz - \frac{i\alpha^2 z}{2kq} - \frac{r^2}{q\omega_0^2}\right) \times \sum_{p=0}^{\infty}\frac{c^p\exp(in\varphi + ip\varphi)}{p!}J_{n+p}\left(\frac{\alpha r}{q}\right). \quad (7.22)$$

The field of Eq. (7.22) generates a paraxial aBG beam at any integer n and complex constant c. For simplicity sake, hereafter, the constant c is assumed to be real and positive: $c \geq 0$. Note, however, that putting c to be complex, $c = |c|\arg c$, or negative, $c < 0$, we find that the intensity distribution generated by the field of Eq. (7.22) is rotated by the angle $\arg c$ about the optical axis. At $c = 0$, a single term in Eq. (7.22) remains to be non-zero at $p = 0$, with the aBG beam changing to a conventional BG beam of Eqs. (7.20) and (7.21).

We consider a linear combination of BG beams in the form defined by Eq. (7.3), which describes a Bessel function with complex argument. Actually, we can write down a reference integral [38] (Eq. 5.7.6.1):

$$\sum_{k=0}^{\infty}\frac{t^k}{k!}J_{k+\nu}(x) = x^{\nu/2}(x-2t)^{-\nu/2}J_\nu\left(\sqrt{x^2-2tx}\right). \quad (7.23)$$

Using (7.23), Eq. (7.22) is rearranged to

$$E_n(r,\varphi,z;c) = \frac{1}{q}\exp\left(ikz - \frac{i\alpha^2 z}{2kq} - \frac{r^2}{q\omega_0^2} + in\varphi\right) \times \left[\frac{\alpha r}{\alpha r - 2cq\exp(i\varphi)}\right]^{n/2} J_n\left\{\frac{1}{q}\sqrt{\alpha r\left[\alpha r - 2cq\exp(i\varphi)\right]}\right\}. \quad (7.24)$$

Relation (7.24) defines a closed form of the complex amplitude of a three-parameter family of paraxial scalar aBG beams, in which the scale and asymmetry degree are, respectively, governed by two continuous parameters, α and c. The Gaussian beam's waist radius ω_0 is assumed to be the same for the entire family of the aBG beams. A combination of the integer n and continuous c defines the value of OAM. The numerator and denominator in Eq. (7.24) take the zero value simultaneously, so that the 0/0 indeterminacy can be resolved. The beam in Eq. (7.24) has a countable number of isolated optical nulls, which generate optical vortices with unit topological charge, except the optical null at the origin which has the topological charge n. To derive the optical nulls of the aBG beam in polar coordinates, the argument of the Bessel function in Eq. (7.24) is equated to the Bessel function's root γ_{np} ($J_n(\gamma_{np})=0$):

$$\alpha^2 r^2 - 2\alpha c q r \exp(i\varphi) = \gamma_{np}^2 q^2. \tag{7.25}$$

Let us separate out the real and imaginary parts of the equation:

$$\begin{cases} \alpha^2 r^2 - 2\alpha c |q| r \cos(\varphi + \Psi) = \gamma_{np}^2 |q|^2 \cos(2\Psi), \\ -2\alpha c |q| r \sin(\varphi + \Psi) = \gamma_{np}^2 |q|^2 \sin(2\Psi), \end{cases} \tag{7.26}$$

where $\Psi = \arctan(z/z_0)$ is Gouy's phase. From (7.26) the coordinates of the optical nulls are given by

$$\begin{cases} \varphi_{np} = \dfrac{1}{2} \arccos\left[\cos(2\Psi) - \dfrac{\gamma_{np}^2}{2c^2} \sin^2(2\Psi) \right], \\ r_{np} = \dfrac{|q|}{\alpha} \sqrt{\gamma_{np}^2 \cos(2\Psi) + 2c^2 \pm 2\sqrt{D}}, \end{cases} \tag{7.27}$$

where $D = (c^2 - \gamma_{np}^2 \sin^2 \Psi)(c^2 + \gamma_{np}^2 \cos^2 \Psi)$. As both relations in Eq. (7.27) suggest, for the optical nulls' coordinates to be real, the following condition needs to be met:

$$\gamma_{np} \sin \Psi \le c. \tag{7.28}$$

From (7.28) it follows that for $c > \gamma_{np}$, the optical nulls with numbers ranging from 0 to p are preserved during propagation, whereas the rest nulls numbered $g = p + 1$, $p + 2, \ldots$ ($c \le \gamma_{ng}$) vanish at a certain distance $z = cz_0/(\gamma_{ng}{}^2 - c^2)^{1/2}$. At $z = 0$, Eq. (7.27) is replaced by

$$\begin{cases} r_{p+} = \alpha^{-1}\left(c + \sqrt{c^2 + \gamma_{np}^2} \right), \quad \varphi_{np} = 2p\pi, \\ r_{p-} = \alpha^{-1}\left(\sqrt{c^2 + \gamma_{np}^2} - c \right), \quad \varphi_{np} = (2p+1)\pi. \end{cases} \tag{7.29}$$

From Eq. (7.29), the transverse intensity of the aBG beam is seen to be asymmetric with respect to the origin, because $r_{p+} > r_{p-}$. It is noteworthy that with growing $c > 0$, the symmetry of the aBG beam also grows, whereas at $c = 0$, the isolated optical nulls of the aBG beam disappear, giving way to the zero-intensity rings of a radially symmetric

BG beam. All the optical nulls of the aBG beam (except the one at $r = 0$) give rise to optical vortices with unit topological charge and opposite signs on the different sides of the origin. Note that the optical nulls have the topological charge −1 at r_{p+} and +1 at r_{p-}. The signs of the optical vortices may be changed by taking the complex conjugate of the expression in Eq. (7.24). From Eq. (7.27), the optical nulls, and hence the entire transverse intensity distribution of the aBG beam, rotate during propagation. However, the rotation pattern is rather sophisticated. In the first relation of Eq. (7.27), the first term in the square brackets can only be discarded when $c \gg 1$, resulting in a transverse diffraction pattern rotating as a whole. In this case, the polar angle varies with distance z as

$$\varphi = \arctan\left(\frac{z}{z_0}\right). \tag{7.30}$$

It follows from Eq. (7.30) that having travelled a distance from $z = 0$ to the Rayleigh range $z = z_0$, the aBG beam makes an anticlockwise turn by $\pi/4$ and then makes a turn of $\pi/4$ on the rest path from $z = z_0$ to $z = \infty$, thus making a total turn of $\pi/2$ over the entire distance travelled. Note that the aBG beam rotates irrespectively of the value of n, meaning that an aBG beam with $n = 0$ will also rotate.

Figure 7.4 depicts the intensity and phase patterns generated by the aBG beam in Eq. (7.24) at different values of the asymmetry parameter c. The numerical

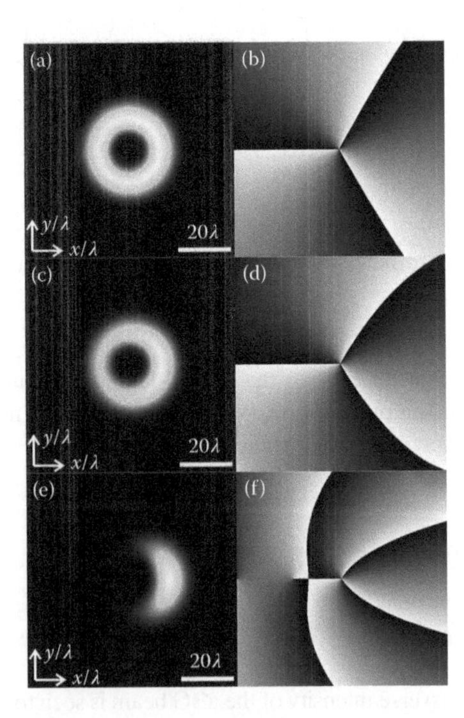

FIGURE 7.4 (a, c, e) Intensity and (b, d, f) phase patterns from the third-order beam of Eq. (7.24) at $n = 3$, $z = 0$, and different values of the asymmetry parameter c: (a, b) 0.1; (c, d) 1; and (e, f) 10.

simulation was conducted for the wavelength $\lambda = 532$ nm, the Gaussian beam's waist radius $w_0 = 10\lambda$, the scale factor $\alpha = 1/(10\lambda)$, and the computation region $-40\lambda \le x$, $y \le 40\lambda$. From Figure 7.4, with increasing $c > 0$, the ring-shape intensity distribution is seen to change into a right-sided semi-crescent. The relationship for an aBG beam able to generate a left-sided semi-crescent can also be deduced. An aBG beam that generates the intensity pattern specularly symmetric to the beam of Eq. (7.24) with respect to the axis $x = 0$, is given by the amplitude:

$$E_n(r,\phi,z=0;c) = \exp\left(-\frac{r^2}{\omega_0^2}\right)\sum_{p=0}^{\infty}\frac{(-c)^p \exp\left[i(n+p)\phi\right]}{p!}J_{n+p}(\alpha r)$$
$$= \exp\left(-\frac{r^2}{\omega_0^2}\right)\left[\frac{\alpha r}{\alpha r + 2c\exp(i\phi)}\right]^{n/2} J_n\left\{\sqrt{\alpha r\left[\alpha r + 2c\exp(i\phi)\right]}\right\}\exp(in\phi). \tag{7.31}$$

From Figure 7.4, the optical nulls located on the axis at $x < 0$ are seen to appear closer to the origin with increasing c: while they are not found in Figure 7.4b and d, an optical null is seen to appear in Figure 7.4f. Such a behavioral pattern is suggested by the second relation of Eq. (7.29) at large $c \gg 1$. At the center of Figure 7.4b, d, and f there is a third-order optical null ($n = 3$).

Shown in Figure 7.5 are the intensity and phase patterns of the third-order aBG beam ($n = 3$) at a relatively large value of the asymmetry parameter $c = 10$ at different

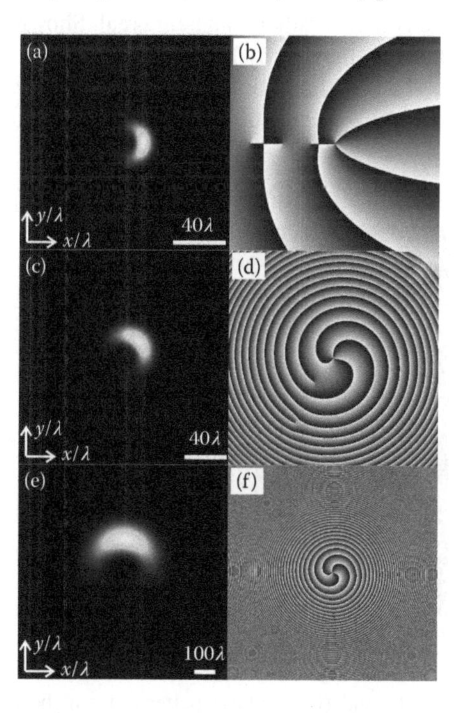

FIGURE 7.5 (a, c, d) Intensity and (b, d, f) phase patterns from the third-order aBG beam of Eq. (7.24) ($n = 3$) at different distances: (a, b) $z = 0$, (c, d) $z = z_0$, and (e, f) $z = 10z_0$.

distances on the optical axis at (a, b) $z = 0$, (c, d) $z = z_0$, and (e, f) $z = 10z_0$. From Figure 7.5, the beam is seen to rotate around the optical axis. The pattern is seen to turn anti-clockwise by $\pi/4$ at the Rayleigh range z_0 (Figure 7.5d), having turned by nearly $\pi/2$ at distance $10z_0$ (Figure 7.5e).

The patterns shown in Figure 7.5 are different in size: (a, b) $-80\lambda \le x, y \le 80\lambda$; (c, d) $-100\lambda \le x, y \le 100\lambda$, and (e, f) $-500\lambda \le x, y \le 500\lambda$. The rest parameters of the numerical simulation are identical to those in Figure 7.4. With increasing z, the optical nulls found in Figure 7.5 on the optical x-axis at $z=0$ also start to rotate anti-clockwise. In Figure 7.5b, three isolated optical nulls are found on the x-axis to the left of the third-order central optical null. From Figure 7.5d, only two optical nulls are seen to be preserved at $z=z_0$, which have turned by 45°, whereas in Figure 7.5f ($z = 10z_0$) the nulls are seen to have merged, forming a second-order optical null, which has turned by nearly 90°.

Inequality (7.28) suggests that at $c \le 1$, all isolated optical nulls on the x-axis (except the one at the origin) will gradually "disappear" as the beam propagates, starting with the most distant nulls ($\gamma_p \gg 1$). The last to "disappear" is the first optical null γ_1 at z equal to

$$z = z_0 \tan\left[\arcsin\left(\frac{c}{\gamma_1}\right)\right]. \tag{7.32}$$

The optical nulls "disappear" in the sense that, considering Eqs. (7.27) and (7.28), their coordinates become complex rather than being real. Shown in Figure 7.6 are the intensity and phase patterns generated by the aBG beam at a relatively small asymmetry parameter, $c=1$. The rest simulation parameters are wavelength, $\lambda = 532$ nm, Gaussian beam's waist radius, $w_0 = 100\lambda$, the scale factor, and $\alpha = 1/(10\lambda)$, the

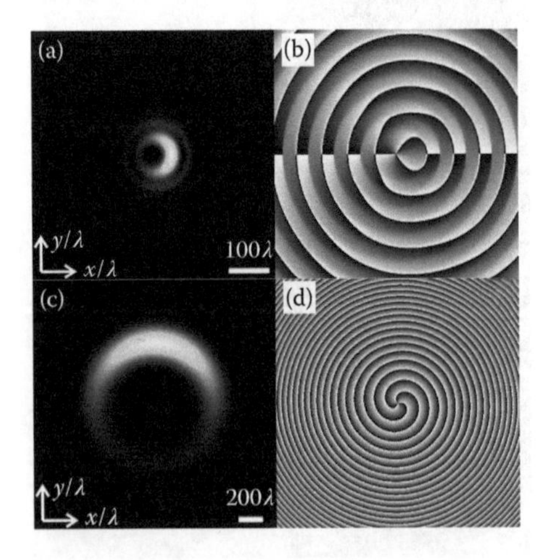

FIGURE 7.6 (a, c) Intensity and (b, d) phase patterns from the third-order aBG beam ($n=3$ and $c=1$) at different distance z: (a, b) 0 and (c, d) z_0.

computation region being $-300\lambda \le x, y \le 300\lambda$ (Figure 7.6a and b) and $-1000\lambda \le x, y \le 1000\lambda$ (Figure 7.6c and d). In this case, the rotation of the aBG beam can no more be described by the simple formula in Eq. (7.30), because the asymmetry parameter c is small. In Figure 7.6, the intensity pattern is seen to have turned anti-clockwise by nearly $\pi/2$ after travelling a distance $z = z_0$. Meanwhile, the isolated optical nulls on the x-axis, which can be observed in Figure 7.6b, "disappear" as the beam propagates: no optical nulls, except the one at the origin, are observed at $z = z_0$ in Figure 7.6d.

Figure 7.7 depicts the intensity and phase of the zero-order aBG beam obtained for the wavelength $\lambda = 532$ nm, Gaussian beam's waist radius $w_0 = 10\lambda$, the scale factor $\alpha = 1/(10\lambda)$, the asymmetry parameter $c = 10$, and the computation domain $-40\lambda \le x, y \le 40\lambda$. An interesting property of the zero-order aBG beam in Figure 7.7 is that it has the intensity maximum close to the optical axis and non-zero OAM. This property can be used for micromanipulation of dielectric microparticles. A particle several times larger than the aBG beam's major intensity maximum in Figure 7.7 can be trapped in the maximum, simultaneously rotating about its axis. From Figure 7.7c, the maximum intensity of the zero-order aBG beam given by

$$E_0(r, \varphi, z = 0; c) = \exp\left(-\frac{r^2}{\omega_0^2}\right) J_0\left\{\sqrt{\alpha r\left[\alpha r - 2c\exp(i\varphi)\right]}\right\}, \tag{7.33}$$

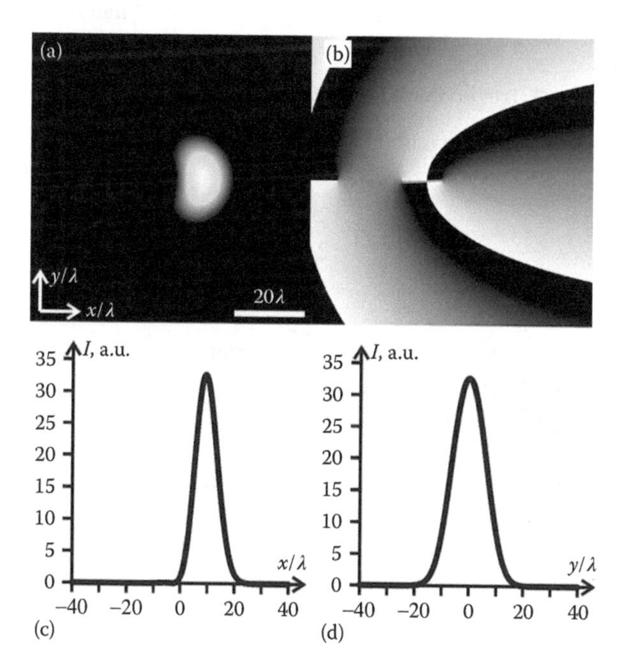

FIGURE 7.7 (a) Intensity and (b) phase patterns of the zero-order aBG beam of Eq. (7.24) ($n = 0$) at $z = 0$ and intensity profiles at (c) $z = y = 0$ and (d) $z = x = 0$.

is seen to be larger than unity and equal to $I_{max} = 32$. This can be explained as follows. Considering that at $\alpha r < 2c$ and $\varphi = 0$, the argument in Eq. (7.33) is purely imaginary, the zero-order aBG beam on this axis interval can be given by

$$E_0(r = x, \phi = 0) = \exp\left(-\frac{x^2}{\omega_0^2}\right) I_0\left\{\sqrt{\alpha x(2c - \alpha x)}\right\}, \tag{7.34}$$

where $I_0(x)$ is a modified zero-order Bessel function. At the ends of the interval $[0, 2c/\alpha]$, the argument of function (7.34) takes a zero value, whereas the Bessel function equals $J_0(0) = I_0(0) = 1$. Therefore, at $x = 0$, the amplitude in Eq. (7.34) equals 1, being equal to $\exp(-4c^2/(\omega_0^2\alpha^2))$ at $x = 2c/\alpha$. On the interval $[0, 2c/\alpha]$, the $I_0(y)$ function reaches its maximum at the point of the maximum argument, which is reached at the interval's midpoint, that is, at $x = c/\alpha$. At the interval's midpoint, the Bessel function equals $I_0(c)$. On the other hand, the Gaussian exponent in (7.34) decreases on the interval from 1 to $x = w_0$. Thus, the maximum of the amplitude in Eq. (7.34) will be found at different points of the interval $[0, 2c/\alpha]$, depending on whether w_0 or c/α is larger. If $c/\alpha < w_0$, the maximum of Eq. (7.34) is found at point $x = c/\alpha$, being equal to $E_0(x = c/\alpha, \phi = 0) = \exp[-c^2/(\omega_0^2\alpha^2)] I_0(c)$. If, on the contrary, $c/\alpha > w_0$, the maximum of (7.34) is at point $x = w_0$ (Figure 7.7c) and equal to $E_0(x = \omega_0, \phi = 0) = e^{-1} I_0\{[\alpha\omega_0(2c - \alpha\omega_0)]^{1/2}\}$. Considering that in Figure 7.7 the latter inequality is observed, the maximal amplitude equals $E_0(x = 10\lambda, \phi = 0) = e^{-1} I_0(\sqrt{19}) \approx 5{,}67$. Thus, the aBG beam's maximal intensity is $I_{max} = \left|E_0(r = 10\lambda, \phi = 0)\right|^2 \cong 32{,}14$. As can be seen from Figure 7.7c and d, the zero-order aBG beam is almost devoid of intensity side-lobes.

Fourier Spectrum of the Asymmetric Bessel-Gaussian Beam

The angular spectrum of a conventional BG beam is known [99]:

$$A_n(\rho, \phi) = (-i)^n \exp\left[-\left(\frac{k\rho\omega_0}{2f}\right)^2 + in\phi\right] I_n\left(\frac{k\alpha\rho\omega_0^2}{2f}\right), \tag{7.35}$$

where:

$I_n(x)$ is a modified Bessel function of the nth order
ρ is the radial coordinate in the Fourier plane
f is the focal length of a lens that forms the spatial spectrum of the BG beam

Taking a linear combination analogous to Eq. (7.22), but composed of functions (7.35), we obtain the spatial spectrum of an aBG beam:

$$A_n(\rho, \phi) = \exp\left[-\left(\frac{k\rho\omega_0}{2f}\right)^2\right] \times \sum_{p=0}^{\infty} \frac{(-ic)^p \exp\left[i(n+p)\phi\right]}{p!} I_p\left(\frac{k\alpha\rho\omega_0^2}{2f}\right). \tag{7.36}$$

Making use of the reference relationship [38]:

$$\sum_{k=0}^{\infty} \frac{t^k}{k!} I_{k+\nu}(x) = x^{\nu/2}(x+2t)^{-\nu/2} I_\nu\left(\sqrt{x^2+2tx}\right), \tag{7.37}$$

the Fourier spectrum of the aBG beam is finally given by

$$\begin{aligned} A_n(\rho,\phi) &= \exp\left[-\left(\frac{k\rho\omega_0}{2f}\right)^2 + in\phi\right] \\ &\times \left(\frac{\xi}{\xi+2ce^{i(\phi-\pi/2)}}\right)^{n/2} I_p\left\{\sqrt{\xi\left[\xi+2c\,e^{i(\phi-\pi/2)}\right]}\right\}, \end{aligned} \tag{7.38}$$

where $\xi = \alpha k\rho\omega_0^2/(2f)$. The angular spectrum in Eq. (7.38) is asymmetric: the modulus of amplitude (7.38) on a circle of fixed radius $\rho = \rho_0$ is maximal at $\varphi = \pi/2$, taking a minimal value at $\varphi = -\pi/2$. The pattern of angular spectrum (7.38) is similar to the amplitude of the aBG beam in Eq. (7.24), with the Bessel function being replaced with a modified Bessel function and the entire pattern rotated by $\pi/2$. Because of this, the asymmetry of spectrum (7.38) is similar to that of the aBG beam in Eq. (7.24).

Orbital Angular Momentum of Asymmetric Bessel-Gaussian Beam

The OAM component J_z (its projection onto the optical axis) and the beam's total intensity I in a plane perpendicular to the optical axis are defined by relationships [96]:

$$J_z = \mathrm{Im}\left\{\iint_{\mathbb{R}^2} E^* \frac{\partial E}{\partial\varphi} r dr d\varphi\right\}, \tag{7.39}$$

$$I = \iint_{\mathbb{R}^2} E^* E r dr d\varphi. \tag{7.40}$$

Substituting the complex amplitude of Eq. (7.22) into Eqs. (7.39) and (7.40) at $z = 0$, we obtain the OAM and the total intensity of the aBG beam:

$$J_z = 2\pi \sum_{p=0}^{\infty} \frac{c^{2p}(n+p)}{(p!)^2} \int_0^{\infty} \exp\left(-\frac{2r^2}{\omega_0^2}\right) J_{n+p}^2(\alpha r) r dr, \tag{7.41}$$

$$I = 2\pi \sum_{p=0}^{\infty} \frac{c^{2p}}{(p!)^2} \int_0^{\infty} \exp\left(-\frac{2r^2}{\omega_0^2}\right) J_{n+p}^2(\alpha r) r dr. \tag{7.42}$$

The constituent integrals in the earlier relations have been known [38]:

$$\int_0^{\infty} x \exp\left(-px^2\right) J_\nu\left(bx\right) J_\nu\left(cx\right) \mathrm{d}x$$
$$= \left(2p\right)^{-1} \exp\left(-\frac{b^2+c^2}{4p}\right) I_\nu\left(\frac{bc}{2p}\right). \tag{7.43}$$

Making use of Eq. (7.43) and dividing (7.41) by (7.42), we deduce a relationship for the OAM normalized with respect to the intensity I:

$$\frac{J_z}{I} = n + \sum_{p=0}^{\infty} \frac{c^{2p} p I_{n+p}\left(y\right)}{\left(p!\right)^2} \left[\sum_{p=0}^{\infty} \frac{c^{2p} I_{n+p}\left(y\right)}{\left(p!\right)^2}\right]^{-1}, \tag{7.44}$$

where $y = \alpha^2 \omega_0^2 / 4$. Attempts to simplify Eq. (7.44) have not met with success. From Eq. (7.44) it follows that OAM of the aBG beams is larger than n, because all terms of the series (7.44) are positive. Thus, with increasing parameter c, the aBG beam becomes more asymmetric and its OAM increases near linearly, as seen from Figure 7.8. Considering that the parameters α and c are real positive numbers, the second summand in Eq. (7.44) can take both integer and fractional positive values. Therefore, from Eq. (7.44) we can also infer that the zero-order aBG beam ($n = 0$) can have any OAM. It is noteworthy that since the scale factors of the Bessel and Gaussian beams given by α and ω_0 are represented as a product in (7.44), different aBG beams for which $\alpha\omega_0$ = const have the same value of OAM (given equal n and c).

Orthogonality of Asymmetric Bessel-Gaussian Beams

Using an approach similar to that used for computing the OAM, we can derive the scalar product of the complex amplitudes of two aBG beams, namely, the nth order aBG beam with parameters α and c and the mth order aBG beam with parameters β and d:

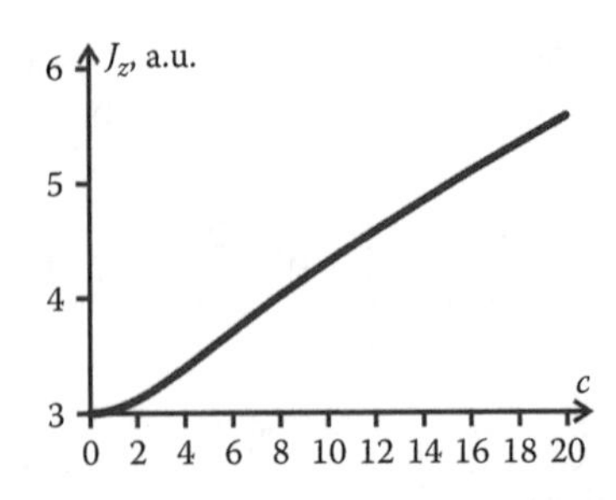

FIGURE 7.8 OAM as a function of asymmetry parameter c at $n = 3$, $w_0 = 10\lambda$, and $\alpha = 1/(10\lambda)$.

$$\left(E_{n\alpha c}, E_{m\beta d}\right) = \frac{\pi\omega_0^2}{2}\left(\frac{d^*}{c}\right)^{\frac{n-m}{2}} \exp\left[-\frac{\omega_0^2}{8}\left(\alpha^2+\beta^2\right)\right]$$

$$\times\sum_{p=0}^{\infty}\frac{\left(cd^*\right)^{p+\frac{|n-m|}{2}}}{p!\left(p+|n-m|\right)!} I_{p+\max(m,n)}\left(\frac{\omega_0^2\alpha\beta}{4}\right), \tag{7.45}$$

where $I_{n-m}(x)$ is a modified Bessel function. From Eq. (7.45) it is seen that as distinct from the Bessel modes [217], the aBG beams are not orthogonal either in terms of the scale factor, the Bessel function order, or the asymmetry parameter. Note that in Eq. (7.45), the asymmetry parameters c and d are assumed to be complex, with * denoting complex conjugation.

Note that with the Gaussian beam's radius tending to infinity, $w_0 \to \infty$, the aBG beams change to the non-diffracting asymmetric elegant Bessel modes [219].

Experiment

The experimental setup is shown in Figure 7.9. The experiment was conducted using a phase spatial light modulator PLUTO VIS. The light from a solid-state laser of wavelength 532 nm was expanded by a collimator and passed through an 8-mm diaphragm. As a result, a uniform intensity distribution was obtained, which can be treated as a plane wave. Then, after passing through a light-splitting cube, the light illuminated the spatial light modulator, was reflected from it, and diverted by the light-splitting cube to a CCD-camera. The computer-generated phase of the aBG beam in combination with that of a parabolic lens of focus 960 mm was transmitted to the half-tone modulator.

The CCD-camera can be moved through a short distance in the close proximity to the focus. Figure 7.10a depicts (a) the computer-generated phase distribution on the modulator (without the lens) and intensity patterns measured by the CCD-camera at distances (b) 850 mm, (c) 900 mm, and (d) 950 mm from the modulator. A 1920×1080 modulator composed of 8-μm sensitive cells was used in the experiment. The phase depicted in Figure 7.10 consisted of 1024×1024 pixels and was generated at the modulator's center. Thus, the size of the generated phase distribution was 8.2 mm.

From Figure 7.10, the modulator-aided optical semi-crescent formed in the converging laser beam's cross-section is seen not only to rotate during propagation (having turned by nearly $\pi/2$ at distance 100 mm) but also to be distorted. The said

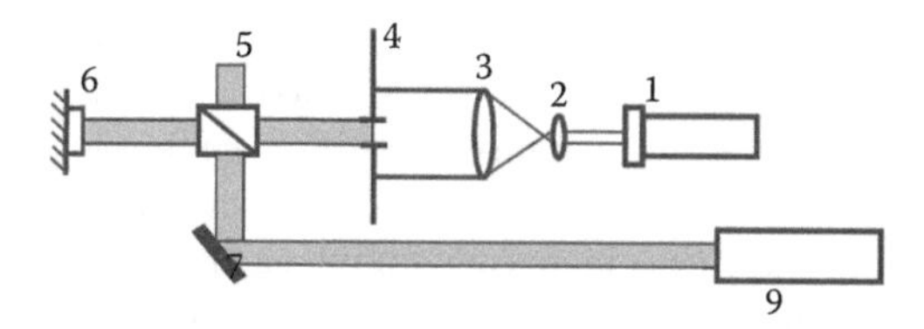

FIGURE 7.9 Experimental setup. 1—solid-state laser of wavelength 532 nm, 2, 3—collimators, 4—a diaphragm, 5—a light-splitting cube, 6—a spatial light modulator PLUTO VIS, 7—a mirror, and 9—a CCD-camera.

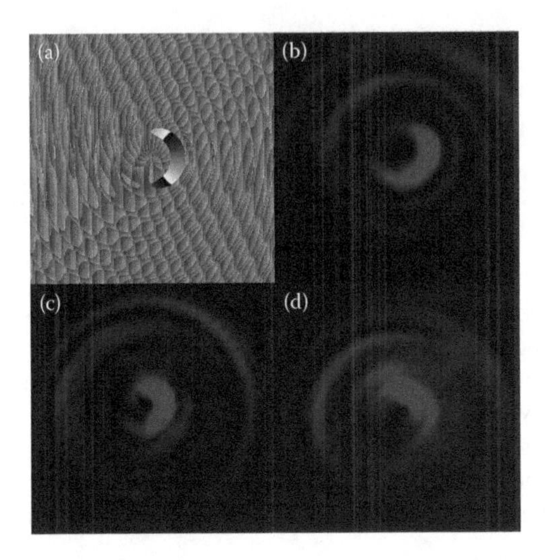

FIGURE 7.10 (a) Computer-generated phase distribution on the modulator PLUTO VIS and intensity patterns measured at the distance of (b) 850 mm, (c) 900 mm, and (d) 950 mm from the modulator.

distortion occurs because the amplitude distribution of the aBG beam in Eq. (7.24) is only partially taken into account in the phase distribution in Figure 7.10a.

Summing up, we have obtained a solution of the paraxial Helmholtz equation to describe a three-parameter family of finite-energy aBG beams, which is described by integer-order Bessel functions of the first kind of complex argument; at $c = 0$, the aBG beams coincide with the conventional BG beams [214]. The intensity distribution generated by the aBG beams has a countable number of isolated optical nulls located on a horizontal line in the original plane ($z = y = 0$). The said optical nulls correspond to the optical vortices with unit topological charges and opposite signs on the different sides of the origin; the topological charge of the optical null located at the origin is equal to the Bessel function order. At large values of the asymmetry parameter $c \gg 1$, the aBG beams rotate as a whole as they propagate in free space, making a $\pi/4$ turn after passing a distance of the Rayleigh range and another $\pi/4$ turn over the entire travel path; at small $c \leq 1$, the beams rotate in a more complex pattern, with the optical vortices (isolated optical nulls) initially found in the beam's transverse plane disappearing at some distance (except the null at the origin) and the beam making a near-($\pi/2$) turn at the distance equal to Rayleigh range. The aBG beams have an asymmetric angular spectrum, which is represented as the product of a Gaussian function by a modified Bessel function with complex argument; the spectrum has asymmetry similar to that of the aBG beam but is turned by $\pi/2$. The aBG beams can have integer and fractional OAM, which increases with increasing number n and shows a near-linear increase with increasing asymmetry parameter c; the zero-order aBG beam has the intensity maximum off-set from the optical axis by c/α and isolated optical nulls, and can carry an arbitrary OAM, depending on the choice of the asymmetry parameter c. The aBG beams are not orthogonal with respect to any of the parameters n, α, and c. Using a liquid crystal light modulator, we have experimentally generated a converging laser beam similar to the aBG beam with a crescent-shape transverse intensity pattern, which has made a near-($\pi/2$) anti-clockwise turn.

7.3 LOMMEL MODES

Among well-known laser beams, a peculiar place belongs to nondiffracting beams. The complex amplitude pattern in their cross-section is structurally preserved, despite diffraction, as the beam propagates over an arbitrary distance along the optical axis. It has been known that three-dimensional (3D) Bessel modes [210] and two-dimensional (2D) Airy beams [165] are nondiffracting. The concept of nondiffracting beams has also been extended on higher-dimension spaces [220]. An optical beam whose angular spectrum of plane waves is described by an infinitely thin circle is also known to be nondiffracting upon propagation in a 3D space. For instance, nondiffracting beams described by a linear combination of Bessel modes have been reported [213]. The complex amplitude of such beams is described by a Mathieu function. aB modes that generate a crescent-shaped transverse intensity pattern have also been proposed [221], with an extended family of the beams later obtained by introducing an extra parameter intended to control the transverse intensity pattern asymmetry [219]. Note that the beams reported in Refs. [222–224] represent a superposition of Bessel beams with identical axial projections of the wave vector. Also, there exist beams described by a linear combination of Bessel modes with different axial projections of the wave vector [222–224]. Such beams allow variations of their transverse diffraction pattern along the optical axis to be controlled. The practical significance of the nondiffracting beams stems from their ability to show stability during propagation in a turbulent atmosphere [225], whereas femtosecond Bessel pulses preserve their shape as they propagate [226].

In this section, we consider a linear combination of Bessel modes with their coefficients fitted so that the beam's complex amplitude is described by a Lommel function of two variables, one of them being complex. The Lommel functions of two variables are not new to optics. By way of illustration, a 3D near-focus optical intensity pattern a spherical monochromatic wave produces after passing through a circular aperture and converging to an axial focus has been analyzed ([163], § 8.8). The desired intensity pattern [163] was obtained using a Fresnel integral expressed through a Lommel function of two variables [227]. The Lommel functions were also used in the description of localized wave pulses in guided media [228,229] and with bandlimited spectrum [230]. In a recent work [231], the Lommel functions were used to describe the lens-aided focusing of a vortex Laguerre-Gaussian beam with zero radial index bounded by a circular aperture. Unlike the conventional Bessel modes [210], the diffraction pattern of the Lommel mode is devoid of a radial symmetry in the form of bright rings and, unlike asymmetric modes [219,221], it is symmetric about two, rather than one, coordinate axes. In this section, the OAM of the Lommel modes is rigorously calculated. It has been found to exceed that of the Bessel mode entering into a linear combination with the least topological charge. Similar to all other nondiffracting beams, the Lommel modes possess an infinite energy and, thus, can be practically implemented only approximately. This work is the follow-up to our previous paper [232]. In the following, we analyze a wider class of Lommel modes with a complex asymmetry parameter c (complexity affects orientation of the intensity pattern), also adding results on the calculation of the side-lobes in the transverse diffraction pattern produced by the Lommel mode and the distortion of the beam's symmetry coming from shifting the amplitude function to the complex plane.

Complex Amplitude of Lommel Modes

We shall study a laser beam whose angular spectrum of plane waves is given by

$$A(\rho,\theta)=\frac{(-i)^n}{\lambda\alpha}\delta\left(\rho-\frac{\alpha}{k}\right)\sum_{p=0}^{\infty}c^{2p}\exp\left[i(n+2p)\theta\right], \tag{7.46}$$

where:

- (ρ, θ) are the polar coordinates in the spectral plane
- $\delta(x)$ is the Dirac delta-function
- $k = 2\pi/\lambda$ is the wave number of light of wavelength λ, with the parameters α, c, and n, respectively, defining the beam's scale, asymmetry, and OAM, as will be shown in the following

Equivalently, using the sum of the geometric progression, Eq. (7.46) can be rewritten as

$$A(\rho,\theta)=\frac{(-i)^n\exp(in\theta)}{\lambda\alpha\left[1-c^2\exp(2i\theta)\right]}\delta\left(\rho-\frac{\alpha}{k}\right), \tag{7.47}$$

From Eq. (7.47), the modulus of the spectrum amplitude is seen to vary along a ring of radius $\rho = \alpha/k$: at real c, the maximum occurs at $\theta = 0, \pi$, with the minimum found at $\theta = \pm\pi/2$. The complex amplitude of the Lommel mode is derived as the following integral transform of the angular spectrum (7.46)

$$E_n(r,\varphi,z)=\int_0^{\infty}\int_0^{2\pi}A(\rho,\theta)\exp\left[ikr\rho\cos(\theta-\varphi)+ikz\sqrt{1-\rho^2}\right]\rho d\rho d\theta, \tag{7.48}$$

where (r, φ, z) are cylindrical coordinates. Substitution of the angular spectrum in Eq. (7.46) into Eq. (7.48) leads to the expression of the complex amplitude represented as a series of Bessel functions:

$$E_n(r,\varphi,z)=\exp\left(iz\sqrt{k^2-\alpha^2}\right)\sum_{p=0}^{\infty}(-1)^p c^{2p}\exp\left[i(n+2p)\varphi\right]J_{n+2p}(\alpha r), \tag{7.49}$$

Using the Lommel function of two variables $U_n(w, z)$ [227]

$$U_n(w,\zeta)=\sum_{p=0}^{\infty}(-1)^p\left(\frac{w}{\zeta}\right)^{n+2p}J_{n+2p}(\zeta), \tag{7.50}$$

we can rewrite Eq. (7.49):

$$E_n(r,\varphi,z) = c^{-n}\exp\left(iz\sqrt{k^2-\alpha^2}\right)U_n\left[c\alpha r\exp(i\varphi),\alpha r\right]. \quad (7.51)$$

Note that unlike aB beams described in [219,221], the Lommel mode's asymmetry parameter c cannot take arbitrary values and is required to be smaller than unity in modulus to provide the convergence of the series in Eqs. (7.46) and (7.49). In Eq. (7.49), the α parameter enters into the argument of the Bessel function, thus characterizing the Lommel mode's scale (bright ring's width) and enabling the same-length axial projection of all constituent Bessel modes of the linear combination in Eq. (7.49). Using (7.49), it can easily be shown that

$$\begin{aligned} |E_n(r,\varphi,z)| &= |E_n(r,-\varphi,z)| = |E_n(r,\pi-\varphi,z)| \\ &= \left|\sum_{p=0}^{\infty}(-1)^p c^{2p}\exp\left[i(n+2p)\varphi\right]J_{n+2p}(\alpha r)\right|. \end{aligned} \quad (7.52)$$

From (7.52) it follows that, unlike aB beams reported in [219,221], the transverse intensity pattern of the Lommel beam in Eq. (7.51) shows symmetry about both the horizontal Oxz-plane and the vertical Oyz-plane, provided that the c parameter takes real or purely imaginary values.

When $c = 0$, all series terms in Eq. (7.49) take zero values, except the first one (at $p = 0$), and the Lommel mode changes to a conventional Bessel mode:

$$E_n(r,\varphi,z) = \exp\left[iz\left(k^2-\alpha^2\right)^{1/2} + in\varphi\right]J_n(\alpha r). \quad (7.53)$$

The Lommel modes in Eq. (7.51) can be rewritten in the Cartesian coordinates:

$$E_n(x,y,z;\alpha,c) = c^{-n}\exp\left(iz\sqrt{k^2-\alpha^2}\right)U_n\left[c\alpha(x+iy),\alpha\sqrt{x^2+y^2}\right]. \quad (7.54)$$

Thus, the Lommel modes in Eqs. (7.51) and (7.54) offer a generalization of the well-known Bessel modes. Note that the Lommel modes in Eq. (7.54) are described by a three-parameter (n, α, c) family of functions, which are proportional to the Lommel function and obtained by solving the Helmholtz equation $(\nabla^2 + k^2)E(x,y,z) = 0$. On the other hand, the Lommel modes represent a superposition of Bessel modes with the same-length projections of the wave vector on the optical axis. Note that while propagating in free space, all Bessel modes of this kind have the same phase velocities and, thus, their linear combination is also a mode. Although one may devise a lot of similar linear combinations with different coefficients but only few of them will be equivalent to the familiar functions. The Lommel modes possess many properties characteristic of Bessel modes. Unlike Bessel modes, the Lommel modes are devoid of circular symmetry, showing instead symmetry about the Cartesian axes. As we show in the following, the Lommel modes carry the OAM that may be continuously varied by smoothly varying the modulus of the parameter $|c| < 1$. Meanwhile, the

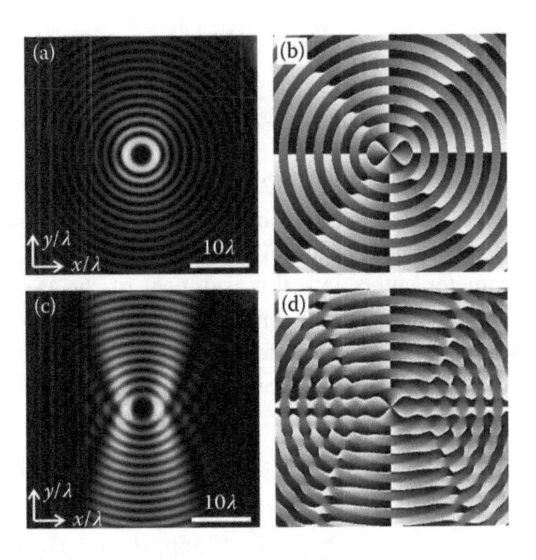

FIGURE 7.11 Transverse (a, c) intensity and (b, d) phase patterns generated by the Lommel beams: (a, b) $c = 0.5i$ and (c, d) $c = 0.9i$.

OAM of the Bessel modes changes in a discrete manner. The argument arg(c) of the asymmetry parameter c defines the angle of rotation of the transverse intensity pattern of the Lommel mode relative to the Cartesian axes. As the argument arg(c) changes, the Lommel mode rotates about the optical axis. Similarly to Bessel modes, the Lommel modes possess an infinite energy and can be practically implemented only approximately, using a bounded aperture, they are diffraction-free and self-reconstructing upon propagation following minor distortions.

Figure 7.11 depicts transverse intensity and phase patterns (at $z = 0$) generated by the Lommel beams with wavelength $\lambda = 532$ nm, topological charge $n = 4$, scale factor $\alpha = k/3$, and the asymmetry parameter (a, b) $c = 0.5i$ and (c, d) $c = 0.9i$. The intensity pattern in Figure 7.11 is shown for the region $-20\lambda \leq x$, $y \leq 20\lambda$. Figure 7.11 shows the results of numerical simulation using Eq. (7.51).

Figure 7.11 suggests that at small absolute values of c the diffraction pattern is similar to that of the Bessel beam but stretched along the x-axis. In the 2D case, when the Lommel beams are propagating in the Oxz-plane, they are similar in shape to the accelerating elliptic modes discussed in Ref. [178]. As the c parameter is further increased, the Lommel beam's asymmetry grows, resulting in two crescents and a central optical null in the transverse plane. In optical micromanipulation, such an intensity pattern is well suited for trapping and guidance of biological microobjects [233].

Generally speaking, the c parameter can take complex values. Note that for the series in Eq. (7.49) to converge to a finite number, the symmetry parameter needs to be less than unity in the absolute value. The argument of the parameter c defines the angle of rotation of the transverse diffraction pattern generated by the Lommel mode (Figure 7.12). For instance, in Figure 7.11 the c parameter is purely imaginary, causing the crescents to become spatially separated along the x-axis. If the c parameter takes a real value, the resulting diffraction pattern will be rotated by 90°

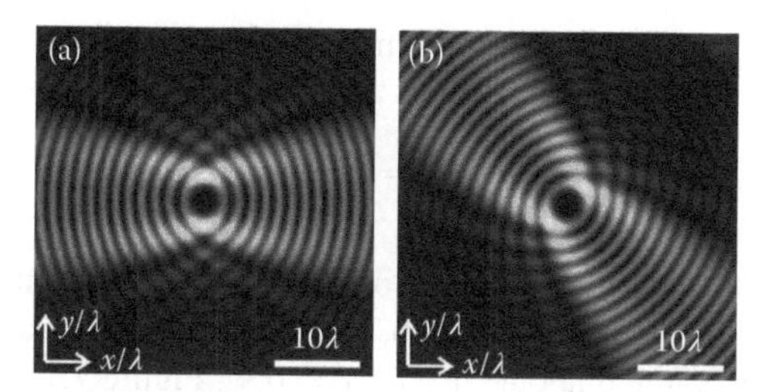

FIGURE 7.12 Transverse intensity patterns generated by the Lommel beams for (a) $c = 0.9$ and (b) $c = 0.9\exp(i\pi/4)$. The rest simulation parameters remain the same as in Figure 7.11.

when compared with that in Figure 7.11c, with the crescents being spatially separated along the y-axis (Figure 7.12a).

With increasing parameter c, not only does the transverse profile asymmetry grow but the intensity contrast on the x- and y-axes increases as well. This is clearly seen from Figure 7.13, which was also derived from Eq. (7.51) for the same parameters as in Figure 7.11.

It is seen from Figure 7.13 that while at $c = 0.5i$ the x-component of the intensity maximum is about 1.5. times that of the vertical component maximum, at $c = 0.9i$ the said ratio reaches 3.5. Note that on the y-axis the side-lobes exceed the first maxima.

Using the asymptotic expansion of the Bessel function for large arguments (Eq. 9.2.1 in [101]):

$$J_\nu(\zeta) = \sqrt{\frac{2}{\pi\zeta}}\left[\cos\left(\zeta - \frac{\nu\pi}{2} - \frac{\pi}{4}\right) + \exp\left(|\mathrm{Im}\,\zeta|\right)O\left(|\zeta|^{-1}\right)\right], \ |\arg\zeta| < \pi, \quad (7.55)$$

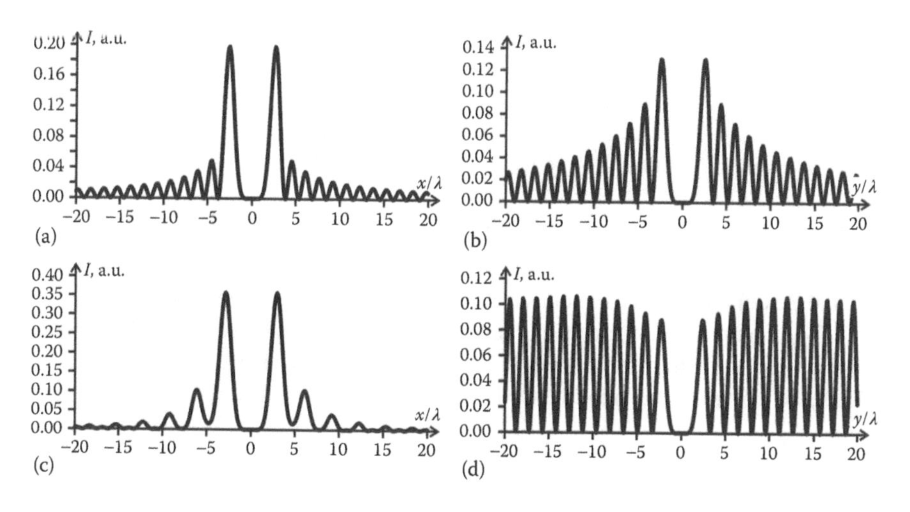

FIGURE 7.13 Intensity profiles at (a, c) $y = 0$ and (b, d) $x = 0$ for Lommel modes: (a, b) $c = 0.5i$ and (c, d) $c = 0.9i$. The wavelengths are shown on the x-axis and the relative intensity—on the y-axis.

where $O(|z|^{-1})$ is the error term of approximation, we can conclude that

$$\left|\frac{E_n(r,\varphi,z)}{E_n(r,\varphi+\pi/2,z)}\right|^2=\left|\frac{1+c^2\exp(i2\varphi)}{1-c^2\exp(i2\varphi)}\right|^2. \tag{7.56}$$

This equation predicts that at the periphery of the diffraction pattern the ratio between intensity maxima in horizontal plane ($y = 0$, $\varphi = 0$) and vertical plane ($x = 0$, $\varphi = \pi/2$) must be 36% for $c = 0.5i$ and about 11% for $c = 0.9i$. Figure 7.13 proves this, since the intensity maxima at the periphery of Figure 7.13b are approximately 3 times higher than those in Figure 7.13a, while the intensity maxima at the periphery of Figure 7.13d are approximately 10 times higher than those in Figure 7.13c.

Orbital Angular Momentum of the Lommel Modes

The OAM projection onto the optical axis, J_z, and the total transverse intensity I of the optical beam are given by [188]

$$J_z=\operatorname{Im}\left\{\iint_{\mathbb{R}^2}E^*\frac{\partial E}{\partial\varphi}r\,\mathrm{d}r\,\mathrm{d}\varphi\right\}=\operatorname{Im}\left\{\lim_{R\to\infty}\int_0^R\int_0^{2\pi}E^*\frac{\partial E}{\partial\varphi}r\,\mathrm{d}r\,\mathrm{d}\varphi\right\}, \tag{7.57}$$

$$I=\iint_{\mathbb{R}^2}E^*E\,r\,\mathrm{d}r\,\mathrm{d}\varphi=\lim_{R\to\infty}\int_0^R\int_0^{2\pi}E^*E\,r\,\mathrm{d}r\,\mathrm{d}\varphi. \tag{7.58}$$

Substituting complex amplitude (7.49) into (7.57) and (7.58) yields

$$J_z=2\pi\lim_{R\to\infty}\sum_{p=0}^{\infty}(n+2p)\left(cc^*\right)^{2p}\int_0^R J_{n+2p}^2(\alpha r)r\,dr. \tag{7.59}$$

$$I=2\pi\lim_{R\to\infty}\sum_{p=0}^{\infty}\left(cc^*\right)^{2p}\int_0^R J_{n+2p}^2(\alpha r)r\,dr. \tag{7.60}$$

The integrals in (7.59) and (7.60) have been known [227] (see 5.54.2):

$$\int J_p^2(\alpha r)r\,\mathrm{d}r=\frac{r^2}{2}\left[J_p^2(\alpha r)-J_{p-1}(\alpha r)J_{p+1}(\alpha r)\right]. \tag{7.61}$$

Using the Bessel function asymptotics at large values of the argument (Eq. 7.55), we find that all the integrals in (7.59) and (7.60) are independent of the Bessel function and equal to $R/(\pi\alpha)$. Then, using numerical series 0.246.1 and 0.246.2 from [135] and dividing (7.59) by (7.60), we derive a relation for the OAM normalized with respect to the intensity:

$$\frac{J_z}{I}=\frac{\sum_{p=0}^{\infty}\left(cc^*\right)^{2p}(n+2p)}{\sum_{p=0}^{\infty}\left(cc^*\right)^{2p}}=n+\frac{2|c|^4}{1-|c|^4}. \tag{7.62}$$

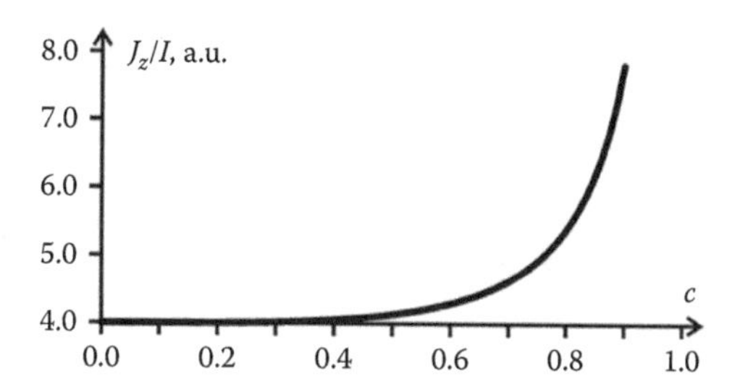

FIGURE 7.14 Normalized OAM vs. the asymmetry parameter c for $n = 4$.

From (7.62) it follows that as the asymmetry parameter $|c| < 1$ is growing, the OAM increases (Figure 7.14). In Figure 7.11a, at $c = 0.5i$, the normalized OAM equals $J_z/I \approx 4.1$, while in Figure 7.11c, at $c = 0.9i$, it is $J_z/I \approx 7.8$.

From Eq. (7.62) it follows that for two Lommel modes with different topological charges n_1 and n_2 one can fit the asymmetry parameters c_1 and c_2 such that the respective OAMs will be the same. This requires the fulfilment of the condition:

$$n_1 - n_2 = \frac{2\left(c_2^4 - c_1^4\right)}{\left(1 - c_1^4\right)\left(1 - c_2^4\right)}. \tag{7.63}$$

Besides, if the numbers n_1 and n_2 have different parity, the modes will also be mutually orthogonal. It is also seen from Eq. (7.62) that for a definite topological charge n of the Lommel mode the OAM can be varied continuously, thus changing the absolute value of the asymmetry parameter c and, thus, causing the transverse intensity pattern to be "deformed." With conventional Bessel or Laguerre-Gauss modes, the OAM can only be varied in a discrete manner, by changing the magnitude of the topological charge n. Considering that the OAM offers an additional degree of freedom in free-space data encoding and transmission (including quantum communications systems) [234], the property of the Lommel mode's OAM to undergo continuous variations may enable the data transmission channel to be essentially multiplexed.

In the following we will consider in more detail the derivation of Eq. (7.62) for the OAM. Substituting complex amplitude (7.49) into (7.57) yields

$$\begin{aligned} J_z &= \mathrm{Im}\left\{ \lim_{R\to\infty} \int_0^R \int_0^{2\pi} E^* \frac{\partial E}{\partial \varphi} r\,\mathrm{d}r\,\mathrm{d}\varphi \right\} \\ &= \mathrm{Im}\left\{ \lim_{R\to\infty} \int_0^R \int_0^{2\pi} \left[\sum_{p=0}^{\infty} (-1)^p \left(c^*\right)^{2p} \exp\left[-i(n+2p)\varphi\right] J_{n+2p}(\alpha r) \right] \right. \\ &\quad \times \left[i(n+2q) \sum_{q=0}^{\infty} (-1)^q c^{2q} \exp\left[i(n+2q)\varphi\right] J_{n+2q}(\alpha r) \right] r\,\mathrm{d}r\mathrm{d}\varphi. \end{aligned} \tag{7.64}$$

The integral over φ yields 2π when $p = q$ and 0 when $p \neq q$. Thus, Eq. (7.64) reduces to

$$J_z = 2\pi \lim_{R\to\infty} \int_0^R \left[\sum_{p=0}^{\infty} (n+2p)\left(cc^*\right)^{2p} J_{n+2p}^2(\alpha r) \right] r\,\mathrm{d}r. \tag{7.65}$$

Using the integral (7.61), we obtain

$$J_z = 2\pi \lim_{R\to\infty} \sum_{p=0}^{\infty} (n+2p)\left(cc^*\right)^{2p} \frac{R^2}{2}\left[J_{n+2p}^2(\alpha R) - J_{n+2p-1}(\alpha R) J_{n+2p+1}(\alpha R)\right]. \tag{7.66}$$

For large values of R, there exists an asymptotic form of the Bessel function at large arguments (Eq. 7.55), therefore Eq. (7.66) leads to

$$\begin{aligned} J_z &= 2\pi \lim_{R\to\infty} \sum_{p=0}^{\infty} (n+2p)\left(cc^*\right)^{2p} \frac{R^2}{2}\frac{2}{\pi\alpha R} \\ &\times \left\{ \cos^2\left[\alpha R - \frac{\pi}{2}(n+2p) - \frac{\pi}{4}\right] \right. \\ &\left. -\cos\left[\alpha R - \frac{\pi}{2}(n+2p-1) - \frac{\pi}{4}\right] \cos\left[\alpha R - \frac{\pi}{2}(n+2p+1) - \frac{\pi}{4}\right] \right\} \\ &= 2\pi \lim_{R\to\infty} \sum_{p=0}^{\infty} (n+2p)\left(cc^*\right)^{2p} \frac{R^2}{2}\frac{2}{\pi\alpha R} = \frac{2}{\alpha} \lim_{R\to\infty} \left[R \sum_{p=0}^{\infty} (n+2p)\left(cc^*\right)^{2p} \right]. \end{aligned} \tag{7.67}$$

Similarly, it can be shown that

$$I = \frac{2}{\alpha} \lim_{R\to\infty} \left[R \sum_{p=0}^{\infty} \left(cc^*\right)^{2p} \right]. \tag{7.68}$$

According to Eqs. (7.67) and (7.68), both the total OAM and total intensity are infinite, but the normalized OAM, that is, the total OAM divided by the total intensity, has a finite value (Eq. 7.61).

Orthogonality of Complex Amplitudes of Lommel Modes

In a similar way to calculating the OAM in the previous section, we may derive a scalar product of two Lommel modes with topological charges n and m, scale factors α and β, and asymmetry parameters c and d:

$$\left(E_{n\alpha c}, E_{m\beta d}\right) = \iint_{\mathbb{R}^2} E_{n\alpha c} E^*_{m\beta d} r dr d\varphi$$

$$= \begin{cases} 2\pi\left(-1\right)^{(n-m)/2} \dfrac{\delta\left(\alpha-\beta\right)}{\alpha} \dfrac{\left(d^*\right)^{n-m}}{1-\left(cd^*\right)^2}, \text{ if } \left(m+n\right) \text{ is even and } n \geq m, \\ 2\pi\left(-1\right)^{(n-m)/2} \dfrac{\delta\left(\alpha-\beta\right)}{\alpha} \dfrac{c^{m-n}}{1-\left(cd^*\right)^2}, \text{ if } \left(m+n\right) \text{ is even and } n \leq m, \\ 0, \text{ if } \left(m+n\right) \text{ is odd.} \end{cases} \tag{7.69}$$

From Eq. (7.69) it is seen that similar to the conventional and aB modes, complex amplitudes of two Lommel modes with different scale factors are orthogonal to each other. From (7.69) it is also seen that unlike aB modes from [219,221], the Lommel modes are divided into two classes, with an even and odd topological charge. The complex amplitudes of the beams belonging to different classes are orthogonal to each other.

Lommel Modes with Complex Shift in the Cartesian Plane

If the solution of the Helmholtz equation $(\nabla^2 + k^2)E(x, y, z) = 0$ is shifted along any Cartesian axis by a constant (including complex) value, the result will also define a solution to the Helmholtz equation. Because of this, alongside the Lommel modes in Eq. (7.54), we may consider modes with the following complex amplitude:

$$\begin{aligned} &E_n\left(x, y, z; \alpha, c, x_0, y_0\right) \\ &= c^{-n} \exp\left(iz\sqrt{k^2-\alpha^2}\right) U_n\left\{c\alpha\left[\left(x-x_0\right)+i\left(y-y_0\right)\right], \alpha\sqrt{\left(x-x_0\right)^2+\left(y-y_0\right)^2}\right\}. \end{aligned} \tag{7.70}$$

Figure 7.15 depicts the transverse intensity distribution generated by the Lommel mode at wavelength $\lambda = 532$ nm, topological charge $n = 4$, scale factor $\alpha = k/3$, the asymmetry parameter (a) $c = 0.9i$: $x_0 = 0.3\lambda i$, $y_0 = 0$ and (b) $x_0 = 0.3\lambda i$, $y_0 = 0.1\lambda i$.

It is seen from Figure 7.15 that imaginary values of the shift parameters x_0 and y_0 allow obtaining diffraction patterns with additional asymmetry. The diffraction patterns in Figure 7.11 have two symmetry axes, x and y, the pattern in Figure 7.15a has only one symmetry axis y, whereas the pattern in Figure 7.15b has no symmetry axis whatever.

Summing up, we have obtained the following results. A new solution of the Helmholtz equation that describes a three-parameter family of nondiffracting non-paraxial Lommel modes has been derived. Their complex amplitude is described by a Lommel function of two variables, with the first variable being complex (Eq. 7.51).

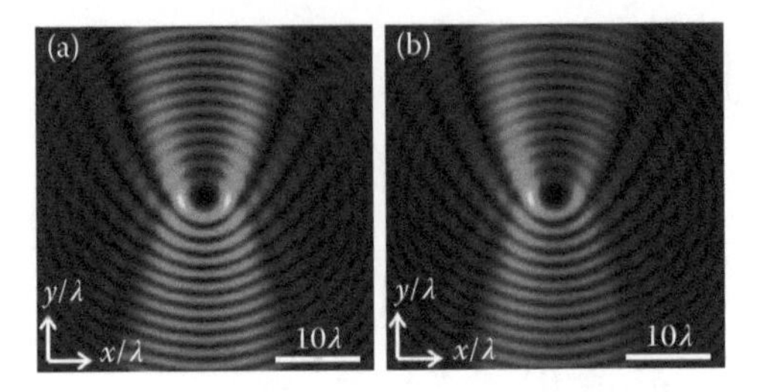

FIGURE 7.15 Transverse intensity patterns generated by the Lommel modes for (a) $x_0 = 0.3\lambda i$, $y_0 = 0$ and (b) $x_0 = 0.3\lambda i$, $y_0 = 0.1\lambda i$.

With increasing modulus of the asymmetry parameter c, the side-lobe intensity of the Lommel modes increases along one coordinate axis and decreases along the other. Complexity of this asymmetry parameter affects orientation of the diffraction pattern. Like any other 3D nondiffracting beams, the Lommel modes have an infinitely-thin ring-shaped angular spectrum of plane waves, which only depends on the polar angle (Eqs. 7.46 and 7.47). Lommel modes carry the OAM that grows linearly with increasing mode number n and nonlinearly with increasing asymmetry parameter c (Eq. 7.62). Thus, in contrast with the Bessel beams, two Lommel modes can have different topological charge at the same OAM and vice versa. Complex amplitudes of the Lommel functions are mutually orthogonal at different scale factor α and nonorthogonal at different asymmetry parameter c; the complex amplitudes of two different-parity beams are orthogonal. The imaginary shift of the complex amplitude along the Cartesian axes leads to additional asymmetry of the diffraction pattern, with the intensity increased in one semi-plane and decreased in the other.

7.4 SUPERPOSITIONS OF ASYMMETRICAL BESSEL BEAMS

In recent papers [219,221,235,236], nonparaxial asymmetrical Bessel modes (aB-modes) and paraxial aBG beams were considered. Transverse intensity distribution of these laser beams has the form of a crescent. In Ref. [237] the aB-modes were investigated experimentally using a digital micromirror device. In [238], by analogy with [219] (i.e., by introducing a complex shift of the Bessel mode), Chebyshev-Bessel asymmetrical beams were considered. In [239] it is proposed to use aB-beams as the acoustic vortex beams. In [240] vector diffraction-free beams are studied with fractional OAM, including asymmetrical (Mathieu and Weber) beams, analogous to beams of [219].

In this section, we show that aB-beams [219] are generated by conventional symmetrical Bessel modes and BG beams by complex shift of the argument of the Bessel function in Cartesian plane. We also consider asymmetrical Bessel beams of the second type, which differ from beams of [219] by the type of complex shift. In addition, we consider superposition of the first (I) and second (II) type aB-beams and we show that despite fractional OAM of the beams of both types, OAM of the sum and difference of the beams does not depend on the asymmetry parameter c and is equal to the topological charge of the Bessel mode n. This implies the possibility of generating nonparaxial modes with different transverse intensity distribution (both symmetrical or asymmetrical), but with the same integer OAM.

Asymmetrical Bessel Modes of the Second Type

Complex amplitude of the aB-mode from [219] reads as

$$E_1(r,\varphi,c)=\left[\frac{\alpha r}{\alpha r-2c\exp(i\varphi)}\right]^{n/2} J_n\left\{\sqrt{\alpha r(\alpha r-2c\exp(i\varphi))}\right\}\exp(in\varphi), \quad (7.71)$$

where:

$J_n(x)$ is the nth order Bessel function of the first kind
(r, φ) are the polar coordinates
α is the scaling factor
c is dimensionless coefficient (complex-valued in general)

It can be shown that the mode (7.71) is generated by conventional Bessel mode

$$E_1(r,\varphi,c=0)=J_n(\alpha r)\exp(in\varphi) \quad (7.72)$$

by complex shift along one Cartesian coordinate. Indeed, light field (7.71) has the following form in Cartesian coordinates:

$$E_1(x,y,c)=\left(\frac{x+iy}{x-iy-2c/\alpha}\right)^{n/2} J_n\left\{\alpha\sqrt{(x+iy)(x-iy-2c/\alpha)}\right\}. \quad (7.73)$$

Complex amplitude (7.73) can be reduced to Eq. (7.72) by the change of variables:

$$\begin{cases} y=y'+ic/\alpha, \\ x=x'+c/\alpha. \end{cases} \quad (7.74)$$

The parameter c in Eq. (7.74) we will assume to be a real positive number. Shift of the mode (7.72) along the horizontal axis (x) is accompanied by its modification since shift along the vertical axis (y) in (7.74) is purely imaginary. This structural change in the beam transverse intensity distribution leads to the fact that instead of the light ring, mode has the form of a weak ellipse ($c < 1$), increasing crescent ($c > 1$) or astigmatic Gaussian beam ($c \gg 1$). It is seen in Eq. (7.74) that absolute values and signs of the shifts are the same for axes x and y. Generally, however, it is not necessary and in order for the mode (7.72) to remain mode, shifts along Cartesian coordinates can be different both in modulus and in sign. But further, for obtaining new Bessel modes by superposition of two different modes, we will constrain ourselves by the two modifications of the beam (7.71). Let us consider a beam with a negative value of the parameter $c' = -c$ ($c > 0$). Then, instead of (7.71) we obtain the complex amplitude of the aB-beam with a shape of decreasing crescent:

$$E_1(r,\varphi,-c)=\left[\frac{\alpha r}{\alpha r+2c\exp(i\varphi)}\right]^{n/2} J_n\left\{\sqrt{\alpha r(\alpha r+2c\exp(i\varphi))}\right\}\exp(in\varphi). \quad (7.75)$$

Now we consider another modification of the Bessel beam (7.71) when shifts along Cartesian coordinates are same in modulus, but opposite in sign:

$$\begin{cases} y = y' + ic/\alpha, \\ x = x' - c/\alpha. \end{cases} \tag{7.76}$$

Thus, we obtain the second-type aB-beam, which also has transverse intensity distribution with a shape of increasing crescent, but shifted from the crescent of the first-type aB-mode with the same value $c > 0$:

$$E_2(x, y, c) = \left(\frac{x + iy + 2c/\alpha}{x - iy} \right)^{n/2} J_n \left\{ \alpha \sqrt{(x - iy)(x + iy + 2c/\alpha)} \right\}. \tag{7.77}$$

In polar coordinates, complex amplitude of the second-type Bessel mode (7.77) reads as

$$E_2(r, \varphi, c) = \left[\frac{\alpha r + 2c \exp(-i\varphi)}{\alpha r} \right]^{n/2} J_n \left\{ \sqrt{\alpha r (\alpha r + 2c \exp(-i\varphi))} \right\} \exp(in\varphi). \tag{7.78}$$

The mirror-like mode of the second type for the mode (7.78) is obtained the same way as the mode (7.75) by replacing the parameter c to $-c$:

$$E_2(r, \phi, -c) = \left[\frac{\alpha r - 2c \exp(-i\phi)}{\alpha r} \right]^{n/2} J_n \left\{ \sqrt{\alpha r (\alpha r - 2c \exp(-i\phi))} \right\} \exp(in\phi). \tag{7.79}$$

Asymmetrical Bessel mode (7.79) has a transverse intensity distribution in a form of decreasing crescent, the same as mode (7.75), but shifted from the original mode (7.72) to the other distance at the same value of c.

OAMs are different for the modes of the first (7.71), (7.75) and second (7.78), (7.79) types:

$$\frac{J_z}{I} = n \pm \frac{c I_1(2c)}{I_0(2c)}, \tag{7.80}$$

where:

$I_n(x)$ is the modified Bessel functions

$I = \iint |E(r, \phi)|^2 \, r dr d\phi$ is the beam power

Equation (7.80) for the first-type aB-mode has been derived in [219]. OAM for the second-type aB-mode is derived the same way. The upper sign in Eq. (7.80) must be chosen for the first-type beams (7.71), (7.75), while the lower sign must be chosen for the second-type beams (7.78), (7.79).

During propagation in free space, complex amplitudes of the aB-beams (7.71), (7.75) and (7.78), (7.79) are multiplied by the same factor $\exp[ikz(1 - \alpha^2)^{1/2}]$.

In order to obtain the corresponding aBG beams of the first and second types, it is necessary to multiply the complex amplitudes (7.71), (7.75) and (7.78), (7.79) by the Gaussian exponential function $\exp(-r^2/w^2)$, where w is the Gaussian beam waist radius.

Linear Combination of the Asymmetrical Bessel Modes

Although laser beams $E_{1,2}(r, \varphi, \pm c)$ have transverse intensity distribution in a form of crescent, superposition of such beams can form a variety of Bessel modes, both symmetrical and asymmetrical. Now we consider a few examples of such superposition of modes $E_{1,2}(r, \varphi, \pm c)$.

Shown in Figure 7.16 are (a) intensity and (b) phase of the sum of two mirror-like asymmetrical first-type Bessel modes:

$$E_+(r,\varphi,z;c) = E_1(r,\varphi,z;c) + E_1(r,\varphi,z;-c). \tag{7.81}$$

The calculation parameters were the following: wavelength $\lambda = 532$ nm, topological charge $n = 3$, asymmetry parameter $c = 1$, scaling factor $\alpha = 1/(10\lambda)$, field size $R = 200\lambda$ (i.e., $-R \le x, y \le R$). It is seen in Figure 7.16 that the resulting mode is symmetrical with respect to the Cartesian axes and its intensity distribution has a form of an ellipse with weak inhomogeneity of intensity.

Shown in Figure 7.17 are distributions of (a) intensity and (b) phase of the sum (7.81), multiplied by the Gaussian exponential function. It is seen that two mirror-like aBG beams with the waist radius of $w = 50\lambda$ at $c = 3$ generate symmetrical elliptic beam, which is almost free of sidelobes and looks like the letter "O."

It is interesting to note that the difference of two mirror-like first-type asymmetrical Bessel modes

$$E_-(r,\varphi,z;c) = E_1(r,\varphi,z;c) - E_1(r,\varphi,z;-c) \tag{7.82}$$

generates intensity (Figure 7.18a) and phase (Figure 7.18b) distribution with almost axial symmetry (weak ellipse) and with the topological charge of the axial optical vortex being incremented by unity. The calculation parameters are the same as in Figure 7.16. It is seen in Figure 7.18b that in the center of the diffraction pattern an axial optical vortex occurs with its topological charge of $n = 4$, although both

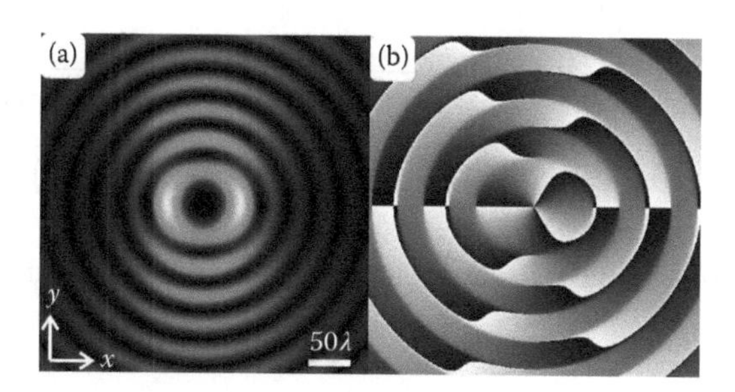

FIGURE 7.16 Distribution of (a) intensity and (b) phase of the sum of two asymmetrical mutually mirror-like Bessel modes (7.81).

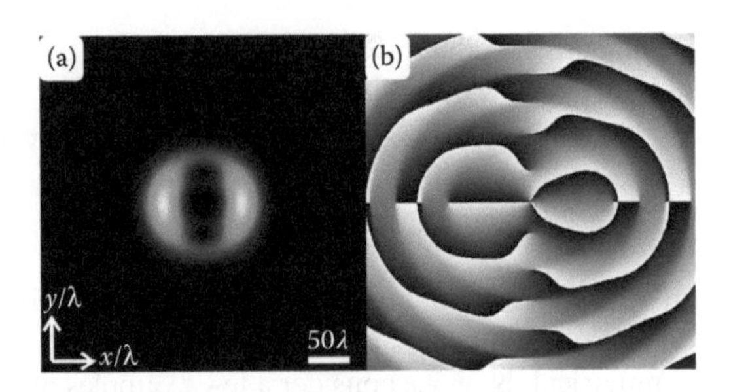

FIGURE 7.17 Distributions of (a) intensity and (b) phase of the sum of two asymmetrical and mutually mirror-like BG beams with the same parameters as in Figure 7.16.

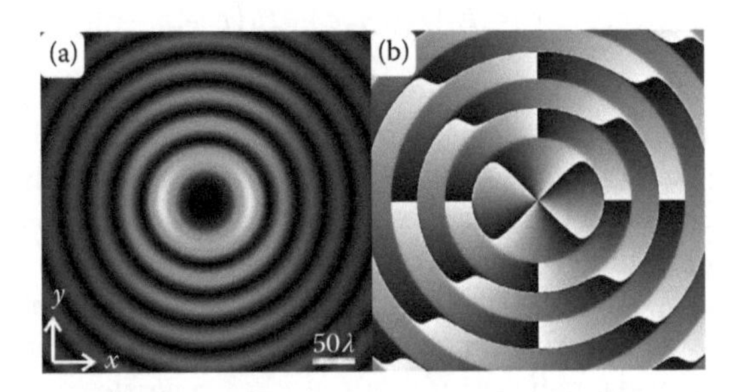

FIGURE 7.18 Distributions of (a) intensity and (b) phase of the difference of two mirror-like asymmetrical Bessel modes (7.82). The calculation parameters are the same as in Figure 7.16.

modes (7.82) have the topological charge of $n = 3$. For the sum of modes (7.81) there is also on-axis optical vortex with the topological charge of $n = 3$ (Figures 7.16b and 7.17b).

Incrementing of the topological charge by unity for the difference of two mirror-like modes (7.71) and (7.75) can be proved. Indeed, if we consider amplitude of the sum (7.81) and the difference (7.82)

$$E_{\pm}(r,\varphi,z) = (\alpha r)^{n/2}\exp(in\varphi) \times \left(\frac{J_n\left\{\sqrt{\alpha r\left[\alpha r - 2c\exp(i\varphi)\right]}\right\}}{\left[\alpha r - 2c\exp(i\varphi)\right]^{n/2}} \pm \frac{J_n\left\{\sqrt{\alpha r\left[\alpha r + 2c\exp(i\varphi)\right]}\right\}}{\left[\alpha r + 2c\exp(i\varphi)\right]^{n/2}} \right), \tag{7.83}$$

in vicinity of the origin, then, using approximation of the Bessel function for small arguments

$$J_n(z) \approx \frac{1}{n!}\left(\frac{z^2}{4}\right)^{n/2}\left[1 + \frac{z^2}{4(n+1)}\right], \tag{7.84}$$

we obtain:

$$E_{\pm}\left(r << \frac{1}{\alpha}, \varphi, z\right) = \frac{1}{n!}(\alpha r)^{n/2}\left(\frac{\alpha r}{4}\right)^{n/2} \exp(in\varphi)$$
$$\times\left\{\left[1+\frac{\alpha r\left[\alpha r - 2c\exp(i\varphi)\right]}{4(n+1)}\right] \pm \left[1+\frac{\alpha r\left[\alpha r + 2c\exp(i\varphi)\right]}{4(n+1)}\right]\right\}, \tag{7.85}$$

i.e.,

$$E_{+} \approx \frac{2}{n!}\left(\frac{\alpha r}{2}\right)^{n}\left[1+\frac{(\alpha r)^2}{4(n+1)}\right]\exp(in\varphi),$$
$$E_{-} \approx \frac{-2c}{(n+1)!}\left(\frac{\alpha r}{2}\right)^{n+1} \exp\left[i(n+1)\varphi\right]. \tag{7.86}$$

It is seen in Eq. (7.86) that for addition of two modes (7.71) and (7.75) with their topological charges of n an optical vortex E_+ occurs near the optical axis with its topological charge of n, while for the difference of modes there is an optical vortex E_- with topological charge of $n+1$.

Both sum and difference of the first-type aB-modes can be expanded into series of the conventional Bessel modes:

$$E_{\pm}(r,\varphi,z;c) = E_1(r,\varphi,z;c) \pm E_1(r,\varphi,z;-c)$$
$$= 2\sum_{p=0}^{\infty} \frac{c^{2p+\delta}\exp\left[i(n+2p+\delta)\varphi\right]}{(2p+\delta)!} J_{n+2p+\delta}(\alpha r), \tag{7.87}$$

where:

$\delta = 0$ for the sum
$\delta = 1$ for the difference

Let us now consider the sum and the difference of two similar modes, but of different types: first type (7.71) and second type (7.78):

$$E_{1+2}(r,\varphi,z) = E_1(r,\varphi,z;c) + E_2(r,\varphi,z;c). \tag{7.88}$$

For small values of the asymmetry parameter ($c = 1$), the sum of modes (7.88) gives almost the same increasing crescent (Figure 7.19a), as each of the added modes separately. Figure 7.19 shows (a) intensity and (b) phase of the sum of two identical asymmetrical Bessel modes of the first and second types (7.88). Calculation parameters are the same as in Figure 7.16.

With increase of the asymmetry parameter ($c = 7$), both crescents generated by the beams described by the terms in Eq. (7.84) are transformed into astigmatic (elongated along the vertical coordinate) Gaussian beams which become further from

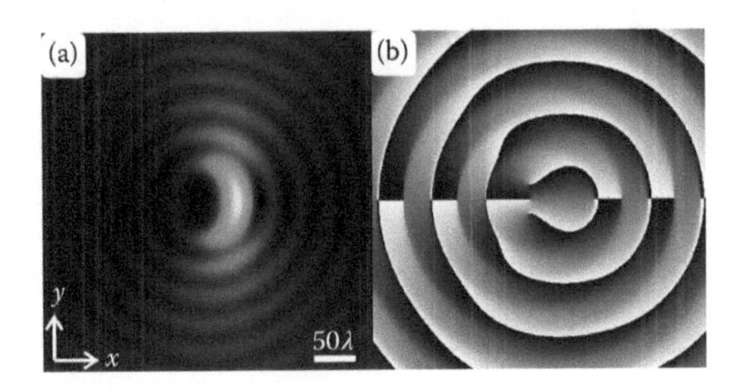

FIGURE 7.19 (a) Intensity and (b) phase of the sum (7.84). Calculation parameters are the same as for Figure 7.16.

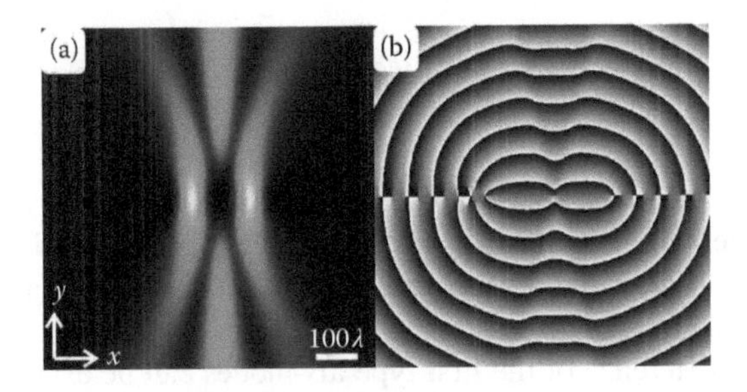

FIGURE 7.20 (a) Intensity and (b) phase of the sum of two identical asymmetrical Bessel modes of the first and second types (7.88). Calculation parameters are the same as in Figure 7.19, but $c = 7$.

each other and form a symmetrical intensity distribution (Figure 7.20a). Shown in Figure 7.20 are the (a) intensity and (b) phase of the sum of two asymmetrical Bessel modes of different types (7.84) at $c = 7$ (other parameters are the same as in Figure 7.16).

Note that the size of areas shown in Figures 7.16 through 7.19 is 200λ, while in Figure 7.20 this size equals to 400λ. The difference of two asymmetrical Bessel modes of the first and second type

$$E_{1-2}\left(r,\varphi,z\right)=E_1\left(r,\varphi,z;c\right)-E_2\left(r,\varphi,z;c\right) \tag{7.89}$$

at small values of the asymmetry parameter ($c = 1$) generates an asymmetrical diffraction pattern (Figure 7.21a), although this pattern has more complicated structure than that for the sum of the same beams (Figure 7.19).

It is seen in Figure 7.21b that along the vertical central line optical vortices occur with alternating topological charges of +1 and −1. Increasing of the asymmetry parameter ($c = 3$) leads to increasing of symmetry of the diffraction pattern of the difference of two beams (7.85) with respect to Cartesian axes. The pattern becomes similar to the letter "X" (Figure 7.22). Shown in Figure 7.22 are the (a) intensity and

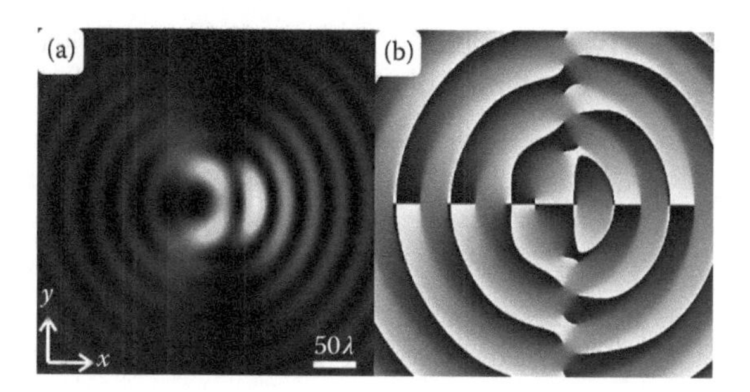

FIGURE 7.21 (a) Intensity and (b) phase of the difference of two identical asymmetrical Bessel modes of the first and second types (7.89). Calculation parameters are the same as in Figure 7.19.

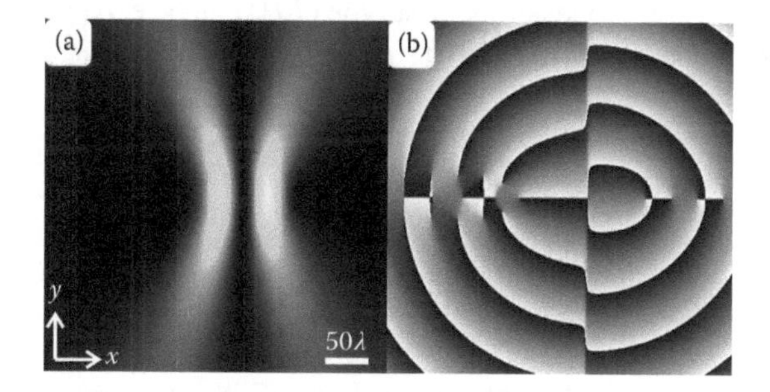

FIGURE 7.22 (a) Intensity and (b) phase of the difference of two identical asymmetrical Bessel modes of the first and second types (7.89). Calculation parameters are the same as in Figure 7.21, but $c = 3$.

(b) phase of the difference between two identical asymmetrical Bessel modes of the first and second types (7.85). Calculation parameters are the same as in Figure 7.21, but $c = 3$. The size of areas in both Figure 7.21 and Figure 7.22 is 200λ. It is seen in Figure 7.22 that there is a line of zero intensity along the vertical axis y (linear dislocation).

It is seen in Figures 7.21b and 7.22b that there is no unitary increase of the topological charge of an optical vortex in the center of the diffraction pattern of the difference between two identical asymmetrical beams Bessel of the first and second type. The topological charge of the optical vortex in the center of Figures 7.21b and 7.22(b) is $n = 3$, as the topological charge of the original beams described by the terms in Eq. (7.89).

Now we consider more general complex shift of the Bessel mode (7.72):

$$\begin{cases} y = y' + ic/\alpha, \\ x = x' + b/\alpha. \end{cases} \tag{7.90}$$

In this case, we obtain a two-parameter (c, b) family of asymmetrical Bessel modes, as opposed to a one-parameter family of modes (7.71):

$$E_3(r,\varphi,c,b)=\left[\alpha r+(c-b)\exp(-i\varphi)\right]^n \times\left[\alpha^2r^2-2\alpha r(b\cos\varphi+ic\sin\varphi)+(b^2-c^2)\right]^{-n/2} \times J_n\left\{\sqrt{\alpha^2r^2-2\alpha r(b\cos\varphi+ic\sin\varphi)+(b^2-c^2)}\right\}\exp(in\varphi). \quad (7.91)$$

If $b=c$, Eq. (7.91) coincides with Eq. (7.71). Modes (7.91) are convenient due to possibility of continuous changing of their shape by changing the parameter c with avoiding the shift of the mode ($b=0$). For modes (7.71) and (7.78), change of their intensity distribution automatically leads to their shift along the axis x.

Asymmetrical Bessel modes of the first type (7.71) can be expanded in a series of unshifted Bessel modes (7.72) [219,221]:

$$E_1(r,\varphi,c)=\left[\frac{\alpha r-2c\exp(i\varphi)}{\alpha r}\right]^{-n/2} J_n\left\{\sqrt{\alpha r(\alpha r-2c\exp(i\varphi))}\right\}\exp(in\varphi) =\sum_{p=0}^{\infty}\frac{c^p\exp\left[i(n+p)\varphi\right]}{p!}J_{n+p}(\alpha r), \quad (7.92)$$

and asymmetrical Bessel modes of the second type (7.78) can be expanded in a series of unshifted Bessel modes (7.72) as well:

$$E_2(r,\varphi,c)=\left[\frac{\alpha r+2c\exp(-i\varphi)}{\alpha r}\right]^{n/2} J_n\left\{\sqrt{\alpha r(\alpha r+2c\exp(-i\varphi))}\right\}\exp(in\varphi) =\sum_{p=0}^{\infty}\frac{c^p\exp\left[i(n-p)\varphi\right]}{p!}J_{n-p}(\alpha r). \quad (7.93)$$

The series (7.92) and (7.93) are used to calculate the OAM (7.80), to calculate the spatial spectrum of modes (7.71) and (7.78), and to calculate the scalar product of two modes with different parameters [219]. For two-parameter modes (7.87) the expansion into series of unshifted Bessel modes was not obtained, so OAM of such modes is hard to calculate.

Orbital Angular Momentum of Superposition of Asymmetrical Bessel Modes of the First and the Second Type

Let us consider superposition of the first-type modes with arbitrary complex coefficients:

$$E_1(r,\varphi,z;c)=C_1E_1(r,\varphi,z;c)+C_2E_1(r,\varphi,z;-c). \quad (7.94)$$

We can calculate the OAM J_z (projection of the OAM onto the optical axis) and total intensity I of the beam in a transverse plane the same way as it is done in [219], that is by using the expressions for the complex amplitudes via the series of Bessel functions (7.92), (7.93). Then the OAM reads as

$$J_z = 2\pi D_1 \lim_{R\to\infty} \sum_{p=0}^{\infty} \frac{(n+p)|c|^{2p}}{(p!)^2} \int_0^R J_{n+p}^2(\alpha r) r dr$$
$$+ 2\pi D_2 \lim_{R\to\infty} \sum_{p=0}^{\infty} (-1)^p \frac{(n+p)|c|^{2p}}{(p!)^2} \int_0^R J_{n+p}^2(\alpha r) r dr. \tag{7.95}$$

where:

$D_1 = |C_1|^2 + |C_2|^2$
$D_2 = C_1^* C_2 + C_1 C_2^*$

Similarly, total energy of the beam is

$$I = 2\pi D_1 \lim_{R\to\infty} \sum_{p=0}^{\infty} \frac{|c|^{2p}}{(p!)^2} \int_0^R J_{n+p}^2(\alpha r) r dr$$
$$+ 2\pi D_2 \lim_{R\to\infty} \sum_{p=0}^{\infty} (-1)^p \frac{|c|^{2p}}{(p!)^2} \int_0^R J_{n+p}^2(\alpha r) r dr. \tag{7.96}$$

Since for arbitrary integer number m and for large values of R the integrals in sums are equal to

$$\int_0^R J_m^2(\alpha r) r dr = \frac{R}{\pi \alpha}, \tag{7.97}$$

an expression for normalized OAM of the light field (7.94) can be obtained:

$$\frac{J_z}{I} = n + |c| \frac{D_1 I_1(2|c|) + D_2 J_1(2|c|)}{D_1 I_0(2|c|) + D_2 J_0(2|c|)}. \tag{7.98}$$

It follows from Eq. (7.98) that for superposition of two first-type modes without phase difference between them (i.e., $C_1 = C_2 = 1$, (7.81)), the normalized OAM reads as

$$\frac{J_z}{I} = n + |c| \frac{I_1(2|c|) + J_1(2|c|)}{I_0(2|c|) + J_0(2|c|)}. \tag{7.99}$$

If there is a phase difference of π in superposition of two first-type modes (i.e., $C_1 = 1$, $C_2 = -1$, (7.82)), however, the normalized OAM reads as

$$\frac{J_z}{I} = n + |c| \frac{I_1(2|c|) - J_1(2|c|)}{I_0(2|c|) - J_0(2|c|)}. \tag{7.100}$$

Now we consider superposition of modes of two different types with arbitrary complex coefficients:

$$E_{12}(r,\varphi,z;c) = C_1 E_1(r,\varphi,z;c) + C_2 E_2(r,\varphi,z;c). \tag{7.101}$$

In this case OAM and total intensity are the following

$$\begin{aligned} J_z &= 2\pi |C_1|^2 \lim_{R\to\infty} \sum_{p=0}^{\infty} \frac{(n+p)|c|^{2p}}{(p!)^2} \int_0^R J_{n+p}^2(\alpha r) r dr \\ &+ 2\pi |C_2|^2 \lim_{R\to\infty} \sum_{p=0}^{\infty} \frac{(n-p)|c|^{2p}}{(p!)^2} \int_0^R J_{n-p}^2(\alpha r) r dr \\ &+ 2\pi n \left(C_1^* C_2 + C_1 C_2^*\right) \lim_{R\to\infty} \int_0^R J_n^2(\alpha r) r dr, \end{aligned} \tag{7.102}$$

$$\begin{aligned} I &= 2\pi |C_1|^2 \lim_{R\to\infty} \sum_{p=0}^{\infty} \frac{|c|^{2p}}{(p!)^2} \int_0^R J_{n+p}^2(\alpha r) r dr \\ &+ 2\pi |C_2|^2 \lim_{R\to\infty} \sum_{p=0}^{\infty} \frac{|c|^{2p}}{(p!)^2} \int_0^R J_{n-p}^2(\alpha r) r dr + 2\pi \left(C_1^* C_2 + C_1 C_2^*\right) \lim_{R\to\infty} \int_0^R J_n^2(\alpha r) r dr. \end{aligned} \tag{7.103}$$

Using (7.102) and (7.103), an expression for normalized OAM can be derived:

$$\frac{J_z}{I} = n + \frac{\left(|C_1|^2 - |C_2|^2\right)|c| I_1(2|c|)}{\left(|C_1|^2 + |C_2|^2\right) I_0(2|c|) + \left(C_1^* C_2 + C_1 C_2^*\right)}. \tag{7.104}$$

In particular, it is seen in Eq. (7.104) that for the cases of addition (7.88) and subtraction (7.89) of modes the OAM is equal to n and does not depend on α and c.

Expression (7.104) can be considered as the main result of this section, since it follows from (7.104) that by addition and subtraction of the complex amplitudes of two aB-modes of the first and second types with the same values of n, α and c, it is possible, by changing the parameters α and c, to derive modes with different transverse intensity distributions, but with the same OAM, equal to n. Therefore, it appears that all modes in Figures 7.19 through 7.22 have the same OAM, equal to $n = 3$.

Thus, we have theoretically and numerically shown that the previously considered asymmetrical Bessel modes [219,221] and BG beams [235,236], can be obtained by the complex shift of the ordinary Bessel modes in Cartesian coordinates. We have also shown that by the complex shift of the Bessel mode, another second-type asymmetrical Bessel mode can be derived (with respect to the first-type modes [221]). Asymmetrical Bessel modes of the first and second types have different OAM. Adding and subtracting the complex amplitudes of the asymmetrical Bessel modes of both types, it is possible to obtain other Bessel modes with different intensity distributions in cross section. In particular, it is possible to obtain symmetrical Bessel modes with their shapes similar to letters "O" and "X." It is also interesting that the difference of two identical mirror-like asymmetrical Bessel modes with topological charge n leads to generation of an optical vortex with a topological charge $n+1$ in the center of the diffraction pattern. The difference or sum of two asymmetrical Bessel modes of the first and second types with the same parameters has the OAM equal to the topological charge n of the original Bessel mode. This OAM does not depend on other parameters: scaling factor α and asymmetry parameter c.

7.5 SHIFTED NONDIFFRACTIVE BESSEL BEAMS

Bessel beams discovered in 1987 [1,210] possess a variety of remarkable properties. They travel without diffraction over a certain distance in free space [210], generate optical tubes or cavities on the optical axis [241,242], and also feature a property of self-healing following a distortion caused by a minor obstacle [243,244]. Bessel beams have an OAM [245,246]. The superposition of Bessel beams can be axially periodic (analog of Talbot effect) [211,247] or experience rotation about the optical axis upon propagation [248,249]. Bessel beams can be generated using digital holograms [241,242,250], a conical diffractive axicon [251,252], a diffractive vortex axicon [41], diffractive optical elements [248,249], and spatial light modulators [50]. It is interesting that astigmatic Bessel beams can be generated by simply tilting the diffractive element or illuminating the diffractive vortex axicon by a tilted beam [95].

Bessel beams have found a variety of applications. One use is for micromanipulation intended for simultaneously trapping several microparticles on the optical axis [253,254] or rotating individual, or a number of, microparticles about the optical axis [31]. Using Bessel beams, it is possible to trap and accelerate individual cooled atoms [255,256]. Recently proposed Hankel-Bessel beams [257] may find use in atmospheric probing because they are immune to atmospheric turbulence [258]. Theoretical analysis of vector Bessel beams was for the first time conducted in [259,260] and analytical relations for the OAM density were derived in [245,246,261]. It should be noted that because the total energy of a Bessel beam is infinite, total OAM is also infinite. Because of this, prior to the present work, no study of OAM of the entire Bessel beams has previously been conducted. As well as defining eigenfunctions for circular billiards, Bessel modes also represent resonant geometric modes that possess OAM [262].

More recently, nonparaxial aB modes [219] and paraxial aBG beams [236] have been proposed. These laser beams have been shown to produce the transverse intensity pattern in the form of a semi-crescent. In Ref. [237], the aB modes were studied experimentally using a digital micromirror array. By analogy with [219], asymmetric Chebyshev-Bessel beams have also been proposed [238]. Previously, study of the superposition of axial Bessel beams has only been reported [247–249,253,261]. The study of the superposition of off-axis laser beams discussed in [262] did not involve Bessel beams.

In this section, we analyze the superposition of off-axis Bessel beams of the same order/topological charge. A general analytical expression for the OAM of the superposition under study has been derived. We show that if the constituent beams of the superposition have real-valued weight coefficients, the total OAM of the superposition of the Bessel beams equals that of an individual non-shifted Bessel beam. This property enables generating nondiffractive beams with different intensity distributions but identical OAM. The superposition of a set of identical Bessel beams centered on an arbitrary-radius circle is shown to be equivalent to an individual constituent Bessel beam put in the circle center. As a result of a complex shift of the Bessel beam, the transverse intensity pattern and OAM of the beam are also shown to change. We show that in the superposition of two or more complex-shifted Bessel beams, the OAM may remain unchanged, while the intensity distribution is changed. Numerical simulation is in good agreement with theory.

Fourier Spectrum of a Shifted Bessel Beam

The complex amplitude of a nonparaxial stationary light field that satisfies the Helmholtz equation has been known to be expressed as the angular spectrum of plane waves [263]:

$$E(x,y,z)=\iint_{\mathbb{R}^2} A(\xi,\eta)\exp\left[ik(\xi x+\eta y)+ikz\sqrt{1-\xi^2-\eta^2}\right]d\xi d\eta, \quad (7.105)$$

where:
- k is the wavenumber of monochromatic light
- $A(\xi,\eta)$ is the complex amplitude of the angular spectrum of plane waves

In polar coordinates (r,φ), Eq. (7.105) takes the form:

$$E(r,\phi,z)=\iint_{\mathbb{R}^2} A(\rho,\theta)\exp\left[ikr\rho\cos(\theta-\phi)+ikz\sqrt{1-\rho^2}\right]\rho d\rho d\theta, \quad (7.106)$$

where (ρ,θ) are the polar coordinates in the Fourier plane. If the beam is shifted from the axis by a Cartesian vector (x_0,y_0), the amplitude of the angular spectrum of plane waves is given by

$$A'(\rho,\theta)=A(\rho,\theta)\exp\left[-ik(\xi x_0+\eta y_0)\right], \quad (7.107)$$

where $A(\rho,\varphi)$ is the amplitude of the angular spectrum of plane waves for the original nonshifted beam.

Note that the shift coordinates (x_0,y_0) may take complex values.

The angular spectrum of a non-shifted n-order Bessel beam has been known [219] to be given by

$$A_n(\rho,\theta)=\frac{(-i)^n}{\alpha\lambda}\exp(in\theta)\delta\left(\rho-\frac{\alpha}{k}\right), \quad (7.108)$$

where:
- $\delta(x)$ is the Dirac delta-function
- α is the scale factor of the non-shifted Bessel mode

$$E_n(r,\varphi,z)=\exp\left(in\varphi+iz\sqrt{k^2-\alpha^2}\right)J_n(\alpha r), \tag{7.109}$$

where $J_n(x)$ is the n-order Bessel function of the first kind. In view of Eq. (7.107), the amplitude of the angular spectrum of the shifted n-order Bessel beam is

$$A'_n(\rho,\theta)=\frac{(-i)^n}{\alpha\lambda}\exp(in\theta)\delta\left(\rho-\frac{\alpha}{k}\right)\exp(-ikx_0\rho\cos\theta-iky_0\rho\sin\theta). \tag{7.110}$$

Relation between the Amplitudes of Spectra of Shifted and Nonshifted Bessel Beams

We shall find coefficients A_{mn} of a series that describes the expansion of the amplitude of the spectrum of a shifted n-order Bessel beam (7.110) in terms of the amplitudes of the spectrum of the nonshifted different-order Bessel beams:

$$A'_n(\rho,\theta)=\delta\left(\rho-\frac{\alpha}{k}\right)\sum_{p=-\infty}^{\infty}A_{pn}\frac{(-i)^p}{\alpha\lambda}\exp(ip\theta). \tag{7.111}$$

Multiplying both sides of Eq. (7.111) by $(\alpha\lambda/\delta)(\rho - \alpha/k)\exp(-im\varphi)$ and integrating over θ from 0 to 2π we obtain:

$$\begin{aligned}&\int_0^{2\pi}\exp\left[i(n-m)\theta\right]\exp(-i\alpha x_0\cos\theta-i\alpha y_0\sin\theta)d\theta\\&=i^n\sum_{p=-\infty}^{\infty}A_{pn}(-i)^p2\pi\delta_{pm}.\end{aligned} \tag{7.112}$$

In the left-hand side of Eq. (7.112), we have

$$\int_0^{2\pi}\exp(in\theta+ia\cos\theta+ib\sin\theta)d\theta=2\pi\left(\frac{ia-b}{\sqrt{a^2+b^2}}\right)^nJ_n\left(\sqrt{a^2+b^2}\right). \tag{7.113}$$

In view of Eq. (7.113), the coefficients in the right-hand side of Eq. (7.112) are

$$A_{mn}=\left(\frac{x_0+iy_0}{\sqrt{x_0^2+y_0^2}}\right)^{n-m}J_{m-n}\left(\alpha\sqrt{x_0^2+y_0^2}\right). \tag{7.114}$$

In particular, if the same-value shift is real on one coordinate and imaginary on the other ($x_0 = c/\alpha$, $y_0 = ic/\alpha$), the coefficients in Eq. (7.114) are simplified to

$$A_m=\begin{cases}\dfrac{c^{m-n}}{(m-n)!}, & m\ge n,\\ \delta_{nm}, & m\le n.\end{cases} \tag{7.115}$$

In this particular case, the amplitude of the angular spectrum of plane waves for the shifted Bessel beam can be represented by a linear combination of Bessel modes:

$$A_n(\rho,\theta) = \delta\left(\rho - \frac{\alpha}{k}\right)\sum_{p=0}^{\infty}\frac{c^p}{p!}\frac{(-i)^{n+p}}{\alpha\lambda}\exp\left[i(n+p)\theta\right], \tag{7.116}$$

with the parameter c defining the degree of asymmetry of the shifted n-order Bessel mode. The angular spectrum of the shifted Bessel mode in Eq. (7.116) is seen to be identical to that of an aB mode [219].

Rearranging Eq. (7.105) with use of Eqs. (7.110) and (7.113), we obtain the relationship for the amplitude of the shifted Bessel beam:

$$E'_n(x,y,z) = \exp\left(iz\sqrt{k^2-\alpha^2}\right)\left[\frac{(x+x_0)+i(y+y_0)}{\sqrt{(x+x_0)^2+(y+y_0)^2}}\right]^n \times J_n\left(\alpha\sqrt{(x+x_0)^2+(y+y_0)^2}\right). \tag{7.117}$$

Orbital Angular Momentum of a Shifted Bessel Beam

Projection of OAM on the optical axis z and total intensity of the laser beam can be found from the relations [219]:

$$iJ_z = \iint_{\mathbb{R}^2} E^*\frac{\partial E}{\partial\phi}rdrd\phi = \left(\frac{2\pi}{k}\right)^2\iint_{\mathbb{R}^2} A^*\frac{\partial A}{\partial\theta}\rho d\rho d\theta, \tag{7.118}$$

$$I = \iint_{\mathbb{R}^2} E^*Erdrd\phi = \left(\frac{2\pi}{k}\right)^2\iint_{\mathbb{R}^2} A^*A\rho d\rho d\theta. \tag{7.119}$$

When calculating Eqs. (7.118) and (7.119) for the shifted Bessel beam, we shall utilize Eq. (7.110). Then, the projection of OAM onto the optical axis is

$$J_z = \frac{\lambda}{\alpha}\delta(0)\left[nI_0(2\alpha r_{0i}) + \frac{\alpha}{r_{0i}}\operatorname{Im}\left(y_0x_0^*\right)I_1(2\alpha r_{0i})\right], \tag{7.120}$$

where:
$r_{0i} = [(\operatorname{Im} x_0)^2 + (\operatorname{Im} y_0)^2]^{1/2}$
$I_0(x)$, $I_1(x)$ are modified Bessel functions, whereas the total power of the beam is

$$I = \frac{\delta(0)}{k\alpha}\int_0^{2\pi}\exp(2\alpha\operatorname{Im} x_0\cos\theta + 2\alpha\operatorname{Im} y_0\sin\theta)d\theta = \frac{\lambda}{\alpha}\delta(0)I_0(2\alpha r_{0i}), \tag{7.121}$$

where $\delta(0)$ is the Dirac delta-function at zero. From Eqs. (7.120) and (7.121), we can infer that although both the projection of OAM onto the optical axis and the power of the shifted Bessel beam are infinite, their ratio is finite:

$$\frac{J_z}{I} = n + \frac{\alpha}{r_{0i}} \operatorname{Im}\left(x_0^* y_0\right) \frac{I_1\left(2\alpha r_{0i}\right)}{I_0\left(2\alpha r_{0i}\right)}. \tag{7.122}$$

From Eq. (7.122) it follows that if the coordinates of the shift vector (x_0, y_0) are both purely real or purely imaginary, the normalized OAM in Eq. (7.122) equals that of a non-shifted Bessel beam:

$$\frac{J_z}{I} = n. \tag{7.123}$$

The normalized OAM of the shifted and non-shifted Bessel beams, Eqs. (7.122) and (7.123), will be different only when the shift is real on one coordinate and imaginary on the other. For instance, assuming $x_0 = b/\alpha$ and $y_0 = ic/\alpha$, the OAM in Eq. (7.122) takes the form:

$$\frac{J_z}{I} = n + b \frac{I_1\left(2|c|\right)}{I_0\left(2|c|\right)}. \tag{7.124}$$

It can be seen that at $b > 0$, the OAM in Eq. (7.124) is larger than that in Eq. (7.123), becoming smaller at $b < 0$. Note that the change of the beam shape is defined by the magnitude of the imaginary shift: the intensity pattern of the shifted beam in Eq. (7.117) takes the form of an ellipse at small $c < 1$, is shaped as a semi-crescent at $c > 1$, finally, taking the form of an astigmatic Gaussian beam at $c >> 1$ [219]. On the x-axis, the center of the Bessel beam in Eq. (7.109) is shifted by $\Delta x = (b - c)/\alpha$.

Orbital Angular Momentum of the Superposition of Shifted Bessel Beams

Let us analyze the superposition of P shifted n-order Bessel beams of Eq. (7.117). Here, the amplitude of angular spectrum of plane waves is given by

$$A(\rho,\theta) = \sum_{p=0}^{P-1} C_p A_{pn}(\rho,\theta), \tag{7.125}$$

where:

$$\begin{aligned} A_{pn}(\rho,\theta) = {} & \frac{(-i)^n}{\alpha\lambda} \exp(in\theta)\delta\left(\rho - \frac{\alpha}{k}\right) \\ & \times \exp\left(-ikx_p\rho\cos\theta - iky_p\rho\sin\theta\right) \end{aligned} \tag{7.126}$$

is the amplitude of angular spectrum of the pth constituent beam shifted by a complex vector with coordinates (x_p, y_p). Based on (7.118) and (7.119), the normalized OAM of the superposition in Eq. (7.125) is given by

$$\frac{J_z}{I} = n - i\alpha \frac{\sum_{p=0}^{P-1}\sum_{q=0}^{P-1} C_p^* C_q \frac{x_p^* y_q - x_q y_p^*}{R_{pq}} J_1\left(\alpha R_{pq}\right)}{\sum_{p=0}^{P-1}\sum_{q=0}^{P-1} C_p^* C_q J_0\left(\alpha R_{pq}\right)}, \tag{7.127}$$

are $J_0(x)$, $J_1(x)$ are the Bessel functions of the zero and first order,

$$\begin{aligned} R_{pq} &= \sqrt{(x_p^* - x_q)^2 + (y_p^* - y_q)^2}, \\ R_{pp} &= 2i\sqrt{\left(\operatorname{Im} x_p\right)^2 + \left(\operatorname{Im} y_p\right)^2}. \end{aligned} \tag{7.128}$$

Although there is an imaginary factor $i\alpha$ in Eq. (7.127), the entire relation is real. This conclusion can be made from the facts that (1) at $p = q$, $|C_p|^2$ in the numerator is real, the magnitudes R_{pq} and $J_1(\alpha R_{pq})$ are purely imaginary, and the difference of two complex conjugated numbers $x_p^* y_p - x_p y_p^*$ is also purely imaginary; and (2) for any p and q, which are not equal to each other, $R_{pq} = R_{qp}^*$, whereas the terms with indices (p, q) and (q, p) also represent the difference of two complex conjugated numbers.

It can be shown that with all constituent Bessel beams in the superposition (7.125) shifted by a real vector (x_p, y_p) and all coefficients C_p assumed to be real, the numerator in (7.127) equals zero, so that the total OAM of the superposition equals that of an individual nonshifted n-order Bessel beam in Eq. (7.123). This is a key finding of the present research. Based on it, it becomes possible to form the most diverse nonparaxial laser beams that would have different transverse intensity distributions but identical OAM of Eq. (7.123), while traveling without diffraction. Examples of such beams are discussed in the following.

From Eq. (7.127), interesting particular cases can be inferred. If $P = 2$, $x_0 = c/\alpha$, $y_0 = ic/\alpha$, $x_1 = -c/\alpha$, $y_1 = ic/\alpha$, then $R_{00} = R_{11} = 2ic/\alpha$, $R_{01} = R_{10} = 0$, a simple expression for the normalized OAM can be derived (with the coefficients in Eq. (7.125) defined by arbitrary complex numbers C_0, C_1):

$$\frac{J_z}{I} = n + \frac{c\left(|C_0|^2 - |C_1|^2\right) I_1\left(2|c|\right)}{\left(|C_0|^2 + |C_1|^2\right) I_0\left(2|c|\right) + 2\operatorname{Re}\left\{C_0^* C_1\right\}}. \tag{7.129}$$

Equation (7.129) suggests that in the superposition of two n-order Bessel beams with complex (purely imaginary) but matching shifts on one axis and equal coefficients $|C_0| = |C_1|$, the normalized OAM of Eq. (7.129) equals OAM of an individual nonshifted Bessel beam of n-order, Eq. (7.123). Thus, assuming $|C_0| = |C_1|$, the transverse intensity pattern of the superposition of two shifted Bessel beams can be varied (because with varying c, the Bessel beam's shape varies), whereas the total OAM remains unchanged.

Superposition of Three Shifted Bessel Beams

In the following, we analyze a superposition of three n-order Bessel beams that are shifted so that their centers are found at the vertices of an equilateral triangle. Thus, in the superposition of Eq. (7.125), $P = 3$, $R_{01} = R_{02} = R_{12}$, the weight coefficients C_0, C_1, C_2 are arbitrary complex numbers, with the coordinates of the complex shift vector assumed to be given by

$$\begin{cases} x_p = R_0 \cos\left(\dfrac{2\pi p}{3}\right) + \dfrac{c}{\alpha}\exp\left(-i\gamma - i\dfrac{2\pi p}{3}\right), \\ y_p = R_0 \sin\left(\dfrac{2\pi p}{3}\right) + i\dfrac{c}{\alpha}\exp\left(-i\gamma - i\dfrac{2\pi p}{3}\right), \end{cases} \tag{7.130}$$

where:

R_0 is the center of a circle on which the singularity centers of the shifted Bessel beams are found

c defines the asymmetry of the shifted Bessel beam

γ is the angle of rotation of the asymmetric shifted Bessel beam

Then, OAM is

$$\frac{J_z}{I} = n + \frac{\pm D_1 \xi I_1(2c) + \mathrm{Im}\left\{D_2\left(\xi \mp ic\sqrt{3}\right)J_1\left(\sqrt{3}\xi \pm ic\right)\right\}}{D_1 I_0(2c) + 2\mathrm{Re}\left\{D_2 J_0\left(\sqrt{3}\xi \pm ic\right)\right\}}, \tag{7.131}$$

with "+" taken for $\gamma = 0$ and "−" taken for $\gamma = \pi$,

$$\begin{cases} D_1 = |C_0|^2 + |C_1|^2 + |C_2|^2, \\ D_2 = C_0^* C_1 + C_1^* C_2 + C_2^* C_0, \end{cases} \tag{7.132}$$

where $\xi = \alpha R_0 \pm c$.

In a particular case of $\gamma = \pi$, $c = \alpha R_0$, the ξ parameter equals zero and Eq. (7.132) is rearranged to

$$\frac{J_z}{I} = n + \frac{c\sqrt{3} I_1(c)\mathrm{Im}\{D_2\}}{D_1 I_0(2c) + 2I_0(c)\mathrm{Re}\{D_2\}}. \tag{7.133}$$

From (7.133), we can infer that assuming real coefficients C_0, C_1, C_2, the term $\mathrm{Im}\{D_2\}$ in (29) equals zero and the OAM of the superposition of three shifted n-order Bessel beams equals OAM of an individual nonshifted Bessel beam of Eq. (7.123).

By way of illustration, Figure 7.23 depicts the calculated (a) intensity pattern and (b) phase of the superposition of three shifted Bessel beams with topological charge $n=3$. The OAM of the superposition is $J_z/I = 3$.

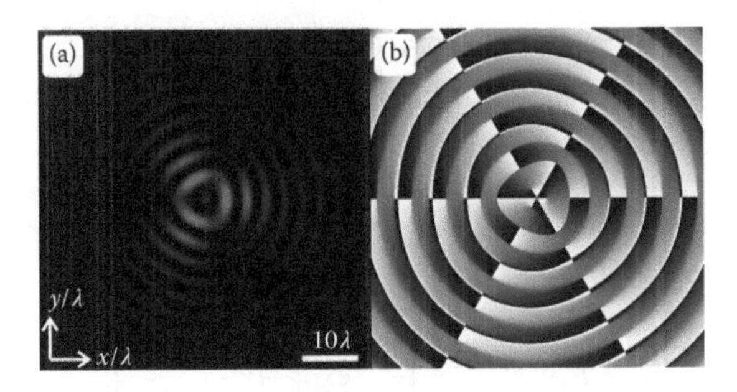

FIGURE 7.23 (a) Intensity and (b) phase of the superposition of three shifted Bessel beams with parameters: $n = 3$, $R_0 = 4\lambda$, $\alpha = 1/\lambda$, $c = 4$, $\gamma = \pi$, vector of weight coefficients, $\mathbf{C} = [1, 1, 1]$. Frame size, $2R = 60\lambda$.

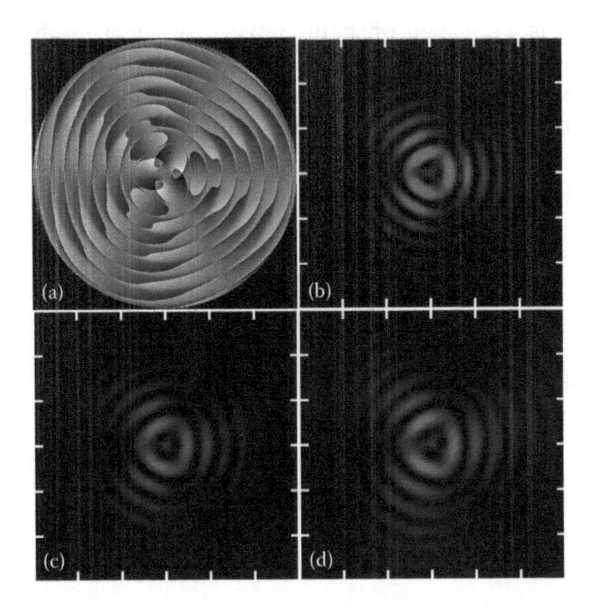

FIGURE 7.24 Encoded phase (Figure 7.23b) to generate a Bessel beam with the transverse intensity pattern shaped as an equilateral triangle and (a) experimentally generated intensity patterns at different distances from the plane $z = 0$: (b) 0 mm; (c) 200 mm; (d) 400 mm. Mesh step, 0.5 mm.

Figure 7.24 depicts the encoded phase (Figure 7.24a) of the superposition of three shifted Bessel beams with topological charge $n = 3$ (Figure 7.23b). The phase was fed to a spatial light modulator SLM PLUTO-VIS (1920×1080 resolution, 8-μm pixel size). Figure 7.24b–d depicts the SLM-aided intensity patterns generated in reflection by the incident wavelength of 633 nm at different distances. From Figure 7.24, the beam is seen to preserve its structure upon propagation, while the intensity pattern is in agreement with the simulation results

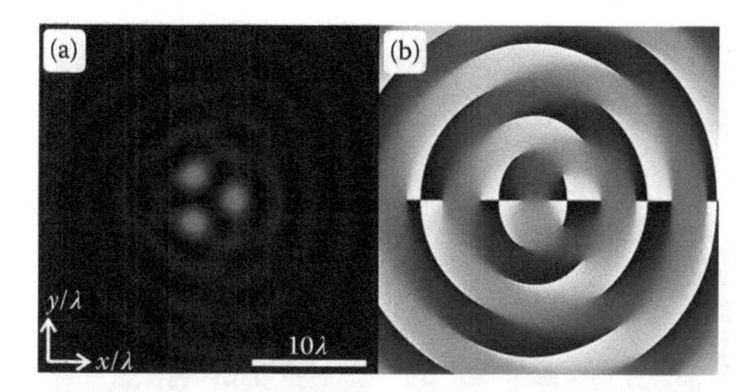

FIGURE 7.25 Calculated (a) intensity and (b) phase patterns generated by the superposition of three shifted Bessel beams with parameters $n = 5$, $R_0 = 8\lambda$, $\alpha = 1/\lambda$, $c = 3$, $\gamma = \pi$, weight coefficient vector, $\mathbf{C} = [1, 1, 1]$. Frame size, $2R = 30\lambda$.

(Figure 7.23a). The intensity was measured with a CMOS-camera MDCE-5A (1/2″, 1280 × 1024 resolution).

Figure 7.25 depicts the calculated phase and intensity distributions generated by the superposition of three shifted Bessel beams with identical weight coefficients, $\mathbf{C} = [1, 1, 1]$, taken at different values of the rest parameters: $n = 5$, $R_0 = 8\lambda$, and $c = 3$. In this case, the diffraction pattern is entirely different, with three bright spots being generated instead of an equilateral triangle. Because in this beam $c \neq \alpha R_0$, OAM cannot be derived from Eq. (7.133). According to Eq. (7.131), the OAM of this beam is fractional and equals

$$\frac{J_z}{I} = 5 + \frac{\mathrm{Im}\left\{\left(5 + i3\sqrt{3}\right) J_1\left(5\sqrt{3} - 3i\right)\right\} - 5 I_1(6)}{I_0(6) + 2\mathrm{Re}\left\{J_0\left(5\sqrt{3} - 3i\right)\right\}} \approx 0.62. \quad (7.134)$$

Figure 7.26 depicts the SLM-aided (a) phase and (b–d) intensity patterns generated at different distances by the superposition of three shifted Bessel beams with parameters $n = 5$, $R_0 = 8\lambda$, $\alpha = 1/\lambda$, $c = 3$, $\gamma = \pi$. From Figure 7.26, the experimental diffraction patterns are seen to be in agreement with the calculated intensity distribution in Figure 7.25(a).

Superposition of Identical Bessel Beams Found at the Vertices of a Regular Polygon

As in the previous case, we shall analyze the superposition of P shifted n-order Bessel beams with their singularity centers found at the vertices of a regular polygon (similar to (7.130)):

$$\begin{cases} x_p = R_0 \cos\left(\dfrac{2\pi p}{P}\right) + \dfrac{c}{\alpha} \exp\left(-i\gamma - i\dfrac{2\pi p}{P}\right), \\ y_p = R_0 \sin\left(\dfrac{2\pi p}{P}\right) + i\dfrac{c}{\alpha} \exp\left(-i\gamma - i\dfrac{2\pi p}{P}\right), \end{cases} \quad (7.135)$$

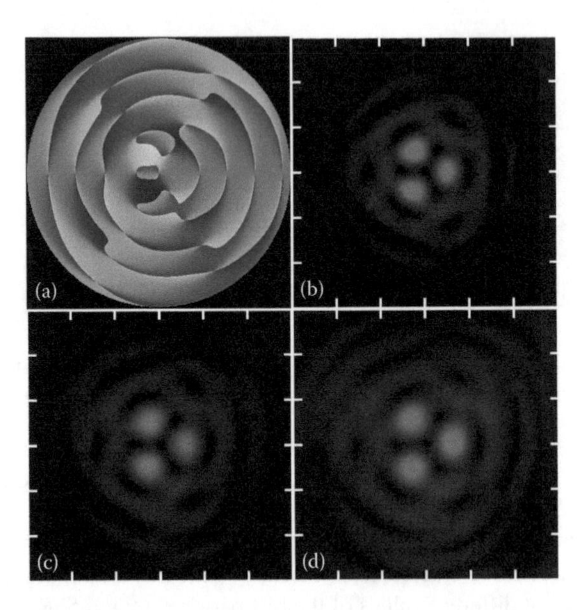

FIGURE 7.26 (a) Encoded phase to generate the superposition of three Bessel beams with the transverse intensity pattern featuring three peaks (Figure 7.25b) and experimentally generated intensity patterns at different distances from the plane $z = 0$: (b) 0 mm; (c) 200 mm; (d) 400 mm. Mesh step, 0.5 mm.

where $p = 0,..., P-1$.

For certainty, assume $\gamma = \pi\ c = \alpha R_0$. Then, we have, instead of (7.135):

$$\begin{cases} x_p = iR_0 \sin\left(\dfrac{2\pi p}{P}\right), \\ y_p = -iR_0 \cos\left(\dfrac{2\pi p}{P}\right). \end{cases} \tag{7.136}$$

The general relationship for the OAM of Eq. (7.128) takes the form:

$$\frac{J_z}{I} = n + \alpha \frac{2R_0 \sum_{p=1}^{P-1} \sum_{q=0}^{p-1} \mathrm{Im}\left\{C_p^* C_q\right\} \sin\left[\dfrac{\pi(p-q)}{P}\right] I_1 \left\{2\alpha R_0 \cos\left[\dfrac{\pi(p-q)}{P}\right]\right\}}{\sum_{p=0}^{P-1} \left|C_p\right|^2 I_0\left(2\alpha r_{pi}\right) + 2\sum_{p=1}^{P-1} \sum_{q=0}^{p-1} \mathrm{Re}\left\{C_p^* C_q J_0\left(\alpha R_{pq}\right)\right\}}. \tag{7.137}$$

With all coefficients C_p assumed to be real, the numerator in Eq. (7.137) equals zero and OAM of the superposition of shifted Bessel beams equals that of an individual nonshifted Bessel beam in Eq. (7.123). By way of illustration, Figure 7.27 depicts the numerically simulated intensity pattern and phase for the superposition of four shifted Bessel beams with topological charge $n = 7$. The normalized

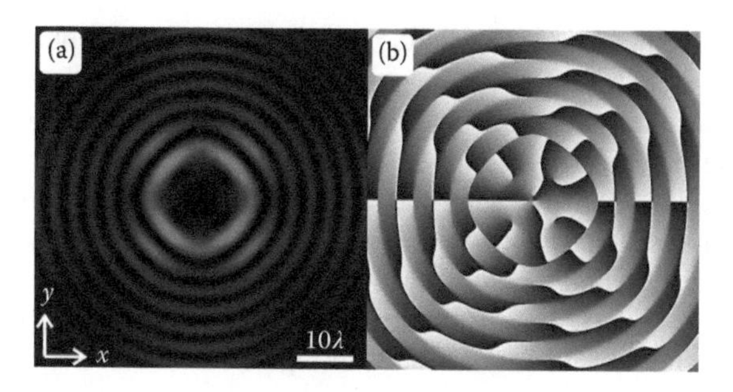

FIGURE 7.27 (a) Intensity and (b) phase of a superposition of four ($P=4$) shifted Bessel beam with parameters $n = 7$, $R_0 = 6\lambda$, $\alpha = 1/\lambda$, $c = 6$, $\gamma = \pi$, $\mathbf{C} = [1, 1, 1, 1]$. Frame size, $2R = 60\lambda$.

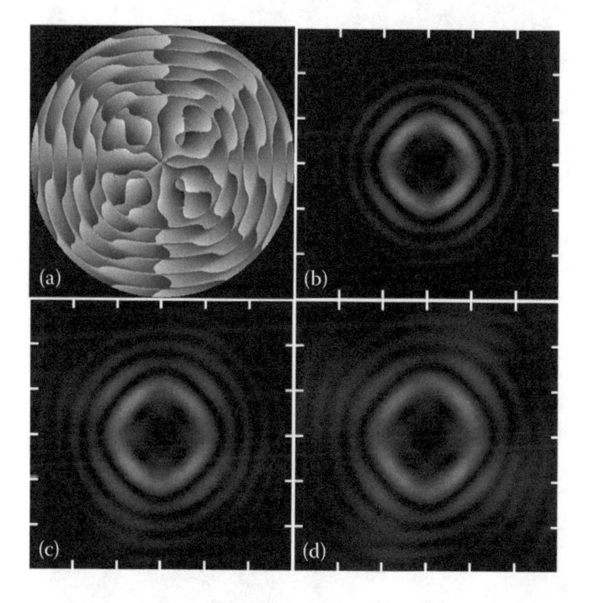

FIGURE 7.28 SLM-aided (a) encoded phase of Figure 7.27b to generate the superposition of four shifted Bessel beams with a square-shaped transverse intensity distribution and experimental intensity patterns at different distances from the plane $z = 0$: (b) 0 mm; (c) 200 mm; (d) 400 mm. Mesh step, 0.5 mm.

OAM of the superposition equals $J_z/I = 7$. As seen in Figure 7.27b, within a main square-like intensity ring there are seven optical vortices with topological charge +1.

Figure 7.28 depicts the SLM-aided encoded (a) phase and (b–d) intensity patterns generated at different distances by the superposition of four shifted Bessel beams (Figure 7.27) with topological charge $n = 7$, $R_0 = 6\lambda$, $\alpha = 1/\lambda$, $c = 6$, $\gamma = \pi$, $\mathbf{C} = [1, 1, 1, 1]$. The diffraction patterns in Figure 7.28 are seen to be in good agreement with the simulated intensity patterns in Figure 7.27a.

Another example in Figure 7.29 shows the (a) intensity and (b) phase of the superposition of six ($P = 6$) shifted Bessel beams with the same topological charge $n = 10$, with their singularity centers found at the vertices of a regular hexagon. The normalized OAM of the superposition equals $J_z/I = 10$.

Figure 7.30 depicts (a) SLM-aided encoded phase and (b–d) intensity pattern generated at different distances by the superposition of six ($P = 6$) shifted Bessel beams (Figure 7.29) with the same topological charge $n = 10$, at $R_0 = 12\lambda$, $\alpha = 1/\lambda$, $c = 12$, $\gamma = \pi$, $\mathbf{C} = [1, 1, 1, 1, 1, 1]$. The diffraction patterns in Figure 7.30 are seen to be in good agreement with the simulation results in Figure 7.29a.

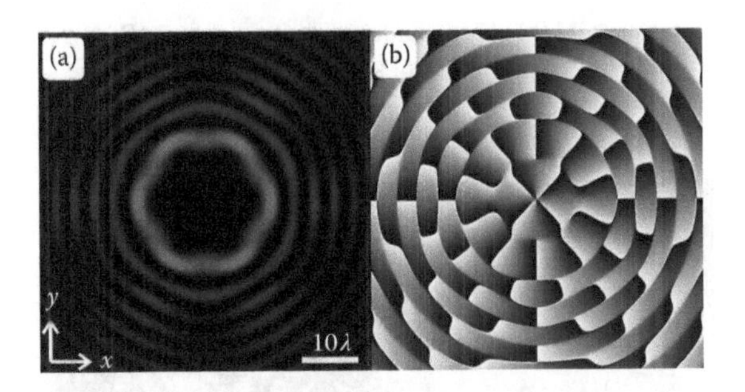

FIGURE 7.29 Simulated (a) intensity and (b) phase of the superposition of six shifted Bessel beams with parameters $P = 6$, $n = 10$, $R_0 = 12\lambda$, $\alpha = 1/\lambda$, $c = 12$, $\gamma = \pi$, $\mathbf{C} = [1, 1, 1, 1, 1, 1]$. Frame size, $2R = 60\lambda$.

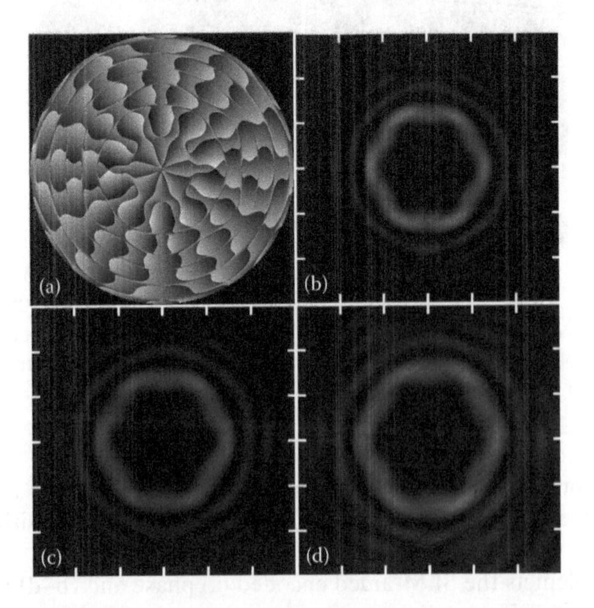

FIGURE 7.30 (a) SLM-aided encoded phase (Figure 7.29a) to generate the superposition of six shifted Bessel beams with a hexagon-shaped transverse intensity distribution and experimental intensity patterns at different distances from the plane $z = 0$: (b) 0 mm; (c) 200 mm; (d) 400 mm. Mesh step, 0.5 μm.

Superposition of a Large Number of Bessel Beams Centred on a Circle

In this section, we show that putting the center of a shifted n-order Bessel beam at each point of a circle of radius R_0 and taking the superposition of an infinite number of such beams with identical weight coefficients, their superposition will generate a conventional nonshifted n-order Bessel beam.

Describing the Bessel function as a series, the nonshifted Bessel beam in Eq. (7.109) can also be presented as a series in the plane $z = 0$:

$$E(x, y, z=0) = \exp(\mathrm{i}\, n\varphi) J_n(\alpha r) = \sum_{p=0}^{\infty} \frac{(-1)^p}{p!(n+p)!} \left(\frac{\alpha}{2}\right)^{n+2p} (x-iy)^p (x+iy)^{n+p}. \tag{7.138}$$

The continuous superposition of the shifted Bessel beams of Eq. (7.138) centred on a circle of radius R_0 is given by

$$E(x, y, z=0) = \sum_{p=0}^{\infty} \frac{(-1)^p}{p!(n+p)!} \left(\frac{\alpha}{2}\right)^{n+2p} \int_0^{2\pi} \left[(x-R_0\cos\theta) - i(y-R_0\sin\theta)\right]^p \times \left[(x-R_0\cos\theta) + i(y-R_0\sin\theta)\right]^{n+p} d\theta. \tag{7.139}$$

In polar coordinates, Eq. (7.139) takes the form:

$$E(r, \varphi, z=0) = \sum_{p=0}^{\infty} \frac{(-1)^p}{p!(n+p)!} \left(\frac{\alpha}{2}\right)^{n+2p} \int_0^{2\pi} \left[r\exp(-i\varphi) - R_0\exp(-i\theta)\right]^p \times \left[r\exp(i\varphi) - R_0\exp(i\theta)\right]^{n+p} d\theta. \tag{7.140}$$

Both integrand terms in Eq. (7.140) can be given as a binomial expansion:

$$E(r, \varphi, z=0) = \sum_{p=0}^{\infty} \frac{(-1)^p}{p!(n+p)!} \left(\frac{\alpha}{2}\right)^{n+2p} \sum_{m=0}^{p} \sum_{k=0}^{n+p} \binom{p}{m} \binom{n+p}{k} (-R_0)^{m+k} r^{2p-m+n-k} \times \exp\left[i(n+p-k)\varphi - i(p-m)\varphi\right] \int_0^{2\pi} \exp\left[i(k-m)\theta\right] d\theta. \tag{7.141}$$

Considering that integral (7.141) over θ takes a non-zero value only at $k = m$, the sum over k drops out and Eq. (7.141) is reduced to

$$E(r, \varphi, z=0) = 2\pi \exp(in\varphi) \sum_{p=0}^{\infty} \frac{(-1)^p}{p!(n+p)!} \left(\frac{\alpha r}{2}\right)^{n+2p} \sum_{m=0}^{p} \binom{p}{m} \binom{n+p}{m} \left(\frac{R_0}{r}\right)^{2m}. \tag{7.142}$$

Changing the order of summation in Eq. (7.142), we obtain:

$$E(r,\varphi,z=0)=2\pi\exp(in\varphi)\left[\sum_{m=0}^{\infty}\left(\frac{R_0}{r}\right)^{2m}\sum_{p=m}^{\infty}\binom{p}{m}\binom{n+p}{m}\frac{(-1)^p}{p!(n+p)!}\left(\frac{\alpha r}{2}\right)^{n+2p}\right]$$
$$=2\pi\exp(in\varphi)\sum_{m=0}^{\infty}\left(\frac{R_0}{r}\right)^{2m}\sum_{p=m}^{\infty}\frac{(-1)^p}{(m!)^2(p-m)!(n+p-m)!}\left(\frac{\alpha r}{2}\right)^{n+2p}. \tag{7.143}$$

Replacing $p - m$ with p and regrouping the terms of both series, we can reduce the earlier relation to Bessel functions:

$$E(r,\varphi,z=0)=2\pi\exp(in\varphi)\left[\sum_{m=0}^{\infty}\frac{(R_0/r)^{2m}}{(m!)^2}\sum_{p=0}^{\infty}\frac{(-1)^{p+m}}{p!(n+p)!}\left(\frac{\alpha r}{2}\right)^{n+2p+2m}\right]$$
$$=2\pi\exp(in\varphi)\left[\sum_{m=0}^{\infty}\frac{(-1)^m}{(m!)^2}\left(\frac{\alpha R_0}{2}\right)^{2m}\right]\left[\sum_{p=0}^{\infty}\frac{(-1)^p}{p!(n+p)!}\left(\frac{\alpha r}{2}\right)^{n+2p}\right] \tag{7.144}$$
$$=2\pi J_0(\alpha R_0)J_n(\alpha r)\exp(in\varphi).$$

Equation (7.144) describes the amplitude of a conventional Bessel beam up to a constant term $2\pi J_0(\alpha R_0)$. Figure 7.31 depicts simulated superpositions of 5, 8, 10, 20, 40, and 60 shifted Bessel beams with topological charge $n=7$, centred on a circle of radius $R_0 = 100\lambda$. For all beams shown in Figure 7.31a–f, simulation parameters were the same: $n = 7$, $R_0 = 100\,\lambda$, $\alpha = 1/\lambda$. Frame size was $2R = 240\lambda$.

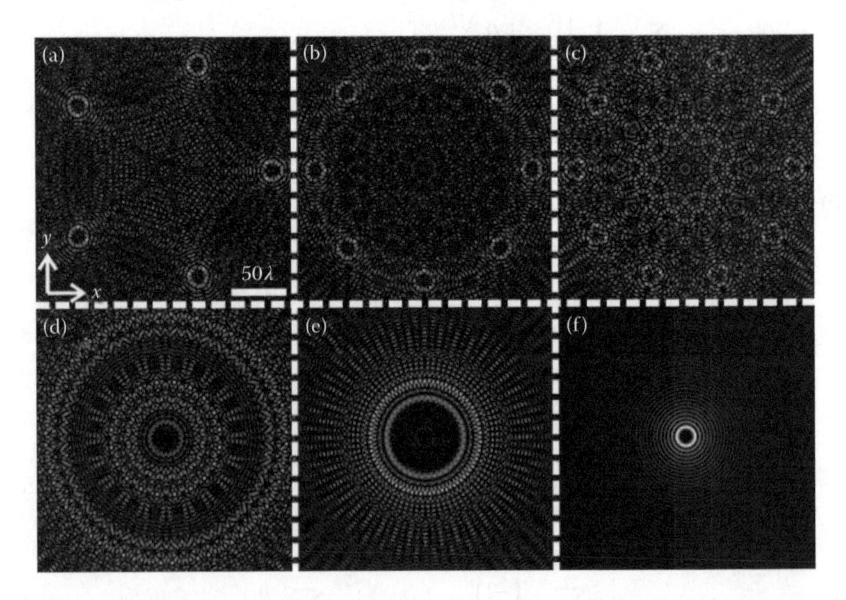

FIGURE 7.31 Transverse intensity distribution of the superposition of P shifted Bessel beams at P = (a) 5, (b) 8, (c) 10, (d) 20, (e) 40, (f) 60.

From Figure 7.31, a near-nonshifted/conventional Bessel mode is seen to be generated already at $P = 60$, with its amplitude defined by Eq. (7.144). All superpositions in Figure 7.31a–f have the same normalized OAM, which is equal to $J_z/I = 7$.

Summing up, the normalized OAM of the superposition of shifted Bessel beams with identical topological charge has been derived in an analytical form, Eq. (7.128). It has been shown that if all weight coefficients of the constituent terms in the superposition of shifted Bessel beams are real, the total OAM of the superposition equals that of an individual nonshifted Bessel beam. Based on this property, nondiffractive beams that have different intensity distributions but the same OAM can be generated. The superposition of a large number of identical Bessel beams centred on an arbitrary-radius circle has been shown to be equivalent to an individual constituent Bessel beam found at the circle center (Eq. 7.144). It has been also shown that a complex shift of a Bessel beam causes changes in the transverse intensity distribution, also changing OAM (Eqs. 7.122, 7.124). In the superposition of two complex-shifted Bessel beams, the OAM may remain unchanged, while the intensity distribution will change, Eq. (7.130).

The experiment has been shown to be in good agreement with theories (Figures 7.24, 7.26, 7.28, and 7.30).

8 Pearcey Laser Beams

8.1 HALF PEARCEY LASER BEAMS

Diffraction is one of the phenomena of the wave nature of light. When a light beam propagates along the optical axis from initial plane to observation plane, light fields from different points of the initial plane interfere with each other, so that in the observation plane the diffraction pattern occurs with its intensity distribution being generally different from the one in the initial plane. Propagation of a light beam in homogeneous medium is described by the Helmholtz equation and its paraxial approximation—Schrödinger type differential equation. Despite diffraction, there are solutions of these equations which describe diffraction-free light fields. First of all, such solutions describe both conventional and recently introduced asymmetrical Bessel beams [210,219,221], Mathieu beams [4], parabolic beams [97], and Airy beams [165]. Bessel beams propagate without diffraction in 3D space while the Airy beams propagate in 2D space. Theoretically, infinite plane waves are diffraction-free as well since their intensity distribution does not change upon propagation from one plane to another. In 3D space, a diffraction-free light field is any field with its angular spectrum of plane waves being nonzero at infinitesimally narrow circle [210]. Along with Bessel and Airy beams, form-invariant paraxial beams are also of interest. Such beams are not diffraction-free, but upon propagation, their transverse intensity distribution changes only in scale, but not structurally. Among the most well known of such beams are Hermite-Gaussian and Laguerre-Gaussian beams [2], as well as hypergeometric (HyG) modes [6]. In Ref. [8] Pearcey beams have been studied as three-dimensional (3D) analogs of Airy beams. Distribution of complex amplitudes of such beams is described by the Pearcey function [264,265], defined as the integral of the complex exponential function with polynomial argument (like the Airy function). Pearcey beams have an angular spectrum of plane waves with phase-only modulation along a parabola. It is shown in [8] that these beams are autofocusing and self-healing after being distorted by obstacles. In recent paper [266] a virtual source is proposed to generate the Pearcey beam.

In this section, we generalize the Pearcey function and consider the half-Pearcey (HP) beams, which are also form-invariant. The conventional Pearcey beam [8] is a superposition of two first-order HP beams. Angular spectrum of plane waves of the HP beams is nonzero at only a half of the parabola. We also consider 2D analogs of the HP beams and show acceleration of their trajectories.

Three-Dimensional Half-Pearcey Beams

Complex amplitude of the paraxial Pearcey beams [8] in initial plane $z = 0$ reads as

$$E(x, y, z=0) = \mathrm{Pe}\left(\frac{x}{x_0}, \frac{y}{y_0}\right) = \int_{-\infty}^{+\infty} \exp\left[is^4 + is^2\left(\frac{y}{y_0}\right) + is\left(\frac{x}{x_0}\right)\right] ds. \quad (8.1)$$

As it is shown in [8], structure of the beam intensity distribution does not change upon propagation. The beam is only shifting along one Cartesian coordinate and scaling along both coordinates (though scaling factors are different along x and y):

$$E(x, y, z) = \frac{1}{(1 - z/z_e)^{1/4}} \mathrm{Pe}\left(\frac{x}{x_0(1 - z/z_e)^{1/4}}, \frac{y - y_0 z/(2kx_0^2)}{y_0(1 - z/z_e)^{1/2}}\right), \quad (8.2)$$

where $z_e = 2ky_0^2$.

At the very end of [8] the generalized beams (8.1) are introduced with their complex amplitude in the initial plane being described by the following function

$$U(x, y) = \int_{-\infty}^{+\infty} \exp\left[i\left(s^{2n} + ys^n + xs^m\right)\right] ds, \quad (8.3)$$

and when $n = 2m$, these beams are also form-invariant. Numbers n and m must be integers, since integral (8.3) is taken over the whole real axis, so for fractional n and m at $s < 0$ the exponential function can tend to infinity (at $s \to -\infty$).

Complex amplitude (8.2) can be derived from (8.1) by using the Fresnel transform with subsequent changing of variables, but without changing of integral limits, since these limits are infinite. However, if the lower limit of integration would be zero instead of infinity, both limits of integration also do not change with the changing of variables. Therefore, along with (8.1) and (8.3) a νth order HP beam can be considered, which also does not change its form upon propagation. Its complex amplitude in the initial plane reads as

$$E(x, y, z=0) = \mathrm{HPe}_\nu\left(\frac{x}{x_0}, \frac{y}{y_0}\right) = \int_{0}^{+\infty} \exp\left[is^{4\nu} + is^{2\nu}(y/y_0) + is^\nu(x/x_0)\right] ds. \quad (8.4)$$

At other arbitrary planes, the complex amplitude of such beams is as follows

$$E(x, y, z) = \frac{1}{(1 - z/z_e)^{1/(4\nu)}} \mathrm{HPe}_\nu\left(\frac{x}{x_0(1 - z/z_e)^{1/4}}, \frac{y - y_0 z/(2kx_0^2)}{y_0(1 - z/z_e)^{1/2}}\right). \quad (8.5)$$

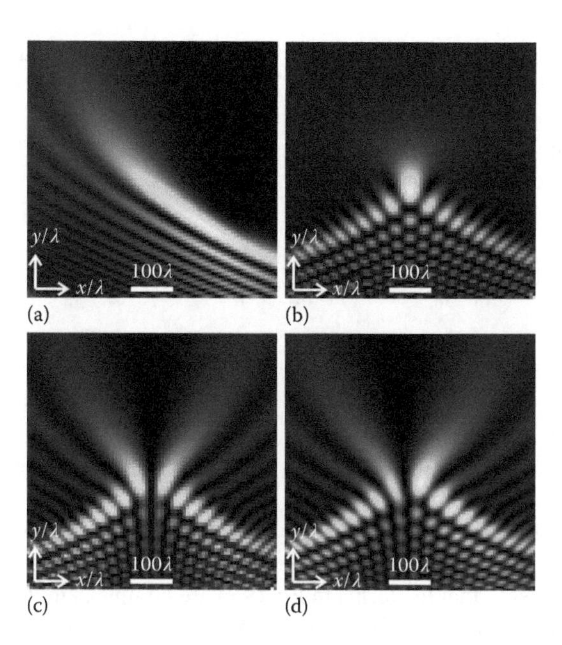

FIGURE 8.1 Intensity distribution of (a) first-order ($\nu = 1$) HP beam (4), (b) full Pearcey beam [8], (c) superposition of two symmetrical HP beams with phase delay of π (4) with zero intensity in plane $x = 0$, and (d) superposition of two symmetrical HP beams with phase delay of $3\pi/4$.

Note that unlike Eq. (8.3), parameter ν in Eq. (8.4) is not necessarily an integer since the integral is taken over only non-negative values of s. The diffraction pattern of beams (8.5) in the transverse plane has the form of inclined light lines with the diffraction thickness (Figure 8.1a).

The definition of the HP function (8.4) is similar to the integral representation of the Airy function, but the powers in the exponent are even higher. Thus, the exponent in the integral oscillates faster than in the case of Airy function. In order to calculate the Airy function, it is convenient to treat it as a sum of two Hankel functions of order 1/3 [267]. When it is difficult or impossible to represent a fast-oscillating integral as the sum of some well-known functions, another approach to cope with integration of such oscillating function is the rotation of integration contour in a complex plane [8]. We also apply this approach here to calculate the function $\mathrm{HPe}_\nu(x/x_0, y/y_0)$, although instead of changing of variables $s \to s'\exp(i\pi/8)$ we used $s \to s'\exp(i\pi/(8\nu))$. This leads to an integral of another function vanishing at infinity. We used the following parameters to calculate the field shown in Figure 8.1: $\lambda = 532$ nm, $x_0 = y_0 = 20\lambda$, $\nu = 1$, size of calculation area $-300\lambda \le x, y \le 300\lambda$.

The HP beam of νth order has the following angular spectrum of plane waves:

$$\tilde{E}\left(k_x, k_y\right) = \begin{cases} \dfrac{x_0 y_0}{\nu}\left(-k_x x_0\right)^{\frac{1-\nu}{\nu}} \exp\left(i k_x^4 x_0^4\right) \delta\left(k_x^2 x_0^2 + k_y y_0\right), \, k_x x_0 < 0, \\ 0, \, k_x x_0 \ge 0. \end{cases} \tag{8.6}$$

Similarly to the angular spectrum of Pearcey beams in [8], the angular spectrum (8.6) is nonzero along the parabola $k_x^2x_0^2 + k_y y_0$. However, instead of the whole parabola, the angular spectrum (8.6) is nonzero only at its half (at $k_x x_0 < 0$).

The Pearcey beam with complex amplitude in the initial plane (8.1) is a superposition of the first-order HP beams (8.4):

$$\mathrm{Pe}\left(\frac{x}{x_0},\frac{y}{y_0}\right)=\mathrm{HPe}_1\left(\frac{x}{x_0},\frac{y}{y_0}\right)+\mathrm{HPe}_1\left(-\frac{x}{x_0},\frac{y}{y_0}\right). \tag{8.7}$$

As full Pearcey beams have infinite energy [8], HP beams have infinite energy as well. By infinite energy of the beam, we mean that its complex amplitude is not square integrable. Physically, when the period of HP-function oscillations becomes less than the wavelength, plane waves become evanescent, making the energy finite.

It is seen in Figure 8.1b that the Pearcey beam [8] is the result of interference of the HP beam of Figure 8.1a and the same beam symmetrically reflected with respect to the plane $x = 0$. Note that the terms in Eq. (8.7) are complex amplitudes of two HP beams in the initial plane and after propagation of both these beams their complex amplitudes are multiplied by the same factor, depending on y_0^2. Therefore, we can use arbitrary coefficients in superposition (8.7) and nevertheless the resulting beam will be form-invariant. In particular, one can consider a light beam which is a superposition of the HP beam of Figure 8.1a and the same beam symmetrically reflected (with respect to the plane $x = 0$) and having phase delay of π. Such a form-invariant beam is similar to the Pearcey beam [8], but in the center of the diffraction pattern there will be zero intensity instead of the maximal intensity. Transverse intensity distribution of such HP beams is shown in Figure 8.1c. In case of another phase delay, the diffraction pattern is similar to Figure 8.1b and c but has no symmetry with respect to axis y. In Figure 8.1d we show the diffraction pattern for phase delay of $3\pi/4$.

The intensity distribution of the full Pearcey beam of Figure 8.1b looks similar to half that of the self-healing caustic beams [268]. Nevertheless, these beams are different. It is written in [268] that a caustic beam can be considered as a cylindrical wave passing through an axicon and that a caustic beam is not invariant on propagation. The Pearcey beams are autofocusing and form-invariant, while distribution of complex amplitudes in a transverse plane, according to Eq. (8.4), cannot be obtained by passing of a cylindrical wave through an axicon.

Figure 8.2 shows the same diffraction patterns as in Figure 8.1 but for the case of fractional beam order $\nu = 1.5$.

It is seen in Figure 8.2 that increasing of the Pearcey function order ν did not lead to significant changes of the diffraction patterns, although intensity peaks in Figure 8.2b are decreasing from the center to the periphery faster than in Figure 8.1b.

The experiment was conducted using a phase spatial light modulator (SLM) PLUTO VIS composed of 1920×1080 sensitive cells (each is 8 μm in size). The phase depicted in Figure 8.3a consisted of 1000×1000 pixels and was generated

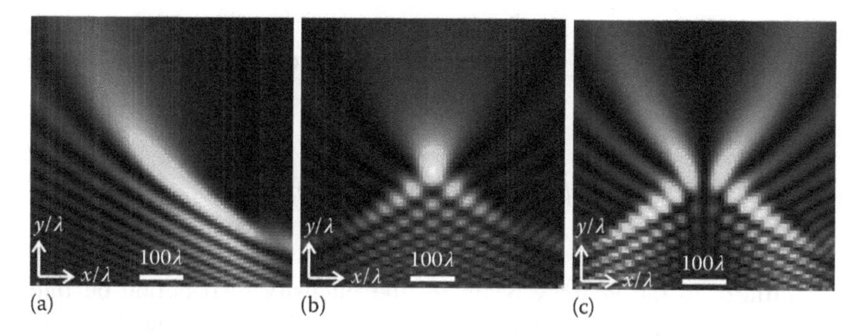

FIGURE 8.2 Intensity distribution of (a) HP beam (4) at $\nu = 1.5$, and (b) superposition of two symmetrical HP beams ($\nu = 1.5$) with phase delay of 0 and (c) with phase delay of π.

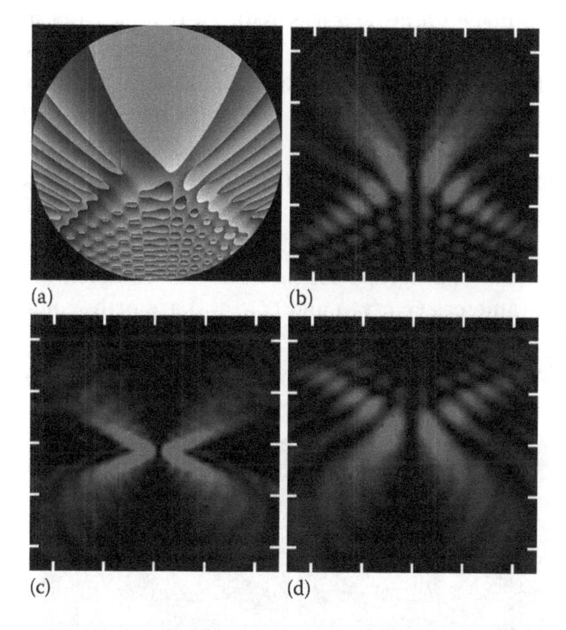

FIGURE 8.3 (a) Computer-generated phase distribution on the SLM PLUTO VIS, intended to generate superposition of two symmetrical HP beams with phase delay of π, and intensity distribution in two transverse planes located at a distance of (b) 50 mm, (c) 400 mm, and (d) 650 mm. The frame size in figures (b), (c), and (d) is 2.5×2.5 mm.

at the modulator's centre. This phase intended to generate superposition of two symmetrical HP beams with phase delay of π. The phase distribution is asymmetric because during encoding the amplitude is taken into account. The SLM was illuminated by a laser beam with a wavelength of 532 nm. Intensity distribution was registered by a CCD-camera with frame size of 2.5×2.5 mm. In Figure 8.3b and c we show intensity distribution at different distances from SLM. It is seen in Figure 8.3b and c that the generated beam is similar to that of Figure 8.1c and propagates without changing its shape.

Note that, during derivation of Eq. (8.5) from Eq. (8.4), change of integration variable can be done only at $z < z_e$. Otherwise, integration limits become imaginary. However, it can be shown that

$$E\left(x, \frac{y_0^3}{x_0^2} + \eta, z_e + \zeta\right) = E^*\left(-x, \frac{y_0^3}{x_0^2} - \eta, z_e - \zeta\right), \tag{8.8}$$

that is, similarly to the Pearcey beam [8] the intensity distribution beyond the plane $z = z_e$ is symmetrical to that in front of the plane $z = z_e$ with regards to the planes $x = 0$ and $y = y_0^3/x_0^2$. It is also seen in Figure 8.3 that after focusing (Figure 8.3c) the intensity pattern in Figure 8.3d is opposite to that of Figure 8.3b.

We also studied experimentally the self-healing effect of the HP beams. In the plane $z = 0$ we placed a cover glass with an opaque area made by a black marker. In this area light was absorbed and thus the beam was distorted. The intensity pattern of the distorted beam at different distances is shown in Figure 8.4. It is seen in Figure 8.4a that one of the two main lobes of the beam is almost absent because of the obstacle. In Figure 8.4b (before the focus) the beam is still asymmetric. Then, the beam is focused (Figure 8.4c) and beyond the focus, the beam has an almost symmetric intensity pattern as if there were no obstacle in the initial plane.

In conclusion to this paragraph let us note that complex amplitudes of 3D HP beams with parameters (x_{01}, y_{01}) and (x_{02}, y_{02}) are orthogonal in cases when $y_{02}/y_{01} \neq x_{02}^2/x_{01}^2$. This can be shown by direct calculation of the dot product of functions $\mathrm{HPe}_\nu(x/x_{01}, y/y_{01})$ and $\mathrm{HPe}_\nu(x/x_{02}, y/y_{02})$ as well as by using the

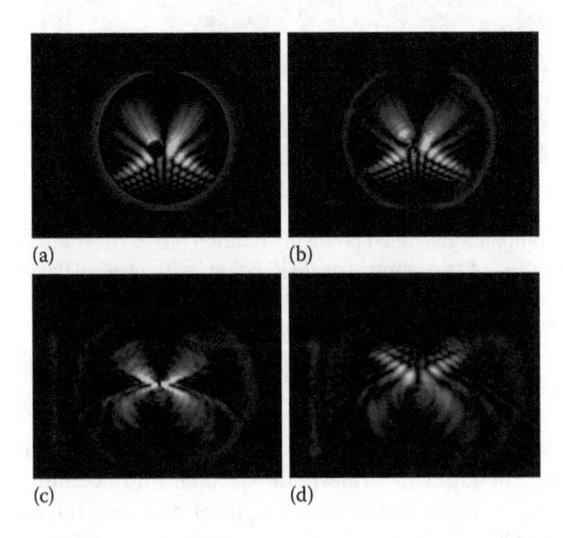

FIGURE 8.4 Intensity patterns of the distorted beam at a distance of (a) 0 mm, (b) 100 mm, (c) 400 mm, and (d) 600 mm.

condition that angular spectra (8.6) are nonzero at different parabolas which are not intersecting with each other.

Two-Dimensional Half-Pearcey Beams

We also considered two-dimensional (2D) HP beams with their initial complex amplitude being similar to Eq. (8.4):

$$E(x,0)=\int_0^{+\infty}\exp\left[is^{\nu}\left(\frac{x}{x_0}\right)+is^{2\nu}\right]ds=\mathrm{HPe}_{\nu}^{2D}\left(\frac{x}{x_0}\right). \tag{8.9}$$

Using the Fresnel transform, it can be shown that at some propagation distance z complex amplitude of the beam (8.9) reads as

$$E(x,z)=\int_0^{+\infty}\exp\left[is^{\nu}\left(\frac{x}{x_0}\right)+i\left(1-\frac{z}{2kx_0^2}\right)s^{2\nu}\right]ds. \tag{8.10}$$

Applying change of integration variable $s=(1-z/z_e)^{-1/(2\nu)}t$ (where $z_e=2kx_0^2$), instead of Eq. (8.10), the complex amplitude of the 2D HP beam is as follows:

$$E(x,z)=\frac{1}{(1-z/z_e)^{1/(2\nu)}}\mathrm{HPe}_{\nu}^{2D}\left(\frac{x}{x_0(1-z/z_e)^{1/2}}\right). \tag{8.11}$$

As in the 3D case, change of integration variable can be done only for $z<z_e$; otherwise, integration limits become imaginary. However, it follows from Eq. (8.10) that

$$E(x,z_e+\zeta)=E^{*}(-x,z_e-\zeta), \tag{8.12}$$

that is, intensity distribution beyond the plane $z=z_e$ is symmetrical to that before the plane $z=z_e$ with regards to the optical axis.

It is seen from Eq. (8.11) that at an arbitrary transverse plane before the focal plane (i.e., $z<z_e$) the intensity maxima of the 2D HP beam are located at points

$$x_{\max}=y_m x_0\sqrt{1-\frac{z}{z_e}}, \tag{8.13}$$

where y_m is the coordinate of the mth maximum of function $\left|\mathrm{HPe}_{\nu}^{2D}(x)\right|^2$. It is easy to see that $(dx_{\max}/dz)(d^2x_{\max}/dz^2)>0$ at $z<z_e$. Therefore, the beam path will be

accelerating before the focal plane. From the symmetry property (8.12) it follows that beyond the plane $z = z_e$ the 2D HP beam propagates with deceleration.

Further, we consider propagation of the 2D HP beams with $\nu = 2$. Let the wavelength be $\lambda = 532$ nm and the beam scaling factor be $x_0 = \lambda$. Therefore, the focal plane is located at a distance of $z = z_e = 4\pi\lambda$. At a distance of $z = 3z_e/4 = 3\pi\lambda$ the beam narrows two times while maximal intensity increases two times as well. At a plane $z = z_e = 4\pi\lambda$ the light field (8.11) narrows infinitely and the focus appears. Intensity distribution in the initial plane for the chosen parameters is shown in Figure 8.5a. We simulated propagation of the 2D HP beam by the finite-difference beam propagation method (BPM). Simulation area was $-80\lambda \leq x \leq 80\lambda$, $0 \leq z \leq 30\lambda$. Shown in Figure 8.5b is intensity distribution in the Oxz plane. Figure 8.5c shows intensity distribution in plane $z = 0$, while Figure 8.5d shows intensity distribution in plane $z = 3\pi\lambda$. Figure 8.5 confirms that the beam has narrowed two times: at $z = 0$ the fourth zero is located approximately at $x \approx -10\lambda$, while at $z = 3\pi\lambda$ this intensity zero is approximately at $x \approx -5\lambda$. Maximal intensity in Figure 8.5d is two times higher than that in Figure 8.5c. In addition,

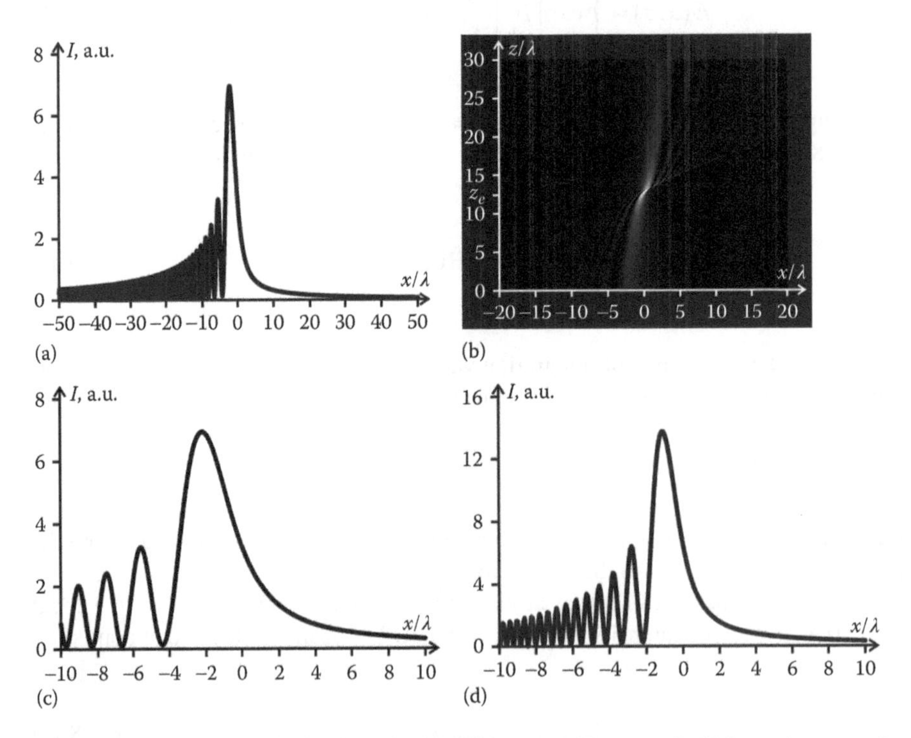

FIGURE 8.5 Intensity distribution of a 2D HP beam (8.11) at $\nu = 2$: (a) intensity at $z = 0$, (b) intensity in Oxz plane, (c) magnified intensity at $z = 0$, and (d) intensity at $z = 3\pi\lambda$.

it is seen that at both distances $z = 0$ and $z = 3\pi\lambda$ intensity of the first sidelobe (i.e., second maximum) is approximately two times smaller than intensity of the major maximum. This also confirms form-invariance of the 2D HP beam.

In conclusion to this paragraph let us note that unlike 3D HP beams, complex amplitudes of 2D HP beams with different parameters x_{01} and x_{02} are not orthogonal to each other.

Thus, we have obtained the following results. We obtained a new solution of the paraxial Helmholtz equation (Schrödinger-type equation). This solution describes a three-parameter family of form-invariant 3D HP laser beams (Eqs. 8.4 and 8.5). We have shown that the full Pearcey beams [8] are superpositions of two symmetrical first order HP beams (Eq. 8.7). Using the linear combination of two HP beams with arbitrary weights, other form-invariant laser beams can be obtained which are neither full Pearcey beams [8] nor HP beams. We have obtained the angular spectrum of plane waves of 3D HP beams (Eq. 8.6). The amplitude of this spectrum is nonzero at half of a parabola. We have obtained the orthogonality condition for the complex amplitudes of 3D HP beams. Functions of complex amplitudes of these beams are orthogonal if their Fourier spectra lie on disjoint parabolas. It was found that 2D HP beams are not orthogonal. Using a spatial light modulator, we generated superposition of two symmetrical HP beams with phase delay of π and demonstrated autofocusing and self-healing properties of such beam. We also obtained a solution of the 2D paraxial Helmholtz equation describing the 2D analog of form-invariant HP beams (Eqs. 8.9 and 8.11). It is shown that similarly to a 3D HP beam, a 2D HP beam has the property of autofocusing. Such beam is also accelerating before the focal plane and decelerating beyond it (Figure 8.5b).

8.2 VORTEX PEARCEY BEAMS

Vortex laser beams that carry an orbital angular momentum (OAM) have been the focus of researchers' attention in optics. They have found uses in Terabit free-space data transmission via channel multiplexing [234], optical trapping and rotation of microparticles [31], and sensing atmospheric turbulence [258]. Vortex beams can be exemplified by the well-known Bessel [210] and Laguerre-Gaussian laser beams [2] as well as HyG modes [6]. Such beams either retain their structure [210] or experience a dimensional scaling during propagation in space [2,6]. To be suitable for micromanipulation problems, the beams need to be focused by means of a microlens. Such beams have a strictly defined intensity profile in the focus. Examples of autofocusing laser beams are provided by Pearcey beams [8]. The Pearcey beams are interpreted in [8] as 3D analogs of the Airy beams [165]. These beams have a complex amplitude distribution described by a Pearcey function [264,265], which is defined as the integral over a complex exponential function of an argument given by a polynomial (like in the Airy function). The angular spectrum associated with such beams is represented by a parabola with phase modulation. These beams have the property of autofocusing, being capable of self-healing following distortion due to an obstacle. A virtual source of light to generate Pearcey beams has been proposed [266]. A general relation for the Pearcey function has been derived and structurally stable HPe beams have been analyzed [269,270]. Standard Pearcey beams [8] are a sum of two HPe beams of the first order. Unlike a parabolic angular spectrum of standard Pearcey beams [8], the angular spectrum of HPe beams is described by a semi-parabola. It is worth noting that Pearcey beams do not carry OAM, since the Pearcey function Pe(x,y) is even with respect to x and arbitrary light field has zero OAM if its complex amplitude $E(x, y)$ is even with respect to at least one Cartesian coordinate, that is, $E(x, y) = E(-x, y)$ or $E(x, y) = E(x, -y)$. It should be also noted that 2D analogs of Pearcey beams that propagate along a curved path, thus showing acceleration, have been introduced [269,270].

In this section, using 2D half Pearcey (HPe) beams, we obtain paraxial vortex autofocusing hypergeometric (AH) beams that carry OAM. With their complex amplitude being proportional to a degenerate HyG function, such beams are close in shape to HyG modes [6]. Taking into account the autofocusing property of Pearcey beams, the beams introduced in this study are also autofocusing and accelerating.

Autofocusing Hypergeometric Beams

In this section, following [269,270], we shall briefly describe 2D HPe beams. The complex amplitude of paraxial 2D HPe beams in the initial plane is given by

$$E(x,0) = \int_0^{+\infty} \exp\left[is^p \left(\frac{x}{x_0}\right) + is^{2p} \right] ds = \mathrm{HPe}_p^{2D}\left(\frac{x}{x_0}\right). \tag{8.14}$$

The complex amplitude of a 2D HPe beam can be given by

$$E(x,z) = \frac{1}{(1-z/z_e)^{1/(2p)}} \mathrm{HPe}_p^{2D}\left(\frac{x}{x_0(1-z/z_e)^{1/2}}\right). \tag{8.15}$$

By twice differentiating (8.15) with respect to z, we find that at $z < z_e$ there holds an inequality $(dx_{max}/dz)(d^2x_{max}/dz^2) > 0$, thus showing that the beam is accelerated prior to autofocusing.

By way of illustration, consider the propagation of a 2D HPe beam at $p = 2$. Assuming the wavelength of light to be $\lambda = 532$ nm and the scale coefficient in (8.14) $x_0 = \lambda$, the focal length is $z_e = 4\pi\lambda$. With the beam becoming twice as thin at a distance of $z = 3z_e/4$, we find that in the case under study $z = 3\pi\lambda$ where the maximum intensity will be doubled. In the plane $z_e = 4\pi\lambda$, the field of Eq. (8.15) is "infinitely" narrowed, producing an autofocus. For the aforementioned parameters, the intensity

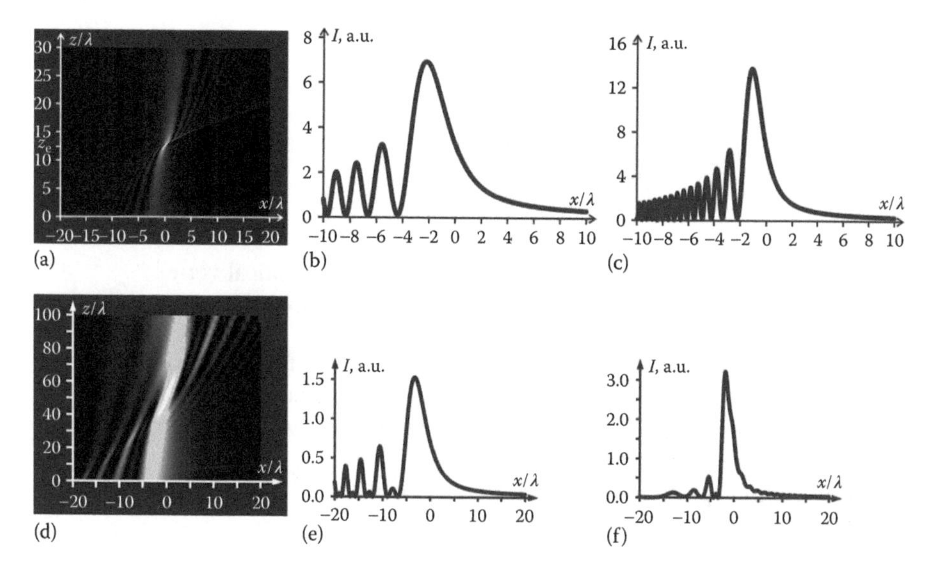

FIGURE 8.6 Intensity distribution of a 2D HPe beam calculated using a BPM-method: (a) intensity pattern in the Oxz-plane of a wide (width/focus ≈ 12.7) HPe beam and its intensity profiles along the x-axis at (b) $z = 0$ and (c) $z = 75\lambda/8 \approx 3z_e/4$, (d) intensity pattern in the Oxz-plane of a paraxial narrow (width/focus ≈ 0.8) HPe beam and its intensity profiles along the x-axis at (e) $z = 0$ and (f) $z = 3z_e/4$.

pattern in the Oxz-plane is shown in Figure 8.6a. The simulation was conducted using a BPM in the domain $-80\lambda \leq x \leq 80\lambda$ (although Figure 8.6a shows only area $-20\lambda \leq x \leq 20\lambda$), $0 \leq z \leq 30\lambda$. The intensity profiles for the pattern in Figure 8.6a are shown at $z = 0$ in Figure 8.6b and at $z = 75\lambda/8$ ($\approx 3\pi\lambda$) in Figure 8.6c. From Figure 8.6, the beam is seen to actually become twice as narrow, with the fourth zero located approximately at -10λ for $z=0$ and at -5λ for $z=75\lambda/8$. The maximum intensity in Figure 8.6c is twice as high as that in Figure 8.6b. Besides, from Figure 8.6b and c the intensity of the second maximum is seen to be approximately half as much as that of the major maximum, additionally testifying to the fact that during propagation the beam is changing in scale while remaining unchanged in structure. Figure 8.6a–c are obtained for highly non-paraxial cases when the input field has a width of 160λ and the beam focuses at a distance of $z_e = 4\pi\lambda$ (since $x_0 = \lambda$). Shown in Figure 8.6d is the intensity pattern in the Oxz plane for paraxial cases when the input field has a width of 40λ, while the scale coefficient in (8.14) is $x_0 = 2\lambda$, thus making the focal distance equal to $z_e = 16\pi\lambda$. In such cases, there are only several lobes in the input plane, and, as can be seen in Figure 8.6e and f, the intensity distribution changes upon propagation.

The complex amplitude of a 3D paraxial monochromatic laser beam can be represented as superposition of 2D HPe beams with each constituent beam being rotated by a polar angle θ and characterized by a phase proportional to the rotation angle, $n\theta$:

$$E_{np}(x,y,z=0)=\frac{(-i)^n}{2\pi}\int_0^{2\pi}e^{in\theta}d\theta\int_0^{\infty}\exp\left[is^{2p}+is^p\left(\frac{x\cos\theta+y\sin\theta}{x_0}\right)\right]ds, \quad (8.16)$$

where:

n is an integer number called a topological charge of an optical vortex
p is a real number that defines the order of the HPe beam

Note that for conventional Pearcey beams [8] $p = 2$ and the integration in (8.16) with respect to the variable s is performed from minus to plus infinity. In Eq. (8.16), beams with different topological charge n are orthogonal, whereas beams with different order p are not orthogonal.

Taking both integrals, Eq. (8.16) is rearranged to

$$E_{np}(r,\phi,z{=}0)$$
$$=\frac{\exp\left[\frac{i\pi}{2}\left(\frac{pn+1}{2p}\right)+in\phi\right]}{2pn!}\Gamma\left(\frac{pn+1}{2p}\right)\left(\frac{r}{2x_0}\right)^n {}_1F_1\left(\frac{pn+1}{2p},n+1,-i\frac{r^2}{4x_0^2}\right), \quad (8.17)$$

where:

(r, φ) are polar coordinates
$\Gamma(x)$ is a γ-function
${}_1F_1(a,b,x)$ is a degenerate HyG function [38].

By taking a Fresnel transform, the complex amplitude of the AH beam in (8.17) at an arbitrary distance z can be given by

$$E_{np}(r,\phi,z)=\frac{\exp\left[\frac{i\pi}{2}\left(\frac{pn+1}{2p}\right)+in\phi\right]}{2pn!\left(1-\frac{z}{z_e}\right)^{1/2p}}\Gamma\left(\frac{pn+1}{2p}\right)\xi^n {}_1F_1\left(\frac{pn+1}{2p},n+1,-i\xi^2\right), \quad (8.18)$$

where:

$$\xi=\frac{r}{2x_0\sqrt{1-\frac{z}{z_e}}},\quad z_e=2kx_0^2.$$

It can be checked that at $z = 0$ Eq. (8.18) can be rearranged to Eq. (8.17). Equation (8.18) suggests that at $n \neq 0$ the on-axis intensity ($r = 0$) will be zero at any z, with the intensity pattern given by an array of concentric rings centered on the optical axis. However, at $n=0$, the beam in (8.18) ceases to have a vortex and annular structure, with the on-axis intensity taking a non-zero value and the complex amplitude given by

$$E_{0p}(r,z)=\frac{\exp\left(\frac{i\pi}{4p}\right)}{2p\left(1-\frac{z}{z_e}\right)^{1/(2p)}}\Gamma\left(\frac{1}{2p}\right){}_1F_1\left(\frac{1}{2p},1,-i\xi^2\right). \tag{8.19}$$

The complex amplitude in (8.19) describes a paraxial wave of infinite energy that converges to an on-axis focus, in a similar way to a spherical or parabolic wave. From (8.18), the argument ξ is seen to tend to infinity at $z = z_e$. Making use of an asymptotic relation at $x \to \infty$

$$x^n\left|{}_1F_1\left(\frac{pn+1}{2p},n+1,-ix^2\right)\right|\approx\frac{1}{x}, \tag{8.20}$$

we find that in the autofocusing plane ($z=z_e$), complex amplitude (8.18) is proportional to

$$E_{np}\left(r,\phi,z=z_e\right)\approx\frac{\exp\left(in\phi\right)}{r}. \tag{8.21}$$

From (8.21), a sharp focus is seen to be formed in the autofocusing plane ($z=z_e$), but a single zero-intensity point is also found in the focus center (if $n\neq 0$). A remarkable property of vortex AH beams is that the autofocusing distance $z_e=2kx_0^2$ is independent of the beam parameters p and n. This means that any superposition of such beams with different values of p and n will be focused at the same distance z_e. This property of vortex beams can be useful for generating a desired intensity pattern and OAM in the autofocusing plane.

Figure 8.7 shows intensity patterns for the radial ($n = 0$) and vortex ($n = 5$) AH beams at different distances from the initial plane: (*a*, *d*) $z=0$, (*b*, *e*) $z=0.75z_e$, and (*c*, *f*) $z=0.999z_e$. At distance $z=3z_e/4$ the diffraction pattern is half as large as that in the plane $z=0$, becoming $10\sqrt{10}$ times smaller at distance $z=0.999z_e$. In Figure 8.7, maximal intensity distribution (in arbitrary units) is (a) 0.821565, (b) 1.64313, (c) 25.9802, (d) 0.0273461, (e) 0.0546921, and (f) 0.864708.

Figure 8.7 was simulated using a formula similar to (8.16)

$$E_{np}\left(x,y,z\right)=\frac{\left(-i\right)^n}{2\pi}\frac{1}{\left(1-z/z_e\right)^{1/(2p)}}\int_0^{2\pi}e^{in\theta}\,\mathrm{HPe}_2^{2D}\left(\frac{x\cos\theta+y\sin\theta}{x_0\sqrt{1-z/z_e}}\right)d\theta, \tag{8.22}$$

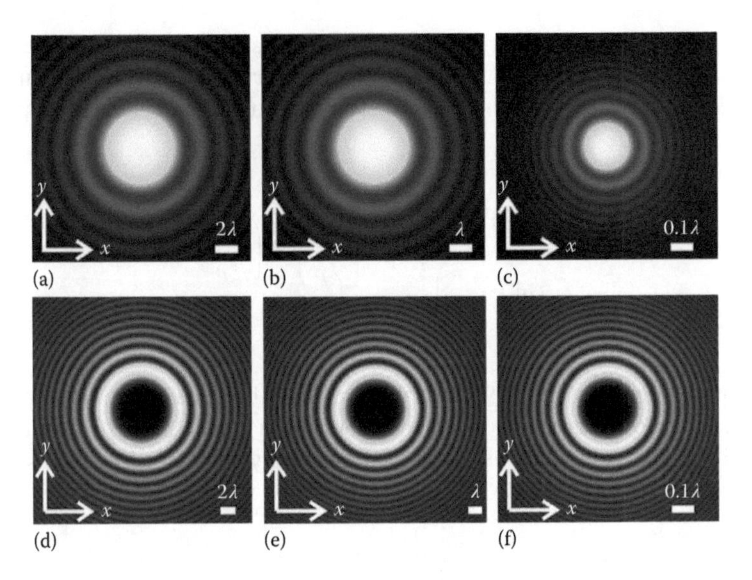

FIGURE 8.7 Intensity pattern of a vortex Pearcey beam at (a, b, c) $n=0$ and (d, e, f) $n=5$ in the planes (a, d) $z=0$, (b, e) $z=3z_e/4$, and (c, f) $z=0.999z_e$.

and an identity relating the Pearcey function at $p=2$ to a Bessel function of the first kind:

$$\mathrm{HPe}_2^{2D}(\xi)=\frac{\pi}{4}\sqrt{\frac{|\xi|}{2}}\exp\left(-\frac{i\xi^2}{8}+\frac{i\pi}{8}\right)\left[J_{-1/4}\left(\frac{\xi^2}{8}\right)+i\,\mathrm{sgn}(\xi)\exp\left(\frac{i\pi}{4}\right)J_{1/4}\left(\frac{\xi^2}{8}\right)\right]. \quad (8.23)$$

This identity is easily derived from several reference integrals (expressions 3.696 in [135]).

It is worth noting that at $p = 1$, vortex AH beams (8.18) are similar in structure to the familiar HyG modes [6] at $\gamma=0$:

$$E_{n\gamma}(r,\phi,z)=\frac{\exp\left[-\frac{i\pi}{4}(n-i\gamma+1)+in\phi\right]}{2\pi n!}\exp\left[\frac{i\gamma}{2}\ln\left(\frac{z}{z_0}\right)\right]\Gamma\left(\frac{n+i\gamma+1}{2}\right)\left(\frac{z_0}{z}\right)^{1/2} \quad (8.24)$$
$$\times\,\xi^{n/2}\,{}_1F_1\left(\frac{n-i\gamma+1}{2},n+1,i\xi\right),$$

where $\xi = kr^2/(2z)$, $z_0=kw^2/2$.

In (8.24), w is a scale factor analogous to x_0 and γ is a real parameter. Note, however, that at $p=1$, Eq. (8.16) describes superposition of parabolic waves, rather than conventional Pearcey beams [8] at $p=2$. Another distinction between (8.18) and (8.24) is that in the initial plane vortex AH beams (8.18) are similar in form to (8.17), whereas HyG modes (8.24) are generated by a complex amplitude [6] given by

$$E_{n\gamma}\left(r,\phi,z=0\right)=\frac{1}{2\pi}\left(\frac{w}{r}\right)\exp\left[i\gamma\ln\left(\frac{r}{w}\right)+in\phi\right]. \tag{8.25}$$

From comparison of (8.21) and (8.25), the vortex AH beams (8.18) in the autofocusing (Fourier) plane are seen to take the form of Eq. (8.21), which is similar to the HyG modes (8.25) in the initial plane (at $\gamma=0$). Note that the function in (8.21) is Fourier-invariant, with its Fourier image being defined as

$$E_n\left(\rho,\theta\right)=\left(-i\right)^{n+1}\frac{\exp\left(in\theta\right)}{\rho},\ \rho>0. \tag{8.26}$$

However, at $n\neq 0$ and $\rho=0$, the amplitude in (8.26) equals zero. It can be inferred that the HyG modes in (8.24) and vortex AH beams in (8.18) differ in an on-axis shift equal to the autofocus distance. An indirect confirmation may be found in a technique for generating accelerating beams from decelerating ones via an on-axis shift [271].

Equating the argument ξ in (8.18) to a value at which the amplitude modulus achieves the maximum ξ_n yields the radius of the brightest ring versus the distance on the optical axis:

$$r_n\left(z\right)=2x_0\xi_n\sqrt{1-\frac{z}{z_e}}. \tag{8.27}$$

From (8.27), the ring radius is seen to equal zero at $z=z_e$. If, similarly to Eq. (8.15), we take the first and second derivatives of the ring radius (8.27) with respect to z, these will have the same sign. This means that propagating along an accelerating path toward focus, the vortex AH beam "steeply" converges to a focus, in a similar way to Airy beams [190].

Autofocusing Hypergeometric-Gaussian Beams

As is the case with Bessel beams [210], beams (8.18) and (8.24) possess an infinite energy. It stands to reason that similarly to explicitly taking a Fresnel transform of function (8.16) with the aid of special functions (8.18), it is possible to take a Fresnel transform of function (8.16) multiplied by a Gaussian exponential function. Hence, we can analyze autofocusing hypergeometric-Gaussian (AHG) beams in the initial plane by multiplying complex amplitude (8.16) by a Gaussian exponential function:

$$F_{np}(x,y,z=0)$$
$$=\frac{(-i)^n}{2\pi}\exp\left(-\frac{x^2+y^2}{\sigma^2}\right)\int_0^{2\pi}e^{in\theta}d\theta\int_0^{\infty}\exp\left[is^{2p}+is^p\left(\frac{x\cos\theta+y\sin\theta}{x_0}\right)\right]ds, \tag{8.28}$$

where σ is the waist radius of the Gaussian beam.

At an arbitrary distance from the initial plane, the complex amplitude of vortex AHG beams that can be derived using a Fresnel transform is

$$F_{np}(r,\varphi,z)=\frac{\exp\left[-\frac{i\pi}{2}(n+1)+in\varphi\right]}{2pn!}\exp\left[-\frac{r^2}{\sigma^2(z)}+\frac{ikr^2}{2R(z)}\right]g^{(1-2p)/p}(z)q^{-1/p}(z)$$
$$\times\left(\frac{pn+1}{2p}\right)\xi^n{}_1F_1\left(\frac{pn+1}{2p},n+1,-i\xi^2\right), \tag{8.29}$$

where:

$$\xi=\frac{r}{2x_0g(z)q(z)},\ g(z)=\sqrt{\frac{z}{z_0}-i},\ q(z)=\sqrt{\frac{z}{z_e}-i\left(\frac{z}{z_0}-i\right)},\ z_0=\frac{k\sigma^2}{2},$$
$$\sigma(z)=\sigma\sqrt{1+\frac{z^2}{z_0^2}},\ R(z)=z\left(1+\frac{z_0^2}{z^2}\right). \tag{8.30}$$

At $\sigma\to\infty$, amplitude (8.29) changes to (8.18). Note that in the denominator of the argument ξ the product $g(z)q(z)$ is a complex number that does not take a zero value at any z. At $z=z_e$, the modulus of the product $g(z)q(z)$ achieves its minimum. That is, as distinct from vortex AH beams (8.18), vortex AHG beams (8.29) have a minimal, but non-zero, radius in the plane of autofocusing. At the same time, vortex AHG beams have a finite energy and can be generated by use of a liquid crystal light modulator.

Also, note that at $p=1$ and $z_e\to\infty$, beams (8.29) become structurally identical to HyG beams [113] at $m=-1$, $\gamma=0$. However, the beams of [113] do not possess an autofocusing property.

Experiment

Figure 8.8a shows the optical setup for the experiment. For the output of the phase distribution of the light field, we used a spatial light modulator SLM PLUTO-VIS

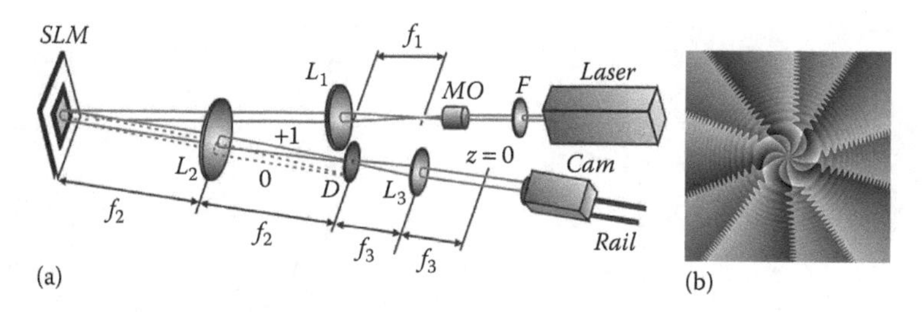

FIGURE 8.8 (a) Optical setup for generation of vortex AH beam: Laser—solid-state laser ($\lambda = 532$ nm), *F*—neutral density filter, *MO*—microobjective (60ˣ, NA = 0.85), L_1, L_2, L_3—lenses with focal distances of $f_1, f_2, f_3 = 150$ mm, *SLM*—spatial light modulator PLUTO VIS, *D*—diaphragm, Cam—CCD-camera LOMO TC-1000. (b) Encoded phase that was used in the experimental setup for generation of vortex AH beam with $p = 2$ and $n = 5$.

(resolution 1920 × 1080 pixels, pixel size is 8 μm). Using the neutral density filter *F*, the power of the source laser beam from a solid-state laser ($\lambda = 532$ nm) was decreased. A microobjective *MO* (60ˣ, NA = 0.85) and a lens L_1 ($f_1 = 350$ mm) were used for collimating the beam and to obtain the beam with the plane wavefront and near-uniform intensity distribution. Lenses L_2 ($f_2 = 350$ mm) and L_3 ($f_3 = 150$ mm) with the diaphragm *D* were used for the high-frequency optical filtering. As a result, the focal plane of the lens L_3 was conjugated with a plane of the SLM (designated as $z = 0$). A camcorder *Cam* LOMO TC-1000 (pixel size of 1.67 μm) was moving on an optical rail and registered generated intensity distributions.

Shown in Figure 8.8b is a 255-level phase function (frame size of 1024 × 1024 pixels) which encodes the complex amplitude (8.17) at $p = 2$ and $n = 5$. This phase (Figure 8.8b) has been used in the SLM. Figure 8.9 shows intensity distributions of the vortex autofocusing AH beam obtained at different propagation distances. The autofocusing distance is $z_e = 45$ mm (Figure 8.9c). It is seen in Figure 8.9 that the beam in a form of concentric rings decreases in diameter up to an autofocus plane and then increases in diameter beyond it.

Figure 8.10 shows experimental intensity distribution registered by a camera at a distance of 20 mm and theoretical intensity distribution calculated by using Eq. (8.18). It is seen in Figure 8.10 that the radii of the light rings are almost the same.

The beam shown in Figures 8.9 and 8.11 shows the dependence of the radius of the first light ring on the propagation distance *z*. The dots correspond to the values obtained from the experimental pictures. It is seen in Figure 8.11 that near the focus the beam narrows with an acceleration. The curve in Figure 8.11 is described by Eq. (8.27) and is the square-root parabola.

Summing up, as a result of our study we have obtained a new solution to the paraxial Helmholtz equation (similar to the Schroedinger equation) to describe a two-parameter family of structurally stable 3D vortex autofocusing accelerating HyG beams (Eq. 8.18). We have shown that while the complex amplitude of AH beams is similar in structure to that of HyG modes [6], the former have an autofocusing property that the latter are devoid of. We have shown that an AH beam is propagating

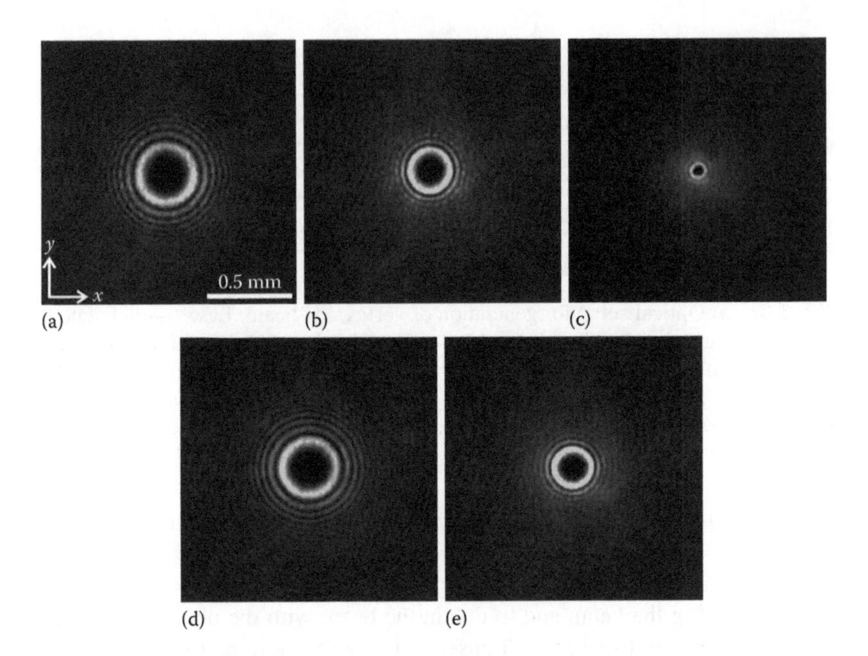

FIGURE 8.9 The experimental light intensity distributions generated by the phase mask (from Figure 8.8b) in the optical setup shown in Figure 8.8a and measured at different distances from the plane $z=0$: (a) 25 mm, (b) 35 mm, (c) 45 mm, (d) 55 mm, and (e) 65 mm. The frame size is the same (1.5 × 1.5) mm.

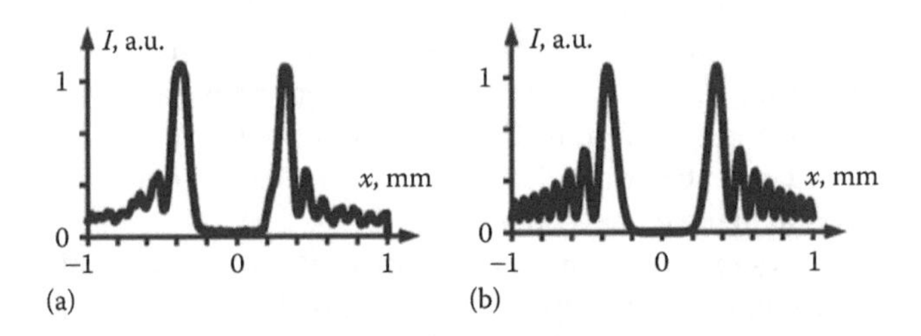

FIGURE 8.10 (a) Experimental intensity distribution registered by a camera at a distance of 20 mm, (b) theoretical intensity distribution calculated by using Eq. (8.18).

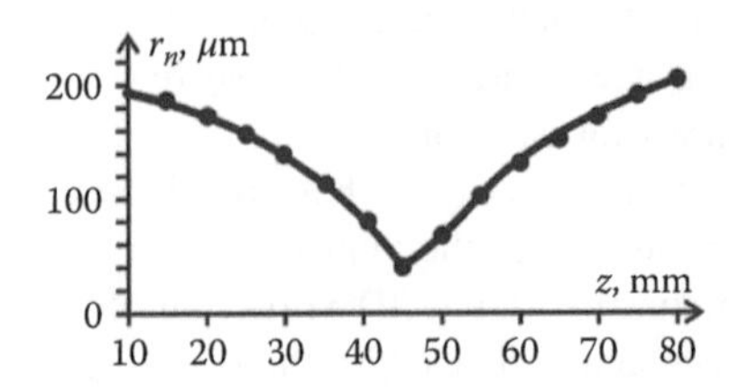

FIGURE 8.11 Radius of the light ring r_n of the beam in Figure 8.9 versus propagation distance z.

toward a focus on an accelerating path, "steeply" converging to the focus, in a similar way to annular Airy beams [271]. We have derived an explicit relation for the complex amplitude of finite-energy vortex AHG beams (Eq. 8.19) that possess an autofocusing property but have a finite radius in the focus, unlike infinite-energy AH beams, which have a zero radius in the focus. The experiment is in good agreement with theory.

9 Asymmetric Gaussian Vortices

9.1 AN ASYMMETRIC GAUSSIAN OPTICAL VORTEX

In Refs. [22,272], A. Zeilinger et al. proposed that photon pairs with entanglement of the orbital angular momentum (OAM) states could be detected in the course of spontaneous parametric down-conversion by illuminating a "fork" hologram (used for generating optical vortices) by a shifted Gaussian beam. Their idea was that as a result of a small shift between the centers a linear combination of a Gaussian and a Laguerre-Gaussian (LG) beam was generated. It has also been shown experimentally [273] that OAM-entangled Stokes and anti-Stokes photons can be generated via four-wave mixing in a hot atomic ensemble and a shift between the centers of the Gaussian beam and the "fork" hologram. There are also studies of the transformation of optical vortices in the classical theory of light. Here, the optical vortex is meant as a laser beam with an isolated intensity null on the optical axis and with a spiral phase, having an integer topological charge. In Ref. [274] it was studied theoretically in which way an axial shift of the center of the Gaussian beam waist from the plane of a spiral phase plate (SPP), combined with a diffractive lens (spiral lens), affected the optical vortex. The propagation of an optical vortex through an opaque screen perforated with multiple pinholes with their centers lying on a circle was experimentally investigated in Ref. [275]. In this case, an nth order optical vortex was shown to split into n first order optical vortices. In [276], the transformation of an optical vortex by applying different ellipticity ratios was reported. This work [276] was done as a continuation of earlier works on the study of elliptic optical vortices [11,60]. In Ref. [277], a method of optical vortex generation was proposed, based on a set of small pinholes in an opaque screen with their centers located on a spiral. Small deformations of the guiding spiral led to the distortion of the optical vortex shape.

Most closely related to the topic of this work are some early papers of the present authors [219,278,279]. In those papers, we studied, both theoretically and experimentally, transformations of optical vortices which occur due to a complex Cartesian shift of the original functions of the complex amplitude. Such a shift leads to asymmetry of the optical vortex: instead of intensity rings (or a "donut") a crescent-shaped intensity appears. In Ref. [219], a complex shift was applied to Bessel beams, while in Ref. [278] it was applied to LG beams. In Ref. [279], an nth order optical vortex was transformed into n first order optical vortices by using an elliptical Gaussian beam illuminating a SPP.

In this section, we theoretically and experimentally study the transformation of the optical vortex proposed in [22,272], by illuminating an amplitude "fork" hologram by a Gaussian beam with its waist center being shifted in the transverse plane from the hologram center. Note that in [22,272] no theory of such a transformation was proposed. In this paper, an optical vortex resulting from the misalignment of the Gaussian beam and the "fork" hologram (or the SPP) centers is called an asymmetric Gaussian optical vortex (AGV). An analytical expression for the OAM of such beams was obtained, showing the OAM to be fractional, that is, the light field is a linear combination of a countable set of optical vortices with integer topological charges. If two photons are in this state, then this state is OAM-entangled. Note that a complex Cartesian shift of Bessel [219] and LG [278] beams leads to the increase of their OAM. The same is true of OAM of an elliptical vortex Hermite-Gaussian beam [279]. In contrast, OAM of the asymmetric Gaussian vortex described here decreases with increasing distance between the centers of the Gaussian beam and the SPP.

Gaussian Beam with a Displaced Optical Vortex

Let us consider a Gaussian beam with the following amplitude in its waist:

$$E_0(x,y)=\exp\left(-\frac{x^2+y^2}{w^2}\right), \tag{9.1}$$

where w is the waist radius. Let an optical vortex with the topological charge n be embedded in the beam (9.1) and shifted from the Gaussian beam center by a distance x_0 along the x-axis. The amplitude of such an optical vortex reads as

$$A_n(x,y)=\left[\frac{(x-x_0)+iy}{w}\right]^n. \tag{9.2}$$

The transmittance (9.2) can be implemented only approximately using an amplitude "fork" hologram [194]. Then, the Fresnel transform allows obtaining the complex amplitude of the AGV at a distance z from the initial plane (plane of the Gaussian beam waist) in cylindrical coordinates:

$$E_n(r,\varphi,z)=w^{-n}\left[q(z)\right]^{-(n+1)}\left[re^{i\varphi}-q(z)x_0\right]^n\exp\left[-\frac{r^2}{w^2(z)}+\frac{ikr^2}{2R(z)}\right], \tag{9.3}$$

where:

$z_0=kw^2/2$ is the Rayleigh range
k is the wavenumber of light
$w^2(z)=w^2[1+(z/z_0)^2]$
$R(z)=z[1+(z_0/z)^2]$, $q(z)=1+iz/z_0$.

From Eq. (9.3), the AGV amplitude is seen to be described by two multipliers. The latter of the multipliers in Eq. (9.3) is a rotationally symmetric Gaussian beam at the propagation distance z from its waist with the beam radius $w(z)$ and the wavefront curvature radius $R(z)$. The multiplier in Eq. (9.3) responsible for the beam asymmetry is shown in the following separately:

$$F = \left[\left(r\cos\varphi - x_0 \right) + i\left(r\sin\varphi - \frac{zx_0}{z_0} \right) \right]^n. \tag{9.4}$$

The intensity distribution of the light field (9.3) is proportional to the square modulus of Eq. (9.4) and has the following form:

$$\begin{aligned} I = \left|E_n(r,\varphi,z)\right|^2 = w^{-2n} \left|q(z)\right|^{-2(n+1)} \exp\left[-\frac{2r^2}{w^2(z)} \right] \\ \times \left[r^2 + \left|q(z)\right|^2 x_0^2 - 2rx_0 \left(\cos\varphi + \frac{z}{z_0}\sin\varphi \right) \right]^n. \end{aligned} \tag{9.5}$$

It is seen in Eq. (9.5), that at $z = 0$ the multiplier (9.4) reads as

$$|F|^2 = \left[r^2 + x_0^2 - 2rx_0 \cos\varphi \right]^n. \tag{9.6}$$

Expression (9.6), together with the Gaussian exponent of Eq. (9.5), describes an inhomogeneous crescent-shaped intensity ring, with an isolated n-order intensity null at point $x = x_0$. The crescent is located so that its maximal and minimal intensity (at a fixed radius r) is achieved at $\varphi = \pi$ and $\varphi = 0$, respectively. At the Fresnel range $z = z_0$ from the beam waist, the multiplier (9.4) responsible for the beam asymmetry takes the form, instead of Eq. (9.6):

$$|F|^2 = \left[r^2 + 2x_0^2 - 2\sqrt{2}rx_0 \cos(\varphi - \pi/4) \right]^n. \tag{9.7}$$

From Eq. (9.7), at the distance $z = z_0$ the crescent is seen to rotate counterclockwise by 45°. The intensity maximum on the crescent is on a line at the angle $\varphi = 5\pi/4$, while the minimum is on a line at the angle $\varphi = \pi/4$. At large distances $z \gg z_0$ from the waist, Eq. (9.7) is rearranged to:

$$|F|^2 = \left[r^2 + \left(\frac{zx_0}{z_0} \right)^2 - 2\frac{zx_0}{z_0} r\sin\varphi \right]^n. \tag{9.8}$$

It is seen in Eq. (9.8) that in the far field ($z \gg z_0$) the crescent is rotated counterclockwise by 90°, and points with the maximal and minimal intensities lie on the lines at $\varphi = -\pi/2$ and at $\varphi = \pi/2$, respectively.

It follows directly from Eq. (9.4) that an isolated intensity null of the AGV has the following coordinates:

$$x = x_0, \quad y = x_0 \frac{z}{z_0}. \tag{9.9}$$

Thus, with increasing propagation distance z, the isolated intensity null, which lies in the initial plane on the horizontal axis at the point $x = x_0$, moves to infinity orthogonally to the horizontal axis.

Orbital Angular Moment of an Asymmetric Gaussian Optical Vortex

In [22,272,273] it was shown experimentally that a shift of the Gaussian beam center from the center of an optical vortex led to the detection of OAM-entangled photons. Thus, it was shown that the AGV must have a fractional OAM. In the following we show that this is true. Since the OAM of a laser beam is conserved upon propagation, it can be calculated in any transverse plane, for example, in the waist plane. It can be shown that the normalized OAM of the AGV (9.3) reads as:

$$\frac{J_z}{W} = n\left[\sum_{l=1}^{n} C_n^l C_{n-1}^{l-1} l!\xi^{2l}\right]\left[\sum_{l=0}^{n}\left(C_n^l\right)^2 l!\xi^{2l}\right]^{-1}, \tag{9.10}$$

where:

$\xi = w/\left(\sqrt{2}x_0\right)$
$C_n^l = n!/l!/(n-l)!$ are the binomial coefficients
$J_z = -i\int_0^\infty \int_0^{2\pi} \bar{E}_n\left(\partial E_n/\partial\varphi\right) r dr d\varphi$ is the axial projection of the OAM
$\bar{E}_n$ is the complex conjugate of the amplitude E_n
$W = \int_0^\infty \int_0^{2\pi} \left|E_n(r,\varphi,z=0)\right|^2 r dr d\varphi$ is the energy of AGV

It is seen in Eq. (9.10) that the normalized OAM of the AGV is generally fractional. At $x_0 = 0$, Eq. (9.10) leads to the well-known result that the OAM of the LG mode [17] equals its topological charge n: $J_z/W = n$.

Eq. (9.10) allows the derivation of simpler expressions for $n = 1$ and $n = 2$, from which particular conclusions about the behavior of the AGV OAM can be made. For example, at $n = 1$ we derive from Eq. (9.10):

$$J_{1z}/W = \left[1+\left(\sqrt{2}x_0/w\right)^2\right]^{-1}. \tag{9.11}$$

This expression coincides with the one obtained previously in Ref. [188]. From Eq. (9.11) it follows that in the absence of the shift ($x_0 = 0$) the OAM is equal to 1. When the shift between the centers of the optical vortex and the Gaussian beam is $x_0 = w/\sqrt{2}$, the beam OAM is equal to 1/2, while at the large shift ($x_0 \to \infty$) the OAM tends to zero. Physically it is justified, since at the sufficiently large distance the optical vortex no longer belongs to the Gaussian beam, and it is located in a region of space where there is no light energy. When $n = 2$, Eq. (9.10) reads as

$$J_{2z}/W = 2\left(\xi^2+\xi^4\right)\left(1/2+2\xi^2+\xi^4\right)^{-1}. \tag{9.12}$$

From Eq. (9.12) it follows that in the absence of the shift, $x_0 = 0$ ($\xi \to \infty$), the OAM is equal to 2, at $x_0 = w/\sqrt{2}$ the OAM is 8/7, while at the large shift, $x_0 \to \infty$ ($\xi = 0$), the OAM is zero. It follows from Eq. (9.10), that for $x_0 \to \infty$ the normalized OAM decreases as $J_z/W \approx \left[wn/\left(\sqrt{2}x_0\right)\right]^2$.

Earlier, we considered asymmetric Gaussian optical vortices that can be generated experimentally by diffraction of a Gaussian beam by an amplitude hologram with a typical "fork" at the point of singularity [22,272]. This case can be described by Eqs. (9.1) through (9.3). However, the amplitude "fork" hologram has low efficiency (a few percent). To increase the efficiency of AGV generation, in the following experiment another setup is used, where a Gaussian beam illuminates a SPP, implemented in a liquid crystal microdisplay. This experimental setup can also be described analytically. The equations in this case are more cumbersome, although the normalized OAM reads as $J_z/W = n\exp(-2x_0^2/w^2)$.

Experiment

Figure 9.1 shows the experimental setup. A light beam of wavelength 532 nm from a solid-state laser was directed via a pinhole *PH* and a lens L_1 onto the display of a spatial light modulator *SLM*, where the phase function of a specified-order SPP was output. Using a set of lenses L_2 and L_3, and a diaphragm *D*, spatial filtering of the phase-modulated laser beam reflected at the SLM display was performed. Further, using a lens L_4, the filtered laser beam was focused onto the CCD array of the CMOS-camera. To obtain interferograms, beam-splitting cubes BS_1 and BS_2 were added to the setup. The first cube split the source beam into two, one of which was directed to the light modulator while the second remained unchanged. Then these two beams were combined into one by the second beam-splitting cube, so that their interference pattern could be observed by the camera. The lens L_5 was used to add spherical wavefront for the Gaussian beam. The Gaussian beam waist diameter was $2w = 1400$ μm, the wavelength was $\lambda = 532$ nm.

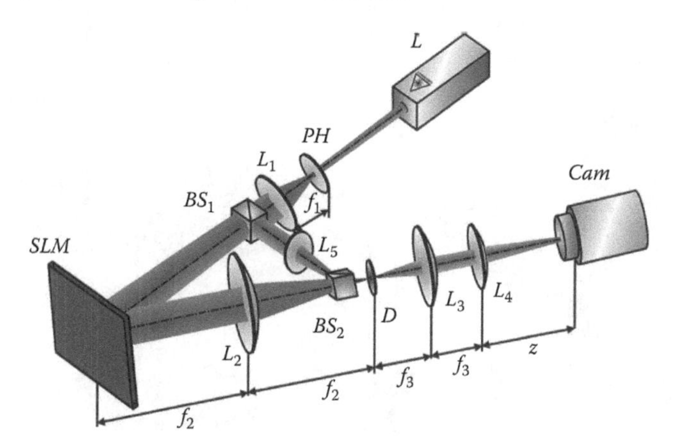

FIGURE 9.1 Experimental setup: *L*—solid-state laser ($\lambda = 532$ nm); *PH*—pinhole (40 μm); L_1, L_2, L_3, L_4 and L_5—lenses with focal distances ($f_1 = 150$ mm, $f_2 = 350$ mm, $f_3 = 150$ mm, $f_4 = 250$ mm, $f_5 = 150$ mm); BS_1, BS_2—beam-splitting cubes; *SLM*—spatial light modulator PLUTO VIS (resolution 1920 × 1080 pixels, pixel size 8 μm); *D*—diaphragm for space filtering; CMOS—video-camera ToupTek U3CMOS08500KPA (pixel size 1.67 μm).

In the experiments, we studied the effect of the shift of the center of the illuminating Gaussian beam from the center of the SPP, which was output onto the light modulator display. Figures 9.2 through 9.4 show generated intensity distributions in the focal plane of a lens with a focal length $f = 250$ mm. There are also shown respective interferograms, obtained by the interference of the studied beams with the Gaussian beam with a spherical wavefront. It should be noted, that because of

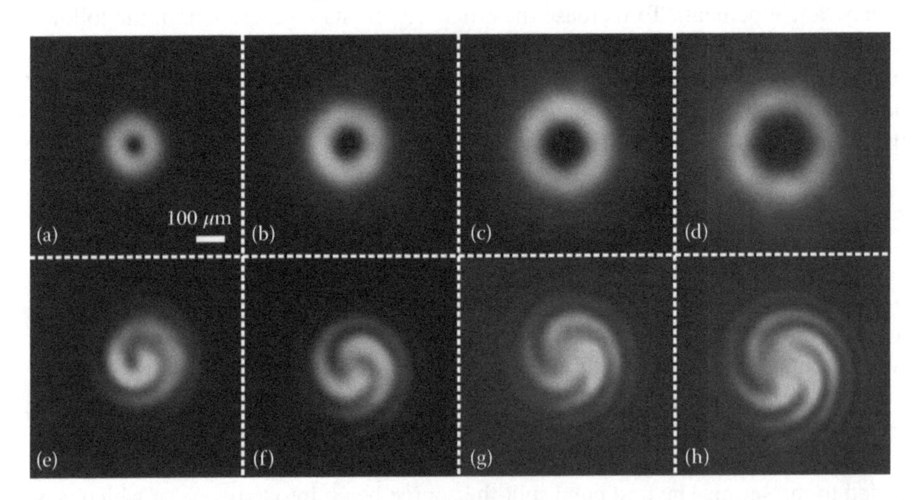

FIGURE 9.2 Centers of the incident beam and the SPP are coinciding ($x_0 = 0$). Top row—obtained intensity distributions for the SPP with the topological charge: (a) $n = 1$, (b) $n = 2$, (c) $n = 3$, (d) $n = 4$. Bottom row—obtained interferograms for the SPP with the topological charge: (e) $n = 1$, (f) $n = 2$, (g) $n = 3$, (h) $n = 4$. Size of the pictures is 750 × 750 μm.

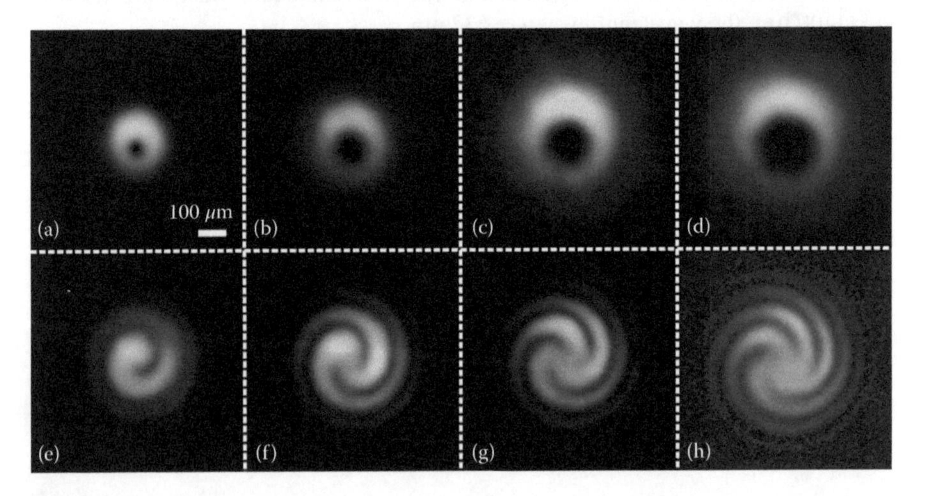

FIGURE 9.3 Shift of the center of the incident beam from the SPP center is $x_0 = 0.125w$. Top row—obtained intensity distributions for the SPP with the topological charge: (a) $n = -1$, (b) $n = -2$, (c) $n = -3$, (d) $n = -4$. Bottom row—obtained interferograms for the AGV with the topological charge: (e) $n = -1$, (f) $n = -2$, (g) $n = -3$, (h) $n = -4$. Size of the pictures is 750 × 750 μm.

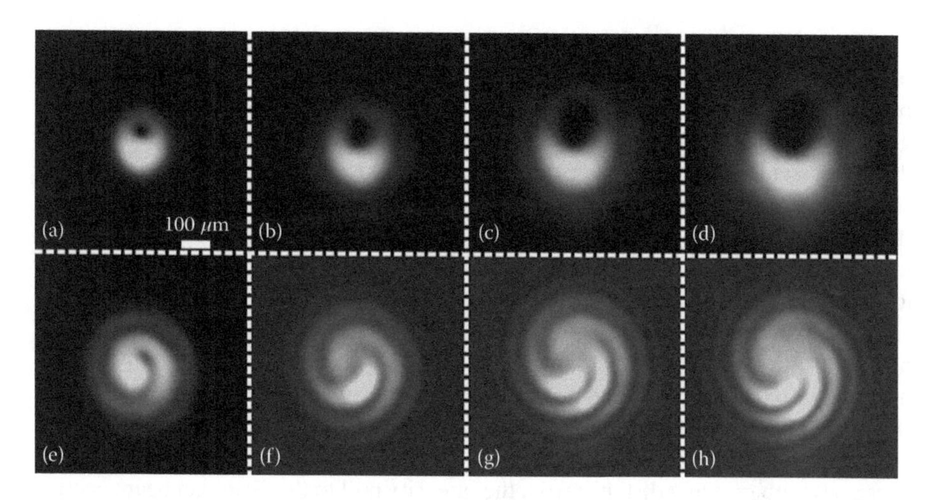

FIGURE 9.4 Shift of the center of the incident beam from the SPP center is $x_0 = 0.250w$. Top row—obtained intensity distributions for the SPP with the topological charge: (a) $n = 1$, (b) $n = 2$, (c) $n = 3$, (d) $n = 4$. Bottom row—obtained interferograms for the SPP with the topological charge: (e) $n = 1$, (f) $n = 2$, (g) $n = 3$, (h) $n = 4$. Size of the pictures is 750×750 μm.

the beam-splitting cube on the path of the laser beam, reflected from the modulator display, the focal plane of the lens L_2 is shifted toward the light modulator. This leads to the convergence of the laser beam, falling onto the lens L_4, so that Fraunhofer diffraction patterns were observed at a distance of $z = 230$ mm from the plane of the lens L_4.

Figure 9.2 shows (a–d) intensity distributions and (e–h) their respective interferograms for the Gaussian optical vortices with the topological charges n ranging from 1 to 4 without the shift between the centers of the SPP and the Gaussian beam. It is seen in Figure 9.2 that the intensity patterns of the optical vortices have rotational symmetry, that is, well-known "donut" patterns with the on-axis intensity null are generated.

Figure 9.3 shows intensity distributions of the Gaussian optical vortices with topological charges n from −1 to −4 with a small shift of the SPP center from the center of the Gaussian beam ($x_0 = 0.125w$). It is seen in Figure 9.3 that all the "donuts" lost their rotational symmetry, although not much. It is also clear that the spiral interferograms changed the directions of their "twists" from "counter-clockwise" to "clockwise" because of the changed sign of the topological charge of the optical vortices.

It is seen in Figure 9.3 that the minimal intensity on the non-uniform bright ring lies on the ray at an angle of −90° to the horizontal axis. This is consistent with Eq. (9.8). Figure 9.4 shows intensity distributions of the Gaussian optical vortices with topological charges n from 1 to 4 at large shift of the SPP center from the center of the Gaussian beam ($x_0 = 0.25w$). It is seen in Figure 9.4 that, first, "crescent" patterns appeared instead of "donut" patterns, and second, because of change of sign of topological charges, the point of minimal intensity on the on-ring inhomogeneous intensity distribution lies on the line at an angle of 90°, unlike the patterns in Figure 9.3, where the minimal intensity on the ring lies on the line at an angle of −90°.

In this section, we theoretically and experimentally studied asymmetric Gaussian optical vortices with the intensity distribution in the form of a crescent. They have been obtained by embedding an optical vortex (isolated intensity null with phase singularity) into the waist of the Gaussian beam, so that coordinates of the intensity null did not coincide with the waist center. It has been shown that as such an asymmetric Gaussian beam propagates, its intensity nulls are shifted to infinity along a straight line, perpendicular to the shift vector. The intensity distribution is rotated so that at the Rayleigh range it is turned by 45°, and in the far field it is turned by 90°, relative to its position in the initial plane. We have derived analytical expressions for the complex amplitudes of such beams generated by a SPP and by an amplitude "fork" hologram. An analytical expression for the OAM was derived. OAM of an AGV has been shown to be fractional and decrease with increasing distance between the intensity null in the waist plane and the Gaussian beam center. Moreover, the higher the order of the optical vortex (the value of its topological charge), the slower the OAM decreases with growing distance from the intensity null to the Gaussian beam center. The decrease of the OAM with increasing distance between the Gaussian beam center and the isolated intensity null of Eq. (9.2), embedded in the Gaussian beam, can be explained by the fact that the optical vortex is embedded to a low-energy region. The experiment is qualitatively consistent with theory. Laser beams that carry fractional OAM can be used to pump a nonlinear crystal for generating OAM-entangled pairs of photons by spontaneous parametric down conversion [22,272]. The AGVs differ from the asymmetric Bessel [219] and LG [278] beams, since with increasing degree of beam asymmetry, the OAM that the latter carry increases linearly [219] and parabolically [278], while the OAM of the AGV decreases with increasing shift between the centers of the beam and the "fork" hologram.

9.2 ELLIPTIC GAUSSIAN OPTICAL VORTICES

Optical vortices that are devoid of radial symmetry and carry a fractional OAM have been studied intensively in the last several years. This interest is due to the fact that optical vortices with fractional OAM have found use in quantum computing for generating the entanglement of the OAM states of photon pairs [22,272,273]. In this way, ultra-high security of quantum communication lines can be achieved.

An optical vortex with fractional OAM can be generated using a variety of techniques. For instance, this can be done by an off-axis shift of a Gaussian beam from the center of a SPP [280]. Another possibility involves generating asymmetric optical vortices with a crescent-shaped intensity pattern [219,278]. However, in this case the microparticle trapped in the beam moves on an open trajectory. It would be of interest to study a situation when the OAM is fractional and the intensity curve in the transverse plane of the vortex is closed. The simplest approach is based on generating an elliptic optical vortex. Transformation of an optical vortex via introducing varying ellipticity was discussed in Ref. [276], which was a follow-up of earlier studies of elliptic optical vortices [11,60]. However, the OAM of elliptic vortices was not considered in Refs. [11,60,276]. Topics dealt with in the present section are most close to those discussed in Ref. [279], which studied an elliptic Hermite-Gaussian vortex, calculating the corresponding OAM and finding it to be fractional. For such a beam to be practically generated, an elliptic Gaussian beam needs to be incident on an amplitude-phase optical element. However, combining the required amplitude and phase in a single element is a challenging task.

In this section, we discuss a scheme that can be simply realized when compared with Ref. [279]. We also derive a relationship to describe the OAM of a Gaussian beam that is embedded with an elliptic optical vortex with n-times degenerate on-axis intensity null (at the Gaussian beam's center). As distinct from Ref. [279] in which the OAM increased with increasing ellipticity of the Gaussian beam, we discuss a situation where the OAM decreases with increasing ellipticity of the embedded optical vortex. Also, note that the optical vortex discussed in Ref. [279] was found to preserve its shape upon propagation up to a scale, with the transverse intensity pattern defined by an ellipse with n isolated intensity nulls. In this section, while featuring an elliptic transverse intensity pattern, the major-axis intensity nulls of the vortex beam are only found in the focal plane of a spherical lens.

Computation of the Orbital Angular Momentum

Assume an isolated n-times degenerate elliptic intensity null at the origin, described as

$$T(x,y)=(\alpha x+iy)^n, \tag{9.13}$$

where:

n is the integer topological charge of the optical vortex

α is a dimensionless parameter that defines the ellipticity of the intensity null:

- if $\alpha < 1$, the major axis is on the x-axis
- if $\alpha > 1$, on the y-axis
- if $\alpha < 0$, the vortex phase rotates clockwise
- if $\alpha > 0$, anticlockwise.

Let the intensity null (9.13) be embedded into the waist of a Gaussian beam, so that the complex amplitude of the light field in the initial plane takes the form:

$$E(x,y,z=0)=(\alpha x+iy)^n \exp\left(-\frac{x^2+y^2}{2w^2}\right), \tag{9.14}$$

where w is the waist radius of the Gaussian beam. We shall seek the OAM and power of the paraxial field using well-known formulae [278]:

$$J_z=\operatorname{Im}\left\{\int_{-\infty}^{\infty}\int_{-\infty}^{\infty}\bar{E}(x,y)\left(x\frac{\partial}{\partial y}-y\frac{\partial}{\partial x}\right)E(x,y)\,dxdy\right\}, \tag{9.15}$$

$$W=\int_{-\infty}^{\infty}\int_{-\infty}^{\infty}\bar{E}(x,y)E(x,y)\,dxdy, \tag{9.16}$$

where:

Im is the imaginary part of a complex number
$\bar{E}$ is a conjugate complex amplitude

For the field in (9.14), (9.15), and (9.16) are rearranged to

$$J_z = \alpha n \int_{-\infty}^{\infty}\int_{-\infty}^{\infty} \exp\left(-\frac{x^2+y^2}{w^2}\right)\left(\alpha^2 x^2 + y^2\right)^{n-1}\left(x^2+y^2\right) dxdy, \tag{9.17}$$

$$W = \int_{-\infty}^{\infty}\int_{-\infty}^{\infty} \exp\left(-\frac{x^2+y^2}{w^2}\right)\left(\alpha^2 x^2 + y^2\right)^{n} dxdy. \tag{9.18}$$

From (9.17) and (9.18) it follows that if the intensity null has zero ellipticity ($\alpha = 1$), we obtain a well-known normalized OAM equal to the topological charge of the vortex [17]:

$$\frac{J_z}{W} = n. \tag{9.19}$$

If $\alpha = -1$, Eq. (9.19) has the opposite sign: $J_z/W = -n$. The integrals in (9.17) and (9.18) can be calculated based on the integral

$$I_m = \int_{-\infty}^{\infty}\int_{-\infty}^{\infty} \exp\left(-\frac{x^2+y^2}{w^2}\right)\left(\alpha^2 x^2 + y^2\right)^{m} dxdy, \tag{9.20}$$

because the OAM in (9.17) and power in (9.18) are connected with (9.20) via simple relations:

$$J_z = -\alpha n \frac{\partial}{\partial\left(w^{-2}\right)} I_{m=n-1}, \quad W = I_{m=n}. \tag{9.21}$$

Integral (9.20) is calculated using an expansion in terms of Newton binomials

$$\left(\alpha x + iy\right)^n = \sum_{l=0}^{n} \frac{n!}{l!(n-l)!}\left(\alpha x\right)^l \left(iy\right)^{n-l} \tag{9.22}$$

and a simple integral

$$\int_{-\infty}^{\infty} x^{2l} \exp\left(-px^2\right) dx = \sqrt{\pi}\, 2^{-l}\left(2l-1\right)!!\, p^{-(2l+1)/2}, \tag{9.23}$$

with the factorial $(2l - 1)!!$ taken over odd integer numbers. Applying (9.22) and (9.23) to (9.20) and accounting for (9.21), (9.17) and (9.18) are rearranged to

$$J_z = \frac{\pi\alpha n^2 w^{2n+2}}{2^{n-1}} A_{n-1}, \quad W = \frac{\pi w^{2n+2}}{2^n} A_n, \tag{9.24}$$

where:

$$A_n = \sum_{l=0}^{n} \frac{n!(2l-1)!!(2n-2l-1)!!}{l!(n-l)!} \alpha^{2l}. \tag{9.25}$$

From (9.24), the normalized OAM of the field in (9.14) takes the form:

$$\frac{J_z}{W} = \frac{2\alpha n^2 A_{n-1}}{A_n}. \tag{9.26}$$

Considering that it is difficult to make specific conclusions from (9.26), in the following we give simplified expressions for the normalized OAM with the topological charges $n = 1, 2, 3$:

$$\left.\frac{J_z}{W}\right|_{n=1} = \frac{2\alpha}{1+\alpha^2}, \tag{9.27}$$

$$\left.\frac{J_z}{W}\right|_{n=2} = \frac{8\alpha\left(1+\alpha^2\right)}{3+2\alpha^2+3\alpha^4}, \tag{9.28}$$

$$\left.\frac{J_z}{W}\right|_{n=3} = \frac{6\alpha\left(3+2\alpha^2+3\alpha^4\right)}{5+3\alpha^2+3\alpha^4+5\alpha^6}. \tag{9.29}$$

From (9.27) to (9.29) the OAM of optical vortex (9.14) is seen to be fractional and smaller than the topological charge n both at $\alpha < 1$ and at $\alpha > 1$. Hence, yet omitting a proof we can suggest that at $\alpha > 0$:

$$\frac{J_z}{W} = \frac{2\alpha n^2 A_{n-1}}{A_n} \le n. \tag{9.30}$$

It can also be shown that the normalized OAM can be expressed more compactly by using the Legendre polynomials:

$$\frac{J_z}{W} = n\frac{P_{n-1}(s)}{P_n(s)}, \tag{9.31}$$

where $s = (1 + \alpha^2)/(2\alpha)$. Using the properties of the Legendre polynomials [38] it can be shown that $P_n(\zeta) > P_{n-1}(\zeta)$ at any $\zeta > 1$. Since the value of s is always not less than unity, this inequality proves the inequality (9.30).

Note that the equality in (9.30) is attained at $\alpha = 1$. The physics behind the decrease of OAM resulting from the replacement of a conventional rotationally symmetric vortex with the elliptic vortex (9.13) may be as follows. In areas where the optical vortex is elongated the amplitude of the Gaussian beam is lower so that the "elongated" optical vortex fragments contribute less to the total OAM. From (9.27) through (9.29), we can infer that at definite vales of α the OAM becomes equal to an integer $m < n$. Therefore, varying the ellipticity of the isolated intensity null (9.13) α from 0 to 1, it is possible to obtain an optical vortex with any normalized OAM in the range from 0 to n.

Computing the Field Complex Amplitude

At an arbitrary distance z, the field's complex amplitude in (9.14) can be calculated using a Fresnel transform. Field (9.14) can be complemented by the transmittance of a thin spherical lens in the paraxial approximation:

$$F(x,y) = \exp\left[-\frac{ik}{2f}\left(x^2 + y^2\right)\right], \tag{9.32}$$

where:
f is the focal length of the thin lens
k is the wavenumber

Then, after passing the spherical lens (9.32), the complex amplitude of the elliptic Gaussian vortex (9.14) takes the integral form:

$$\begin{aligned} E(\xi,\eta,z) = \left(\frac{-ik}{2\pi z}\right) \int_{-\infty}^{\infty}\int_{-\infty}^{\infty} E(x,y,z=0)F(x,y) \\ \times \exp\left[\frac{ik}{2z}\left\{(x-\xi)^2 + (y-\eta)^2\right\}\right] dxdy. \end{aligned} \tag{9.33}$$

Integral (9.33) can be calculated as the sum in (9.32) and two reference relationships [12]:

$$\begin{aligned} &\int_{-\infty}^{\infty} x^m \exp\left(-Ax^2 - Bx\right) dx \\ &= \sqrt{\pi}\left(\frac{i}{2}\right)^m A^{-(m+1)/2} \exp\left(\frac{B^2}{4A}\right) H_m\left(\frac{iB}{2\sqrt{A}}\right), \end{aligned} \tag{9.34}$$

$$\sum_{l=0}^{m} \frac{m!t^l}{l!(m-l)!} H_l(X) H_{m-l}(Y) = \left(1+t^2\right)^{m/2} H_m\left(\frac{tX+Y}{\sqrt{1+t^2}}\right), \tag{9.35}$$

where $H_m(x)$ is a Hermite polynomial. Then, Eq. (9.33) reduces to ($\alpha > 1$):

$$E(\xi,\eta,z)=\left(\frac{-1}{2}\right)^n\left(\frac{-ik}{2z}\right)\left(\frac{\sqrt{2}w}{1+iz_0/z}\right)^{n+2} \times\exp\left[-\frac{\xi^2+\eta^2}{P^2}+\frac{ik}{2S}\left(\xi^2+\eta^2\right)\right]\left(1-\alpha^2\right)^{n/2}H_n\left(Q\left(R+iI\right)\right), \tag{9.36}$$

where:

$$P=\frac{\sqrt{2}wz}{z_0}\sqrt{1+\frac{z_0^2}{z_1^2}},\ S=z\left\{1+\left[\frac{z}{z_1}\left(1+\frac{z_1^2}{z_0^2}\right)\right]^{-1}\right\}^{-1},\ z_1=\frac{zf}{z-f},\quad z_0=kw^2, \tag{9.37}$$

$Q=kw/(\sqrt{2}z)(\alpha^2-1)^{-1/2}(1+z_0^2/z_1^2)^{-1/4}$, $I=\eta\cos\psi-\alpha\xi\sin\psi$, $R=-\alpha\xi\cos\psi-\eta\sin\psi$, $\psi=(1/2)\arctan(z_0/z_1)$.

From (9.36) and (9.37) it follows that the argument of the Hermite polynomial takes real values only on a straight line in the transverse plane, satisfying the equation:

$$\eta=\alpha\xi\,\mathrm{tg}(\psi). \tag{9.38}$$

From (9.38) it follows that at $z=0$, $z_1=0$ and $\psi=\pi/2$, that is, the straight line (9.38) is vertical. Conversely, at $z=f$, $z_1\to\infty$ and $\psi=0$, which means that the line is horizontal. At other distances, $0<z<f$, the line (9.38) is gradually rotated from the vertical to horizontal position. It is on the line of real-valued arguments that the roots of the Hermite polynomial (intensity nulls of field (9.36)) are found. Having equated the real part of the Hermite polynomial in (9.36) to the root value γ_n, where $H_n(\gamma_n)=0$, and taking account of (9.38), we derive an equation for the coordinates of the intensity nulls on the horizontal ξ-axis (the intensity null coordinates on the vertical axis follow from (9.38)):

$$\xi_n=-\frac{\sqrt{2}\gamma_n z\sqrt{\alpha^2-1}\left(1+\frac{z_0^2}{z_1^2}\right)^{1/4}}{\alpha kw}\cos(\psi). \tag{9.39}$$

Equation (9.39) suggests that at $\alpha=1$, function (9.36) has a single intensity null, $\gamma_n=0$. At small z and $\alpha>1$, the intensity nulls (9.39) are located close to each other. The maximal spacing between the nulls (9.39) is in the lens focus at $z=f$:

$$\xi_n=-\frac{\sqrt{2}\gamma_n f\sqrt{\alpha^2-1}}{\alpha kw}. \tag{9.40}$$

From (9.40) it can be inferred that the spacing between the nulls of field (9.36) located on a horizontal line in the focal plane of spherical lens (9.32) depends on the optical vortex's ellipticity, α. At $\alpha = 1$, the spacing between the intensity nulls equals zero, whereas at $\alpha \to \infty$ it is maximal, becoming equal to

$$\xi_n = -\frac{\sqrt{2}\gamma_n f}{kw}. \tag{9.41}$$

It is worth noting that by choosing the inverse value $\beta = 1/\alpha$ as the ellipticity of the optical vortex (9.13), Eq. (9.40) is rearranged to

$$\xi_n = -\frac{\sqrt{2}\gamma_n f\sqrt{1-\beta^2}}{kw}, \tag{9.42}$$

from which it follows that at $\beta = 1$ the spacing between the intensity nulls equals zero (i.e., there is just one intensity null at the origin in the lens focal plane), whereas at $\beta = 0$ the spacing between the nulls is maximal and defined by (9.41).

Experiments on Generating an Elliptic Gaussian Beam

In the experiment, a linearly polarized Gaussian beam of waist diameter $2w = 2.7$ mm was near orthogonally incident on an elliptic SPP with the transmittance

$$V(r,\varphi) = \exp(in\varphi), \tag{9.43}$$

where $\varphi = \arctan[(c_y y)/(c_x x)]$. The ratio $c_y/c_x = \beta = 1/\alpha$ was taken to equal 0.1; 0.2; 0.4; 0.6; 0.8; and 1.0. Note that although function (9.43) is different from function (9.13), both optical vortices have the same topological charge, given the same n and α. This makes both vortices behave in a similar way.

An experimental setup is shown in Figure 9.5. A solid-state laser L ($\lambda = 532$ nm) was used as a light source, generating a fundamental Gaussian beam. The laser light was expanded and collimated by sequentially passing through a 40-μm pinhole *PH* and lens L_1 ($f_1 = 250$ mm), before hitting the display of a modulator *SLM* (PLUTO VIS, 1920 × 1080 resolution, and 8-μm pixel size). The diaphragm D_1 was utilized to single out the central bright ring from surrounding bright and dark rings resulting from

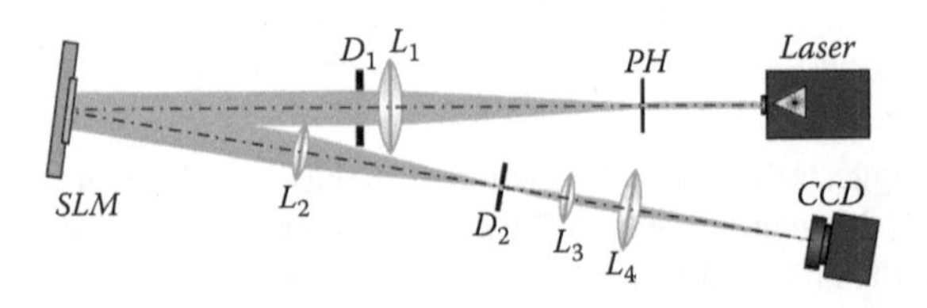

FIGURE 9.5 An experimental setup: L is a solid-state laser ($\lambda = 532$ nm); PH is a 40-μm pinhole; L_1, L_2, L_3, and L_4 are lenses with focal lengths $f_1 = 250$ mm, $f_2 = 350$ mm, $f_3 = 150$ mm, and $f_4 = 500$ mm; D_1 and D_2 are diaphragms; *SLM* is a spatial light modulator PLUTO VIS; and *CCD* is a video-camera LOMO TC-1000.

diffraction by the pinhole. Then, using lenses L_2 ($f_2 = 350$ mm) and L_3 ($f_3 = 150$ mm) and diaphragm D_2 the phase-modulated laser beam reflected at the modulator's display was spatially filtered. Lens L_4 ($f_4 = 500$ mm) was used to focus the laser beam on the matrix of the CCD-camera LOMO TC 1000 (3.34×3.34-μm pixel size).

Shown in Figure 9.6 are phase functions of the SPPs with different ellipticity β and corresponding intensity distributions generated in the focus of lens L_4. The SPP carries a topological charge of $n = 1$. A minor deviation of the intensity

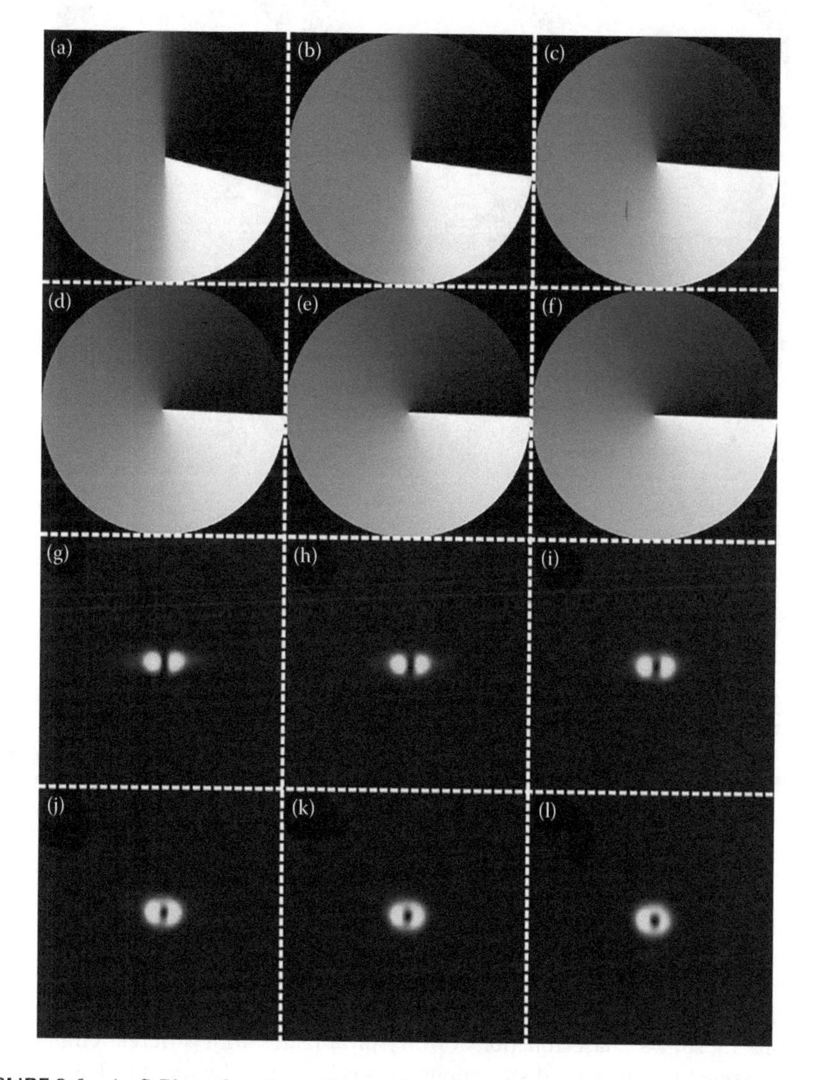

FIGURE 9.6 (a–f) Phase functions of the optical elements and (g–l) corresponding intensity distributions in the focus of lens L_4 when using an SPP with $n = 1$ and the ratio β taking values of (a, g) 0.1; (b, h) 0.2; (c, i) 0.4; (d, j) 0.6; (e, k) 0.8; and (f, l) 1.0. The intensity patterns are 900 μm × 900 μm in size.

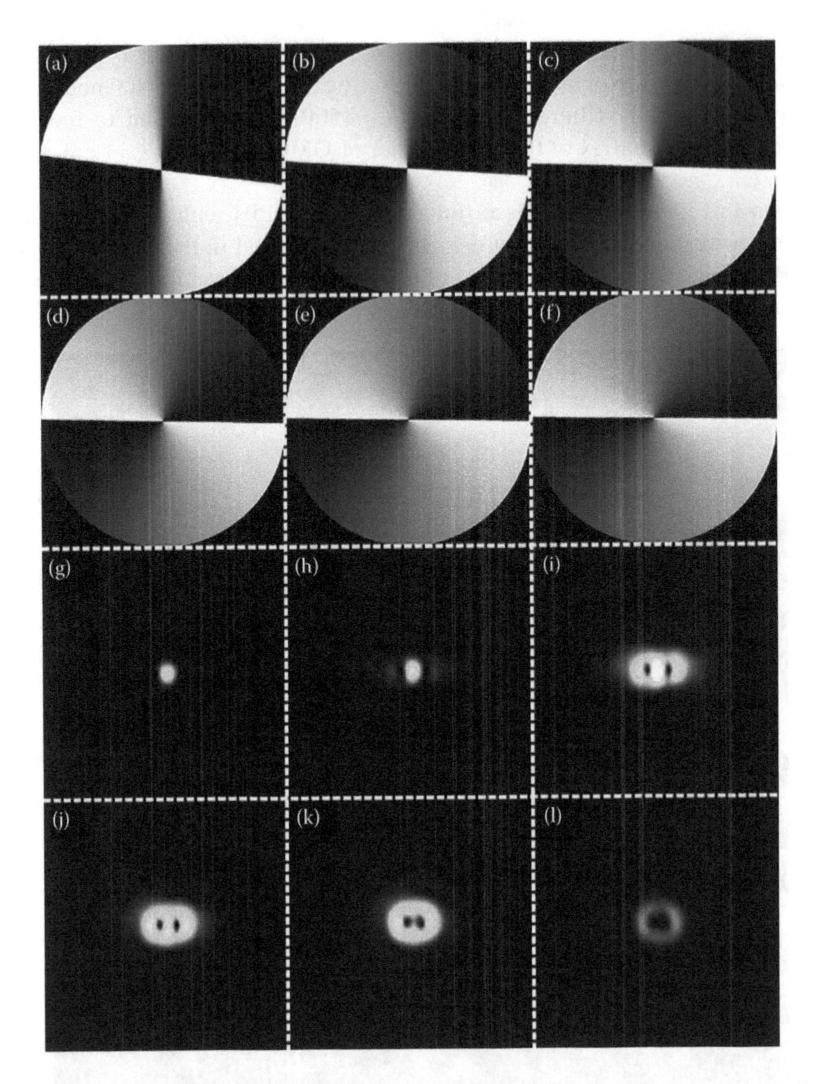

FIGURE 9.7 (a–f) Phase functions of the elements and (g–l) corresponding intensity distributions in the focus of lens L_4 for an SPP with the topological charge $n = 2$ and with the ratio β taking values of (a, g) 0.1; (b, h) 0.2; (c, i) 0.4; (d, j) 0.6; (e, k) 0.8; and (f, l) 1.0. The intensity patterns are 900 μm × 900 μm in size.

distribution observed at $\beta = 1$ from a perfect ring is due to the minor deviation of the incident light from normal.

Figure 9.7 shows phase functions of the spiral plates with different ellipticity β and corresponding intensity distributions they generate in the focus of lens L_4. The SPP carries a topological charge of $n = 2$. A minor deviation of the resulting intensity

TABLE 9.1
Normalized OAM Derived from (9.28) for an Elliptic Gaussian Vortex with $n = 2$ and Varying Ellipticity β

Figure 9.7	g	h	i	j	k	L
β	0.1	0.2	0.4	0.6	0.8	1.0
J_z/W	0.26	0.53	1.11	1.56	1.91	2.00

distribution observed at $\beta = 1$ from a perfect ring is due to the minor deviation of the incident beam from normal.

Note that the relations (9.27) through (9.29) are invariant to the substitution of parameters: $\alpha \to \beta$. Hence, substituting in (9.28) β for α, we get the OAM of the optical vortices shown in Figure 9.7g–l, as presented in Table 9.1. From Figure 9.7, the spacing between two adjacent intensity nulls is seen to decrease (according to (9.42)) from Figure 9.7g–l, while the OAM on the contrary increases from Figure 9.7g–l (Table 9.1, row 3).

It is interesting that Table 9.1 suggests that in the range 0.2–0.4 there is a β value such that the elliptic beam has a unit OAM. Nonetheless, such an optical vortex has two intensity nulls in the lens focal plane, rather than having a single null, which is the case with all similar optical vortices in Figure 9.7.

Figure 9.8 depicts phase functions of the SPPs with different ellipticity β and corresponding intensity distributions generated in the focus of lens L_4. The SPP has $n = 3$. A minor deviation of the intensity distribution generated at $\beta = 1$ from a perfect ring is due to the minor deviation of the incident light from normal.

Table 9.2 gives values of the normalized OAM derived from (9.29) for an elliptic Gaussian beam with $n = 3$ and different values of β. Figure 9.8 shows that the spacing between three intensity nulls is decreasing (in accordance with (9.42)) from Figure 9.8g–l, while OAM, on the contrary, is increasing from Figure 9.8g–l (Table 9.2, row 3). In Figure 9.8g and h is clearly seen a single intensity null because two other intensity nulls are found in the Gauss beam's low-intensity region. The nulls are getting well discernible starting from Figure 9.8i. In Figure 9.8l, the three intensity nulls get merged, forming the intensity null of a conventional radially symmetric optical vortex.

Figure 9.9 depicts phase functions of the SPP with varying ellipticity β and corresponding intensity distributions generated in the focus of lens L_4. The SPP has $n = 4$. A minor deviation of the intensity distribution generated at $\beta = 1$ is due to the minor deviation of the incident light from normal.

Four intensity nulls in the lens focus are clearly discernible only in Figure 9.9i and j. In Figure 9.9g, the nulls are not yet seen because they are located in the Gaussian beam's low-intensity region. In Figure 9.9h just two intensity nulls can be seen, with the two others found in the low-intensity region. In Figure 9.9k and l the situation is different: being located close to each other, the nulls are hardly discernible.

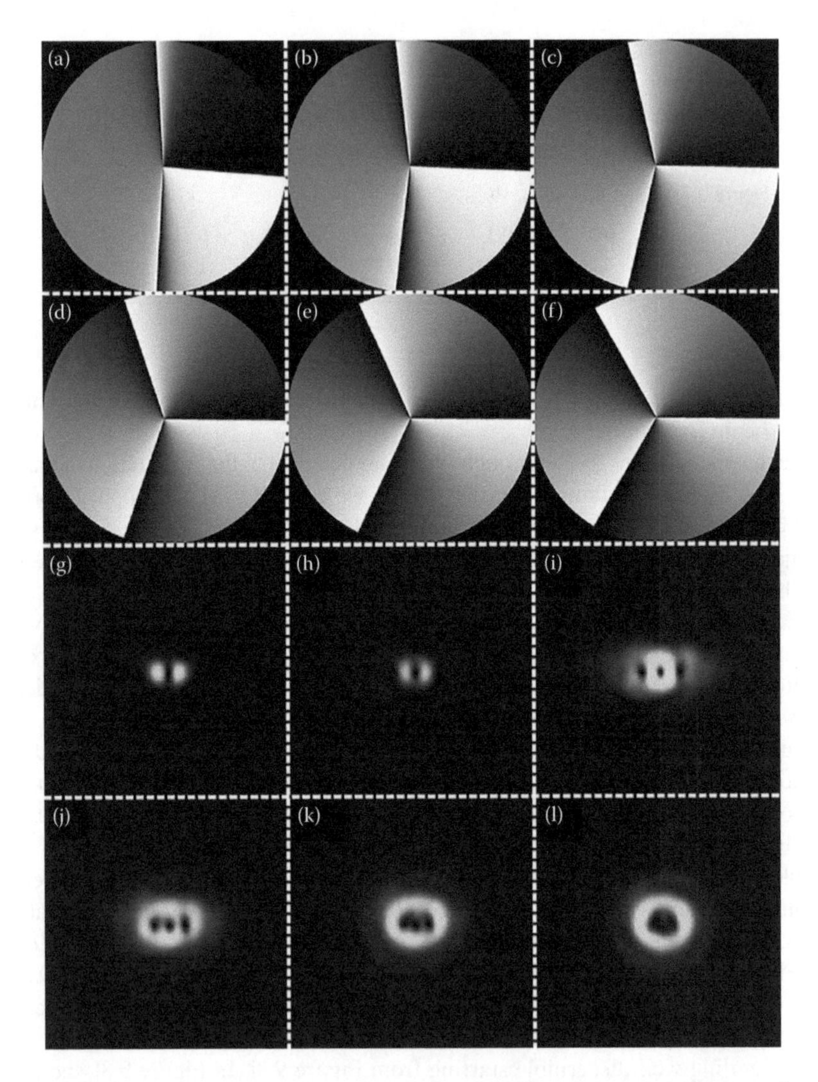

FIGURE 9.8 (a–f) Phase functions of the elements and (g–l) corresponding intensity distributions in the focus of lens L_4 when using an SPP with $n = 3$ and with the ratio β taking values (a, g) 0.1; (b, h) 0.2; (c, i) 0.4; (d, j) 0.6; (e, k) 0.8; and (f, l) 1.0. The intensity patterns are 900 μm × 900 μm in size.

TABLE 9.2
Normalized OAM Derived from (9.29) for an Elliptic Gauss Vortex with $n = 3$ and Varying Ellipticity β

Figure 9.7	g	h	i	j	k	l
β	0.1	0.2	0.4	0.6	0.8	1.0
J_z/W	0.36	0.72	1.47	2.20	2.97	3.00

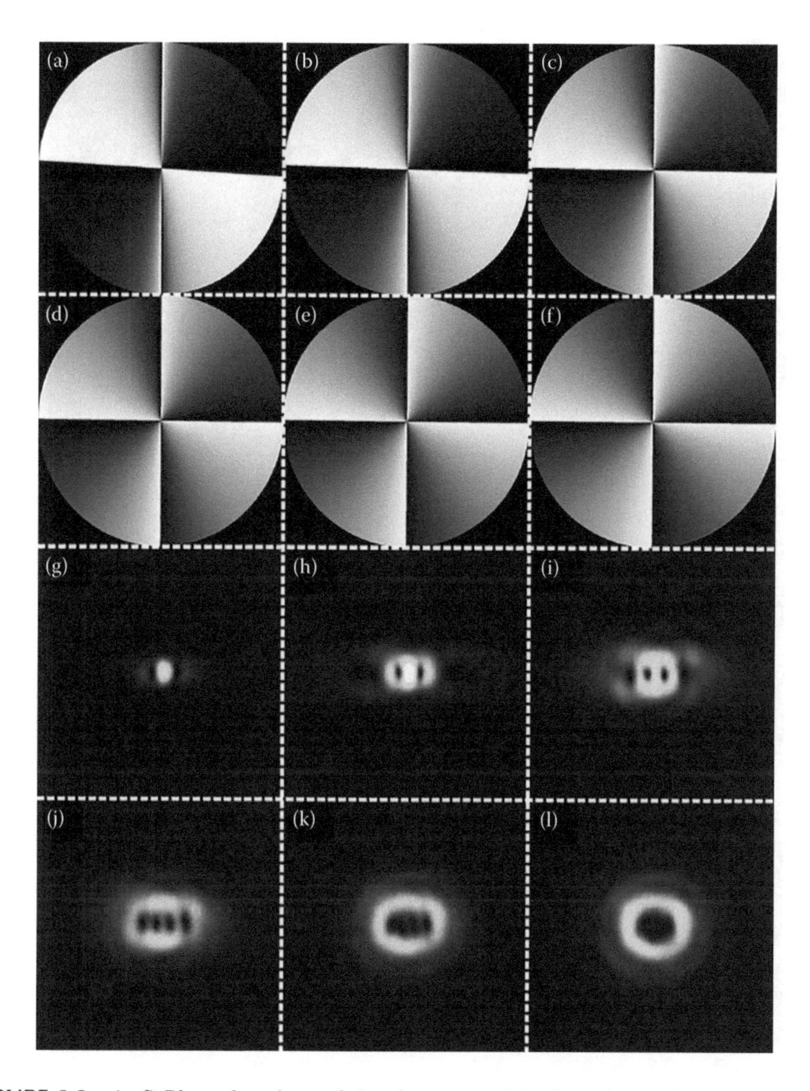

FIGURE 9.9 (a–f) Phase functions of the elements and (g–l) corresponding intensity distributions in the focus of lens L_4 when using an SPP with $n = 4$ and with the ratio β taking values (a, g) 0.1; (b, h) 0.2; (c, i) 0.4; (d, j) 0.6; (e, k) 0.8; and (f, l) 1.0. The intensity patterns are 900 μm × 900 μm in size.

Figure 9.10 depicts intensity distributions of an elliptic Gauss vortex with the topological charge $n = 4$ and ellipticity $\beta = 0.6$ registered at different distances from the initial plane (prior to and behind the focal plane of lens L_4).

Figure 9.10 suggests that, in compliance with (9.38), while remaining in quadrants II and IV, the major axis of the transverse intensity ellipse of the optical vortex rotates upon propagation, changing from the initial vertical to the horizontal position in the lens focal plane (Figure 9.10g). Conversely, after passing the focus, the major axis of the intensity ellipse rotates from the horizontal to vertical position,

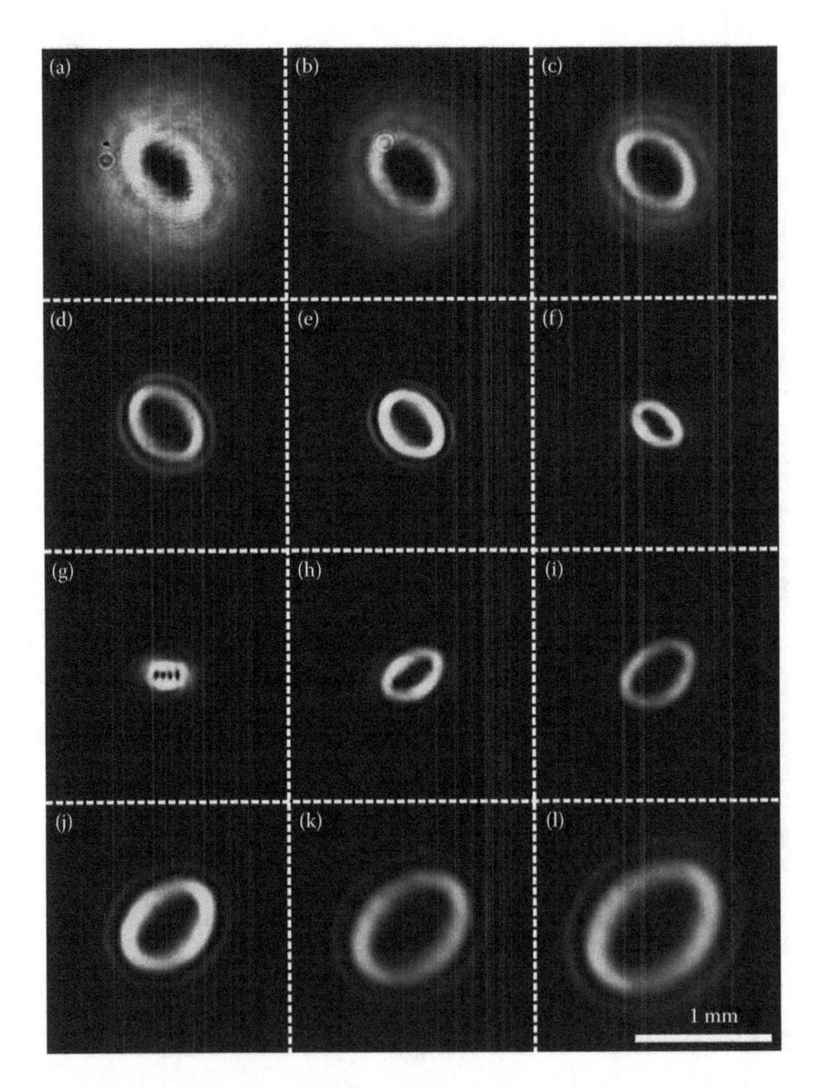

FIGURE 9.10 Intensity patterns generated at different distances from the focal plane of lens L_4 (f_4 = 150 mm) when using an elliptic SPP with $n = 4$ and $\beta = 0.6$: (a) 100 mm, (b) 125 mm, (c) 150 mm, (d) 175 mm, (e) 200 mm, (f) 225 mm, (g) 250 mm (focus), (h) 275 mm, (i) 300 mm, (j) 325 mm, (k) 350 mm, and (l) 375 mm. The intensity patterns are 1800 μm × 1800 μm in size.

remaining in quadrants I and III. The tilt of the straight line in (9.38) changes its sign because of changing sign of the variable $z_1 = zf/(z - f)$ after passing the focus, leading to the change of sign of the angle $\psi = \text{arctg}(z_0/z_1)/2$. In accordance with (9.39), the spacing between four intensity nulls on the major axis of the intensity ellipse increases, reaching its maximum (9.40) in the lens focal plane (Figure 9.10g). Because of this, four intensity nulls are clearly discernible just in the focal plane in Figure 9.10g. After passing the focus, as the spacing decreases, the nulls become hardly discernible.

Summing up, the following results have been obtained. Explicit closed relationships for the complex amplitude and normalized OAM of a conventional Gaussian beam implanted with an elliptic optical vortex with an n-times degenerate intensity null at the Gaussian beam's center have been deduced. The elliptic Gaussian beam has been shown to carry a fractional OAM whose maximal value is equal to the vortex topological charge n attained at zero ellipticity of the vortex. The major axis of the intensity ellipse has been found to rotate, making an angle of 90° while propagating from the initial plane to the focal plane of a spherical lens. There are n intensity nulls on the major axis of the intensity ellipse, with the spacing between them varying both during propagation of the HGV and with varying ellipticity. The inter-null spacing is maximal in the focal plane, given the same ellipticity. When ellipticity is zero, all nulls get merged into a single on-axis, n-times degenerate intensity null. Such a beam has been experimentally generated via illuminating an SPP by a conventional Gaussian beam. Although strictly speaking, the transmittance of such an SPP in (9.43) is different from the complex amplitude of an elliptic intensity null in (9.13), experimental results are in qualitative agreement with theory. This can be explained by the fact that functions in (9.13) and (9.43) have the same phase distributions.

It is worth noting that if, on the contrary, we assume an elliptic Gaussian beam implanted with a conventional radially symmetric n-times degenerate intensity null at the Gaussian beam's center, such a beam has an integer OAM equal to the vortex's topological charge n whatever the Gaussian beam ellipticity.

9.3 CONTROLLING ORBITAL ANGULAR MOMENTUM OF AN OPTICAL VORTEX BY VARYING ITS ELLIPTICITY

Optical vortices without circular symmetry and with a fractional OAM have been intensively studied in the last few years. This is duc to applicability of the fractional-OAM optical vortices in quantum informatics to detected the OAM-entangled photons [22,272,273]. At the same time, maximal degree of quantum link security is achieved.

There are several ways to generate a fractional-OAM optical vortex. For example, it can be done by shifting the Gaussian beam center from the center of the SPP [280,281], or by using a non-integer 2π phase step SPP [34,282], or by superposition of light modes with different values of integer topological charge [204,283]. Similar expansion of the fractional-OAM Bessel beams has also been studied theoretically in [284] and experimentally in [285]. Vortices with the fractional OAM can also be obtained by generating asymmetric optical vortices with the crescent-shape intensity distributions [219,278]. However, in this case the trajectory of microparticles motion will be split. So, it is interesting to study the case when the OAM is fractional and the light intensity curve is closed in the transverse plane. The easiest way to do this is to generate an elliptical optical vortex. In [276], the transformation of an optical vortex is investigated by embedding into it a different degree of ellipticity. The work [276] continues earlier works on the study of elliptical optical vortices [11,60]. However, the OAM of elliptic vortices was not investigated in [11,60,276]. The closest to this section is Ref. [279]. In [279], an elliptical vortex Hermite-Gaussian beam was considered and its OAM was calculated, which turned out to be fractional. However, in order to generate such a beam in practice, it is necessary to use an elliptical Gaussian beam as the incident onto the amplitude-phase optical element, but combining the given amplitude and phase in one element is a difficult task.

In this section, a simple formula is obtained for the OAM of an elliptic Gaussian vortex with an n-fold degenerate intensity null on the optical axis (in the Gaussian beam center). It follows from this formula that depending on the type of the optical vortex ellipticity, the increasing of ellipticity can make the OAM to increase or decrease or remain unchanged. It turns out that if an elliptical vortex is embedded into the center of a circularly symmetric Gaussian beam, then the OAM of the whole beam, normalized to its energy, is fractional and it does not exceed the vortex topological charge. In addition, the OAM decreases with increasing of the vortex ellipticity. If both the Gaussian beam and the optical vortex have a matched (similar or equal) ellipticity, then the normalized OAM of the whole beam exceeds its topological charge, and it increases with the increasing ellipticity. We already studied an elliptic optical vortex embedded into a Gaussian beam [286], but in [286] we considered only a circular Gaussian beam. Here we study a more general case when both the vortex and the Gaussian beam are elliptic (with different ellipticity).

Orbital Angular Momentum of an Elliptic Gaussian Beam with an Embedded Intensity Null

We consider an isolated n-fold degenerate elliptic intensity null embedded in the center of the waist of an elliptic Gaussian beam:

$$E_n(x,y)=\left(\frac{x}{a}+i\frac{y}{b}\right)^n \exp\left(-\frac{x^2}{2w^2}-\frac{y^2}{2\sigma^2}\right), \tag{9.44}$$

where:

- n is the integer topological charge of the optical vortex
- a and b are positive values, which determine the ellipticity of the optical vortex (n-fold degenerate isolated intensity null)
- w and σ are the waist radii of the elliptical Gaussian beam along the Cartesian axes

The OAM and the beam power (energy) of the paraxial field (9.44) are determined by the well-known equations [219,278]:

$$J_z=\operatorname{Im}\int_{-\infty}^{\infty}\int_{-\infty}^{\infty}\bar{E}_n(x,y)\left(x\frac{\partial}{\partial y}-y\frac{\partial}{\partial x}\right)E_n(x,y)\,dxdy, \tag{9.45}$$

$$W=\int_{-\infty}^{\infty}\int_{-\infty}^{\infty}\bar{E}_n(x,y)E_n(x,y)\,dxdy, \tag{9.46}$$

where:

Im means imaginary part of a complex number

$\bar{E}$ is the conjugated complex amplitude

Then, instead of (9.45) and (9.46) we get for the field (9.44):

$$\frac{J_z}{W} = n\left[ab\frac{\sigma^2 - w^2}{a^2\sigma^2 - b^2w^2} + w\sigma\frac{a^2 - b^2}{a^2\sigma^2 - b^2w^2}\frac{P_{n-1}(x)}{P_n(x)}\right], \tag{9.47}$$

where:

$x = [(w\sigma)/(2ab)]\,[(a/w)^2 + (b/\sigma)^2]$

$P_n(x)$ is the Legendre polynomial [101]

From Eq. (9.47), interesting partial cases follow, which show how one can control the OAM of a laser beam in a wide range by using the beam ellipticity.

Partial Cases

1. If the optical vortex (9.44) is circularly symmetric ($a = b$), while the Gaussian beam has an elliptical waist ($w \neq \sigma$), then Eq. (9.47) leads to:

$$\frac{J_z}{W} = n. \tag{9.48}$$

This means that the OAM of a circularly symmetric optical vortex, embedded into the center of the waist of an elliptical Gaussian beam, equals the topological charge. So, even if the shape of the real amplitude function (9.44) depends on the azimuth angle φ, it does not affect the OAM. The OAM value is affected only by the phase component of the beam complex amplitude (9.44), which in this case is represented by single angular harmonic $\exp(in\varphi)$, where φ is the azimuthal angle in the polar coordinates (r, φ).

2. If an elliptical optical vortex ($a \neq b$) is embedded into the center of the waist of a circularly symmetric Gaussian beam ($w = \sigma$), then instead of Eq. (9.47) we get:

$$\frac{J_z}{W} = nP_{n-1}(y)\left[P_n(y)\right]^{-1},\; y = \left(a^2 + b^2\right)\left(2ab\right)^{-1}. \tag{9.49}$$

From Eq. (9.49) it follows that if only vortex part of the beam is elliptical then the OAM is fractional and it is less than the topological charge n. Indeed, since $a^2 + b^2 \geq 2ab$, then $y \geq 1$, while it is known [101] that $P_{n-1}(y) \leq P_n(y)$ at arbitrary $y \geq 1$ for any order $n > 0$. The equality is achieved at $a = b$ (i.e., for circularly symmetric optical vortex), since $P_n(1) = 1$ for any $n > 0$. In order to explain why the OAM (9.49) is less than n, we transform the amplitude (9.44) by using the Newton binomial:

$$E_n(r,\varphi) = \left[\frac{r(a+b)}{2ab}\right]^n \exp\left(-\frac{r^2}{2w^2}\right)\sum_{k=0}^{n}\frac{n!}{k!(n-k)!}\left(\frac{b-a}{b+a}\right)^k e^{i(n-2k)\varphi}. \tag{9.50}$$

It is seen in Eq. (9.50) that $2n$ different-weight angular harmonics $\exp(\pm ik\varphi)$ contribute to the field amplitude. It can be shown that the OAM of the sum of n angular harmonics with weights, normalized by the full energy, is less than the OAM of one angular harmonic with the maximal number n. Indeed, let us consider a light field with the complex amplitude consisting of the sum of a finite number of the angular harmonics: $F(r,\varphi)=\sum_{k=0}^{k=n} A_k(r)\exp(ik\varphi)$. According to Eqs. (9.45) and (9.46), the normalized OAM of such field reads as $J_z/W=\sum_{k=0}^{k=n} k\tilde{I}_k$, where $I_k=\int_0^{\infty}|A_k(r)|^2\,r dr$ and $\tilde{I}_k=I_k\left(\sum_{k=0}^{k=n} I_k\right)^{-1}$. This leads to an obvious inequality: $J_z/W=\sum_{k=0}^{k=n} k\tilde{I}_k \le n\sum_{k=0}^{k=n}\tilde{I}_k=n$.

Thus, we have shown that an elliptical vortex, embedded in the center of the waist of a circularly symmetric Gaussian beam, contains a finite number of angular harmonics, making the normalized OAM to be less than the topological charge. It also follows from Eq. (9.49) that the greater is the degree of ellipticity of the vortex, that is, the larger the ratio a/b differs from unity, the smaller is the OAM of the light beam (9.44). This can be explained physically, since the OAM density of the elliptical vortex is greater at the elongated parts of the ellipse, which, at the same time, are farthest from the center of the Gaussian beam, and in which the light intensity is minimal. In addition, decreasing of the OAM of an elliptical vortex embedded in the center of the waist of a circularly symmetric Gaussian beam can be explained by analyzing the distribution of the OAM density normalized to the intensity:

$$j_z=\mathrm{Im}\left(\bar{E}_n(r,\varphi)\frac{\partial E_n(r,\varphi)}{\partial\varphi}\right),\quad I=\left|E_n(r,\varphi)\right|^2. \tag{9.51}$$

In this case, the normalized OAM density reads as

$$\frac{j_z}{I}=\frac{n}{ab}S^2(\varphi),\quad I=r^{2n}\exp\left(-\frac{r^2}{w^2}\right)S^{-2n}(\varphi), \tag{9.52}$$

where:

(r,φ) are the polar coordinates

$S(\varphi)=(a^{-2}\cos^2\varphi+b^{-2}\sin^2\varphi)^{-1/2}$ is the radius of vortex ellipse in the direction of the polar angle φ

It is seen in Eq. (9.52) that if $a>b$, then at $\varphi=0$ or $\varphi=\pi$ the normalized OAM density (9.52) is maximal and equals $j_z/I=an/b$, while at $\varphi=\pi/2$ or $\varphi=3\pi/2$ the OAM density is minimal and equals $j_z/I=bn/a$. So, in the parts of the ellipse with the maximal curvature the normalized OAM density is also maximal and exceeds the topological charge, while in the parts with the minimal curvature the OAM density is also minimal and is less than n. However, the total OAM depends on the summary OAM density. Since the intensity distribution (9.52) $I(r,\varphi)$ depends on φ, we cannot integrate the OAM density j_z/I in Eq. (9.52) only over the azimuthal angle φ, and therefore the total normalized OAM for the optical vortex (9.44) at $w = \sigma$ and $a \neq b$ is as follows:

$$\frac{J_z}{W} = \frac{n}{ab}\left[\int_0^{2\pi} S^{2(1-n)}(\varphi)\,d\varphi\right]\left[\int_0^{2\pi} S^{-2n}(\varphi)\,d\varphi\right]^{-1} \leq n. \tag{9.53}$$

Equation (9.53) can be expressed via the ratio of the Legendre polynomials and coincides with Eq. (9.49). It can be derived from the following equation:

$$P_n\left(\frac{a^2+b^2}{2ab}\right) = \frac{(ab)^n}{2\pi}\int_0^{2\pi} S^{-2n}(\varphi)\,d\varphi, \tag{9.54}$$

which follows directly from the integral representation of the Legendre polynomials [101].

3. Now we consider the case of matched ellipticities of the vortex and of the Gaussian beam, that is, $a\sigma = bw$. The argument of the Legendre polynomials in Eq. (9.47) becomes equal to unity, making the polynomials themselves also equal to unity. After resolving the 0/0 uncertainty we can get instead of Eq. (9.47):

$$\frac{J_z}{W} = n\left(\frac{aw+b\sigma}{a\sigma+bw}\right) = n\left(\frac{w^2+\sigma^2}{2w\sigma}\right) = n\left(\frac{a^2+b^2}{2ab}\right) \geq n. \tag{9.55}$$

It follows from Eq. (9.55) that the normalized OAM of the light field with matched ellipticities of the vortex and of the Gaussian beam is always larger than the value of the topological charge. The greater is the degree of ellipticity, that is, the larger the ratio $a/b = w/\sigma$ differs from unity, the greater is the OAM (9.55) as compared to n. It also follows from Eq. (9.55) that when the degree of ellipticity of the vortex and of the Gaussian beam is the same, that is, $a = w$, $b = \sigma$, then the OAM (9.55) does not change [287]. The normalized OAM density in this case coincides with Eq. (9.52), but the intensity distribution has elliptical symmetry:

$$\frac{j_z}{I} = n(ab)^{-1} S^2(\varphi),$$
$$I = r^{2n} S^{-2n}(\varphi)\exp\left[-\left(\frac{a^2}{w^2}\right)S^{-2}(\varphi)\,r^2\right]. \tag{9.56}$$

Introducing elliptical polar coordinates (r', φ'): $x = ar'\cos\varphi'$, $y = br'\sin\varphi'$, the total normalized OAM can be obtained by integration of the OAM density solely over elliptical azimuthal angle:

$$\frac{J_z}{W} = \frac{n}{2\pi}\int_0^{2\pi}\left(\frac{a}{b}\cos^2\varphi' + \frac{b}{b}\sin^2\varphi'\right)d\varphi' = n\frac{a^2+b^2}{2ab} \geq n. \tag{9.57}$$

4. In an experiment, for generation of vortex beams one usually uses phase optical elements, e.g., SPP with the transmittance function $T(\varphi) = \exp(in\varphi)$ [24]. For generating the elliptical vortex an elliptical SPP should be used with the following transmittance:

$$T(\varphi) = \exp\left\{ in \arctan\left[\left(\frac{ay}{bx} \right) \right] \right\}. \tag{9.58}$$

Then the complex amplitude of the elliptical Gaussian vortex immediately behind the SPP (9.58) reads as

$$E_{1n}(x, y) = \exp\left[-in \arctan\left(\frac{ay}{bx} \right) - \frac{x^2}{2w^2} - \frac{y^2}{2\sigma^2} \right]. \tag{9.59}$$

It can be shown that if the ellipticities of the Gaussian beam and of the phase of the spiral plate (9.58) are matched (i.e., $bw = a\sigma$ for the field (9.59)), then the normalized OAM coincides with (9.55), (9.57) and is equal to

$$\frac{J_z}{W} = n\left(a^2 + b^2\right)\left(2ab\right)^{-1} \geq n. \tag{9.60}$$

Increasing of the OAM (9.60) for an elliptical Gaussian vortex, compared to a circularly symmetric Gaussian vortex, follows from the OAM density. For the field (9.59), the OAM density (9.51), normalized to the intensity, reads as

$$\begin{gathered} \frac{j_z}{I} = n\left(ab\right)^{-1}\left(a^2 \cos^2 \varphi' + b^2 \sin^2 \varphi'\right), \\ I = \exp\left(-r'^2 p^{-2}\right), \quad p = bw = a\sigma, \end{gathered} \tag{9.61}$$

where:
$r' = (b^2x^2 + a^2y^2)^{-1/2}$
$\varphi' = \arctan[(ay)/(bx)]$

Since in Eq. (9.61) the intensity does not depend on the elliptical azimuthal angle φ', the total normalized OAM can be obtained by the integration of Eq. (9.61) over the angle φ', and it exactly coincides with the expression (9.57).

5. It is shown earlier that two different elliptical Gaussian optical vortices (9.44) and (9.59) have the same normalized OAM (9.60), if ellipticities of the Gaussian beam and of the vortex are matched ($bw = a\sigma$). Therefore, it can be suggested that this holds in other cases. For example, if an elliptic vortex is embedded into the center of the waist of a circularly symmetric Gaussian beam, then the OAM of such beam is less than the topological charge and equals (9.49) and (9.53). It can be supposed that if an elliptical SPP is illuminated by a circularly symmetric Gaussian beam, then the OAM of such beam is also less than the topological charge and equals (9.49) and (9.53). However, it is not so. Now we determine the OAM for a beam with the complex amplitude (9.49) when a circularly symmetric Gaussian

beam is incident onto an elliptical SPP (9.58). Let $w = \sigma$ in Eq. (9.59). Then the amplitude of such field immediately behind the SPP is equal to

$$E_{1n}(x, y) = \exp\left[-in\arctan\left(\frac{ay}{bx}\right) - \frac{x^2 + y^2}{2w^2}\right]. \tag{9.62}$$

It can be shown that the normalized OAM for the field (9.62) is equal to the topological charge (9.48): $J_z/W = n$. This means that changing of the ellipticity of the SPP (9.58) does not lead to changing of the OAM of the optical vortex, if such a SPP is combined with a circularly symmetric Gaussian beam. This can be explained, since the normalized OAM density of the beam (9.62) is "inverse" to the expression (9.61) and coincides with the OAM density (9.56):

$$\frac{j_z}{I} = n(ab)^{-1} S^2(\varphi), \quad I = \exp\left(-r^2/w^2\right), \tag{9.63}$$

where:
$r = (x^2 + y^2)^{1/2}$
$\varphi = \arctan(y/x)$

Since the intensity in Eq. (9.63) does not depend on the angle φ, the total normalized OAM of the field (9.62) can be obtained by the integration of the OAM density (9.63) over the angle:

$$\frac{J_z}{W} = \frac{nab}{2\pi} \int_0^{2\pi} \frac{d\varphi}{b^2 \cos^2 \varphi + a^2 \sin^2 \varphi} = n. \tag{9.64}$$

6. If the SPP (9.58) without ellipticity ($a = b$) is illuminated by an elliptical Gaussian beam ($\sigma \neq w$), then similarly to the embedding of a circularly symmetric intensity null into the center of the waist of an elliptical Gaussian beam the normalized OAM equals the topological charge and coincides with Eq. (9.48). In this case the normalized OAM density coincides with the total normalized OAM and reads as

$$\frac{j_z}{I} = \frac{J_z}{W} = n. \tag{9.65}$$

Numerical Simulation

Figure 9.11 shows intensity distributions of incident circular and elliptical Gaussian beams (wavelength $\lambda = 532$ nm), phase distributions of circular and elliptical SPPs with $n = 10$, and intensity distributions of diffracted beams at propagation distance $z = 500\lambda$, obtained by numerical evaluation of the Fresnel transform. It is seen that the intensity patterns are quite different. The calculated normalized OAM is $J_z/W = 9.985$ (Figure 9.11a–c), 20.924 (Figure 9.11d–f), 9.916 (Figure 9.11g–i), and 9.953 (Figure 9.11j–l), while the corresponding theoretical values are 10, 21.25, 10, and 10, respectively.

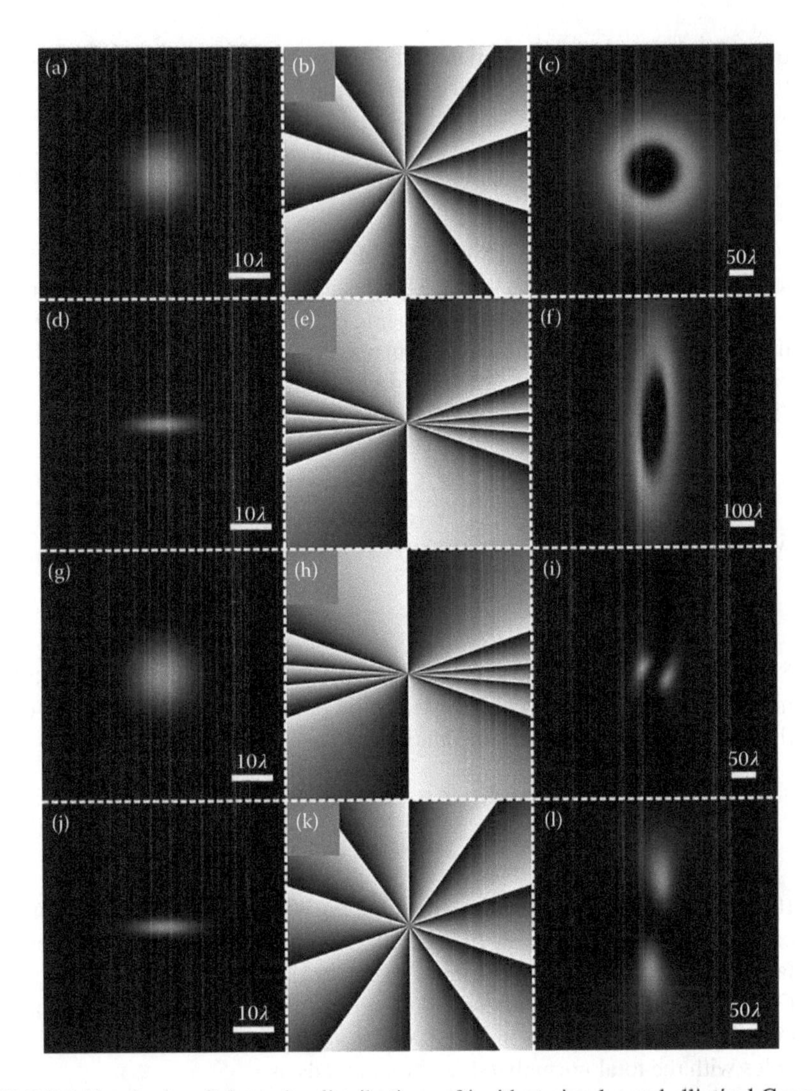

FIGURE 9.11 (a, d, g, j) Intensity distributions of incident circular and elliptical Gaussian beams, (b, e, h, k) phase distributions of circular and elliptical 10th order SPPs, and (c, f, i, l) intensity distributions of diffracted beams at propagation distance $z = 500\lambda$, obtained by numerical evaluation of the Fresnel transform. Horizontal waist radius of all incident Gaussian beams is $w = 8\lambda$, vertical waist radius is (a, g) $\sigma = 8\lambda$ and (d, j) $\sigma = 2\lambda$. Ellipticity factors of the SPP are (b, k) $a/b = 1$ and (e, h) $a/b = 4$. Depicted area is (a, b, d, e, g, h, j, k) $-30\lambda \le x, y \le 30\lambda$, (c, i, l) $-250\lambda \le x, y \le 250\lambda$, and (f) $-500\lambda \le x, y \le 500\lambda$.

In conclusion, let us summarize the obtained results. It is rigorously shown that combining the type and degree of ellipticity of a Gaussian optical vortex allows wide-range varying of its OAM. A general formula is obtained for the normalized total OAM of an elliptic vortex embedded into the waist of an elliptic Gaussian beam. The OAM of such a beam is expressed in terms of the ratio of the Legendre polynomials. It follows from the obtained formula that if the optical vortex has circular symmetry and the Gaussian beam is elliptical, then the OAM is equal to the topological charge. If, on the contrary, the optical vortex has elliptical symmetry and the Gaussian beam is circular, then the OAM is less than the topological charge. In addition, if both the optical vortex and the Gaussian beam have the same elliptical symmetry, then the OAM exceeds the topological charge. We also obtain expressions for an optical vortex generated by using an elliptical SPP placed in the waist of an elliptical Gaussian beam. It is shown that if a Gaussian beam is elliptical and a SPP is conventional with circular symmetry, or conversely, an ordinary circularly symmetric Gaussian beam is incident onto an elliptical SPP, then in both cases the normalized OAM equals the topological charge. If ellipticities of the Gaussian beam and of the SPP are matched (or the same), then the normalized OAM exceeds the topological charge. In those cases where the OAM is larger (or respectively smaller) than the topological charge, the more the ratio of the ellipse's axes differs from unity, the larger (or respectively smaller) is the value of the OAM. Elliptic Gaussian optical vortices can be used for optical trapping and accelerated or decelerated motion of microscopic particles along an elliptical trajectory. In addition, such fractional-OAM laser beams can be used in the spontaneous parametric down-conversion for pumping nonlinear crystals to generate pairs of OAM-entangled photons. This increases the already high cryptographic strength of the quantum communication system. In addition, entangled (correlated) photons can be used to generate quantum images with superresolution [288].

9.4 ASYMMETRIC LAGUERRE-GAUSSIAN BEAMS

LG modes comprise a well-studied class of light fields. The transverse intensity profile of these fields is invariant to the propagation in a uniform medium and shows a radial symmetry. The LG modes have found use in the areas such as optical micromanipulation, quantum optics, and optical communications. Each mode of the class is characterized by two indices—radial and azimuthal, the latter defining the OAM.

Despite long history and a considerable bulk of research dealing with LG modes, articles concerned with the study of their properties [289–293], generation [294–296], and uses [23,297–301] have been actively published.

For instance, the propagation of composite vortex beams generated by coaxial superposition of LG beams with the identical location and size of the waist was discussed in [289]. Fields composed of equidistant arrays of solitary or tandem low-intensity spots located on diffractive rings have been generated. The physical meaning of the radial index of LG modes was looked into in Ref. [290], whereas Ref. [291] has studied in which way the three-dimensional intensity distribution of sharply focused LG beams depends on the homogeneous polarization (linear and circular) and topological charge. The polarization type was shown to have a greatest effect on the longitudinal E-field component, with the total intensity pattern showing largest variations versus polarization when using the first-order vortex phase. The non-paraxial propagation of LG modes in the presence of an aperture was discussed in Ref. [292]. The diffraction by the aperture was shown to essentially distort the near-field pattern, while having an unessential effect on the far-field intensity (unless the aperture blocked off a substantial proportion of the beam). Properties of light fields that carried the OAM, had no radial symmetry, and were affected by a harmonic potential were analyzed in Ref. [293]. A technique for generating LG modes in a cavity of a solid-state laser was proposed in Ref. [294]. Generation of lower-order LG modes with the ability to control the topological charge by means of a solid-state laser was discussed in Ref. [295]. LG modes with non-zero radial index were generated by means of spiral zone plates in Ref. [296]. In Ref. [297] an operator was considered which was linked with the radial index in the Laguerre-Gauss modes of a two-dimensional harmonic oscillator in cylindrical coordinates. In Ref. [298], the replacement of a conventional Gaussian beam with a LG beam was shown to result in a reduced Doppler width in the absorption spectrum of rubidium-85 and rubidium-87 atoms. The reduction of the thermal noise effect on gravitational wave detectors with use of the LG modes was discussed in Ref. [299]. Spin-orbital coupling of ultra-cold atoms with the aid of LG beams was reported in Ref. [300]. The interaction of a LG beam with an atom or a diatomic molecule was studied in Ref. [301]. The transfer of the OAM between the mass center and internal motion of a sufficiently cooled atom or molecule has been shown to take place. A three-dimensional off-axis optical trap for dielectric submicron microbeads created with a single LG beam was described in Ref. [23]. The classical communication by means of LG modes at a 3-km distance in a turbulent atmosphere was demonstrated in Ref. [19]. The quantum communication with entangled twisted photons was described in Ref. [302]. Note that in addition to laser beams, electron beams [303] and even neutron beams [304] can carry the OAM.

The previous review of the latest publications relating to LG beams suggests that not only do they find new applications but also form a basis for constructing advanced light fields that have been studied theoretically so far. Alongside looking into the properties of various superpositions of the familiar laser beams, it is possible to obtain novel beams of interest by simply performing a complex-valued shift of their complex amplitude in the Cartesian plane. It is known that if a paraxial point source is shifted along the optical axis by an imaginary distance, then instead of a parabolic wave, a Gaussian beam is generated [305]. Asymmetric diffraction-free Bessel modes that produce a crescent-shaped transverse intensity pattern have been generated in a similar way [219]. However, Bessel beams have infinite energy and therefore can be physically realized only approximately. In addition, the dependence of the OAM on the asymmetry parameter is linear for Bessel beams [219], while for the asymmetric Laguerre-Gaussian (aLG) beams under study this dependence is parabolic.

In this section, also making use of a complex-valued shift in the Cartesian plane, we conduct theoretical and experimental studies of aLG beams. Like in conventional LG beams—and as distinct from Bessel beams—their transverse intensity pattern consists of a finite number of diffraction rings, which, however, have a non-uniform intensity distribution. As the aLG mode propagates in a uniform medium, the intensity of the peripheral ring increases. The OAM and power of the aLG beam is calculated analytically. In particular, we analyze aLG beams with zero radial index that have a crescent-shaped transverse intensity pattern. For such beams, an analytic relation for the coordinates of the intensity peak is derived and the diffraction pattern is shown to rotate upon propagation in a uniform medium. An aLG beam with zero radial index is generated using a spatial light modulator and experimentally shown to rotate in space during propagation. A feasibility to generate the superposition of misaligned beams that has a near-Gaussian intensity pattern and rotates as a whole during propagation in space is demonstrated.

Elliptical beams [306] have been known to constitute the most general family of paraxial laser beams. Circular beams [14] and Ince-Gaussian beams [90] are partial cases of the elliptical beams. At definite parameters these beams carry a non-zero OAM, but these beams can be reduced neither to LG modes nor to elegant or aLG beams. In addition, the Ince polynomials do not have a closed form, making it difficult to derive analytical expressions for them. For example, a closed expression for the OAM of an elliptical vortex [306] cannot be obtained.

Complex Amplitude of an Asymmetric Laguerre-Gaussian Beam

The complex amplitude of a conventional LG beam in polar coordinates in the initial plane is given by [307]

$$E_{mn}(r,\varphi,z=0)=\left(\frac{\sqrt{2}r}{w_0}\right)^n L_m^n\left(\frac{2r^2}{w_0^2}\right)\exp\left(-\frac{r^2}{w_0^2}+in\varphi\right), \tag{9.66}$$

where:

(r, φ, z) are cylindrical coordinates
w_0 is the Gaussian beam waist
n is the topological charge of an optical vortex
$L_m^n(x)$ is an adjoint Laguerre polynomial

If the beam is shifted by x_0 along the x-coordinate and by y_0 along the y-coordinate (x_0 and y_0 can take complex values), the beam amplitude in the Cartesian coordinates takes the form:

$$E_{mn}(x,y,z=0)=\left(\frac{\sqrt{2}}{w_0}\right)^n\left[(x-x_0)+i(y-y_0)\right]^n\exp\left(-\frac{s^2}{w_0^2}\right)L_m^n\left(\frac{2s^2}{w_0^2}\right), \tag{9.67}$$

where $s^2 = (x - x_0)^2 + (y - y_0)^2$.

When propagating in free space at an arbitrary distance z, the beam complex amplitude takes the form:

$$\begin{aligned}E_{mn}(x,y,z)=&\frac{w(0)}{w(z)}\left[\frac{\sqrt{2}}{w(z)}\right]^n\left[(x-x_0)+i(y-y_0)\right]^n L_m^n\left[\frac{2s^2}{w^2(z)}\right]\\&\times\exp\left[-\frac{s^2}{w^2(z)}+\frac{iks^2}{2R(z)}-i(n+2m+1)\zeta(z)\right],\end{aligned} \tag{9.68}$$

where:

$$w(z) = w_0\sqrt{1+\left(\frac{z}{z_R}\right)^2},\quad R(z) = z\left[1+\left(\frac{z_R}{z}\right)^2\right],\quad \zeta(z) = \arctan\left(\frac{z}{z_R}\right), \tag{9.69}$$

$z_R = kw_0^2/2$ is the Rayleigh range, and $k = 2\pi/\lambda$ is the wavenumber of light of wavelength λ.

If the shifts x_0 and y_0 are not real-valued, then the magnitudes s^2, $w(z)$, and $R(z)$ do not have the same physical meaning as for real x_0 and y_0, no more, respectively, denoting a distance from the optical axis, a beam width, and a wavefront curvature radius. Besides, unlike conventional LG beams, the transverse intensity pattern of such a beam is not radially symmetric. Figure 9.12 depicts simulated intensity patterns of beam (9.68) at different distances for the following values of parameters: wavelength, λ = 532 nm, waist radius, $w_0 = 10\lambda$, beam index $(m, n) = (8, 7)$, transverse shifts $x_0 = 0$, $y_0 = \lambda\, i$, and the on-axis distances are: $z = 0\lambda$ (Figure 9.12a), $z = z_R/4 = 25\lambda$ (Figure 9.12b), $z = z_R/2 = 50\lambda$ (Figure 9.12c), $z = z_R = 100\lambda$ (Figure 9.12d), $z = 3z_R/2 = 150\lambda$ (Figure 9.12e), and $z = 5z_R/2 = 250\lambda$ (Figure 9.12f). The computational domain size is $2R$, where $R = 60\lambda$ (Figure 9.12a and b), 70λ (Figure 9.12c), 80λ (Figure 9.12d), 100λ (Figure 9.12e), and 150λ (Figure 9.12f). At the aforementioned parameters, the Rayleigh range is $z_R = 100\pi\lambda$.

Figure 9.12 suggests that upon propagation the first crescent-shaped ring nearly turns into a ring, although this is not the case for the peripheral rings. As the aLG beam propagates, the energy is also seen to be redistributed from the central to peripheral crescents.

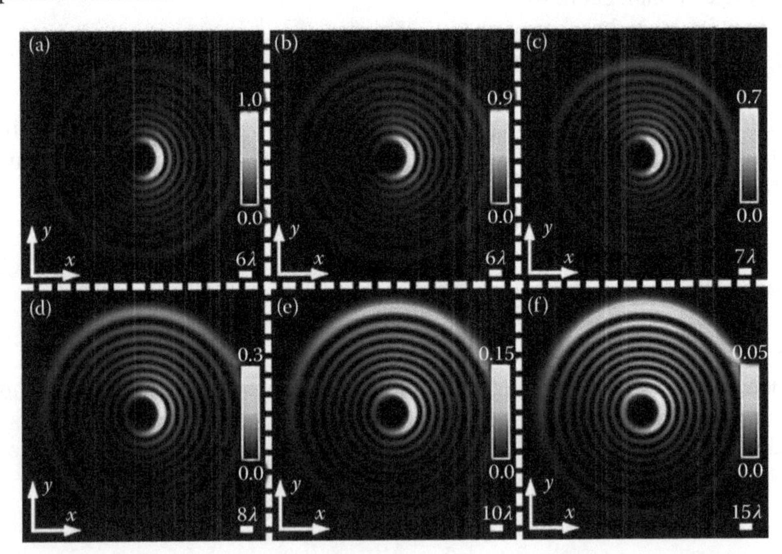

FIGURE 9.12 The transverse intensity pattern of an aLG beam in different planes for the parameters: wavelength λ = 532 nm, waist radius $w_0 = 10\lambda$, beam index $(m, n) = (8, 7)$, transverse shifts $x_0 = 0$ and $y_0 = \lambda$i, the on-axis distances are (a) $z = 0\lambda$, (b) $z = z_R/4 = 25\lambda$, (c) $z = z_R/2 = 50\lambda$, (d) $z = z_R = 100\lambda$, (e) $z = 3z_R/2 = 150\lambda$, and (f) $z = 5z_R/2 = 250\lambda$.

Power of a Shifted Laguerre-Gaussian Beam

The power of a paraxial light beam can be expressed through both the complex amplitude E and the angular spectrum of plane waves, A:

$$W = \int_{-\infty}^{+\infty}\int_{-\infty}^{+\infty} E^* E dx dy = \lambda^2 \int_{-\infty}^{+\infty}\int_{-\infty}^{+\infty} A^* A d\alpha d\beta, \tag{9.70}$$

where:

$$A(\alpha,\beta) = \lambda^{-2} \int_{-\infty}^{+\infty}\int_{-\infty}^{+\infty} E(x,y,0)\exp\left[-ik(\alpha x+\beta y)\right] dx dy. \tag{9.71}$$

For an aLG beam, the power can be more conveniently calculated in the spectral plane. First, we derive a relation for the angular spectrum of a non-shifted beam:

$$\begin{aligned} A(\rho,\theta) &= \frac{1}{\lambda^2}\left(\frac{\sqrt{2}}{w_0}\right)^n \\ &\times \int_0^\infty r^n \exp\left(-\frac{r^2}{w_0^2}\right) L_m^n\left(\frac{2r^2}{w_0^2}\right)\left\{\int_0^{2\pi} \exp\left[in\varphi - ik\rho r\cos(\varphi-\theta)\right] d\varphi\right\} r dr \\ &= (-i)^n \frac{k^2}{2\pi}\left(\frac{\sqrt{2}}{w_0}\right)^n \exp(in\theta) \int_0^\infty r^{n+1} \exp\left(-\frac{r^2}{w_0^2}\right) L_m^n\left(\frac{2r^2}{w_0^2}\right) J_n(k\rho r) dr, \end{aligned} \tag{9.72}$$

where (r, φ) and (ρ, θ) are polar coordinates in the initial plane and in the Fourier plane, respectively (ρ is a dimensionless coordinate).

We shall make use of a reference integral (Ref. [135], Eq. 7.421.4):

$$\int_0^\infty x^{\nu+1} \exp(-\beta x^2) L_n^\nu(\alpha x^2) J_\nu(xy) dx = \frac{(\beta-\alpha)^n y^\nu}{2^{\nu+1}\beta^{\nu+n+1}} \exp\left(-\frac{y^2}{4\beta}\right) L_n^\nu\left[\frac{\alpha y^2}{4\beta(\alpha-\beta)}\right]. \tag{9.73}$$

In view of (9.73), the angular spectrum of plane waves of a LG beam is

$$A(\rho,\theta) = C_0 \rho^n L_m^n\left[\frac{(kw_0\rho)^2}{2}\right] \exp\left[-\frac{(kw_0\rho)^2}{4} + in\theta\right], \tag{9.74}$$

where:

$$C_0 = (-i)^n (-1)^m \frac{(kw_0)^{n+2}}{2^{2+n/2}\pi}. \tag{9.75}$$

For the shifted beam in Eq. (9.67), the angular spectrum of plane waves takes the form:

$$A(\rho,\theta)=C_0\rho^n \exp\left[-\frac{(kw_0\rho)^2}{4}+in\theta\right] \times L_m^n\left[\frac{(kw_0\rho)^2}{2}\right]\exp\left[-ik\rho(x_0\cos\theta+y_0\sin\theta)\right]. \tag{9.76}$$

Making use of Eq. (9.76), the power of the aLG beam is given by

$$W=2\pi\lambda^2|C_0|^2\int_0^\infty \rho^{2n+1}\exp\left[-\frac{(kw_0\rho)^2}{2}\right]\left\{L_m^n\left[\frac{(kw_0\rho)^2}{2}\right]\right\}^2 J_0(2ikD_0\rho)d\rho, \tag{9.77}$$

where $D_0 = [(\mathrm{Im}\, x_0)^2 + (\mathrm{Im}\, y_0)^2]^{1/2}$.

Integral (9.77) can be calculated using a reference integral (Eq. 2.9.12.14 in Ref. [38]), which, following a numerical check-up, reads as

$$\int_0^\infty x^{(\gamma+\delta)/2}e^{-cx}J_{\gamma+\delta}\left(b\sqrt{x}\right)L_\mu^\gamma(cx)L_\nu^\delta(cx)dx = \frac{(-1)^{\mu+\nu}}{c^{\gamma+\delta+1}}\left(\frac{b}{2}\right)^{\gamma+\delta}\exp\left(-\frac{b^2}{4c}\right)L_\nu^{\gamma+\mu-\nu}\left(\frac{b^2}{4c}\right)L_\mu^{\delta-\mu+\nu}\left(\frac{b^2}{4c}\right), \tag{9.78}$$

where Re $c > 0$, Re $(\gamma + \lambda) > -1$, $|\arg b| < \pi$. In (9.78), we change the integration variable $x \to x^2$ and put $\mu = m$, $\gamma = n$, $\delta = -n$, $\nu = m + n$, $c = (kw_0)^2/2$, $b = 2ikD_0$. Then, considering the identity $L_\mu^{-\sigma}(x)\equiv\left[(\mu-\sigma)!/\mu!\right](-x)^\sigma L_{\mu-\sigma}^\sigma(x)$, we obtain the beam power:

$$W=\frac{\pi w_0^2}{2}\frac{(m+n)!}{m!}\exp\left(\frac{2D_0^2}{w_0^2}\right)L_{m+n}^0\left(-\frac{2D_0^2}{w_0^2}\right)L_m^0\left(-\frac{2D_0^2}{w_0^2}\right). \tag{9.79}$$

Although being proportional to the Laguerre polynomials, the beam power cannot take negative or zero values. The reason is that $2(D_0/w_0)^2 \geq 0$, whereas the Laguerre polynomials are always positive in the non-positive domain:

$$L_m(-\xi)=\sum_{k=0}^m\frac{(-1)^k}{k!}C_m^k(-\xi)^k=\sum_{k=0}^m C_m^k\frac{\xi^k}{k!}=1+\underbrace{\sum_{k=1}^m C_m^k\frac{\xi^k}{k!}}_{\geq 0}\geq 1, \tag{9.80}$$

where C_m^k are binomial coefficients.

From (9.80) and the presence of the factor $\exp[2(D_0/w_0)^2]$ in (9.79), we can infer that a complex-valued shift always results in an increased power of the beam.

In a particular case, when the beam is shifted by a real distance, the parameter D_0 takes a zero value and the power equals $[\pi w_0^2/2][(m+n)!/m!]$, which is coincident with the power of LG beams reported in [307] within a constant.

The increase in the power of the aLG beam with increasing asymmetry has no physical meaning. However, we shall use Eq. (9.79) in the next paragraph to calculate the normalized OAM.

Orbital Angular Momentum of a Shifted Laguerre-Gaussian Beam

Let us derive a relation for the projection of OAM of an aLG beam on the optical axis. Note that the rest projections of a paraxial beam equal zero. This is also convenient to do using the angular spectrum of plane waves:

$$J_z = -i\lambda^2 \iint_{\mathbb{R}^2} A^* \frac{\partial A}{\partial \theta} \rho d\rho d\theta. \tag{9.81}$$

Substituting (9.76) in (9.81) yields:

$$\begin{aligned} J_z = &-i\lambda^2 |C_0|^2 \iint_{\mathbb{R}^2} \rho^{2n} \exp\left[-\frac{(kw\rho)^2}{2}\right] \left\{L_m^n\left[\frac{(kw\rho)^2}{2}\right]\right\}^2 \\ &\times \exp\left[-in\theta + ik\rho\left(x_0^* \cos\theta + y_0^* \sin\theta\right)\right]\left[in + ik\rho\left(x_0 \sin\theta - y_0 \cos\theta\right)\right] \\ &\times \exp\left[in\theta - ik\rho\left(x_0 \cos\theta + y_0 \sin\theta\right)\right] \rho d\rho d\theta. \end{aligned} \tag{9.82}$$

We can single out a term proportional to the power:

$$\begin{aligned} J_z = &\, nW + k\lambda^2 |C_0|^2 \int_0^\infty \rho^{2n+2} \exp\left[-\frac{(kw\rho)^2}{2}\right] \left\{L_m^n\left[\frac{(kw\rho)^2}{2}\right]\right\}^2 d\rho \\ &\times \int_0^{2\pi} \left(x_0 \sin\theta - y_0 \cos\theta\right) \exp\left\{2k\rho\left[(\operatorname{Im} x_0)\cos\theta + (\operatorname{Im} y_0)\sin\theta\right]\right\} d\theta. \end{aligned} \tag{9.83}$$

The inner integral is expressed through Bessel functions [308]. Then, (9.83) takes the form:

$$\begin{aligned} J_z = &\, nW - 2\pi k\lambda^2 |C_0|^2 \int_0^\infty \rho^{2n+2} \exp\left[-\frac{(kw\rho)^2}{2}\right] \left\{L_m^n\left[\frac{(kw\rho)^2}{2}\right]\right\}^2 \\ &\times \left\{(ix_0 + y_0)\left[\frac{(\operatorname{Im} x_0) - (\operatorname{Im} y_0)}{2iD_0}\right] + (ix_0 - y_0)\left[\frac{(\operatorname{Im} x_0) - (\operatorname{Im} y_0)}{2iD_0}\right]^{-1}\right\} iI_1(2kD_0\rho) d\rho \\ = &\, nW + 4\pi^2\lambda |C_0|^2 \frac{\operatorname{Im}(x_0^* y_0)}{D_0} \int_0^\infty \rho^{2n+2} \exp\left[-\frac{(kw\rho)^2}{2}\right] \left\{L_m^n\left[\frac{(kw\rho)^2}{2}\right]\right\}^2 I_1(2kD_0\rho) d\rho. \end{aligned} \tag{9.84}$$

The integral in (9.84) can be calculated in a way similar to the beam power integral by expressing it as a derivative with respect to D_0:

$$J_z = nW - 4\pi^2\lambda|C_0|^2 \frac{\mathrm{Im}(x_0^* y_0)}{2kD_0}\frac{\partial}{\partial D_0}\left(\int_0^\infty \rho^{2n+1}\exp\left[-\frac{(kw\rho)^2}{2}\right]\right.$$
$$\left.\times\left\{L_m^n\left[\frac{(kw\rho)^2}{2}\right]\right\}^2 I_0(2kD_0\rho)d\rho\right). \tag{9.85}$$

The integral in (9.85) can be taken, being coincident with the power integral:

$$J_z = nW - 4\pi^2\lambda|C_0|^2 \frac{\mathrm{Im}(x_0^* y_0)}{2kD_0}\frac{\partial}{\partial D_0}\left\{\frac{2^n}{(kw_0)^{2n+2}}\frac{(m+n)!}{m!}\exp\left(\frac{2D_0^2}{w_0^2}\right)\right.$$
$$\left.\times L_{m+n}\left(-\frac{2D_0^2}{w_0^2}\right)L_m\left(-\frac{2D_0^2}{w_0^2}\right)\right\} \tag{9.86}$$
$$= nW - \frac{\pi w_0^2}{4}\frac{(m+n)!}{m!}\frac{\mathrm{Im}(x_0^* y_0)}{D_0}\frac{\partial}{\partial D_0}\left\{\exp\left(\frac{2D_0^2}{w_0^2}\right)L_{m+n}\left(-\frac{2D_0^2}{w_0^2}\right)L_m\left(-\frac{2D_0^2}{w_0^2}\right)\right\}.$$

After taking the derivative, the normalized OAM takes the form:

$$\frac{J_z}{W} = n + \frac{2\,\mathrm{Im}(x_0^* y_0)}{w_0^2}\left[\frac{L_m^1\left(-\frac{2D_0^2}{w_0^2}\right)}{L_m\left(-\frac{2D_0^2}{w_0^2}\right)} + \frac{L_{m+n}^1\left(-\frac{2D_0^2}{w_0^2}\right)}{L_{m+n}\left(-\frac{2D_0^2}{w_0^2}\right)} - 1\right]. \tag{9.87}$$

The physical meaning of the second term in the right-hand side of Eq. (9.87) can be better understood at $n = m = 0$. In this case, the laser beam transforms into an asymmetric Gaussian beam. The first term in Eq. (9.87) is equal to zero ($n = 0$), while the second one reads as $4\,\mathrm{Im}(x_0^* y_0)/w_0^2$. It means that if the shift of the Gaussian beam is real (i.e., $x_0 = y_0 = aw_0$), then the Gaussian beam is just displaced and not deformed, that is, $J_z/W = 0$. If both shifts are purely imaginary along both coordinates (i.e., $x_0 = y_0 = iaw_0$), then the shape of the Gaussian beam is distorted, while we still have $J_z/W = 0$. Only when the shift is real along one coordinate and imaginary along the other (i.e., $ix_0 = y_0 = iaw_0$), then the Gaussian beam is displaced, with its shape being distorted and its acquiring the OAM: $J_z/W = 4a^2$. The aLG beam, at any other values of n and m, is similar.

From (9.87), the normalized OAM is seen to be independent of the wavelength and fully defined by the ratio of the shifts to the waist radius, that is, by x_0/w_0 and y_0/w_0.

The increase or decrease of the normalized OAM can be shown to be fully determined by the sign of the magnitude $\mathrm{Im}(x_0^* y_0)$, because the expression in the square brackets in (9.87) is always larger than or equal to 1.

The normalized OAM depends at once on several parameters, namely, two indices of the Laguerre polynomial and the real and imaginary shifts in the Cartesian coordinates. Let us analyze a particular case of purely imaginary magnitude $x_0^* y_0$, whereas the complex shifts have the same absolute value. Let $x_0 = aw_0\exp(iv)$, $y_0 = ix_0$, where a is a real number (i.e., $a = |x_0/w_0|$). Then, the normalized OAM is given by

$$\frac{J_z}{W} = n - \xi\left[\frac{L_m^1(\xi)}{L_m(\xi)} + \frac{L_{m+n}^1(\xi)}{L_{m+n}(\xi)} - 1\right], \tag{9.88}$$

where $\zeta = -2a^2$. The a parameter can be referred to as an asymmetry parameter of the aLG beam. Equation (9.88) suggests that at $|\zeta| >> 1$ the normalized OAM approximately equals $J_z/W \approx n - \zeta$, depending quadratically on the asymmetry parameter a, as is confirmed by the graph in Figure 9.13 plotted using Eq. (9.88) for $m = 3$ and $n = 5$.

From Figure 9.13 it seen that as distinct from conventional LG beams, the OAM of aLG beams varies in a continuous manner, taking integer and fractional values.

From (9.87) it follows that if $x_0^* y_0$ is a real number, in a similar way to radially symmetric optical vortices, the normalized OAM is coincident with the topological charge n, although the beam has no radial symmetry (if x_0 and y_0 are purely imaginary). Thus, Figure 9.14 depicts intensity patterns in the plane $z = 0$ for the aLG beams with the following parameters: wavelength $\lambda = 532$ nm, waist radius $w = 2\lambda$, beam index $(m, n) = (3, 5)$, transverse shifts $x_0 = 0.01wi$ and $y_0 = 0.01wi$ (Figure 9.14a), $x_0 = 0.05wi$ and $y_0 = 0.05wi$ (Figure 9.14b), $x_0 = 0.1wi$ and $y_0 = 0.1wi$ (Figure 9.14c), $x_0 = 0.2wi$ and $y_0 = 0.2wi$ (Figure 9.14d), $x_0 = 0.5wi$ and $y_0 = 0.5wi$ (Figure 9.14e), $x_0 = 2wi$ and $y_0 = 2wi$ (Figure 9.14f). The computational domain size is $2R$, where $R = 10\lambda$. Equation (9.87) suggests that in all pictures in Figure 9.14, OAM should be equal to 5. From the numerical simulation, OAM was found to equal 4.999 (Figure 9.14a–d) and 4.998 (Figure 9.14e and f).

All the beams in Figure 9.14 are seen to be different in shape, showing near radially symmetric diffraction rings in Figure 9.14a, an intensity crescent enclosed by peripheral rings in Figure 9.14b and c, a crescent with disintegrated peripheral rings in Figure 9.14d, a crescent with no peripheral rings in Figure 9.14e, and an elliptic

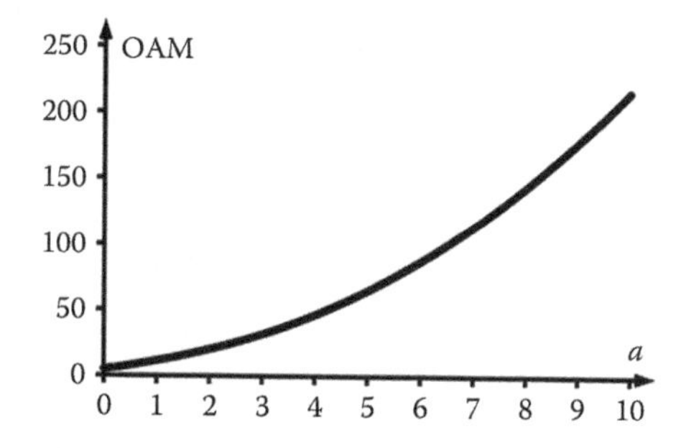

FIGURE 9.13 Normalized OAM vs. normalized asymmetry parameter a ($n = 5$).

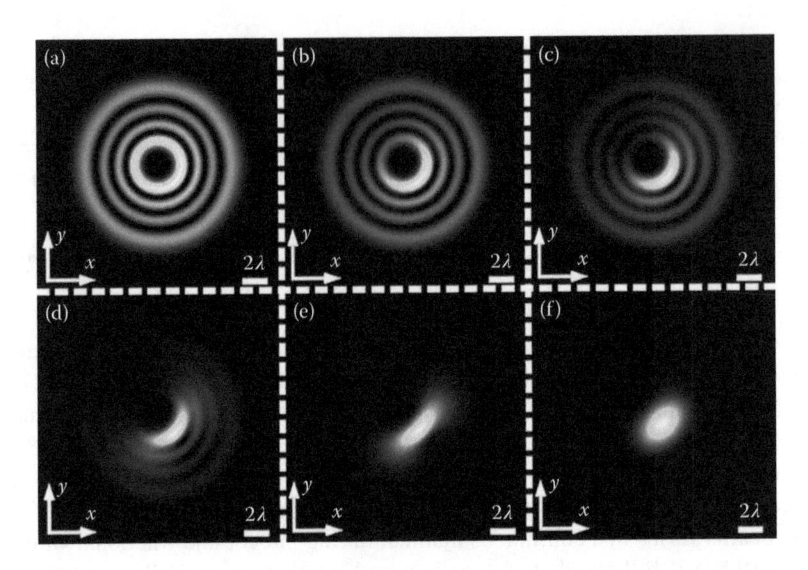

FIGURE 9.14 Intensity patterns in the plane $z = 0$ for the aLG beams at the following parameters: wavelength $\lambda = 532$ nm, waist radius $w = 2\lambda$, beam index $(m, n) = (3, 5)$, at transverse shifts: (a) $x_0 = 0.01wi$ and $y_0 = 0.01wi$, (b) $x_0 = 0.05wi$ and $y_0 = 0.05wi$, (c) $x_0 = 0.1wi$ and $y_0 = 0.1wi$, (d) $x_0 = 0.2wi$ and $y_0 = 0.2wi$, (e) $x_0 = 0.5wi$ and $y_0 = 0.5wi$, and (f) $x_0 = 2wi$ and $y_0 = 2wi$.

intensity spot in Figure 9.14f. However, despite being different in shape, all these beams have the same OAM.

Note that although the beam in Figure 9.14f looks like an elliptical Gaussian beam, this beam has the index $(m, n) = (3, 5)$ and its OAM equals 5. For the explanation of the shape of the beam in Figure 9.14f we use Eq. (9.76). From Eq. (9.76) it follows that for purely imaginary large shifts along both coordinates ($x_0 = y_0 = 2wi$), the second exponent in Eq. (9.76) grows the fastest with increasing ρ at $\theta = \pi/4$. Thus, only a proportion of the spectrum in (9.76) in the first quadrant near the angle $\theta = \pi/4$ gives an effective contribution to the field of Eq. (9.68). Therefore, the field (9.68) is effectively generated only in the fourth quadrant with its center positioned at the angle $\theta = -\pi/4$. Since only a small proportion of spectrum (9.76) contributes to field (9.68), there are no narrow bright rings in Figure 9.14f that are seen in Figure 9.14a.

Paraxial Laguerre-Gaussian Beams in the Form of a Rotating Crescent

Equation (9.68) is essentially simplified at $m = 0$, with the diffraction pattern having a single ring. Then, the intensity takes the form:

$$I_{0n}(x,y,z) = |E|^2 = \frac{w^2(0)}{w^2(z)}\left[\frac{\sqrt{2}}{w^2(z)}\right]^n \exp\left[2\frac{(\mathrm{Im}\,x_0)^2 + (\mathrm{Im}\,y_0)^2}{w^2(z)}\right]$$
$$\times\left[(u + \mathrm{Im}\,y_0)^2 + (v - \mathrm{Im}\,x_0)^2\right]^n \exp\left\{-\frac{2(u^2+v^2)}{w^2(z)} + \frac{2k\left[(\mathrm{Im}\,x_0)u + (\mathrm{Im}\,y_0)v\right]}{R(z)}\right\}. \tag{9.89}$$

where $u = x - \operatorname{Re} x_0$, $v = y - \operatorname{Re} y_0$.

Considering that the intensity cannot be negative, the intensity zeros are its minima. From (9.89) it is seen that as the aLG beam propagates in a uniform space at $n > 0$, the central intensity minimum is observed at the point $(x_{min}, y_{min}) = (\operatorname{Re} x_0 - \operatorname{Im} y_0, \operatorname{Re} y_0 + \operatorname{Im} x_0)$, forming a phase singularity. Setting the partial Cartesian derivatives of intensity (9.24) equal to zero, we can show that at $n > 0$ the location of the intensity peak is a function of the distance z traveled and rotates about the minimum:

$$\frac{y_{max} - \operatorname{Re} y_0 - \operatorname{Im} x_0}{x_{max} - \operatorname{Re} x_0 + \operatorname{Im} y_0} = \frac{(\operatorname{Im} y_0)z - (\operatorname{Im} x_0)z_R}{(\operatorname{Im} x_0)z + (\operatorname{Im} y_0)z_R}, \tag{9.90}$$

where (x_{max}, y_{max}) are the coordinates of the maximum intensity point.

Let us introduce a new coordinate system which is shifted and rotated around the initial system by an angle defined by the distance traveled, z:

$$\begin{pmatrix} \xi \\ \eta \end{pmatrix} = \begin{pmatrix} \cos\varphi & \sin\varphi \\ -\sin\varphi & \cos\varphi \end{pmatrix} \begin{pmatrix} x - \operatorname{Re} x_0 + \operatorname{Im} y_0 \\ y - \operatorname{Re} y_0 - \operatorname{Im} x_0 \end{pmatrix}, \tag{9.91}$$

where:

$$\varphi = \arctan\left[\frac{(\operatorname{Im} y_0)z - (\operatorname{Im} x_0)z_R}{(\operatorname{Im} x_0)z + (\operatorname{Im} y_0)z_R}\right],$$

$$\cos\varphi = \frac{(\operatorname{Im} x_0)z + (\operatorname{Im} y_0)z_R}{D_0\sqrt{z^2 + z_R^2}}, \tag{9.92}$$

$$\sin\varphi = \frac{(\operatorname{Im} y_0)z - (\operatorname{Im} x_0)z_R}{D_0\sqrt{z^2 + z_R^2}}.$$

It follows from Eq. (9.92), that if $\operatorname{Im}(y_0) = \operatorname{Im}(x_0)$, then in the initial plane ($z = 0$) the light crescent is rotated by the angle $-\pi/4$, which is confirmed by Figure 9.14.

In the new coordinate system (ξ, η), intensity (9.89) takes the form:

$$I_{0n}(\xi,\eta,z) = |E|^2 = \frac{w^2(0)}{w^2(z)}\left[\frac{\sqrt{2}}{w^2(z)}\right]^n (\xi^2+\eta^2)^n \exp\left[-\frac{2(\xi^2+\eta^2)}{w^2(z)} + \frac{2kD_0}{\sqrt{z^2+z_R^2}}\xi\right]. \tag{9.93}$$

Such a function can be shown to have three stationary points. The first point is a minimum at $\xi = 0$, $\eta = 0$, which corresponds to the aforementioned point (x_{min}, y_{min}). The second point is a maximum with the coordinates

$$\begin{cases} \xi_{max} = \dfrac{1}{2}\dfrac{w(z)}{w(0)}\left(D_0 + \sqrt{D_0^2 + 2nw_0^2}\right), \\ \eta_{max} = 0, \end{cases} \tag{9.94}$$

and the third one is a saddle point with the coordinates

$$\begin{cases} \xi_{\text{saddle}} = \dfrac{1}{2}\dfrac{w(z)}{w(0)}\left(D_0 - \sqrt{D_0^2 + 2nw_0^2}\right), \\ \eta_{\text{saddle}} = 0, \end{cases} \tag{9.95}$$

where there is an intensity maximum with respect to the variable ξ and a minimum with respect to the variable η. The intensity maximum with respect to ξ means that the point is found on a bright ring, whereas the minimum with respect to η is where the minimal intensity is found on the ring.

It follows from Eqs. (9.93) and (9.94), that the intensity decreases from the maximum by e times in the points, which are located on a ring at the following angles from the intensity maximum

$$\theta_e = \pm 2\arcsin\left(\frac{1}{\sqrt{8n}}\sqrt{\sqrt{1+2n\frac{w_0^2}{D_0^2}}-1}\right). \tag{9.96}$$

This means that increasing of the asymmetry parameter leads to decreasing of the length of the arc of the light crescent. If $D_0 << w_0$, then the intensity does not decrease e times on the whole ring. If $D_0 >> w_0$, then, conversely, the intensity drops e times at $\theta_e \approx \pm w_0/(2^{1/2}D_0)$.

From Eqs. (9.93) and (9.94) it also follows that having travelled over a distance z, the maximum intensity drops by a factor of $1 + (z/z_R)^2$.

In the initial coordinate system, the intensity maximum point has the coordinates:

$$\begin{cases} x_{\max} = \text{Re}\, x_0 - \text{Im}\, y_0 + \dfrac{(\text{Im}\, x_0)z + (\text{Im}\, y_0)z_R}{2z_R}\left(1+\sqrt{1+\dfrac{2nw_0^2}{D_0^2}}\right), \\ y_{\max} = \text{Re}\, y_0 + \text{Im}\, x_0 + \dfrac{(\text{Im}\, y_0)z - (\text{Im}\, x_0)z_R}{2z_R}\left(1+\sqrt{1+\dfrac{2nw_0^2}{D_0^2}}\right). \end{cases} \tag{9.97}$$

From (9.97) it follows that the maximum intensity point $(x_{\max}, y_{\max})$ is rotating about the point $(x_{\min}, y_{\min})$, making an angle of α_0 at the distance

$$z = z_R \tan(\alpha_0). \tag{9.98}$$

Figure 9.15 shows transverse intensity patterns of the beam in Eq. (9.68) with zero radial index $m = 0$ at different planes for the following parameters: wavelength $\lambda = 532$ nm, waist radius $w_0 = 5\lambda$ (with the Rayleigh range equal to $z_R = 25\pi\lambda$), the topological charge $n = 8$, transverse shifts $x_0 = 0.25\lambda = w_0/20$ and $y_0 = 0.25\lambda\, i = iw_0/20$, on-axis distances $z = 0$ (Figure 9.15a), $z_R \tan(\pi/12)$ (Figure 9.15b), $z_R \tan(\pi/4) = z_R$ (Figure 9.15c), and $z_R \tan(5\pi/12)$ (Figure 9.15d). The beam centre is at the origin

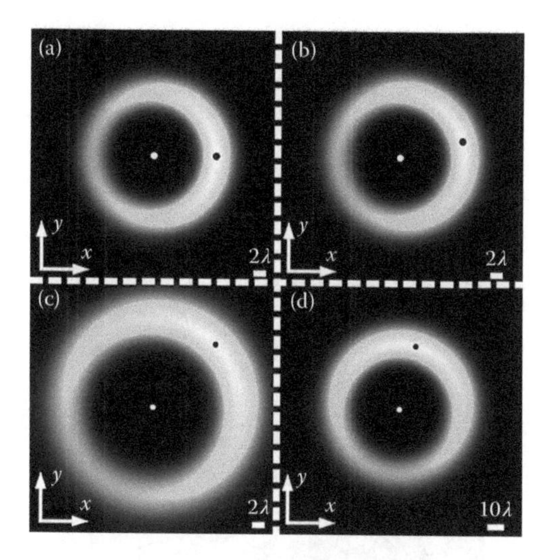

FIGURE 9.15 Transverse intensity patterns of beam (9.68) with zero radial index $m = 0$ at different distances for the following parameters: wavelength, $\lambda = 532$ nm, waist radius, $w_0 = 5\lambda$, topological charge, $n = 8$, transverse shifts $x_0 = 0.25\lambda = w_0/20$ and $y_0 = 0.25\lambda\ i = iw_0/20$, the on-axis distances are (a) $z = 0$, (b) $z_R \tan(\pi/12)$, (c) $z_R \tan(\pi/4) = z_R$, and (d) $z_R \tan(5\pi/12)$. The computational domain size is $2R$, where $R =$ (a) 20λ, (b) 20λ, (c) 20λ, and (d) 75λ. The black dot shows the location of the intensity maximum derived from (9.97).

of coordinates: $(x_{\min}, y_{\min}) = (0, 0)$. The computational domain size is $2R$, where $R = 20\lambda$ (Figure 9.15a), 20λ (Figure 9.15b), 20λ (Figure 9.15c), and 75λ (Figure 9.15d). The black dot marks the location of the intensity maximum, which was calculated using (9.97). The white dot marks the location of the minimal intensity or a phase singularity point.

From (9.93) through (9.95) it follows that the ratio of the maximal intensity on the diffraction ring to a minimal one (in the saddle point) is defined by the relation:

$$\frac{I_{0n,\max}}{I_{0n,\text{saddle}}} = \left(\frac{\sqrt{D_0^2 + 2nw_0^2} + D_0}{\sqrt{D_0^2 + 2nw_0^2} - D_0} \right)^{2n} \exp\left(\frac{2D_0}{w_0^2} \sqrt{D_0^2 + 2nw_0^2} \right). \tag{9.99}$$

From (9.99), the asymmetry of the beam is seen to increase with increasing shift D_0. What this means is that with increasing imaginary shifts $\text{Im}(x_0)$ and $\text{Im}(y_0)$ a disintegration of the bright ring occurs, with the intensity pattern appearing as an off-centre bright spot rotating upon propagation about a phase singularity point $(x_{\min}, y_{\min})$. The term "disintegration" is quoted because the intensity nowhere becomes zero in the bright ring, as the denominator in (9.99) never turns zero at $n > 0$.

Figure 9.16 depicts the simulation results for the transverse intensity pattern and the phase of the same beam as was depicted in Figure 9.15, but with larger shifts: $x_0 = 5\lambda = w_0$ and $y_0 = 5\lambda\ i = iw_0$. The rest parameters remained unchanged. The black dot marks the intensity maximum calculated from (9.97). Color bars confirm that upon propagation the intensity drops by a factor of $1 + (z/z_R)^2$.

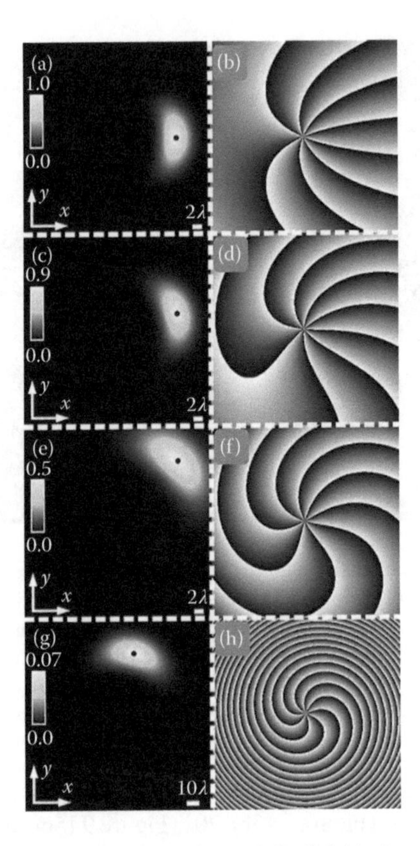

FIGURE 9.16 (a, c, e, g) Transverse intensity and (b, d, f, h) phase patterns of beam (9.68) with zero radial index $m = 0$ at different distances z for the following parameters: wavelength, $\lambda = 532$ nm, waist radius, $w_0 = 5\lambda$, topological charge, $n = 8$, transverse shifts, $x_0 = 5\lambda = w_0$ and $y_0 = 5\lambda\, i = iw_0$, on-axis distances are (a, b) $z = 0$, (c, d) $z_R \tan(\pi/12)$, (e, f) $z_R \tan(\pi/4) = z_R$, and (g, h) $z_R \tan(5\pi/12)$. The computational domain size is $2R$, where $R =$ (a–f) 20λ and (g, h) 75λ. The black dot marks the intensity maximum derived from (9.96). In the phase patterns black and white colors, respectively, mark $-\pi$ and $+\pi$.

From Figure 9.16, an oblong focal spot is seen to be rotated by respective angles of $\pi/12$, $\pi/4$, and $5\pi/12$, whereas the phase singularity center remains at the origin of coordinates.

At $m = 0$, an aLG beam is similar to an asymmetric Bessel-Gaussian (aBG) beam [236]. Both beams have the same rotation velocity defined by (9.98). Note, however, that with an aBG beam being more difficult to define analytically, only relationships for the coordinates of intensity zeros were deduced in [236], whereas the coordinates of intensity maxima similar to (9.97) were not analytically derived. As distinct from a closed-form relation for the OAM of an aLG beam in Eq. (9.87), the OAM of an aBG beam is defined via modified Bessel functions series. Besides, the OAM of an aLG beam is quadratically dependent on the asymmetry parameter, whereas the OAM of an aBG beam is described by a near-linear function [236].

Experimental Generation of an Asymmetric Laguerre-Gaussian Beam Using a Spatial Light Modulator

The experimental optical setup is shown in Figure 9.17. The output beam of a solid-state laser L ($\lambda = 532$ nm) was expanded using a system composed of lenses L_1 ($f_1 = 250$ mm) and L_2 ($f_2 = 500$ mm). The expanded laser beam with radius of about 1.1 mm illuminated the display of a spatial light modulator SLM (PLUTO VIS, 1920 × 1080 resolution, with 8-μm pixels). The input of the SLM display composed of a phase function generated by the superposition of the encoded phase function of the initial element used to generate the aLG beam and a linear phase mask (see an inset in Figure 9.17). The aim was to separate spatially the first and zero diffraction orders, with the nonmodulated wave being reflected to the latter. Using lenses L_3 ($f_3 = 350$ mm) and L_4 ($f_4 = 150$ mm), the laser beam reflected to the first order was guided to a lens L_5 ($f_5 = 280$ mm), which focused the aLG beam onto a *CMOS* array of the *LOMO TC-1000* video-camera (3664 × 2740 resolution, 1.67-μm pixels). The CMOS-camera was mounted on an optical rail to travel along and register the intensity pattern at different distances from lens L_5. The diaphragm D served to filter out the zero-diffraction order.

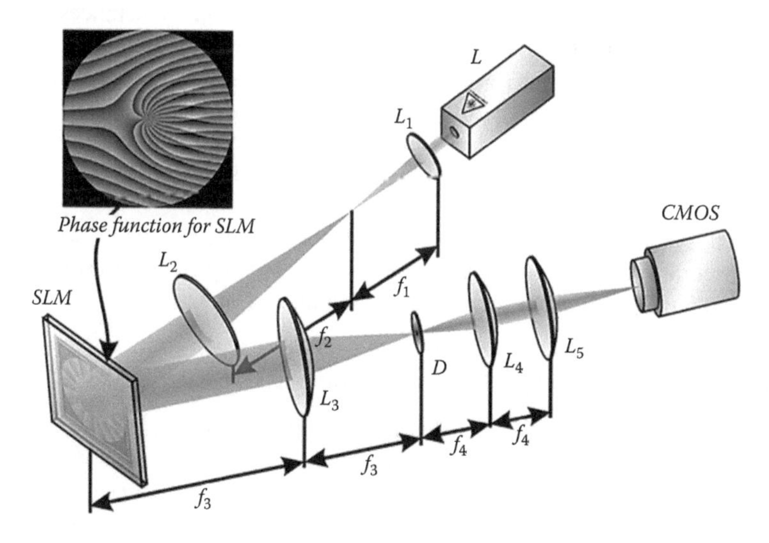

FIGURE 9.17 Experimental setup for generating aLG beams: L—a solid-state laser ($\lambda = 532$ nm); L_1, L_2, L_3, L_4, and L_5—lenses with foci $f_1 = 250$ mm, $f_2 = 500$ mm, $f_3 = 350$ mm, $f_4 = 150$ mm, and $f_5 = 280$ mm; SLM—a spatial light modulator PLUTO VIS (1920 × 1080 resolution and 8-μm pixels); D—a diaphragm serving as a spatial filter; and CMOS—a video-camera LOMO TC-1000 (3664 × 2740 resolution and 1.67-μm pixels).

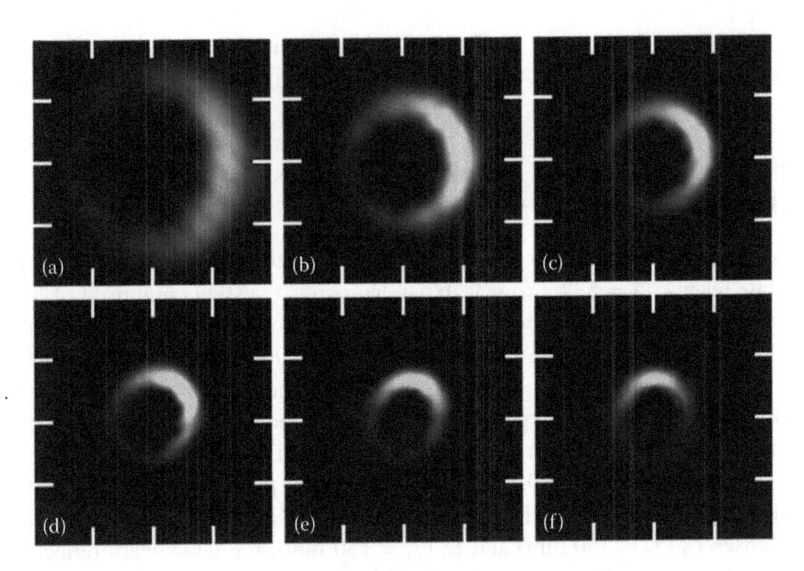

FIGURE 9.18 Intensity patterns for an aLG beam, generated at different distances from the plane of lens L_5: (a) 0 mm, (b) 100 mm, (c) 150 mm, (d) 200 mm, (e) 250 mm, and (f) 280 mm. The scale size is 500 μm. Parameters are: $w_0 = 1$ mm, $n = 8$, $x_0 = 0.2w_0$, $y_0 = 0.2w_0 i$.

Figure 9.18 depicts the intensity patterns registered at different distances from the surface of lens L_5. While registering the pattern at distance 0 mm, the lens L_5 was temporarily removed from the setup. The depicted images show the crescent-shaped beam to rotate about the axis with increasing distance from the lens. The beam generated in the focus of lens L_5 is seen to be rotated by 90° with respect to that in the plane of lens L_5. In addition, due to focusing, the transverse size of the beams is also reduced.

Figure 9.19 shows the intensity patterns calculated using Eq. (9.68), with regard for the lenses, at the parameters of the experiment. From Figure 9.19, the experimental and calculated patterns are seen to be in a qualitative agreement.

Rotating Superpositions of Asymmetric Laguerre-Gaussian Beams

As the imaginary shifts x_0 and y_0 further increase, the beam's asymmetry increases to the extent that the intensity on the ring gets concentrated near a maximum point, defined by (9.96). Rather than being a crescent, the intensity pattern looks more like a Gaussian beam, which is shifted from the origin and rotated by an angle of $\pi/2$ upon propagation.

Actually, according to Eq. (9.93), in the initial plane of the rotated coordinate system (9.91), the intensity is distributed by the law:

$$I_{0n}(\xi,\eta,z=0)=\left(\frac{\sqrt{2}}{w_0^2}\right)^n \exp\left(\frac{2D_0^2}{w_0^2}\right)\left(\xi^2+\eta^2\right)^n \exp\left\{-\frac{2}{w_0^2}\left[\left(\xi-D_0\right)^2+\eta^2\right]\right\}. \quad (9.100)$$

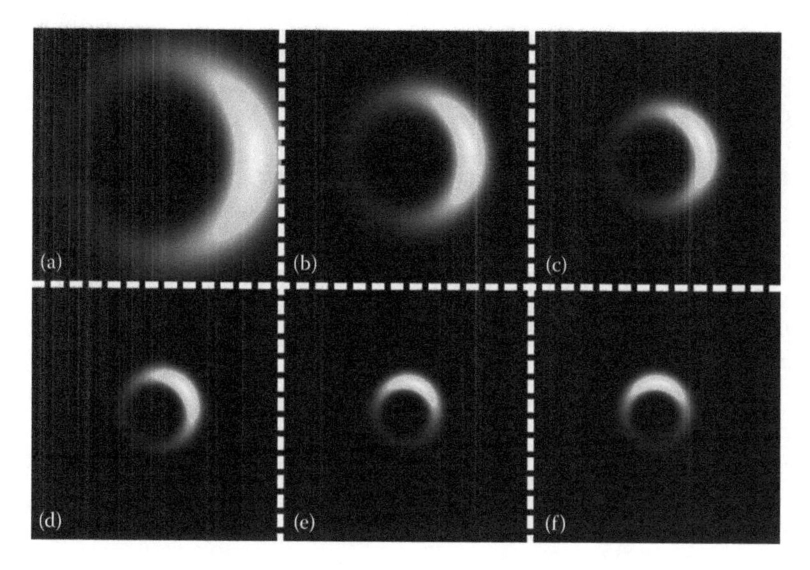

FIGURE 9.19 Intensity patterns of an aLG beam calculated using Eq. (9.68) (with regard for the lenses) at the parameters used in Figure 9.18.

In the other planes at distance z, the intensity in Eq. (9.99) takes the form:

$$I_{0n}(\xi,\eta,z)=\frac{w^2(0)}{w^2(z)}\left[\frac{\sqrt{2}}{w^2(z)}\right]^n\exp\left(\frac{2D_0^2}{w_0^2}\right)$$
$$\times\left(\xi^2+\eta^2\right)^n\exp\left(-\frac{2}{w^2(z)}\left\{\left[\xi-D(z)\right]^2+\eta^2\right\}\right). \tag{9.101}$$

where $D(z)/D_0 = [1 + (z/z_R)^2]^{1/2} = w(z)/w(0)$.

That is, at $D_0 >> w_0$ the power component $(\xi^2 + \eta^2)^n$ weakly affects the intensity pattern, with the remaining exponential component in (9.101) corresponding to the intensity pattern of a Gaussian beam with waist radius $w(z)$, shifted by distance $D(z) \approx \xi_{max}$ from the origin (where the phase singularity is found). Hence, the superposition of beams (9.100) with different shifts D_0, will appear as misaligned Gaussian beams, which are rotated by the same angle defined by (9.98) as they propagate. Also, note that according to (9.101), with increasing distance z from the initial plane, there will be an $[1 + (z/z_R)^2]^{1/2}$ times increase in the off-axis shift of the intensity maxima, which is proportional to the Gaussian beam expansion. What this means is that as they propagate, the beams hardly interfere with each other and the diffraction pattern remains unchanged.

Figure 9.20 illustrates the propagation of two aLG beams, with their complex amplitude given by

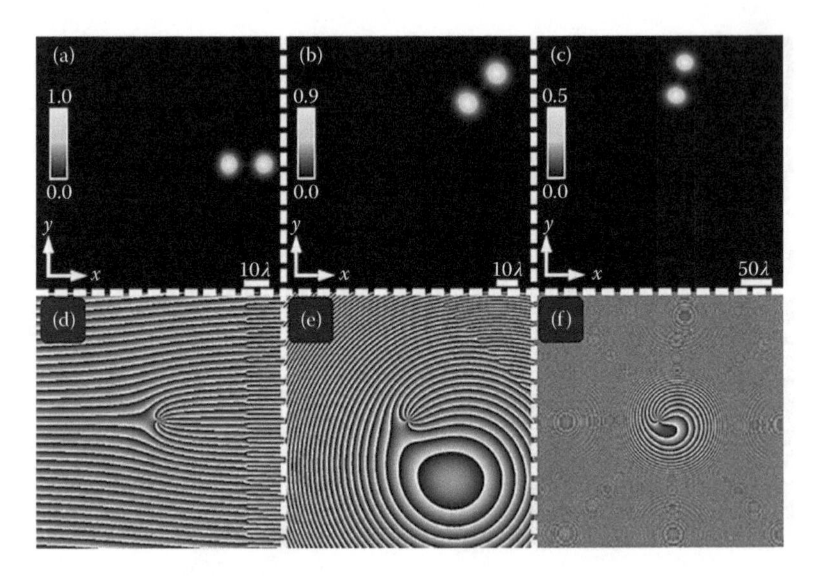

FIGURE 9.20 (a–c) Intensity and (d–f) phase patterns of the aLG beams (9.101) at the parameters ($N = 2$): wavelength $\lambda = 532$ nm, waist radius $w_0 = 5\lambda$, beam index $(m, n) = (0, 8)$, the transverse shifts $x_0 = 5w_0$, $y_0 = 5w_0 i$ (for the first beam) and $x_0 = 8w_0$, $y_0 = 8w_0 i$, (the second beam), the on-axis distances (a, d) $z = 0$, (b, e) $z = z_R \tan(\pi/4)$, and (c, f) $z = z_R \tan(5\pi/12)$. The computational domain size is $2R$, where $R =$ (a, d) 50λ, (b, e) 60λ, and (c, f) 200λ. In the phase patterns, black and white colors, respectively, mark $-\pi$ and $+\pi$.

$$E(x, y, z) = \frac{w(0)}{w(z)}\left[\frac{\sqrt{2}}{w(z)}\right]^n \sum_{j=1}^{N} C_j \left[(x - x_{0j}) + i(y - y_{0j})\right]^n$$
$$\times \exp\left[-\frac{s_j^2}{w^2(z)} + \frac{iks_j^2}{2R(z)} - i(n+1)\zeta(z)\right], \tag{9.102}$$

where:

$N = 2$ for two beams

(x_{0j}, y_{0j}) are the shifts of the jth beam

$s_j = [(x - x_{0j})^2 + (y - y_{0j})^2]^{1/2}$

The C_j coefficients were fitted so as to attain the same maximum intensity for all constituent beams. The rest parameters of the simulation were as follows: $\lambda = 532$ nm, waist radius, $w_0 = 5\lambda$ (Rayleigh range $z_R = 25\pi\lambda$), beam index, $(m, n) = (0, 8)$, the transverse shifts of the first beam, $x_0 = 5w_0$, $y_0 = 5w_0 i$, of the second beam, $x_0 = 8w_0$, $y_0 = 8w_0 i$, the on-axis distances, $z = 0$ (Figure 9.20a and d), and $z = z_R \tan(\pi/4)$ (Figure 9.20b and e), $z = z_R \tan(5\pi/12)$ (Figure 9.20c and f). The computational domain size was $2R$, where $R = 50\lambda$ (Figure 9.20a and d), 60λ (Figure 9.20b and e), and 200λ (Figure 9.20c and f).

From Figure 9.20, the beam similar to the superposition of two Gaussian beams is seen to remain nearly unchanged upon propagation (within a scale), being rotated by an angle of $\pi/4$ (Figure 9.20b and e) and $5\pi/12$ (Figure 9.20c and f).

Summing up, we have proposed a generalization of well-studied LG modes. aLG beams have no modal properties, showing an asymmetric intensity pattern in a plane perpendicular to the propagation axis. As an aLG beam propagates in a uniform space, the asymmetry of the first diffraction ring decreases, with the energy being redistributed to peripheral rings. The number of diffraction rings is coincident with that of a conventional (symmetric) LG beam. The power transferred by aLG beams and the projection of their OAM onto the optical axis have been analytically derived. The normalized OAM has been found to increase quadratically with increasing asymmetry parameter, which is defined as the ratio of the Cartesian shift to the Gaussian beam radius. Conditions for the normalized OAM becoming equal to the topological charge (as is the case for conventional LG beams) have been derived. A particular case of the aLG beams with zero radial index that have a crescent-shaped transverse intensity pattern has been discussed. A relation to describe the coordinate of the intensity maximum has been deduced and the intensity crescent has been shown to rotate during propagation in space. An aLG beam with zero radial index has been generated using a liquid-crystal SLM. The crescent-shaped transverse intensity pattern has been experimentally shown to rotate upon propagation. A feasibility to generate misaligned superpositions of aLG beams with a near-Gaussian intensity distribution that rotate as a whole upon propagation has been demonstrated. The crescent-shaped aLG beams can find uses for trapping and manipulating biological objects (cells) [233], because in this case the cell is less exposed to heat than compared with traps based on a symmetric Gaussian beam. The aLG beams will also be useful in quantum communications systems to form the entanglement of the OAM states of photons. Considering that the aLG beams can carry both integer and fractional OAM, the latter corresponds to the entanglement of the OAM state of a photon [22]. That is, if the aLG-beam is used as a pumping laser beam in the spontaneous parametric down-conversion, two photons should appear with the entangled OAM.

Note that any solution of the paraxial equation can be shifted by complex distances along the Cartesian coordinates. This leads to a new solution. Therefore, it would be interesting to study other beams in the same way: asymmetric Hermite-Gaussian [188], asymmetric Ince-Gaussian [90], and the like.

9.5 ASTIGMATIC LASER BEAMS WITH A LARGE ORBITAL ANGULAR MOMENTUM

Laser beams with an OAM are currently actively investigated because of the wide use of such beams for optical trapping and rotation of microscopic particles [309] and cold atoms [310], as well as in phase contrast microscopy [47], stimulated emission depletion microscopy [311], and optical classical [234] and quantum [22] informatics.

An interesting laser beam which has the OAM but is free of isolated intensity nulls with phase dislocations has been considered in [312]. In [312], the OAM of an elliptic Gaussian beam focused by a cylindrical lens has been calculated. The idea of using the cylindrical lens for assignment of the OAM to a beam was first introduced in Ref. [195]. It has been shown experimentally in [195] that after passing a cylindrical lens at certain propagation distance and at certain conditions an OAM-free Hermite-Gaussian beam transforms to a LG beam which has the OAM.

In Refs. [203,277,313–322] there are attempts to obtain as large as possible values of the OAM. In Ref. [313], it is proposed to increase the OAM by using an array of Gaussian vortex beams, whose centers lie on a circle and whose axes and the common optical axis are crossed lines. It is shown in Ref. [313] that the OAM of such composite beam can reach 204 per photon. In Ref. [277], instead of the Gaussian beams it is proposed to use small holes in an opaque screen, which act like point sources. If these sources are located along a spiral then together they generate a vortex beam with the OAM. In practice, in Ref. [277] a beam has been generated with the OAM equal to 3. It is shown in Ref. [314] that tight focusing of an optical vortex with high topological charge leads to decreased contrast or visibility of sidelobes. In Ref. [314], a beam with the OAM per photon equal to 15 was practically focused. An interesting method for determining the topological charge of an optical vortex by using an annular diffraction grating has been proposed in Ref. [315]. It has been shown experimentally that this way allows to determine the topological charge of ± 25. In Ref. [316], by using the three-wave mixing in a nonlinear Kerr medium, vortex harmonics with the OAM up to 30 per photon have been experimentally generated. In Ref. [317], a perfect optical vortex with a topological charge of 90 has been generated by using a digital multi-element mirror (with the number of micromirrors 1024×768). In Ref. [318], an optical vortex with the topological charge of 200 has been generated by using a liquid-crystal spatial light modulator (the number of elements 1900×1200). It allowed rotating 1.4-μm-diameter microparticles at a speed of 500 μm/sec. In Ref. [319], also by using a light modulator (the number of pixels 1920×1080), entangled pairs of photons with the OAM per photon of ± 300 have been generated. In Ref. [320], an optical vortex with a topological charge of 100 was experimentally generated by using a spiral phase mirror fabricated in an aluminum plate by direct machining with a diamond tool. The same authors [203] created a spiral mirror capable to generate optical vortices with the topological charge of 1020. They used an advanced technology on a 75-mm-diameter aluminum substrate with a roughness of 3 nm. In Ref. [203], it was also shown interferometrically that an optical vortex that has been generated by a mirror has a topological charge of 5050. In Ref. [321], using electronic lithography in a PMMA resist, an 80-μm-diameter hologram with a resolution of 35 nm and a relief height of 25 nm was fabricated. It allowed generating a vortex electron beam with energy of 0.5–1 eV and with a topological charge of 1000. Finally, photons entangled by both the OAM and polarization have been generated by a spiral aluminum mirror with a diameter of about 50 mm for a wavelength of 810 nm [322]. What is more, the quantum OAM of photons was ± 10010. This is the maximum value of the OAM received so far.

In this section, following Ref. [312], we show that an elliptic Gaussian beam after passing a cylindrical lens, whose axis does not coincide with the axes of an elliptical Gaussian profile, rotates upon propagation and at this it is not a vortex laser beam, although its full OAM can reach large values by varying the waist radii of the elliptical Gaussian beam and the focal length of the cylindrical lens. We calculate the coefficients of the angular harmonics expansion of the complex amplitude of such an astigmatic Gaussian beam. An exact formula is also obtained for the OAM in a form of a series of the Legendre functions of the second kind.

Vortex-Free Beam with the Orbital Angular Momentum

For the convenience of the reader, the first four formulas in this section coincide with Ref. [312]. Vortex laser beams with the OAM are usually considered within the paraxial limit. Such beams have singularity points—isolated intensity nulls with undefined phase. Isophase wavefront surface around such point has a spiral shape. However, it turns out that there are simple light fields that have the OAM and do not have the isolated intensity nulls with the vortex phase. We consider a Gaussian elliptic beam with a cylindrical lens [312] in its waist, whose generatrix is rotated in the waist plane by an angle α. The complex amplitude of light immediately after the cylindrical lens reads as

$$E(x,y) = \exp\left(-\frac{x^2}{w_x^2} - \frac{y^2}{w_y^2}\right)\exp\left(-\frac{ikx^2\cos^2\alpha}{2f} - \frac{iky^2\sin^2\alpha}{2f} - \frac{ikxy\sin 2\alpha}{2f}\right), \quad (9.103)$$

where w_x and w_y are the waist radii of the Gaussian beam along the Cartesian axes, f is the focal length of the thin cylindrical lens, whose generatrix is rotated by an angle α with respect to the vertical axis y (the lens is rotated counterclockwise), k is the wave number of light. The normalized OAM within the paraxial limit is given by the expressions [312] (up to the constants):

$$J_z = \mathrm{Im}\int_{-\infty}^{\infty}\int_{-\infty}^{\infty}\bar{E}(x,y)\left(x\frac{\partial E(x,y)}{\partial y} - y\frac{\partial E(x,y)}{\partial x}\right)dxdy, \quad (9.104)$$

$$W = \int_{-\infty}^{\infty}\int_{-\infty}^{\infty}\bar{E}(x,y)E(x,y)\,dxdy, \quad (9.105)$$

where:

- J_z is the axial component of the OAM vector
- W is the energy (power) density of light
- Im is the imaginary part
- $\bar{E}$ is the complex conjugate to the amplitude (9.103)

Substituting Eq. (9.103) into Eqs. (9.104) and (9.105), we derive a simple expression for the normalized OAM of the light field (9.103):

$$\frac{J_z}{W} = \left(\frac{k\sin 2\alpha}{8f}\right)\left(w_y^2 - w_x^2\right). \quad (9.106)$$

It is seen in Eq. (9.106) that the OAM equals zero if the Gaussian beam is circular ($w_x = w_y$) or the lens is not inclined to the vertical axis ($\alpha = 0$). For the inclination angle of 45°, the OAM (4) is maximal, all other conditions being the same. It is also seen in Eq. (9.106) that the OAM of the beam (9.103) is generally fractional, although it can be integer too. The smaller is the focal length of the cylindrical lens and the larger is the ellipticity of the beam (9.103), the greater is the OAM. The OAM sign is determined by whether the Gaussian beam is elongated along the x or y axis in its waist. The advantage of the beam (9.103) is that it can be generated without any additional elements, without a light modulator or a SPP or a fork hologram. Only two lenses are needed to generate it. The first one generates an elliptic Gaussian beam, while the second one creates its OAM.

Now we estimate the OAM for specific values of the variables in Eq. (9.106). A Gaussian beam is considered to be paraxial if its waist radii exceed its wavelength. Let the waist radii be equal to $w_x = 2$ mm and $w_y = 1$ mm, while the focal length is $f = 10$ mm, wavelength is $\lambda = 0.5$ μm, and the tilt angle of the lens is 45° ($\alpha = \pi/4$). Then the OAM in Eq. (9.106) equals 471.24.

The Elliptic Gaussian Beam is Rotating after Passing the Cylindrical Lens

In the following, in difference with Ref. [312], we show that the elliptic Gaussian beam is rotating after passing the cylindrical lens. Let's derive equations to describe propagation of the beam (9.103) and show that no isolated intensity nulls appear on propagation, that is, the beam (9.103) is not a vortex or a singular beam [203,277,313–322]. The Fresnel transform of the complex amplitude (9.103) reads as

$$E(\xi,\eta,z) = \frac{-ik}{z\sqrt{p(z)q(z)}} \exp\left[A(z)\xi^2 + B(z)\eta^2 + C(z)\xi\eta\right], \tag{9.107}$$

where:

$$\begin{aligned} A(z) &= \frac{ik}{2z} - \frac{k^2}{4z^2 p(z)} + \frac{k^4 \sin^2 2\alpha}{64 f^2 z^2 p^2(z) q(z)}, \\ B(z) &= \frac{ik}{2z} - \frac{k^2}{4z^2 q(z)}, \quad C(z) = \frac{ik^3 \sin 2\alpha}{8fz^2 p(z) q(z)}, \\ p(z) &= \frac{1}{w_x^2} + \frac{ik}{2z_x}, \quad q(z) = \frac{1}{w_y^2} + \frac{ik}{2z_y} + \frac{k^2 \sin^2 2\alpha}{16 f^2 p(z)}, \\ z_x &= \frac{zf}{z\cos^2\alpha - f}, \quad z_y = \frac{zf}{z\sin^2\alpha - f}. \end{aligned} \tag{9.108}$$

It is seen in Eq. (9.107) that the Gaussian beam (9.103) preserves its Gaussian shape on propagation but changes its scale and rotates. Eq. (9.107) is simplified significantly for $\alpha = \pi/4$ and $z = 2f$, since $z_x \to \infty$ and $z_y \to \infty$ for these values:

$$E(\xi,\eta,z=2f)=-2i\gamma^{-1}\exp\left[\frac{ik}{4f}\left(\xi^2+\eta^2\right)-\frac{\xi^2}{w_y^2\gamma^2}-\frac{\eta^2}{w_x^2\gamma^2}+\frac{ik\xi\eta}{2f\gamma^2}\right], \quad (9.109)$$

where:

$$\gamma=\left(1+\frac{16f^2}{k^2w_x^2w_y^2}\right)^{1/2}. \quad (9.110)$$

It is seen in Eq. (9.109) that at the distance $z = 2f$ the elliptic Gaussian beam (9.103) is rotated by 90° and widened since $\gamma > 1$.

Vortex Beam after Two Crossed Cylindrical Lenses

It is shown that using of two crossed cylindrical lenses with the focal lengths, equal in modulus but opposite in sign, increases the OAM two times.

Now instead of the field (9.103) we consider a more general light field, when an elliptical Gaussian beam passes through two crossed cylindrical lenses with different focal lengths f_x, f_y, whose axes are inclined at an angle α to the horizontal axis. The complex amplitude of such a field in the initial plane ($z = 0$) reads as

$$E(x,y)=\exp\left(-\frac{x^2}{w_x^2}-\frac{y^2}{w_y^2}-\frac{ikx'^2}{2f_x}-\frac{iky'^2}{2f_y}\right), \quad (9.111)$$

where:

$$\begin{cases} x' = x\cos\alpha + y\sin\alpha, \\ y' = y\cos\alpha - x\sin\alpha, \end{cases} \quad (9.112)$$

Then, substitution of the field (9.111) into Eqs. (9.104) and (9.105) gives the following normalized OAM:

$$\frac{J_z}{W}=\frac{k}{8}\left(\frac{1}{f_y}-\frac{1}{f_x}\right)\left(w_x^2-w_y^2\right)\sin 2\alpha. \quad (9.113)$$

Comparison of Eq. (9.113) with Eq. (9.106) shows that if the focal lengths of the cylindrical lenses have different signs (one lens is converging and the other is diverging) then the OAM can be increased two times compared to using only one cylindrical lens.

Elliptical Gaussian Beam after ABCD System

In this paragraph expressions are obtained that describe propagation of the field (9.103) through an ABCD system. A general formula is also obtained for the OAM density of the astigmatic Gaussian beam at any distance from the waist plane.

Now we consider propagation of the field (9.111) through an ABCD-system. Then the complex amplitude in the output plane reads as:

$$E_2(\xi,\eta)=\frac{-ik}{B\sqrt{G}}\exp\left[\frac{ikD}{2B}\left(\xi^2+\eta^2\right)-\frac{k^2}{B^2}\frac{P_{xx}\eta^2+P_{yy}\xi^2-P_{xy}\xi\eta}{G}\right], \tag{9.114}$$

where:

$$\begin{aligned}
P_{xx}&=\frac{1}{w_x^2}+\frac{ik}{2f_x}\cos^2\alpha+\frac{ik}{2f_y}\sin^2\alpha-\frac{ikA}{2B},\\
P_{yy}&=\frac{1}{w_y^2}+\frac{ik}{2f_x}\sin^2\alpha+\frac{ik}{2f_y}\cos^2\alpha-\frac{ikA}{2B},\\
P_{xy}&=\frac{ik}{2}\left(\frac{1}{f_x}-\frac{1}{f_y}\right)\sin 2\alpha,\quad G=4P_{xx}P_{yy}-P_{xy}^2.
\end{aligned} \tag{9.115}$$

The intensity distribution of the field (9.111) in the output plane of the ABCD-system is as follows:

$$I_2(\xi,\eta)=\frac{k^2}{B^2|G|}\exp\left[-\frac{2k^2}{B^2|G|^2}\Psi(\xi,\eta)\right], \tag{9.116}$$

where $\Psi(\xi,\eta)=\mathrm{Re}\left\{P_{yy}G^*\right\}\xi^2+\mathrm{Re}\left\{P_{xx}G^*\right\}\eta^2-\mathrm{Re}\left\{P_{xy}G^*\right\}\xi\eta$.

It is seen in Eq. (9.116) that the intensity distribution looks like an elliptic Gaussian beam, as in the initial plane (9.111). If there is only one cylindrical lens with the focal length f_x and $f_y\to\infty$, $\alpha=\pi/4$, $A=1$, $B=z$, then, using Eq. (9.114), an equation can be obtained for the OAM density in the polar coordinates (r,φ) at an arbitrary propagation distance z:

$$\frac{j_z}{|E_2|^2}=kr^2\frac{2z_{0x}z_{0y}\left[f_xz^2+z_{0x}z_{0y}\left(z-f_x\right)\right]\cos 2\varphi+f_x\left(z_{0y}^2-z_{0x}^2\right)z\left(z-2f_x\right)\sin 2\varphi}{4\left[f_xz^2+z_{0x}z_{0y}\left(z-f_x\right)\right]^2+\left(z_{0x}+z_{0y}\right)^2z^2\left(z-2f_x\right)^2}, \tag{9.117}$$

where $z_{0x}=kw_x^2/2$, $z_{0y}=kw_y^2/2$ are the Rayleigh distances. In particular, for the double focal length the following simple expression for the OAM density can be derived from Eq. (9.117):

$$\frac{j_z}{|E_2|^2}=\frac{k}{2f_x}\frac{z_{0x}z_{0y}}{4f_x^2+z_{0x}z_{0y}}r^2\cos 2\varphi. \tag{9.118}$$

From Eq. (9.117) it follows that the normalized OAM density depends on the distance to the lens z. The numerator is proportional to z^2, while the denominator is proportional to z^4. This means that with increasing distance the OAM density (9.117) decreases parabolically, similarly to the decreasing intensity of the Gaussian beam due to its divergence.

Cylindrical Lens is Placed in Arbitrary Plane

This paragraph contains an expression for the full OAM of an astigmatic Gaussian beam (9.103) if the cylindrical lens is placed in arbitrary transverse plane (not only in the waist plane).

Now we consider an elliptic Gaussian beam (9.103) not in the waist plane, but in any other plane at a distance z from the waist. The complex amplitude at a distance z from the waist of the beam (9.103) reads as

$$E(x,y,z)=\left[q_x(z)q_y(z)\right]^{-1/2}\exp\left(-\frac{x^2}{w_x^2 q_x(z)}-\frac{y^2}{w_y^2 q_y(z)}\right), \tag{9.119}$$

where:

$$q_x(z)=1+\frac{iz}{z_{0x}},\quad q_y(z)=1+\frac{iz}{z_{0y}}, \tag{9.120}$$

and the Rayleigh distances are the same as in Eq. (9.117).

If in the field (9.119) a cylindrical lens is placed with a focal length f, whose generatrix is rotated in the transverse plane by an angle α, then the complex amplitude of light immediately after the cylindrical lens is as follows

$$\begin{aligned} E(x,y,z)=&\left[q_x(z)q_y(z)\right]^{-1/2}\exp\left(-\frac{x^2}{w_x^2 q_x(z)}-\frac{y^2}{w_y^2 q_y(z)}\right)\\ &\times\exp\left(-\frac{ikx^2\cos^2\alpha}{2f}-\frac{iky^2\sin^2\alpha}{2f}-\frac{ikxy\sin 2\alpha}{2f}\right). \end{aligned} \tag{9.121}$$

The normalized OAM of the beam (9.121) reads as

$$\frac{J_z}{W}=\left(\frac{k\sin 2\alpha}{8f}\right)\left(w_y^2|q_y|^2-w_x^2|q_x|^2\right). \tag{9.122}$$

It is seen in Eq. (9.122) that the sign and the magnitude of the OAM depend on the distance z from the waist of the elliptical Gaussian beam to the plane of the cylindrical lens. This is more clearly seen if we write Eq. (9.122) in a different form:

$$\frac{J_z}{W}=\left(\frac{k\sin 2\alpha}{8f}\right)\left(w_y^2-w_x^2\right)\left(1-\frac{4z^2}{k^2w_x^2w_y^2}\right). \tag{9.123}$$

It is seen in Eq. (9.123) that if the cylindrical lens is placed at the distance $z = kw_xw_y/2$, then the OAM equals zero, since at this distance the elliptical Gaussian beam has a circular transverse section. If $z < kw_xw_y/2$ then the OAM is positive and if $z > kw_xw_y/2$ then the OAM is negative. At large values of z the modulus of the OAM increases with z parabolically. We note, however, that for the previously used parameters of the Gaussian beam ($w_x = 2$ mm, $w_y = 1$ mm, $\lambda = 0.5$ μm) the distance along which the OAM drops from its maximum value to zero is $z = 4\pi$ meters. Therefore, to achieve the maximal OAM the cylindrical lens should be placed into the waist of the elliptic Gaussian beam.

Generation of an Elliptic Gaussian Beam

In Ref. [312], an elliptic Gaussian beam has been generated by using two cylindrical lenses. However, one cylindrical lens is sufficient for generation of converging or diverging elliptic Gaussian beam. Now we consider this in detail. Let a cylindrical lens with the curvature along the x-axis and with focal length f_1 be placed into the waist of a conventional circular Gaussian beam with the waist radius w. Then the complex amplitude of the elliptic Gaussian beam at a distance z behind the cylindrical lens reads as

$$E(x,y,z)=\frac{1}{\sqrt{q_0(z)q_1(z)}}\exp\left(-\frac{x^2}{w^2q_2(z)}-\frac{y^2}{w^2q_0(z)}\right), \tag{9.124}$$

where:

$$q_0(z)=1+\frac{iz}{z_0},\quad q_1(z)=q_0(z)-\frac{z}{f_1},$$
$$q_2(z)=q_1(z)\left(1+\frac{iz_0}{f_1}\right)^{-1},\quad z_0=\frac{kw^2}{2}. \tag{9.125}$$

If a cylindrical lens with the focal length f is placed into the light field (9.124) and rotated by an angle α, then the complex amplitude immediately behind the lens is

$$E(x,y,z)=\left(q_0(z)q_1(z)\right)^{-1/2}\exp\left(-\frac{x^2}{w^2q_2(z)}-\frac{y^2}{w^2q_0(z)}\right)$$
$$\times\exp\left(-\frac{ikx^2\cos^2\alpha}{2f}-\frac{iky^2\sin^2\alpha}{2f}-\frac{ikxy\sin 2\alpha}{2f}\right). \tag{9.126}$$

The normalized OAM of the beam (9.126) reads as

$$\frac{J_z}{W}=\left(\frac{kw^2\sin 2\alpha}{8f}\right)\left[\frac{|q_0|^2}{\mathrm{Re}(q_0)}-\frac{|q_2|^2}{\mathrm{Re}(q_2)}\right]=\left(\frac{kw^2\sin 2\alpha}{8f}\right)\left(\frac{z}{f_1}\right)\left(2-\frac{z}{f_1}\right). \tag{9.127}$$

It is seen in Eq. (9.127) that the OAM tends to zero at $f_1 \to \infty$, since the beam (9.126) tends to the conventional Gaussian beam. It is also seen in Eq. (9.127) that increasing of the distance z between the first cylindrical lens with the focal length f_1 and the second cylindrical lens with the focal length f allows unlimited increasing of the OAM of a laser beam. At $z = 0$ and $z = 2f_1$ the OAM (9.106) is also equal to zero since the Gaussian beam has a circular shape. The OAM (9.127) has maximal positive value at $z = f_1$, that is, when the second cylindrical lens is placed in the focus of the first one. The instead of Eq. (9.127) we get ($z = f_1$)

$$\frac{J_z}{W} = \left(\frac{kw^2 \sin 2\alpha}{8f} \right). \tag{9.128}$$

For $w_y = w$ and $w_x = 0$, Eq. (9.128) coincides with the expression (9.106) for the OAM. At $z > 2f_1$ the OAM (9.127) changes its sign (becomes negative) and is increasing (in modulus) with increasing distance z.

Orbital Angular Momentum of an Astigmatic Beam

In this section another expression is obtained for the normalized OAM of the beam (9.103). It is based on the angular harmonics expansion of the amplitude of the field (9.103). We write the amplitude (9.103) in the polar coordinates (r, φ)

$$E(r,\varphi) = \exp\left(-ar^2 \cos^2 \varphi - br^2 \sin^2 \varphi - icr^2 \sin 2\varphi\right), \tag{9.129}$$

as a series of the angular harmonics:

$$E(r,\varphi) = \sum_{n=-\infty}^{\infty} C_n(r) \exp(in\varphi), \tag{9.130}$$

where:

$$C_n(r) = (2\pi)^{-1} \int_0^{2\pi} E(r,\varphi) \exp(-in\varphi) \, d\varphi, \tag{9.131}$$

and

$$\begin{aligned} a &= w_x^{-2} + ik(2f)^{-1} \cos^2 \alpha, \\ b &= w_y^{-2} + ik(2f)^{-1} \sin^2 \alpha, \\ c &= k(4f)^{-1} \sin 2\alpha. \end{aligned} \tag{9.132}$$

The integral (9.131) can be written as

$$C_n(r) = (2\pi)^{-1} e^{-D} \int_0^{2\pi} \exp\left(-in\varphi - A\cos 2\varphi - iB \sin 2\varphi\right) d\varphi, \tag{9.133}$$

where:

$A = [(a - b)/2]r^2$
$B = cr^2$
$D = [(a + b)/2]r^2$

After replacing $t = 2\varphi$ and for even numbers $n = 2m$, instead of Eq. (9.133) we get:

$$\begin{aligned} C_n(r) &= (2\pi)^{-1} e^{-D} \int_0^{2\pi} \exp(-imt - A\cos t - iB\sin t)\,dt \\ &= (-i)^m e^{-D+im\theta} J_m(F) = e^{-D}\left(\frac{F}{A-B}\right)^m J_m(F), \end{aligned} \tag{9.134}$$

where:

$F = (B^2 - A^2)^{1/2}$
$\text{tg}\,\theta = iB/A$

For the odd numbers of the angular harmonics $n = 2m + 1$ the coefficients (9.134) are equal to zero:

$$\begin{aligned} 2\pi C_{2m+1}(r) &= e^{-D} \int_0^{2\pi} \exp\left[-i(2m+1)\varphi - A\cos 2\varphi - iB\sin 2\varphi\right] d\varphi \\ &= e^{-D} \left\{ \int_0^{\pi} \exp\left[-i(2m+1)\varphi - A\cos 2\varphi - iB\sin 2\varphi\right] d\varphi \right. \\ &\quad \left. + \int_0^{\pi} \exp\left[-i(2m+1)(\varphi+\pi) - A\cos 2(\varphi+\pi) - iB\sin 2(\varphi+\pi)\right] d\varphi \right\} \\ &= e^{-D} \int_0^{\pi} \left[1 + (-1)^{2m+1}\right] \exp\left[-i(2m+1)\varphi - A\cos 2\varphi - iB\sin 2\varphi\right] d\varphi = 0. \end{aligned} \tag{9.135}$$

Then, for the coefficients (9.133) we get:

$$C_n(r) = \begin{cases} e^{-D}\left(\dfrac{F}{A-B}\right)^{n/2} J_{n/2}(F), \; n = 2m, \\ 0, \; n = 2m+1. \end{cases} \tag{9.136}$$

It is seen in Eq. (9.136) that the coefficients at the angular harmonics with the positive and negative numbers are different in modulus ($n = 2m$):

$$\begin{aligned} \left|C_{2m}(r)\right|^2 &= e^{-2\,\mathrm{Re}\,D} \left|\frac{A+B}{A-B}\right|^m \left|J_m(F)\right|^2, \\ \left|C_{-2m}(r)\right|^2 &= e^{-2\,\mathrm{Re}\,D} \left|\frac{A-B}{A+B}\right|^m \left|J_m(F)\right|^2. \end{aligned} \tag{9.137}$$

The normalized OAM for the expansion (9.108) can be represented as

$$\frac{J_z}{W}=\frac{\sum_{n=-\infty}^{\infty} n\bar{C}_n}{\sum_{n=-\infty}^{\infty} \bar{C}_n}, \tag{9.138}$$

where $\bar{C}_n=\int_0^\infty |C_n(r)|^2\, rdr$.

Further, we suppose $\alpha=\pi/4$ to make the difference $a-b$ real and to remove the modulus sign of the squared Bessel function. Coefficients in the sum (9.134) for the even numbers read as

$$\bar{C}_{2m}=\left|\frac{a-b+2c}{a-b-2c}\right|^m \int_0^\infty \exp\left[-\mathrm{Re}(a+b)r^2\right] J_m^2\left(\frac{r^2}{2}\sqrt{4c^2-(a-b)^2}\right) rdr. \tag{9.139}$$

In the general case, when $\alpha\neq\pi/4$, instead of Eq. (9.139) a more tedious expression is obtained:

$$\begin{aligned}\bar{C}_{2m}&=\int_0^\infty e^{-2\,\mathrm{Re}\,D}\left|\frac{A+B}{A-B}\right|^m \left|J_m(F)\right|^2 rdr\\ &=\left|\frac{a-b+2c}{a-b-2c}\right|^m \int_0^\infty e^{-\mathrm{Re}(a+b)r^2} J_m\left(Gr^2\right) J_m\left(G^* r^2\right) rdr\\ &=\frac{1}{2}\left|\frac{a-b+2c}{a-b-2c}\right|^m \int_0^\infty e^{-\mathrm{Re}(a+b)u} J_m(Gu) J_m\left(G^* u\right) du.\end{aligned} \tag{9.140}$$

where $G=[4c^2-(a-b)^2]^{1/2}/2$.

The integral in Eq. (9.139) can be evaluated by using a reference integral [38]

$$\int_0^\infty e^{-pr^2} J_m^2\left(qr^2\right) rdr=(2\pi q)^{-1} Q_{|m|-1/2}\left(1+\frac{p^2}{2q^2}\right). \tag{9.141}$$

To evaluate the integral in Eq. (9.140), another reference integral can be used [38]:

$$\int_0^\infty e^{-px} J_\nu(bx) J_\nu(cx)\, dx=\frac{1}{\pi\sqrt{bc}} Q_{\nu-1/2}\left(\frac{p^2+b^2+c^2}{2bc}\right), \tag{9.142}$$

where $Q_\nu(x)$ is the spherical Legendre function or the Legendre function of the second kind ($x>1$):

$$Q_\nu(x)=\sqrt{\pi}\,(2x)^{-\nu-1}\,\Gamma(\nu+1)\,\Gamma^{-1}(\nu+3/2)\,{}_2F_1\left(\frac{\nu+1}{2},\frac{\nu+2}{2},\nu+\frac{3}{2},x^{-2}\right), \tag{9.143}$$

where:

$\Gamma(x)$ is the Gamma function

${}_2F_1(a,b,c,x)$ is the hypergeometric function [101]

Then, Eq. (9.139) reads as

$$\bar{C}_{2m} = \pi^{-1}\left(\frac{c+a-b}{c-a+b}\right)^m \left[c^2-(a-b)^2\right]^{-1/2} Q_{|m|-1/2}\left[1+\frac{2(a+b)^2}{c^2-(a-b)^2}\right], \quad (9.144)$$

or in the more general case of the tilt angle $\alpha \neq \pi/4$ instead of Eq. (9.144) we get:

$$\bar{C}_{2m} = \frac{1}{2\pi|G|}\left|\frac{a-b+2c}{a-b-2c}\right|^m Q_{|m|-1/2}\left(\frac{\mathrm{Re}^2(a+b)+2\mathrm{Re}G^2}{2|G|^2}\right). \quad (9.145)$$

Let $|G|^2 \approx 0$. Then the argument of the Legendre function in Eq. (9.145) tends to infinity and from the asymptotic of the Legendre function at $x >> 1$ [101]

$$Q_{|\nu|}(x) \approx \sqrt{\pi}(2x)^{-|\nu|-1}\Gamma(|\nu|+1)\Gamma^{-1}(|\nu|+3/2) \quad (9.146)$$

an expression follows for the expansion coefficients in the formula for the OAM (9.138):

$$\bar{C}_{2m} = \frac{1}{2^{2|m|+1}\sqrt{\pi}}\frac{\Gamma(|m|+1/2)}{\Gamma(|m|+1)}\frac{|a-b+2c|^{|m|+m}|a-b-2c|^{|m|-m}}{\mathrm{Re}^{2|m|+1}(a+b)}. \quad (9.147)$$

We note that the argument of the Legendre function in Eq. (9.144) is complex. Further, for the certainty, let the cylindrical lens to be rotated again by the angle α equal to 45°. When the Gaussian beam is circular ($w_y = w_x$), then $a - b = 0$ in Eq. (9.144) and the coefficients (9.144) at the angular harmonics are equal for the numbers m and $-m$, that is, the OAM (9.138) equals zero. If $2c = b - a$, that is, $f^{-1} = z_{0y}^{-1} - z_{0x}^{-1}$, all the coefficients (9.144) are equal to zero at $m > 0$, while at $m < 0$ they are not equal to zero:

$$\bar{C}_{2m} = \begin{cases} 0, m > 0, \\ \dfrac{f}{k\sqrt{\pi}}\dfrac{\Gamma(|m|+1/2)}{\Gamma(|m|+1)}\left(\dfrac{w_y^2+w_x^2}{w_x^2-w_y^2}\right)^{-2|m|-1}, m \leq 0. \end{cases} \quad (9.148)$$

The OAM (9.138) in this case is non-zero. In the opposite case, when $2c = a - b$ or $f^{-1} = z_{0x}^{-1} - z_{0y}^{-1}$, all the coefficients at $m < 0$ are equal to zero, while at $m > 0$ they are non-zero:

$$\bar{C}_{2m} = \begin{cases} \frac{f}{k\sqrt{\pi}} \frac{\Gamma(m+1/2)}{\Gamma(m+1)} \left(\frac{w_x^2 + w_y^2}{w_y^2 - w_x^2} \right)^{-2m-1}, m \geq 0, \\ 0, m < 0. \end{cases} \quad (9.149)$$

In this case, the OAM (9.138) is also non-zero. It is seen in Eq. (9.149) that with increasing number the coefficients $|\bar{C}_n|$ of the series (9.138) decrease to zero. For large values of m, instead of Eq. (9.149) it can be written ($m > 0$) that:

$$\bar{C}_{2m} \approx \frac{f}{k\sqrt{\pi}} \left(\frac{w_y^2 - w_x^2}{w_y^2 + w_x^2} \right)^{2m+1}. \quad (9.150).$$

Let for the certainty $a - b > 0$ or $w_y > w_x$, and let $c > a - b$, that is, $f^{-1} > z_{0y}^{-1} - z_{0x}^{-1}$. Then, the coefficients (9.144) with the positive numbers $m > 0$ will exceed the coefficients with the negative numbers: $\bar{C}_{2m} > \bar{C}_{-2m}$. This means that the angular harmonics $\exp(i2m\varphi)$ with the positive numbers give the larger contribution into the full OAM, which is therefore positive. In the opposite case, when $a - b < 0$ or $w_y < w_x$ and $f^{-1} < z_{0y}^{-1} - z_{0x}^{-1}$, the OAM is negative since $\bar{C}_{2m} < \bar{C}_{-2m}$. These conclusions are in concordance with the expression (9.106).

So, we have shown that an elliptical Gaussian beam with an astigmatism (9.103) has the OAM (9.138), which consists of contributions of only even angular harmonics with both positive ($2m$) and negative ($-2m$) topological charge, although these contributions are not equal ($\bar{C}_{2m} \neq \bar{C}_{-2m}$). For $m = 0$, the coefficient $\bar{C}_0$ is non-zero and equals

$$\bar{C}_0 = \frac{1}{\pi\sqrt{c^2 - (a-b)^2}} Q_{-1/2}\left[1 + \frac{2(a+b)^2}{c^2 - (a-b)^2}\right]. \quad (9.151)$$

The value of the Legendre function in Eq. (9.151) can be obtained via the complete elliptic integral [101]:

$$Q_{-1/2}(x) = \sqrt{\frac{2}{1+x}} K\left(\sqrt{\frac{2}{1+x}}\right),$$
$$K(t) = \int_0^1 \left[(1-x^2)(1-tx^2)\right]^{-1/2} dx. \quad (9.152)$$

Numerical Simulation

Using the Fresnel transform, intensity and phase distributions of the field (9.103) were computed for several propagation distances from the cylindrical lens. The following parameters were used: wavelength $\lambda = 532$ nm, Gaussian beam waist radii $w_x = 20\lambda$ and $w_y = 10\lambda$, cylindrical lens focal length $f = 100\lambda$, inclination angle of the

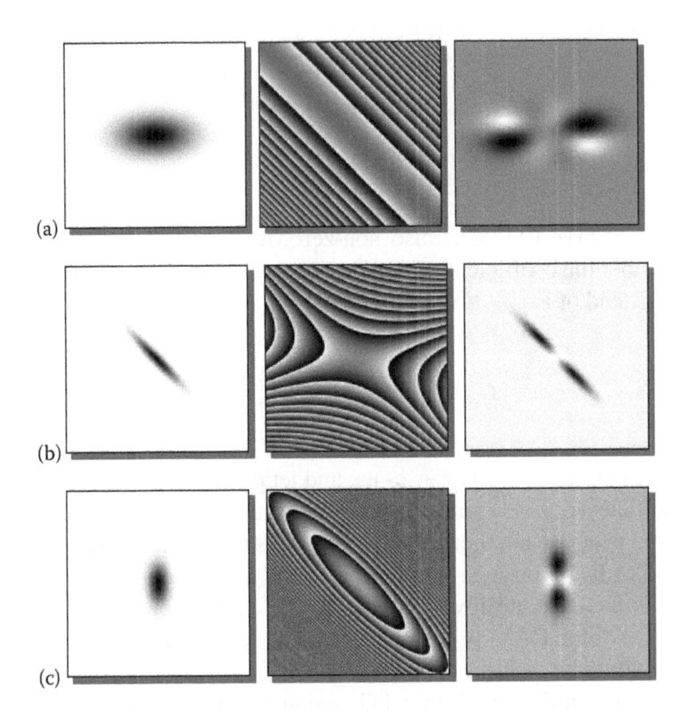

FIGURE 9.21 Distributions of intensity (left column), phase (middle column) and OAM density (right column) of the field (9.103) at different distances from the initial plane: (a) $z = \lambda$ ($R = 40\lambda$); (b) $z = f$ ($R = 40\lambda$); (c) $z = 2f$ ($R = 80\lambda$).

lens from the Cartesian coordinates $\alpha = \pi/4$, and computation area $-R \le x, y \le R$. Normalized OAM density of the field (9.103) was computed by using the expression: $j_z = kI(x, y)(2f)^{-1}\left(y^2 - x^2\right)$.

Figure 9.21 shows distributions of intensity, phase, and OAM density of the elliptical Gaussian beam (9.103) for the different distances after the cylindrical lens. It is seen in Figure 9.21 that the elliptic Gaussian beam rotates on propagation after passing the cylindrical lens. The OAM density rotates with the beam synchronously, while the total OAM is certainly preserved. It is also seen that at the double focal distance the Gaussian beam is turned by 90° (Figure 9.21c, left column) with respect to its initial position (Figure 9.21a), as predicted by Eq. (9.109).

Now we verify numerically the previously obtained expressions. We calculate the OAM by using the initial expression (9.106) and by using the angular harmonics expansion (9.138). The calculation parameters are the following: the wavelength λ = 532 nm, tilt angle of the cylindrical lens $\alpha = \pi/4$, waist radii of the elliptic Gaussian beam $w_x = 20\lambda$ and $w_y = 400\lambda$, focal length of the cylindrical lens $f = 1/(1/z_{0x} - 1/z_{0y}) \approx 1260\lambda$. Figure 9.22 shows the distribution of the absolute values of the coefficients $\left|\bar{C}_n\right|$ (continuous curve). For the other waist radii $w_x = 10\lambda$, $w_y = 100\lambda$ and for the other focal length of the cylindrical lens $f = 1/(1/z_{0x} - 1/z_{0y}) \approx 317\lambda$, the distribution of the absolute values of the coefficients $\left|\bar{C}_n\right|$ in Eq. (9.120) is shown also in Figure 9.22 (dashed line).

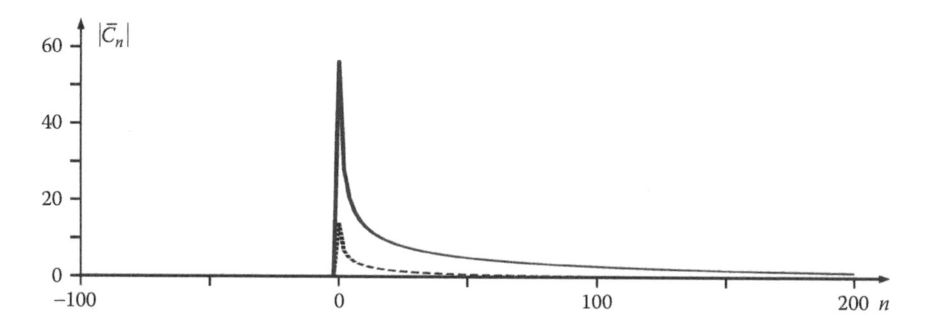

FIGURE 9.22 Distribution of the coefficients $|\bar{C}_n|$ (only even numbers, since for the odd numbers they are equal to zero) for two different beams and focal lengths of the cylindrical lens: $w_x = 20\lambda$, $w_y = 400\lambda$, $f = 1260\lambda$ (continuous curve); $w_x = 10\lambda$, $w_y = 100\lambda$, $f = 317\lambda$ (dashed curve).

For the first case (continuous curve in Figure 9.22), the OAM, calculated by Eq. (9.106), is equal to 99.500625, while the OAM calculated by Eq. (9.138) equals 99.48454. For the second case (dashed curve in Figure 9.22), the OAM calculated by Eq. (9.106) equals 24.5025, while the OAM calculated by Eq. (9.138) is equal to 24.5025. In the second case there is a coincidence up to four decimal points. In Eq. (9.138), 2000 coefficients $|\bar{C}_n|$ were used for the calculation. It is seen in Figure 9.22 that the less is the OAM, the less is the value of the non-zero angular harmonics expansion coefficients $|\bar{C}_n|$ (the dashed curve is below the continuous curve).

In this section we obtained the following results. For an astigmatic Gaussian beam generated by focusing of an elliptical Gaussian beam with a cylindrical lens, an expression is obtained for the normalized OAM in the form of a series of Legendre functions of the second kind with half-integer orders. If the parameters of the elliptic Gaussian beams and of the cylindrical lens are such that $f^{-1} = z_{0y}^{-1} - z_{0x}^{-1}$ then only the angular harmonics with the negative numbers ($2m < 0$) make a contribution to the OAM. Conversely, if $f^{-1} = z_{0x}^{-1} - z_{0y}^{-1}$ then only the angular harmonics with the positive numbers ($2m > 0$) contribute to the OAM. In all cases, the largest coefficient in the angular harmonics expansion of the astigmatic field is the zero coefficient. However, since this coefficient is multiplied by zero (zero topological charge), it does not contribute to the OAM. Because of the large contribution of the zero angular harmonic to the amplitude of the astigmatic field, there are no isolated intensity nulls, and therefore there are no points with phase singularities. Such an astigmatic beam looks like a vortex-free beam, although it has the non-zero OAM.

10 Perfect Vortices

10.1 AN OPTIMAL PHASE ELEMENT FOR GENERATING A PERFECT OPTICAL VORTEX

For the first time, a perfect optical vortex (POV) that does not change its radius with changing topological charge was proposed in [323]. Theoretically, such a perfect vortex can be described by a series expanded in terms of Bessel functions of the same order and different scale. In practice, the series is truncated, leading to errors in generating a "perfect" optical vortex. In [323], a perfect vortex was generated experimentally using a phase optical element containing a finite set of concentric rings. The finite-width thickness of the rings was approximated by a delta function, while the phase of light within the rings varied linearly with the azimuthal angle and in direct proportion to the topological charge. However, with the amplitude in each ring assumed to be different, the ring widths proportional to the amplitude are also assumed to be different. As a result, one has to reconcile contradictory requirements—on the one hand, the ring width is supposed to be minimal to approximate the delta-function, and on the other hand, the width of each ring is supposed to be proportional to a pregiven amplitude. In [323], such contradictions led to low-quality results both in simulation (Figure 10.3 in [1]) and in experiments (Figure 10.6 in [323]). In [324], a different approach was taken to generating a POV with the aid of a conical axicon and a spiral phase plate. However, the relationship that the authors claimed to describe the amplitude of the resulting POV (Eq. (10.1) in [324]) offered a very coarse approximation. In Ref. [325] the POV is used to generate pairs of entangled photons in spontaneous parametric down-conversion (SPDC). In [325] a POV is generated by an axicon, as in [324]. In [325], studying the axicon-aided generation of POVs, the authors showed that it was possible to vary the POV radius by simply moving the axicon, thus eliminating the need to utilize lenses with different focal length. In Ref. [326], an inverse approach is considered for the generation of POVs. The initial narrow ring is imaged by a 4*f*-system. The disadvantage of this approach is obvious: it does not allow generating a POV in the focus of a high-numerical-aperture objective for optical trapping and rotation of microparticles. In [327], a spatial light modulator was described, with the phase mask generated as a combination of an axicon and a spiral phase plate. A Bessel-Gaussian beam generated by the mask then travelled through a lens that performed a Fourier transform. However, such an approach also relied on an inaccurate formula to describe diffraction by a spiral axicon. In [328] rather than using an axicon, it was proposed that POVs should be generated using an amplitude-phase optical element whose profile approximated a Bessel mode of the nth order. Note that the amplitude-phase filter reported in [328] is most close to that we propose in this work.

In this section, we aim to derive accurate relationships to describe POVs generated using different techniques, demonstrating that, weak as it may be, there is a dependence between the POV doughnut intensity and the topological charge. We compare the following techniques for VOP generation: (1) using an amplitude-phase field described by a finite-radius Bessel mode; (2) using an optimal phase of the electromagnetic field, as described in [329]; and (3) using an electromagnetic field whose phase is formed with a conical axicon and a spiral phase plate [324,325]. All three approaches are different from that proposed in [323]. We aim to demonstrate that an optimal optical element discussed in [329] is best suited for generating a POV.

Generating the Perfect Optical Vortex Using an Amplitude-Phase Optical Element

A POV [323] is described by the complex amplitude given by:

$$E_0(\rho,\theta)=\delta(\rho-\rho_0)\exp(in\theta), \tag{10.1}$$

where:

$\delta(x)$ is the Dirac delta-function
(ρ, θ) are the polar coordinates in the Fourier plane of a spherical lens
n is the integer topological charge of the optical vortex

From (10.1), the radius of an infinitely thin ring ρ_0 is seen to be independent of the topological charge n. The POV in (10.1) can be generated using an ideal Bessel mode, which at the focal length of the spherical lens is given by

$$F_0(r,\varphi)=J_n(\alpha r)\exp(in\varphi), \tag{10.2}$$

with the dimension parameter α defining the scale of the Bessel function of the nth order and first kind, $J_n(x)$. This follows from the property of Bessel functions to be orthogonal on the entire real axis [330]:

$$\alpha\int_0^\infty J_m(\alpha r)J_m\left(\frac{k\rho r}{f}\right)rdr=\delta\left(\alpha-\frac{k\rho}{f}\right), \tag{10.3}$$

where:

k is the wavenumber of monochromatic, coherent light
f is the focal length of the lens.

From (10.3), the radius of a ring with the maximum intensity of the POV defined by (10.1) can be derived in the form:

$$\rho_0=\frac{\alpha f}{k}. \tag{10.4}$$

When generating field (10.2), there arise two problems. First, with the Bessel function being alternating, for it to be optically implemented, one needs to combine an amplitude mask whose transmission is proportional to $|J_n(\alpha r)|$ with a phase mask whose transmission is given by

$$\operatorname{sgn} J_n(\alpha r) = \begin{cases} 1, \ J_n(\alpha r) > 0, \\ -1, \ J_n(\alpha r) < 0. \end{cases} \tag{10.5}$$

Second, in practice, the light field in (10.2) needs to be bounded by a circular aperture of radius R. These are two reasons why the POV gets distorted.

In the following, we analyze a technique for generating a POV by means of function (10.2), which is bounded by a circular aperture of radius R:

$$F_1(r,\varphi) = \operatorname{circ}\left(\frac{r}{R}\right) J_n(\alpha r) \exp(in\varphi), \tag{10.6}$$

where:

$$\operatorname{circ}\left(\frac{r}{R}\right) = \begin{cases} 1, R \geq 0, \\ 0, R < 0. \end{cases}$$

For the light field in (10.6), we can derive its complex amplitude in the Fourier plane of an ideal spherical lens of focus f using a Fourier transform in cylindrical coordinates:

$$\begin{aligned} E_1(\rho,\theta) &= (-i)^{n+1}\left(\frac{k}{f}\right) e^{in\theta} \int_0^R J_n(\alpha r) J_n\left(\frac{kr\rho}{f}\right) r dr \\ &= (-i)^{n+1}\left(\frac{kR}{f}\right) e^{in\theta}\left[\frac{\alpha J_{n+1}(\alpha R) J_n(xR) - xJ_n(\alpha R) J_{n+1}(xR)}{\alpha^2 - x^2}\right], \end{aligned} \tag{10.7}$$

where $x = k\rho/f$. It is worth noting that although field (10.7) is essentially different from the desired POV in (10.1), the radius of the maximum intensity ring for this filed is the same as in (10.4), being independent of the topological charge. Actually, from (10.4) we can infer that the radius of the POV maximum intensity ring should take a value at which the denominator in (10.7) turns to zero. However, at $x = \alpha$ the numerator also becomes equal to zero. The 0/0 uncertainty in (10.7) can be resolved at $x = \alpha$ (in the bright ring of the POV maximum intensity) using the reference integral [38]:

$$\int_0^R J_n^2(\alpha r) r dr = \frac{R^2}{2}\left[J_n^2(\alpha R) - J_{n-1}(\alpha R) J_{n+1}(\alpha R)\right]. \tag{10.8}$$

Then, at $x = \alpha$ and in view of Eq. (10.8), Eq. (10.7) can be replaced with

$$E_1\left(\rho = \alpha f k^{-1}, \theta\right) = (-i)^{n+1}\left(\frac{kR^2}{2f}\right)e^{in\theta}\left[J_n^2(\alpha R) - J_{n-1}(\alpha R)J_{n+1}(\alpha R)\right]. \quad (10.9)$$

From (10.9), putting α and R constant, the maximum intensity on the POV ring is seen to depend on the topological charge n. Note, however, that at large αR, the intensity becomes nearly independent of n, because based on asymptotic properties of a Bessel function, we can write:

$$\begin{aligned} J_n^2(\alpha R) - J_{n-1}(\alpha R)J_{n+1}(\alpha R) &\approx \left(\frac{2}{\pi\alpha R}\right)\Big[\cos^2(\alpha R - n\pi/2 - \pi/4) \\ &\quad -\cos(\alpha R - (n+1)\pi/2 - \pi/4)\cos(\alpha R - (n-1)\pi/2 - \pi/4)\Big] \\ &= \left(\frac{2}{\pi\alpha R}\right)\left[\cos^2(\alpha R - n\pi/2 - \pi/4) + \sin^2(\alpha R - n\pi/2 - \pi/4)\right] = \left(\frac{2}{\pi\alpha R}\right). \end{aligned} \quad (10.10)$$

From (10.10) it follows that at large αR, the maximum intensity on the POV ring asymptotically tends to a value independent of the topological charge:

$$I_1\left(\rho = \frac{\alpha f}{k}\right) = |E_1|^2 = \left(\frac{kR}{\pi\alpha f}\right)^2. \quad (10.11)$$

To be able to derive accurate characteristics of a POV, the scale of the Bessel function in (10.5) can be chosen in a special way by putting $\alpha R = \gamma_{n,\nu}$, where $\gamma_{n,\nu}$ is the ν-th root of the Bessel function: $J_n(\gamma_{n,\nu}) = 0$. Then, instead of (10.7), we have:

$$E_1(\rho,\theta) = (-i)^{n+1}\left(\frac{kR^2}{f}\right)e^{in\theta}\left[\frac{\gamma_{n,\nu}J_{n+1}(\gamma_{n,\nu})J_n(xR)}{\gamma_{n,\nu}^2 - (xR)^2}\right]. \quad (10.12)$$

From (10.12), the amplitude function zeros (dark rings of the POV intensity) are seen to coincide with the zeros of a Bessel function of the nth order, with the rings' radii being equal to each other:

$$\rho_{n,\mu} = \frac{\gamma_{n,\mu} f}{kR}, \ \mu \neq \nu, \quad (10.13)$$

Putting $\mu = \nu$, we obtain a bright ring whose maximum intensity is given by

$$I_1\left(\rho = \frac{\gamma_{n,v} f}{kR}\right) = \left|E_1\right|^2 = I_0 J_{n+1}^4\left(\gamma_{n,v}\right), \tag{10.14}$$

where $I_0 = [kR^2/(2f)]^2$ is the intensity at the center of the Airy disc, which is defined as the bright central spot of the Fraunhofer diffraction pattern from a unit-amplitude plane wave diffracted by a circular aperture of radius R. Equation (10.14) can be shown to follow from (10.9) at $\alpha R = \gamma_{n,v}$. Actually, in view of (10.9) and putting $\alpha R = \gamma_{n,v}$, Eq. (10.12) changes to

$$\begin{aligned} E_1\left(\rho = \frac{\gamma_{n,v} f}{kR}\right) &= \left(-i\right)^{n+1} e^{in\theta}\left(\frac{kR^2}{2f}\right)\left[-J_{n-1}\left(\gamma_{n,v}\right)J_{n+1}\left(\gamma_{n,v}\right)\right] \\ &= \left(-i\right)^{n+1} e^{in\theta}\left(\frac{kR^2}{2f}\right)J_{n+1}^2\left(\gamma_{n,v}\right). \end{aligned} \tag{10.15}$$

From (10.15) we can directly obtain (10.14). From (10.14), the maximum intensity on the POV ring is seen to depend on n and $\gamma_{n,v}$. Rather than being chosen arbitrarily, the root values are supposed to satisfy the following conditions for the Bessel function scale α and the aperture radius R:

$$\alpha R = \gamma_{n,v} = \gamma_{m,\mu}. \tag{10.16}$$

Hence, for different topological charges n and m the roots of the Bessel functions need to be chosen so as to ensure the fulfilment of condition (10.16). In this case, the ring of maximum intensity retains its radius with varying topological charge. Note that the Bessel function has no coincident roots, with all roots being mutually intermittent. Nonetheless, two close roots can always be found. From (10.12), the width of the maximum intensity ring can be derived. For the Bessel function of the nth order, the width equals the distance between two roots with the numbers $v - 1$ and $v + 1$:

$$\Delta\rho_0 = \frac{\left(\gamma_{n,v+1} - \gamma_{n,v-1}\right)f}{kR} \approx \frac{2\pi f}{kR}. \tag{10.17}$$

Note that at the full-width half maximum (FWHM) of intensity, the width of the POV ring is twice as small: FWHM $= \pi f/(kR)$.

Let us evaluate the efficiency of generating a POV based on function (10.6). If a unit-intensity plane wave is incident on a circular aperture of radius R, the incident power on the optical element with transmission (10.6) is proportional to the circle area: $W_0 = \pi R^2$. After passing the element with transmission function (10.6), the power of light is defined by the right-hand side of (10.8) multiplied by 2π. Considering Eq. (10.10) for large αR, we have $W_1 = 2R/\alpha$. Hence, the efficiency of generating the

POV by means of (10.6) equals $\eta = W_1/W_0 = 2/(\pi\alpha R)$. This formula suggests that with increasing aperture radius the efficiency is decreasing.

Generating the Perfect Optical Vortex Using an Optimal Phase Optical Element

We refer to an optical element as optimal if it redistributes the largest proportion of the incident energy to a bright ring of designed radius. Such an element has been described [7], with its transmission being given by

$$F_2(r,\varphi) = \mathrm{circ}\left(\frac{r}{R}\right)\mathrm{sgn}\, J_n(\alpha r)\exp(in\varphi). \tag{10.18}$$

The sign and aperture functions in (10.18) were defined in (10.5) and (10.6). The amplitude of light generated by the optical element with transmission function (10.18) in the focus of a spherical lens has been derived for $n = 0$ [329]. In the following, we derive the relationship for the amplitude at any integer n. We proceed from the assumption that a circle of radius R can be broken down into N rings with radii r_m, with the sign of function (10.5) alternating within the rings:

$$r_m = \frac{\gamma_{n,m}}{\alpha},\ m = 1,2,\ldots,N,\quad r_N = R. \tag{10.19}$$

The complex amplitude in the lens focus from the optical element described by transmission function (10.18) can be expressed as a sum of contributions from each ring above ($r_0 = 0$):

$$E_2(\rho,\theta) = (-i)^{n+1}\left(\frac{k}{f}\right)e^{in\theta}\sum_{m=0}^{N-1}(-1)^m\int_{r_m}^{r_{m+1}} J_n\left(\frac{k\rho r}{f}\right) r\,dr. \tag{10.20}$$

Putting $\rho = \alpha f/k$ in (10.20), the argument of the Bessel function in (10.20) becomes independent of the physical parameters f and k, and equal to αr. We may put, without loss of generality, that the scale factor α is such that the Bessel function takes a zero value at the edge of the aperture, that is, $\alpha R = \gamma_{n,N}$. In this case, the integrals in (10.20) and the factor $(-1)^m$ take positive values for even m, with both becoming negative for odd m. Hence, at $\rho_0 = \gamma_{n,N} f/(kR)$ all constituent terms in the sum are positive, producing the maximum contribution to the field on the ring of the POV radius. Thus, we can infer that the phase element in (10.18) is actually optimal in terms of generating the maximum intensity on the ring of the given radius $\rho_0 = \gamma_{n,N} f/(kR)$. For the radius of the POV ring to be independent of the topological charge, close roots of the Bessel function need to be chosen, in a way similar to (10.16): $\gamma_{n,N} = \gamma_{m,M}$. Based on the earlier remarks concerning Eq. (10.20), we can infer that the intensity on the POV ring is given by (at $\alpha R = \gamma_{n,N}$):

$$I_2\left(\rho=\frac{\alpha f}{k}\right)=\left|E_2\right|^2=\left(\frac{k}{f}\right)^2\left(\int_0^R\left|J_n(\alpha r)\right|rdr\right)^2. \tag{10.21}$$

Next, we can derive an explicit relation for the intensity in the focus of a spherical lens from the initial field in (10.18). In each ring confined by the radii (r_m, r_{m+1}) of Eq. (10.19), the radial transmission of function (10.18) is constant, alternating in sign from ring to ring. Because of this, the Fraunhofer diffraction of a plane wave of field (10.18) in each ring can be described by a complex amplitude [57]:

$$\begin{aligned}E_m(\rho,\theta)&=(-i)^{n+1}\left(\frac{k}{f}\right)e^{in\theta}\int_{r_m}^{r_{m+1}}J_n\left(\frac{k\rho r}{f}\right)rdr=(-i)^{n+1}\left(\frac{1}{n!(n+2)}\right)e^{in\theta}\\&\times\left[\left(\frac{kr_{m+1}^2}{f}\right)x_{m+1}^n\,{}_1F_2\left(\frac{n+2}{2},\frac{n+4}{2},n+1,-x_{m+1}^2\right)\right.\\&\left.-\left(\frac{kr_m^2}{f}\right)x_m^n\,{}_1F_2\left(\frac{n+2}{2},\frac{n+4}{2},n+1,-x_m^2\right)\right],\end{aligned} \tag{10.22}$$

where:

$m = 1, 2,\ldots, N-1$

$x_m = k\rho r_m/(2f) = k\rho\gamma_{n,m}/(2\alpha f)$

${}_1F_2(a, b, c, x)$ is a hypergeometric (HyG) function [38]

The total light field in the spherical lens focus from the initial field in (10.18) is obtained by summing up contributions (10.22) from all annular apertures (10.19) in a similar way to (10.20):

$$\begin{aligned}E_2(\rho,\theta)&=-(-i)^{n+1}\left[\frac{k}{n!(n+2)\alpha^2 f}\right]e^{in\theta}\\&\times\left[2\sum_{m=1}^{N-1}(-1)^m\gamma_{n,m}^2x_m^n\,{}_1F_2\left(\frac{n+2}{2},\frac{n+4}{2},n+1,-x_m^2\right)\right.\\&\left.-(-1)^N\gamma_{n,N}^2x_N^n\,{}_1F_2\left(\frac{n+2}{2},\frac{n+4}{2},n+1,-x_N^2\right)\right],\end{aligned} \tag{10.23}$$

where $x_N = k\rho R/(2f)$. After substituting the POV radius $r_0 = \gamma_{n,N}f/(kR)$ into (10.23), the term in the second square brackets becomes independent of the physical parameters (α, k, f), because all arguments are proportional to the Bessel function roots: $x_m = k\rho r_m/(2f) = k\rho\gamma_{n,m}/(2\alpha f) = \gamma_{n,m}/2$. Therefore, from (10.23) the relationship between the maximum intensity on the POV ring and the said parameters is derived in the form:

$$I_2\left(\rho=\frac{\alpha f}{k}\right)=\left|E_2\right|^2\sim\left(\frac{k}{\alpha^2 f}\right)^2. \tag{10.24}$$

Considering that $\alpha R = \gamma_{n,N}$, Eqs. (10.24) and (10.11) are similar. It is worth noting that for each term in (10.23), values of the HyG function arguments on the POV ring of maximum intensity depend on the roots of a Bessel function of the nth order ($x_m = \gamma_{n,m}/2$). Hence, by changing the POV topological charge we change roots and arguments of the Bessel function. This leads to a change in the modulus of amplitude in (10.23) and a consequent change in the intensity on the POV ring. The magnitude of such a change can only be evaluated via modeling. With Eq. (10.18) describing a phase optical element, the change of the topological charge n does not cause the change of the aperture radius R or the POV radius. Hence, the change of intensity (10.21) of field (10.23) on the ring $r_0 = \gamma_{n,N}f/(kR)$ implies that the energy is redistributed between the POV ring and side-lobes. Note that we can directly infer from (10.21) that the intensity on the POV ring depends on n, because the integral takes different values for the different-order Bessel functions.

Generating a Perfect Optical Vortex Using a Conical Axicon

Generation of the POV using a combination of a conical axicon and a spiral phase plate was reported in [324]. Note that the study in [324] described only the experimental part, omitting the theoretical background. In this section we are going to fill the gap. Let, instead of Eqs. (10.6) and (10.18), an optical element be described by a complex transmission function in the form:

$$F_3(r,\varphi)=\operatorname{circ}\left(\frac{r}{R}\right)\exp(i\alpha r+in\varphi). \tag{10.25}$$

The optical element in (10.25) was proposed for the first time in [41] as an element intended to generate light pipes. Rather than defining the scale parameter of the Bessel function, in the present context α is interpreted as an axicon parameter: $\alpha = k \sin \psi$. Note, however, that the axicon $\exp(i\,\alpha\, r)$ approximately generates a zero-order Bessel function whose scale factor is just equal to α. This can be inferred from the following well-known relation [330]:

$$\exp(i\alpha r)=\sum_{m=-\infty}^{\infty} i^m J_m(\alpha r). \tag{10.26}$$

In [324,325] the light field in (10.25) was assumed to generate a POV in the focus of a spherical lens, with the POV amplitude distribution given by

$$E_3(\rho,\theta) \sim \exp\left[-\frac{(\rho-\rho_0)^2}{\Delta\rho^2}\right]\exp(in\theta). \tag{10.27}$$

However, such an assumption is a harsh simplification because the complex amplitude of the POV in the spherical lens focus is defined by a Fourier transform of function (10.25), taking a much more complex form than that in (10.27). Let us prove this. The Fraunhofer diffraction of light field (10.25) is described by the following relation [80]:

$$E_3(\rho,\theta) = (-i)^{n+1}\left(\frac{k}{f}\right)e^{in\theta}\int_0^R \exp(i\alpha r)J_n\left(\frac{k\rho r}{f}\right)rdr. \tag{10.28}$$

Equation (10.28) suggests that if the radius of the maximum intensity ring is defined by Eq. (10.4), the scale factors of the integrand functions in (10.28) become equal to each other. Hence,

$$\begin{aligned} E_3\left(\rho=\frac{\alpha f}{k},\theta\right) &= (-i)^{n+1}\left(\frac{k}{f}\right)e^{in\theta}\int_0^R \exp(i\alpha r)J_n(\alpha r)rdr \\ &= (-i)^{n+1}\left(\frac{k}{f}\right)e^{in\theta}\sum_{m=-\infty}^{\infty} i^m\int_0^R J_m(\alpha r)J_n(\alpha r)rdr. \end{aligned} \tag{10.29}$$

The second equality in (10.29) has been derived with regard for (10.26). Equation (10.29) suggests that with the scales of the integrand function being maximally matched, the POV generated in the lens focus using the optical element in (10.25) has the maximum intensity on radius (10.4). The intensity in (10.29) is much higher than that on the POV ring in (10.11) generated with an amplitude-phase element in (10.6), because the latter is formed by the contribution from a single term of the series in (10.29) at $m = n$.

On the other hand, the intensity on the POV ring (10.29) is still lower than that on the POV ring (10.21), which is generated by means of the optimal phase element in (10.18), because:

$$\begin{aligned} I_3\left(\rho=\frac{\alpha f}{k}\right) &= |E_3|^2 = \left(\frac{k}{f}\right)^2\left|\int_0^R \exp(i\alpha r)J_n(\alpha r)rdr\right|^2 \\ &< \left(\frac{k}{f}\right)^2\left|\int_0^R |J_n(\alpha r)|rdr\right|^2 = I_2\left(\rho=\frac{\alpha f}{k}\right). \end{aligned} \tag{10.30}$$

From (10.30), the intensity I_3 is seen to depend on the order n of the Bessel function at constant α, f, R. Note that while Eqs. (10.18) and (10.25) both define phase optical elements with the same radius R, the corresponding intensities on the same-radius rings (10.4) are still different, as evident from (10.30). This means that the POV ring in (10.29) is wider than the POV ring in (10.20). Physically, this can be explained as follows. The intensity ring in (10.20) is generated by the optical element in (10.18) containing a binary phase annular grating whose main +1 and −1 diffraction orders both contribute to each point of the POV ring (10.20). Thus, the entire annular aperture of radius R contributes to the intensity ring and, similarly to (10.7), the ring's width is approximately equal to (10.17) or the diffraction limit FWHM = $\pi f/(kR)$. In the meantime, the POV ring (10.29) is generated by the axicon (10.25), with just half the aperture, ranging from 0 to R, contributing to each point on the ring. Because of this, ring (10.29) is approximately twice as wide as rings (10.20) and (10.7). This is confirmed by the numerical simulation.

To express field amplitude (10.28) in an explicit form, we can expand the integrand exponential function into a Taylor series and obtain:

$$E_3(\rho,\theta) = (-i)^{n+1}\left(\frac{k}{f}\right)e^{in\theta}\sum_{m=0}^{\infty}\frac{(i\alpha)^m}{m!}\int_0^R r^m J_n\left(\frac{k\rho r}{f}\right)rdr. \tag{10.31}$$

The integrals under the sum sign in (10.31) can be found in a reference book [38], so that (10.31) takes the form [80]:

$$\begin{aligned}E_3(\rho,\theta) = {} & \frac{(-i)^{n+1}}{n!}\left(\frac{kR^2}{f}\right)e^{in\theta}x^n\sum_{m=0}^{\infty}\frac{(i\alpha R)^m}{(m+n+2)m!}\\ & \times {}_1F_2\left(\frac{m+n+2}{2},\frac{m+n+4}{2},n+1,-x^2\right),\end{aligned} \tag{10.32}$$

where $x = k\rho R/(2f)$, ${}_1F_2(a, b, c, x)$ is a HyG [38]. Note that whereas the POV from the optimal element (10.18) is described by a finite sum of HyG (10.23), the POV from the axicon (10.25) is described by series (10.32) composed of the HyG functions, having nothing in common with relation (10.27) from [324]. The maximum intensity radius of the POV ring is defined by Eq. (10.4). As before, for the radius of the POV ring to remain unchanged with changing topological charge n, closely set roots need to be chosen: $\alpha R = \gamma_{n,N} \approx \gamma_{m,M}$. On the maximum intensity ring, the argument x in (10.32) equals $x = \alpha R/2 = \gamma_{n,N}/2$.

Therefore, both the argument x and the entire relation in (10.32) are independent of the physical parameters k, f, R. Hence, in a similar way to (10.24), the relationship between the maximum intensity on the POV (10.30) and physical parameters can be given in the form:

$$I_3\left(\rho = \frac{\alpha f}{k}\right) = |E_3|^2 \sim \left(\frac{kR^2}{f}\right)^2. \tag{10.33}$$

Relationship (10.33) holds for all three optical elements under study in Eqs. (10.6), (10.18), and (10.25).

Thus, in this section we showed that the generation of a POV with the aid of a bounded-aperture spiral axicon has the following features: (1) the on-ring intensity is lower than that from optimal element (10.18) but higher than that from amplitude element (10.6); (2) the on-ring intensity depends on the topological charge: Eq. (10.30); (3) the ring is approximately twice as wide as in the previous techniques studied; and (4) the radius of the maximum intensity ring is nearly independent of the topological charge, since $\gamma_{n,N} \approx \gamma_{m,M}$.

Simulation Results

In this section we discuss simulation results when generating the POV using three optical elements that were analyzed in the previous sections. These results are in agreement with the theoretical predictions discussed in Sections "Generating the Perfect Optical Vortex using an Amplitude-Phase Optical Element," "Generating the POV using an Optimal Phase Optical Element," and "Generating a POV using a Conical Axicon." The simulation parameters are: wavelength of light $\lambda = 532$ nm, circular aperture radius $R = 20\lambda$, focal length of an ideal spherical lens $f = 100\lambda$. Note that Bessel function's scale factor α was chosen so as to satisfy the equality $\alpha R = \gamma_{1,20} = 63.6114$, where $\gamma_{1,20}$ is the 20th root of a Bessel function ($\nu = 20$) of the first order ($n = 1$). The POV was simulated for two topological charges: $n = 1$ and $n = 14$. The rest parameters were kept unchanged. Note that for the Bessel function of order $n = 14$ we chose the 14th root ($\nu = 14$), because $\gamma_{14,14} \approx \gamma_{1,20} = 63.6114$. Figure 10.1 depicts absolute values of two Bessel functions, $|J_1(\gamma_{1,20}x/R)|$ and $|J_{14}(\gamma_{14,14}x/R)|$, within radius R. From Figure 10.1 both functions are seen to take a zero value at $x = R$.

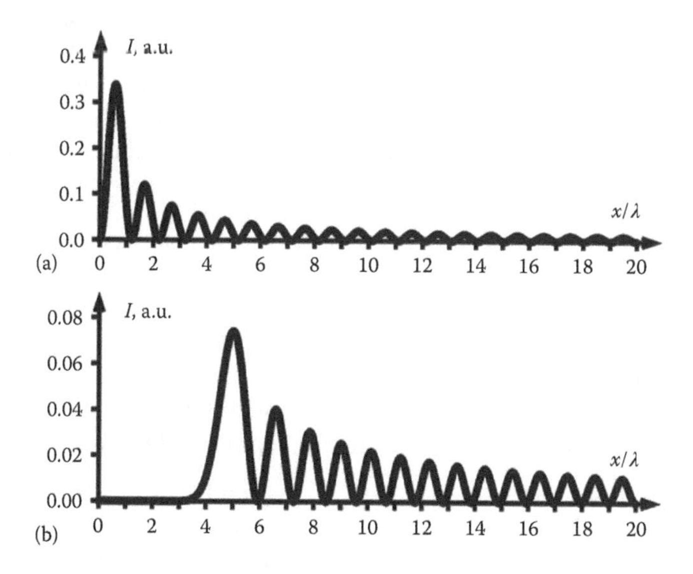

FIGURE 10.1 Absolute values of the Bessel functions: (a) $|J_1(\gamma_{1,20}x/R)|$ and (b) $|J_{14}(\gamma_{14,14}x/R)|$ within radius R.

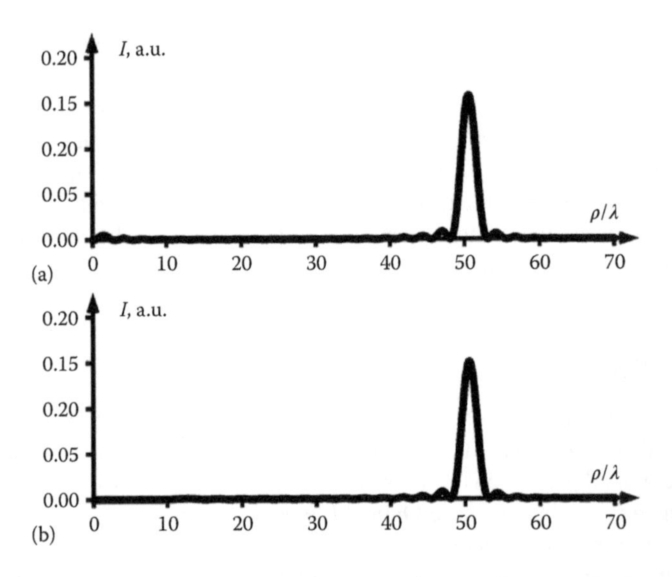

FIGURE 10.2 POV intensity profiles at (a) $n = 1$ and (b) $n = 14$ for initial optical filed (10.6). On the y-axis are arbitrary units and on the x-axis is radius in wavelengths.

Figure 10.2 depicts intensity profiles of the POV in the Fourier plane of a spherical lens obtained on the assumption of an initial light field with complex amplitude (10.6) for topological charges $n = 1$ and $n = 14$. The POV characteristics calculated for the initial field (10.6) are given in Table 10.1. From Table 10.1, the POV radius is seen not to change with changing topological charge n. For the above-described simulation parameters, the POV radius derived from (10.4) is $\rho_0 = \alpha f/k \approx 50.62\lambda$. This value of the radius is just 3% different from that in Table 10.1. With the topological charge having increased by an order of magnitude, there is only a 5% decrease in the maximum intensity of the POV. Note that according to (10.11) the maximum intensity in Figure 10.2 at the chosen simulation parameter is supposed to equal $I_1(\rho_0) = [kR/(\pi\alpha f)]^2 \approx 0.015816$. This value agrees well with the intensity value in Table 10.1 (with a 0.1% difference). Considering that both the ring radius and the optical element's aperture radius R have remained unchanged, the ring

TABLE 10.1
Comparison of the POV Parameters with the Initial Optical Field in (10.6) at Different Topological Charge *n*

Topological Charge	$n = 1$	$n = 14$
Radius of maximum intensity ring, ρ_0, λ	50.781563	50.781563
Maximum intensity in relative units, I_{max}	0.0157968	0.0150522
Ring thickness at half-intensity, FWHM, λ	2.244489	2.244489

width is also supposed to remain unchanged. From Table 10.1 the ring width is actually seen to remain unchanged with changing topological charge of the optical vortex. According to (10.17), the ring width at the above-chosen simulation parameters is supposed to be equal to FWHM = $5/2\lambda$. This value is 11% different from that given in Table 10.1.

Let us analyze the generation of a POV using the optimal phase element in (10.18). Figure 10.3 depicts POV intensity profiles at (a) n = 1and (b) n = 14 from the initial optical field in (10.18). Table 10.2 gives calculated parameters of the POV in Figure 10.3. From Table 10.2, the ring radius is seen to insignificantly decrease as compared with that in Figure 10.2 (less than by 0.3%). With a 14-fold increase in the topological charge, the radius is seen to remain unchanged. The intensity on the ring is seen to be nearly 100 times higher than that of the POV in Figure 10.2. This intensity can be derived from (10.21). Note that with the

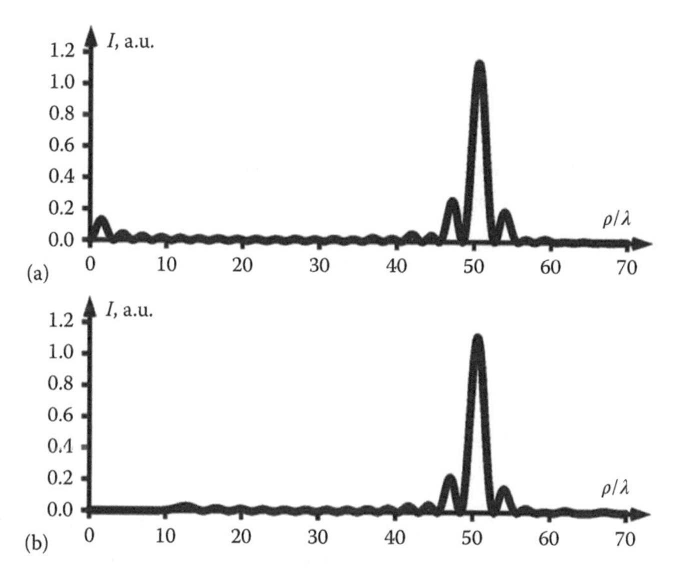

FIGURE 10.3 Intensity profiles of the POV at (a) $n = 1$ and (b) $n = 14$ from initial light field (10.18). On the y-axis are arbitrary units and on the x-axis is radius in terms of wavelengths.

TABLE 10.2
Comparison of Parameters of the POV from Optical Phase Element (10.18) at Different Topological Charges *n*

Topological Charge	$n = 1$	$n = 14$
Radius of maximum intensity ring, ρ_0, λ	50.641283	50.641283
Maximum intensity in relative units, I_{max}	1.140685	1.1181689
Ring thickness at half-intensity, FWHM, λ	1.9639279	1.9639279

topological charge having increased by a factor of 14, the intensity on the ring is seen to decrease just by 2%. Note that in this case, the ring is about 14% thinner than that in Figure 10.2. The ring width remains unchanged with changing topological charge of the optical vortex. From Figure 10.3 the side-lobes are seen to increase.

Let us now analyze the POV generated with spiral axicon (10.25). Figure 10.4 shows intensity profiles for the POV from initial field (10.25) at (a) $n = 1$ and (b) $n = 14$, with the calculated parameters of the POV given in Table 10.3.

From Figure 10.4 and Table 10.3, the width of the POV ring generated using spiral axicon (10.25) in the spherical lens focus is approximately 2.5 times that of the ring in Figure 10.2.

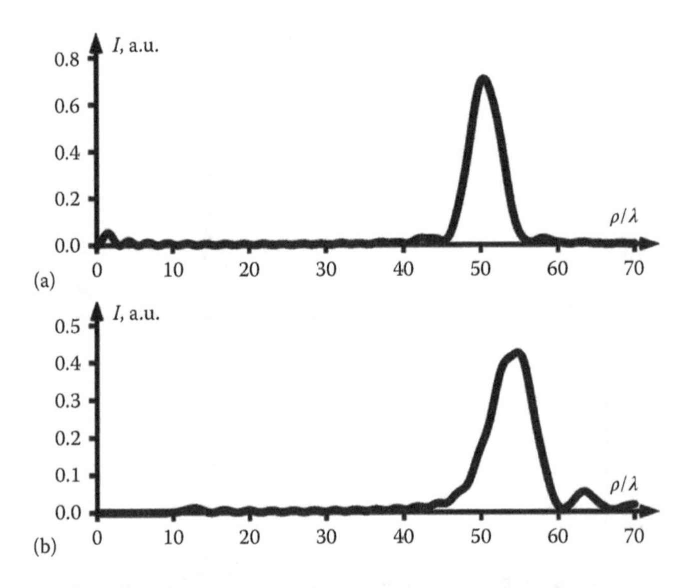

FIGURE 10.4 Intensity distribution for the POV from initial field (10.25) at (a) $n = 1$ and (b) $n = 14$. On the y-axis are arbitrary units and on the x-axis is radius in wavelengths.

TABLE 10.3
Comparison of Parameters of the POV Generated Using Spiral Axicon (10.25) at Different n

Topological Charge	$n = 1$	$n = 14$
Radius of maximum intensity ring, ρ_0, λ	50.501002	54.849699
Maximum intensity in relative units, I_{max}	0.7070332	0.4249419
Ring thickness at half-intensity, FWHM, λ	4.9098196	6.5931864

Also, note that with a 14-fold increase in the topological charge, the ring thickness has increased by a factor of 1.3.

An increase in the ring width (Figure 10.4) with increasing n is seen to result in a decrease in the intensity. Table 10.3 shows that with a 14-fold increase of n, there is a 1.7-fold decrease in the maximum intensity on the POV ring (Figure 10.4). Moreover, the radius of the maximum intensity ring increases by 8%.

Summing up, the simulation has shown that out of three optical elements for generating a POV, the optimal phase element in (10.18) offers the best results, providing the narrowest ring (FWHM = $1.96\lambda = 0.39\lambda f/R$) with the maximum intensity, which is 1.6 times higher than that from the spiral axicon in (10.25).

EXPERIMENT

The experimental study of the optical elements to generate POV was conducted using an optical arrangement in Figure 10.5. The fundamental Gaussian beam was generated by a solid-state laser L of wavelength $\lambda = 532$ nm. The laser beam was expanded and collimated using a system *PH* composed of a 40-μm pinhole and a lens L_1 ($f_1 = 250$ mm) before illuminating the display of a spatial light modulator *SLM* (PLUTO VIS, 1920 × 1080 resolution, with 8-mμ pixels). Using a diaphragm D_1, the central bright spot was isolated from surrounding dark and bright rings caused by diffraction by the pinhole. Using a set of lenses L_2 ($f_2 = 350$ mm) and L_3 ($f_3 = 150$ mm), and a diaphragm D_2, spatial filtering of the phase-modulated laser beam reflected at the SLM display was performed. Using a lens L_4 ($f_4 = 500$ mm), the laser beam was focused on the *CCD* array of a video-camera LOMO TC 1000 (of pixel size 1.67 × 1.67 μm). When generating the POV we utilized phase masks shown in Figure 10.6, which were output to the SLM display. The non-modulated beam reflected at the display and the phase-modulated beam were spatially separated using a linear phase mask superimposed on the initial phase mask.

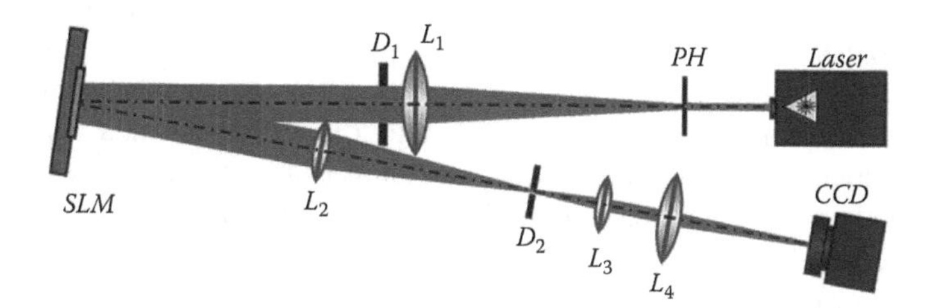

FIGURE 10.5 An experimental setup: L—a solid-state laser ($\lambda = 532$ nm); PH—a 40-μm pinhole; L_1, L_2, L_3, and L_4 are lenses with focal lengths $f_1 = 250$ mm, $f_2 = 350$ mm, $f_3 = 150$ mm, and $f_4 = 500$ mm; D_1 and D_2—diaphragms; *SLM*—a spatial light modulator PLUTO VIS; and *CCD*—a video-camera LOMO TC-1000.

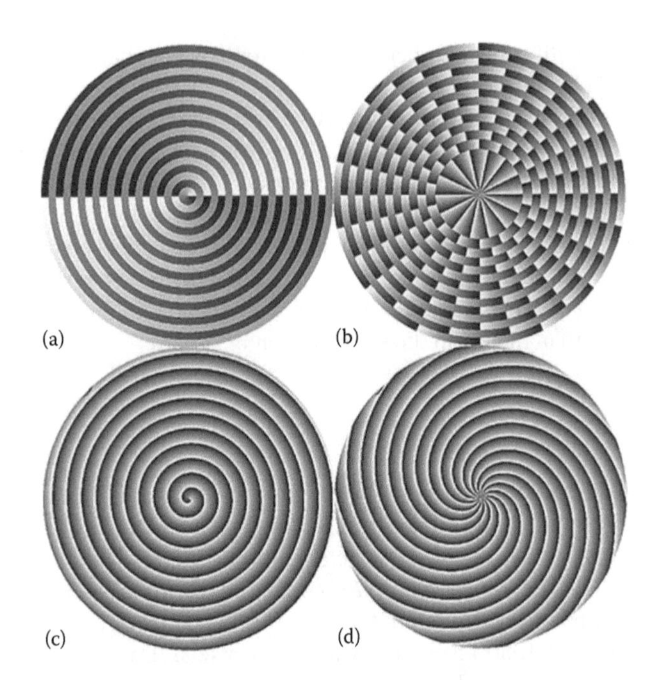

FIGURE 10.6 Phase masks for optical elements to generate a POV with topological charge (a, c) $n = 1$ and (b, d) $n = 14$. Figures (a, b) depict optimal phase elements and Figures (c, d) are for spiral axicons.

Shown in Figure 10.7 are intensity distributions in the focus of lens L_4 generated using phase masks corresponding to the optimal phase elements with $n = 1$ and $n = 14$. Values of the parameters of the resulting POV are given in Table 10.4. Figure 10.8 depicts intensity distributions in the focus of lens L_4 generated using phase masks corresponding to the spiral axicons with $n = 1$ and $n = 14$. Values of the parameters of the resulting POV are given in Table 10.5.

Thus, the analysis of the experimentally measured parameters of the POV suggests that their relative values are in good agreement with simulation results.

Figure 10.9 depicts interferograms resulting from interference of the POVs under analysis with a Gaussian beam with spherical wavefront. The interferograms are seen to have a spiral structure, thus corroborating that the generated POVs carry the orbital annular momentum. Note that the number of spirals correlates with the topological charge of the corresponding POV. By analogy with the results reported in [325], the spiral structure is not observed at the center of the interference pattern due to the large dark core of the perfect vortices.

We have derived explicit analytical relationships to describe the complex amplitude of a light field for perfect optical vortices generated using three different optical elements: an amplitude-phase element with transmission proportional to a Bessel mode, an optimal phase element, and a spiral axicon. It has been shown that all three optical elements are capable of generating same-radius diffraction rings, with the radius weakly depending on the topological charge of the optical vortex. The optimal

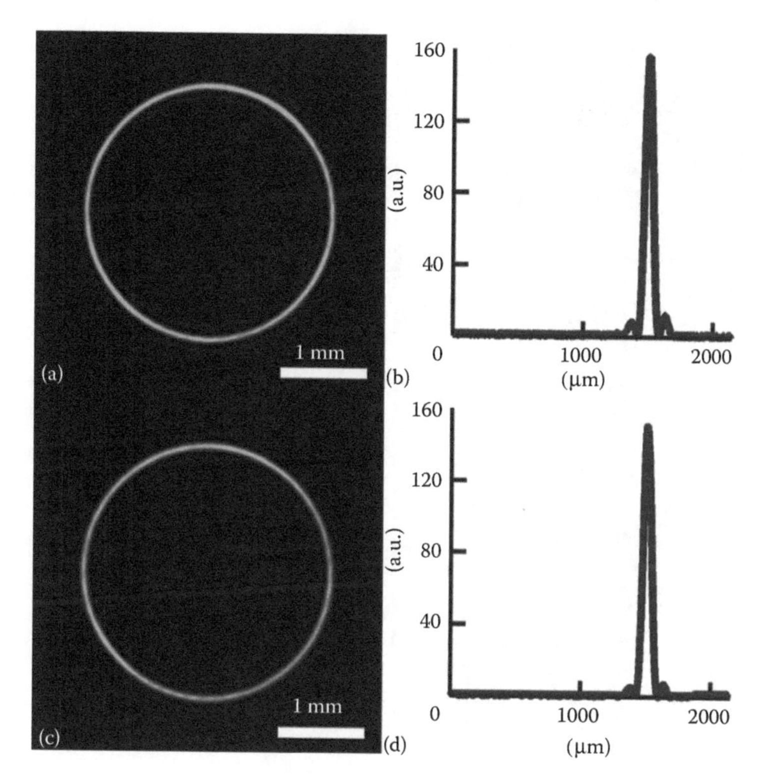

FIGURE 10.7 Intensity pattern of a POV (left column) and respective profiles depicted from the center to the edge (right column) obtained using an optimal phase element with (a, b) $n = 1$ and (c, d) $n = 14$.

TABLE 10.4
Comparison of POV Parameters Obtained Using an Optimal Phase Element with $n = 1$ and $n = 14$

Topological Charge	$n = 1$	$n = 14$
Radius of the maximum intensity ring, μm	1491.0 ± 2.0	1496.5 ± 2.0
Maximum intensity, in relative units	156.0 ± 0.5	151.0 ± 0.5
Ring thickness at half-intensity, μm	70.0 ± 2.0	73.0 ± 2.0

phase element has been shown to provide the highest intensity on the ring, all other factors being the same. The intensity on the ring from the optimal phase element has been found to be weakly related with the topological charge (Table 10.4). At the same time, we have shown by simulation (Table 10.3) and experiment (Table 10.5) that using the spiral axicon the intensity drops nearly by a factor of two with increasing topological charge from $n = 1$ to $n = 14$. The vortex-axicon-aided intensity ring

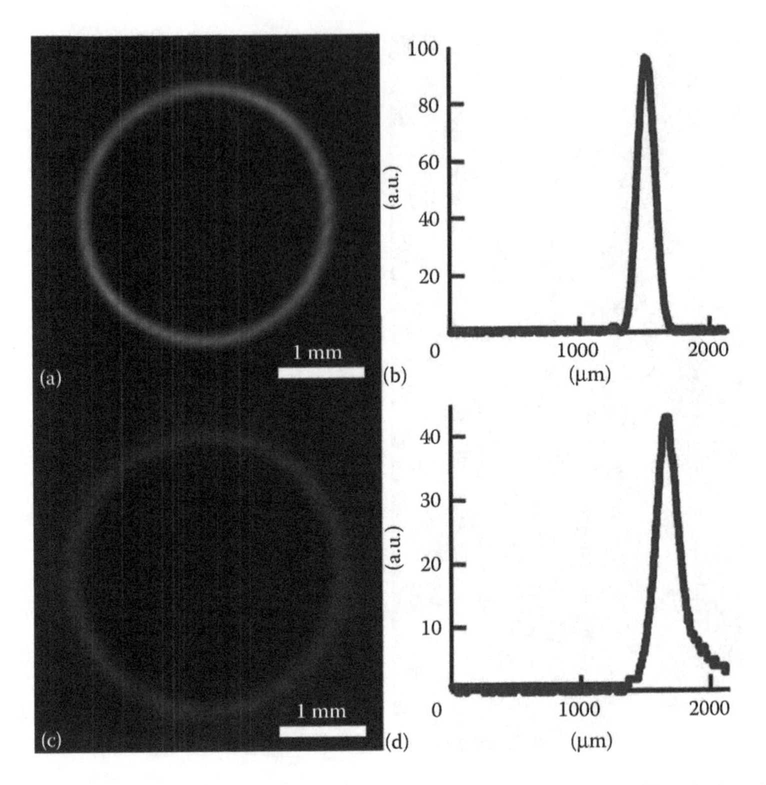

FIGURE 10.8 Intensity pattern of a POV (left column), respective profiles depicted from the center to the edge (right column) for a spiral axicon with (a, b) $n = 1$ and (c, d) $n = 14$.

TABLE 10.5
Comparison of POV Parameters Obtained Using a Spiral Axicon with $n = 1$ and $n = 14$

Topological Charge	$n = 1$	$n = 14$
Radius of the maximum intensity ring, μm	1498.0 ± 2.0	1655.0 ± 2.0
Maximum intensity, in relative units	96.0 ± 0.5	43.0 ± 0.5
Ring thickness at half-intensity, μm	158.0 ± 2.0	206.0 ± 2.0

is approximately twice as wide in comparison with the two other rings analyzed. Summing up, optimal element (10.18), which was first proposed for a different application in [329], is best suited for generating a POV. The simulation results corroborate theoretical predictions, with the experimental results being in agreement with theory and the simulation results.

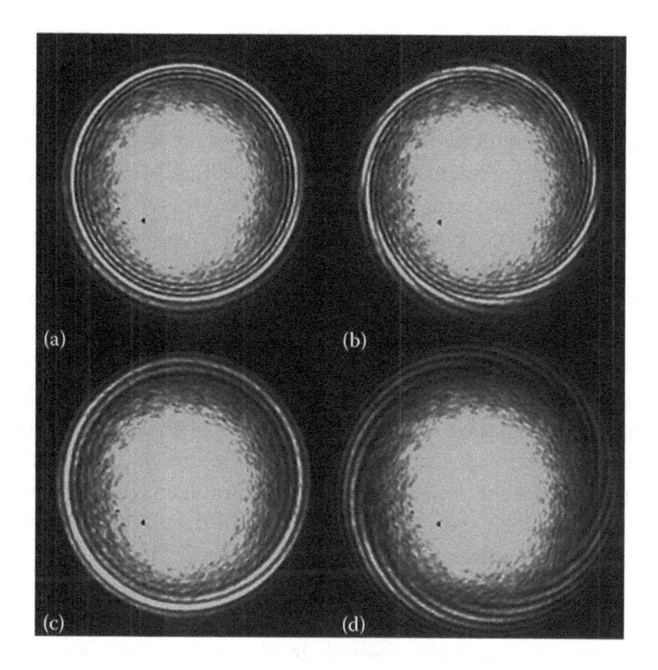

FIGURE 10.9 Interferograms resulting from interference of the POVs with a Gaussian beam with spherical wavefront when using an optimal phase element with the topological charge (a) $n = 1$ and (b) $n = 14$, and when using a spiral axicon with the topological charge (c) $n = 1$ and (d) $n = 14$.

10.2 ELLIPTIC PERFECT OPTICAL VORTICES

In 2013, the concept of the POV has been introduced [323]. A POV has a shape of an infinitely thin ring with a radius independent of the topological charge. Since then, perfect vortices were studied in many works. In [324], microparticle dynamics within a perfect vortex were analyzed. An improved technique for generating the POV based on the width-pulse approximation of Bessel functions was proposed in [328]. In [327], a POV was generated by Fourier transforming of the Bessel-Gauss beams. In [317], POVs with topological charge up to 90 were generated by a digital micromirror device. In [325], the POV was generated the same way as in [327] but adjusting the POV radius was done by varying the separation between the lens and the axicon. Using such vortices in SPDC, non-collinear interaction of SPDC photons was studied. In [331], the POV beams were scattered through a rough surface and random non-diffracting fields were generated, useful for authentication in cryptography. In [332], POVs were used in high-resolution plasmonic structured illumination microscopy. In [333], POV was used for single cells refractive index measurements. Paper [334] is devoted to in situ determining of the topological charge of a POV. The work [326] introduces perfect vectorial vortex beams, which have intensity profiles independent of the polarization order and the topological charge. In our paper [335], we considered three optical elements to generate POVs: (1) an amplitude-phase element with a transmission proportional to the Bessel function, (2) an optimal phase element with a transmission function equal to the signum function of the Bessel function, and (3) a spiral axicon. The doughnut intensity was shown to be highest when using an optimal phase element, while for the spiral axicon the light ring was twice wider.

So, in all the papers that we are aware of, there are studies of circular POVs. On the other hand, optical vortices are studied long ago and generalized elliptic vortices were proposed [11,60,212,276,279,336–339]. Obviously, an elliptic perfect optical vortex (EPOV) can be observed in the far field of the elliptical Bessel beam studied in [54,340,341]. For generation of a light ellipse, the initial light field in [54] has a transmittance function of the axicon and spiral phase plate but stretched along one dimension. In [340], Bessel functions with their arguments having elliptic locus were used to generate the mask, which was recorded using holographic technique. In [341], diffraction-free beams with elliptic Bessel envelope were studied in periodic media. The papers [22,23] were devoted mainly to the experimental generation of the elliptic Bessel beams (and consequently narrow light ellipse in the far field). Neither the OAM density nor the summary OAM of such vortices have been studied. Power efficiency of the proposed techniques and the thickness of the obtained light ellipses have not been considered as well.

Here we bring together these two research areas—perfect vortices and non-circular vortices. We generalize the concept of perfect optical vortices and consider a POV with an elliptic shape. We obtain analytical expression for its complex amplitude when it is generated from the elliptic Bessel beam, bounded by an elliptic aperture. An estimation of ellipse thicknesses is obtained. We obtain an exact analytical expression for the total OAM of the EPOV, as well as for the OAM density. The density is shown to be higher on the smaller side of the EPOV. Horizontal and vertical diameters of the EPOV are independent of the topological charge, similarly to the circular POV. We propose a phase optical element which generates in the far field a light ellipse with several times smaller thickness than an elliptical axicon with a spiral phase plate. Using such an element, EPOVs of several topological charges are generated experimentally using a spatial light modulator. EPOVs can find application for movement of microscopic particles along an ellipse with acceleration, as well as for generation of OAM-entangled photons.

Phase Optical Element

We consider the Bessel beam stretched along one dimension. Its complex amplitude is

$$E_{\text{in}}(x,y)=J_m\left[\alpha\sqrt{x^2/a^2+y^2/b^2}\right]\exp\left[im\arg\left(x/a+iy/b\right)\right], \tag{10.34}$$

where:

- (x, y) are the Cartesian coordinates in the transverse plane
- a and b are normalizing factors which define the ellipticity of the Bessel beam
- m is the integer topological charge of the elliptic vortex
- α is the scaling factor (proportional to the cone angle for the conventional Bessel beam) which defines the beam width

If such a Bessel beam is generated in the front focal plane of a lens with the focal distance f, then the complex amplitude in the rear focal plane is defined by the Fourier transform, which leads to the following expression:

$$E_{\text{out}}(r,\varphi)=(-i)^{m+1}\alpha^{-1}\exp(im\varphi)\delta\left(r-\alpha f/(kab)\right), \tag{10.35}$$

where:

$k = 2\pi/\lambda$ is the wavenumber of light with the wavelength λ

$r = [(u/b)^2 + (v/a)^2]^{1/2}$

$\varphi = \arg(u/b + iv/a)$

(u, v) are the Cartesian coordinates in the rear focal plane of the lens.

It is seen that the diffraction pattern looks like an infinitely thin light ellipse with horizontal and vertical diameters equal to $2\alpha f/(ka)$ and $2\alpha f/(kb)$, respectively.

Now we consider the same elliptic Bessel beam, but bounded by an elliptic aperture. Its complex amplitude is

$$\begin{aligned} E_{\text{in}}(x,y) &= J_m\left(\alpha\sqrt{x^2/a^2+y^2/b^2}\right) \\ &\times\exp\left[im\arctan\left((ay)/bx\right)\right]\operatorname{circ}\left(x^2/a^2+y^2/b^2\right), \end{aligned} \tag{10.36}$$

where $\operatorname{circ}(x) = 1$ at $x \le 1$ and $\operatorname{circ}(x) = 0$ at $x > 1$.

In the rear focal plane, Fourier transform of Eq. (10.36) gives the following complex amplitude:

$$\begin{aligned} E_{\text{out}}(r,\varphi) &= (-i)^{m+1}(k/f)abe^{im\varphi}\left(\alpha^2-\zeta^2\right)^{-1} \\ &\times\left[\alpha J_{m+1}(\alpha)J_m(\zeta)-\zeta J_m(\alpha)J_{m+1}(\zeta)\right], \end{aligned} \tag{10.37}$$

where $\zeta = kabr/f$. Since the intensity distribution depends only on ζ, the intensity on the light ellipse is constant. Putting $\alpha = \gamma_{m,\nu}$ ($\gamma_{m,\nu}$ is the νth root of the mth order Bessel function), we get $E_{\text{out}}(r,\varphi) \propto (\gamma_{m,\nu}^2 - \zeta^2)^{-1} J_m(\zeta)$. Therefore, the FWHM of the maximum intensity ellipse is approximately equal to $\Delta r_0 = (2kab)^{-1}(\gamma_{m,\nu+1} - \gamma_{m,\nu-1})f \approx \lambda f/(2ab)$. On the horizontal axis $\text{FWHM}_x = \lambda f/(2a)$, while on the vertical axis $\text{FWHM}_y = \lambda f/(2b)$.

For a paraxial beam with a complex amplitude $E(r', \varphi')$, the OAM density reads as $j_z = \operatorname{Im}\{E^* \partial E/\partial \varphi'\}$, where (r', φ') are the polar coordinates in the transverse plane. For the beam (10.34), we get the following expression for the OAM density (in the elliptic variables (r, φ)):

$$j_z(r,\varphi) = m\alpha^{-2}\delta^2\left(r-\frac{\alpha f}{kab}\right)\left(\frac{b}{a}\cos^2\varphi+\frac{a}{b}\sin^2\varphi\right). \tag{10.38}$$

It is seen that at $a \neq b$ maximal OAM density is concentrated on the smaller side of the light ellipse, while the minimal OAM density is $[\max\{a, b\}/\min\{a, b\}]^2$ times smaller and is located on the larger side.

The beam power and the OAM of the EPOV read as $W = (\lambda f/\alpha)\delta(0)$ and $J_z = m\lambda f/(2\alpha)(b/a + a/b)\delta(0)$ respectively. Thus, the normalized OAM is

$$J_z/W = (m/2)(b/a + a/b). \tag{10.39}$$

Since $b/a + a/b \geq 2$, the normalized OAM of the EPOV is always greater than the topological charge.

The EPOV can be generated approximately by the Fraunhofer diffraction of a beam with a following phase-only complex amplitude, like in [335]:

$$\begin{aligned} E_0(x, y) = &\operatorname{sgn}\left[J_m\left(\alpha\sqrt{x^2/a^2 + y^2/b^2}\right)\right] \\ &\times \exp\left[im\arctan\left((ay)/bx\right)\right]\operatorname{circ}\left(x^2/a^2 + y^2/b^2\right). \end{aligned} \tag{10.40}$$

The EPOV can also be generated by using an elliptic axicon. It was used in [54] to generate elliptic Bessel beams, which in the rear focal plane of the lens look like narrow light ellipses. After passing an elliptical axicon with an elliptical spiral phase plate (elliptical spiral axicon), the complex amplitude of the field reads as

$$\begin{aligned} E_0(x, y) = &\operatorname{circ}\left(x^2/a^2 + y^2/b^2\right) \\ &\times \exp\left[i\alpha\sqrt{x^2/a^2 + y^2/b^2} + im\arctan\left((ay)/bx\right)\right]. \end{aligned} \tag{10.41}$$

Numerical Simulation

Figure 10.10 shows intensity distribution of the EPOV obtained by using the elliptic phase-only Bessel beam (10.40) and by the spiral axicon (10.41) with the topological charge $m = 7$. Parameters of calculations: scaling factor $\alpha = 30$, ellipticity factors $a = 15\lambda$, $b = 5\lambda$, lens focal distance $f = 50\lambda$, calculation area $-60\lambda \leq x, y \leq 60\lambda$. Theoretical values of the horizontal and vertical radii are respectively $\alpha f/(ka) = 15.915\lambda$ and $\alpha f/(kb) = 47.746\lambda$. For the field (10.40), numerically obtained values are respectively 15.94λ and 47.80λ, while for the spiral axicon (10.41) these radii are 17.25λ and 51.75λ, that is, have larger difference with the theoretical values. The horizontal thickness ($FWHM_x$) is 1.35λ for the field (10.40) and 3.70λ for the field (10.41). The vertical thickness ($FWHM_y$) is respectively 4.05λ and 11.50λ. The increase of the ellipse thickness for the axicon leads to decrease of the maximal intensity. For the field (10.40) it is 340 a.u., while for the field (10.41) it is 148 a.u. Therefore, in the following experimental research we use the element (10.40).

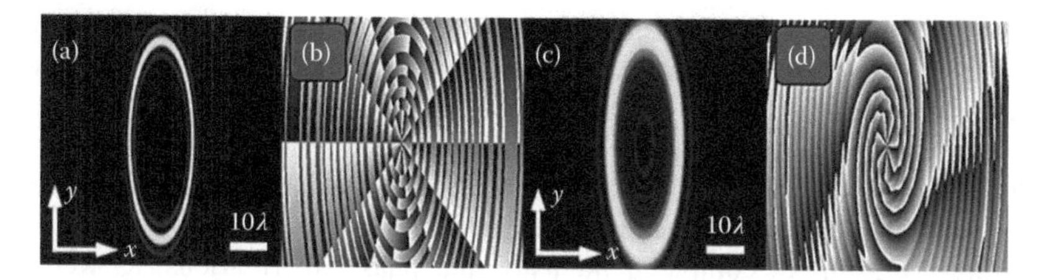

FIGURE 10.10 Distributions of the (a, c) intensity and (b, d) phase of the EPOV with the topological charge $m = 7$, obtained by (a, b) Fraunhofer diffraction of the elliptic phase-only Bessel beam (10.40) and (c, d) by using the elliptic spiral axicon (10.41).

Experiment

For the experimental study of elements that generate elliptic perfect optical vortices, we used an optical setup shown in Figure 10.11a. A fundamental Gaussian beam from a solid-state frequency doubled Nd:YAG-laser L ($\lambda = 532$ nm) was expanded and collimated by sequentially passing through a 40-μm pinhole PH and lens L_1 ($f_1 = 250$ mm), before hitting the display of a modulator SLM (PLUTO VIS, 1920 × 1080 resolution, and 8-μm pixel size). The diaphragm D_1 was utilized to single out the central bright ring from surrounding bright and dark rings resulting from diffraction by the pinhole. Then, using lenses L_2 ($f_2 = 350$ mm) and L_3 ($f_3 = 150$ mm) and diaphragm D_2 the phase-modulated laser beam reflected at the modulator's display was spatially filtered. Lens L_4 ($f_4 = 140$ mm) was used to focus the laser beam on

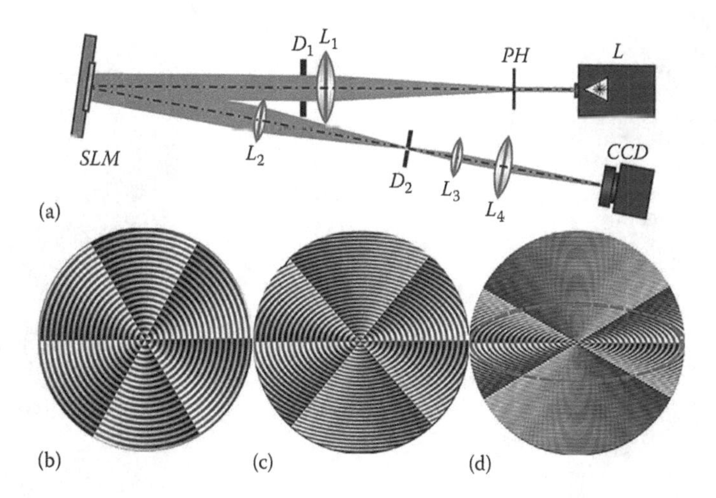

FIGURE 10.11 (a) An experimental setup: L is a solid-state frequency doubled Nd:YAG-laser ($\lambda = 532$ nm); PH is a 40-μm pinhole; L_1, L_2, L_3 and L_4 are lenses with focal lengths $f_1 = 250$ mm, $f_2 = 350$ mm, $f_3 = 150$ mm, and $f_4 = 140$ mm; D_1 and D_2 are diaphragms; SLM is a spatial light modulator PLUTO VIS; and CCD is a video-camera LOMO TC-1000. (b–d) Phase elements for generating the EPOVs with the topological charge $m = 3$ and with the ellipticity parameter equal to (b) $a/b = 1$, (c) $a/b = 1.5$, and (d) $a/b = 3$.

the matrix of the *CCD*-camera LOMO TC 1000 (1.67 × 1.67-μm pixel size). Shown in Figure 10.11(b–d) are phase elements for the spatial light modulator that we used in our experiments to generate various EPOVs. The dashed ellipse in Figure 10.11d shows the elliptic aperture described in Eq. (10.36). For spatial separation of non-modulated and phase-modulated beams reflected at the modulator, we used a linear phase mask additionally with the initial phase mask.

Figure 10.12 shows computed intensity and phase distributions, as well as intensity distributions obtained experimentally in the focus of the lens L_4 by using phase elements, which generate EPOVs with topological charges 3, 7, and 20, and with the ellipticity parameter $a/b = 3$. Figure 10.13 shows the same intensity and phase distributions, but for the ellipticity parameter $a/b = 1.5$. Scaling factor α equals 120.

It is seen in both Figures 10.12 and 10.13 that a narrow light ellipse is generated. For Figure 10.12a, we computed the efficiency which is about 60%. This is due to the sidelobes which are almost invisible in Figure 10.12a, but amount about 20% of the maximal intensity and can be seen distinctly in the intensity cross-section (Figure 10.14).

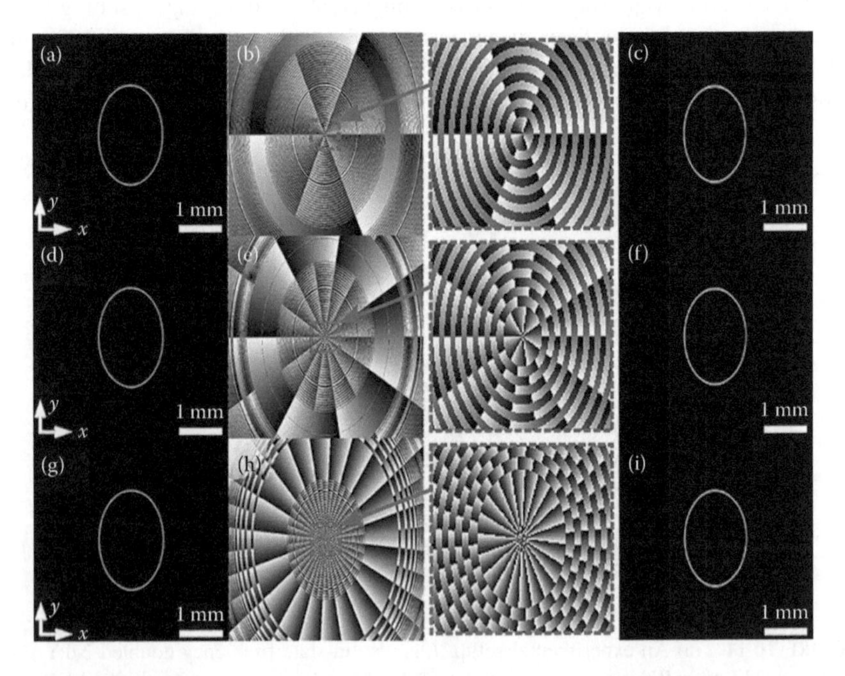

FIGURE 10.12 (a, d, g) Computed intensity and (b, e, h) phase distributions, as well as (c, f, i) experimentally generated intensity distributions of EPOVs with different topological charges m when the ellipticity parameter (ratio between x- and y-diameters of the ellipse) is $a/b = 3$: (a) $m = 3$, (b) $m = 7$, (a) $m = 20$. Frame size is 4500 × 4500 μm. Insets in Figure 10.12 (b, e, h) show magnified areas bounded by dashed line.

FIGURE 10.13 (a, d, g) Computed intensity and (b, e, h) phase distributions, as well as (c, f, i) experimentally generated intensity distributions of EPOVs with different topological charges m when the ellipticity parameter is $a/b = 1.5$: (a) $m = 3$, (b) $m = 7$, (a) $m = 20$. Frame size is 4500×4500 μm. Insets in Figure 10.13 (b, e, h) show magnified areas bounded by dashed line.

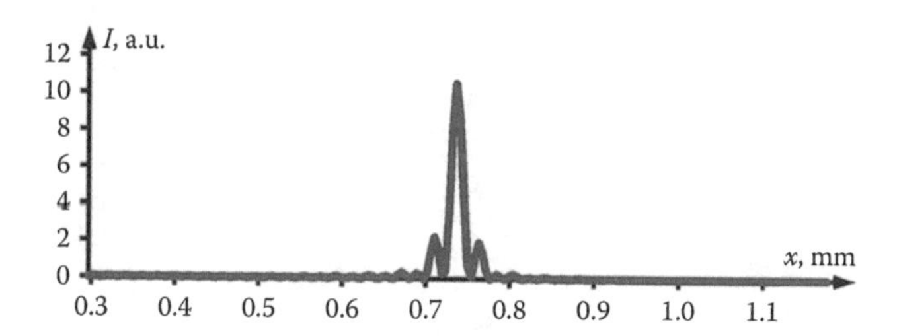

FIGURE 10.14 Horizontal cross-section of the intensity distribution of the EPOV in Figure 10.12a (x is the distance from the centre of the ellipse).

For the EPOV in Figure 10.12g–i ($m = 20$), when $a/b = 3$, the computed normalized OAM (10.37) is $J_z/W = 31.13$, while the theoretical value is $J_z/W = 33.33$. When $a/b = 1.5$ (Figure 10.13g–i), the computed normalized OAM is $J_z/W = 20.83$, while the theoretical value is $J_z/W = 21.67$ (Figure 10.15).

Note that theoretical and numerical values are close to each other, although they are not exactly the same. One of the reasons is that the theoretical value is obtained for the EPOV generated by the infinite elliptic Bessel beam, while the computed value is obtained for the EPOV generated by an initial field with its complex amplitude equal to the sign of bounded elliptic Bessel function (10.41). Another reason is that

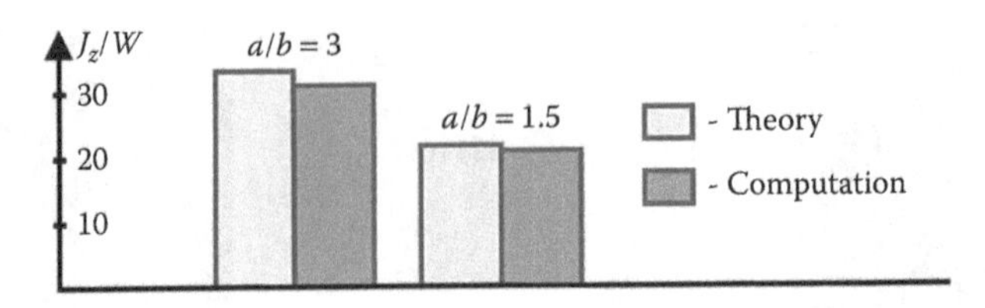

FIGURE 10.15 Theoretical and computed normalized OAM J_z/W of the EPOVs from Figures 10.12 and 10.13.

numerical calculation of the OAM is done on a figure with narrow ring and therefore the sampling can affect the accuracy. It can also be seen that in the experimental patterns (Figure 10.12c, f, and i and Figure 10.13c, f, and i) the intensity in the top and bottom parts of the ellipse exceeds that in the left and right parts. This is because the experimental patterns were generated by using the phase elements from Figure 10.11b–d without an elliptic aperture shown in Figure 10.11d. For the on-ellipse intensity to be uniform, the light outside the elliptic aperture should be blocked and since it was not, it leads to the increasing intensity in the top and bottom areas.

The topological charge of the experimentally generated EPOVs was measured interferometrically. Figure 10.16 shows computed and experimentally measured

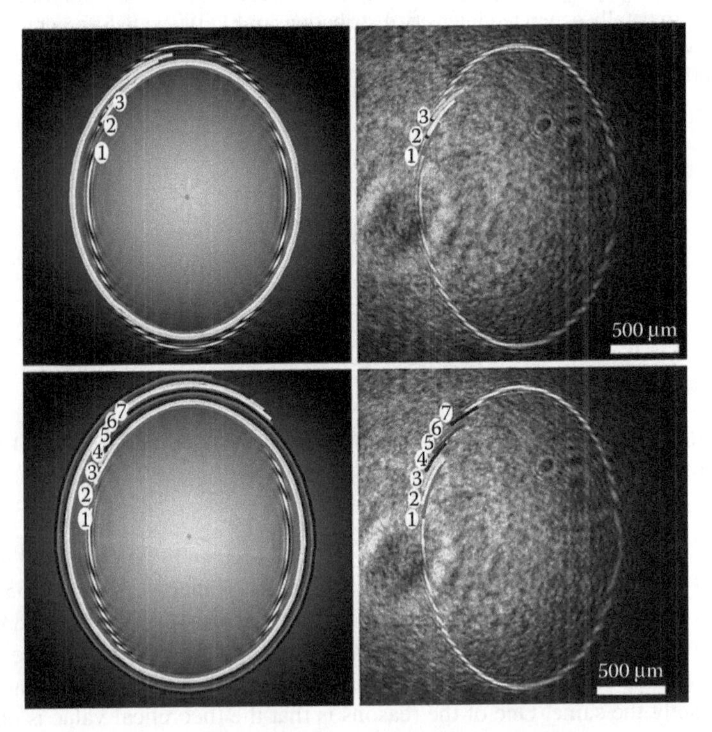

FIGURE 10.16 Interferogram of an EPOV with topological charge $m = 3$ and with ellipticity parameter $a/b = 1.5$, obtained by interference of the EPOV with a diverging Gaussian beam: left column – simulation, right column – experiment. Branches of the spirals are marked by the numbers 1–3 and 1–7.

interferograms of the EPOVs with the topological charges $m = 3$ and $m = 7$ (the ellipticity parameter is $a/b = 1.5$). These patterns were obtained by interference of the EPOVs and a diverging Gaussian beam. It is known that the interference between the helical wavefront and the spherical wavefront results in spiral fringes. Intersections of such spirals with the light ellipse form fringes that are seen in the interferograms in Figure 10.16. Thus, Figure 10.16 confirms the non-zero topological charge of the generated beams. Despite the large number of fringes, all are lying on only a few spirals (i.e., one spiral can pass several fringes). Each spiral is marked by a different color (or by a different index) and their number corresponds to the topological charge of the EPOV. Using such an approach, it was experimentally confirmed that topological charges of generated beams are different, while their sizes are the same.

We studied an EPOV which can be approximately generated by the Fraunhofer diffraction of bounded pure-phase optical elements (10.40) (sign of Bessel function) and (10.41) (spiral axicon). The element (10.40) appears more effective than the spiral axicon (10.41). Compared to the spiral axicon (10.41), the light ellipse generated by the proposed element (10.40) is more than twice narrower, while its intensity is respectively more than twice higher. As is the case for the conventional circular POVs, horizontal and vertical diameters of the EPOV are independent of the topological charge. The obtained exact analytical expression for the OAM of the EPOV shows that it can be fractional and at fixed topological charge it exceeds the OAM of the conventional circular POV. The obtained expression for the OAM density shows that it is higher on the smaller side of the EPOV and is lower on the larger side. The ratio between the maximal and minimal OAM density equals the squared ratio between the ellipse diameters. EPOVs of several topological charges are generated experimentally using a spatial light modulator. Independence of their size of the topological charge, which was determined interferometrically, was confirmed. Such EPOVs can find applications in optical trapping and elliptic rotation of microscopic particles, as well as generation of OAM-entangled photons.

11 Hankel Optical Vortices

11.1 VORTEX HANKEL LASER BEAMS

Laser beams described by scalar complex amplitudes derived as exact solutions of a nonparaxial Helmholtz equation have been well studied in optics. These include well-known plane and spherical waves [163], as well as more recently proposed Bessel modes [1], Mathieu beams [98], parabolic laser beams [97], Hankel-Bessel beams [257], and asymmetric Bessel modes [219]. In Refs. [342,343] we also described Hankel beams of the first, second, and third types with their complex amplitude being proportional to the Hankel function.

Relations to describe projections of the electric and magnetic field strength (E and H vectors) of an electromagnetic wave have been deduced [344]. The projections of the E and H vectors were expressed as three integral relations in the form an expansion in terms of the angular spectrum of plane waves. Based on the electromagnetic field projections, relations for the Poynting vector and the angular momentum (AM) density were derived. It is worth noting that at the time being, closed analytic relations for the AM density have been reported only for paraxial Laguerre-Gaussian beams [17] and vector Bessel beams [246].

This section is a continuation of our work [342] about scalar Hankel beams and therefore we briefly describe here some theory from Ref. [342]. In this section, we propose new vectorial nonparaxial vortex beams with their complex amplitude described by a Hankel function of semi-integer order, prompting us to call them vectorial Hankel beams. We call these beams laser beams since their complex amplitude satisfies the Helmholtz equation for monochromatic field. In addition, we do not consider here any incoherence or partial coherence, that is, we suppose that the light beam is fully coherent. These beams differ from the Hankel-Bessel beams of Refs. [257,343] since the latter beams are scalar and their complex amplitude is proportional to multiplication of the Hankel and Bessel functions. A lower (zero)-order Hankel beam is shown to describe a spherical wave. We derive explicit analytical expressions for amplitudes of all six components of the linearly polarized vectorial Hankel beam, which allow obtaining analytical expressions for the Poynting vector and the AM. The laser beams under analysis can be generated using a liquid crystal microdisplay, showing promise for applications in optically trapping and rotating dielectric microobjects due to their nonparaxiality and absence of sidelobes.

Scalar Hankel Laser Beams

Let us consider the following differential equation:

$$\nabla^2 E - \gamma^2 k^2 E = 0, \tag{11.1}$$

where $k \in \mathbb{R}, \gamma \in \mathbb{C}$. At $\gamma = \pm i$, this equation coincides with the Helmholtz equation. It is easy to see that the general solution of Eq. (11.1) can be found in the following form:

$$E(x,y,z) = \iint_{\mathbb{R}^2} A(\xi,\eta) \exp\left[ik(x\xi + y\eta) - kz\sqrt{\gamma^2 + \xi^2 + \eta^2} \right] \mathrm{d}\xi \mathrm{d}\eta, \quad (11.2)$$

where:

(x, y, z) are the Cartesian coordinates

$A(\zeta,\eta)$ is an arbitrary function

In cylindrical coordinates (r, φ, z) the solution (11.2) reads as

$$E(r,\varphi,z) = \iint_{\mathbb{R}^2} A(\rho,\theta) \exp\left[ikr\rho\cos(\theta - \varphi) - kz\sqrt{\rho^2 + \gamma^2} \right] \rho \mathrm{d}\rho \mathrm{d}\theta. \quad (11.3)$$

If the function $A(\rho, \theta)$ is given by $A(\rho)\exp(in\theta)$, where n is an integer non-negative number, then, instead of (11.3), the solution of Eq. (11.1) can be written via an integral with the Bessel functions $J_n(x)$:

$$E(r,\varphi,z) = 2\pi \mathrm{i}^n \exp(in\varphi) \int_0^\infty A(\rho) \exp\left(-kz\sqrt{\rho^2 + \gamma^2} \right) J_n(kr\rho) \rho \mathrm{d}\rho. \quad (11.4)$$

Let us choose the function $A(\rho)$ in the following form:

$$A(\rho) = \frac{\rho^n}{\sqrt{\rho^2 + \gamma^2}}. \quad (11.5)$$

Then Eq. (11.4) reads as

$$E(r,\varphi,z) = 2\pi \mathrm{i}^n \exp(in\varphi) \int_0^\infty \frac{\rho^{n+1}}{\sqrt{\rho^2 + \gamma^2}} \exp\left(-kz\sqrt{\rho^2 + \gamma^2} \right) J_n(kr\rho) \mathrm{d}\rho. \quad (11.6)$$

We use the following reference integral (expression 2.12.10.10 in [38]):

$$\int_0^\infty x^{\nu+1} \frac{\exp\left(-p\sqrt{x^2 + z^2} \right)}{\sqrt{x^2 + z^2}} J_\nu(cx) \mathrm{d}x = \sqrt{\frac{2}{\pi}} c^\nu z^{\nu+1/2} \left(p^2 + c^2 \right)^{-(2\nu+1)/4} K_{\nu+1/2}\left(z\sqrt{p^2 + c^2} \right), \quad (11.7)$$

where:

Re $p >$ |Im c|, Re $z > 0$, Re $\nu > -1$

$K_\nu(\xi)$ is the modified Bessel function of the second kind

In our case, the first and the third conditions are fulfilled since $n \geq 0$ and Re $p = kz >$ |Im c| = |Im (kr)| = 0. Further, we suppose that Re $\gamma > 0$. Then Eq. (11.6) can be written in an explicit form:

$$E(r,\varphi,z) = 2\mathrm{i}^n \gamma^{n+1/2} \sqrt{\lambda} r^n \exp(in\varphi) R^{-n-1/2} K_{n+1/2}(\gamma kR), \tag{11.8}$$

where $R^2 = r^2 + z^2$. Using the relation between the modified Bessel function of the second kind and the Hankel function $H_\nu^{(1)}(x) = J_\nu(x) + \mathrm{i}Y_\nu(x)$ ($Y_\nu(x)$ is a Neumann function), we obtain:

$$E(r,\varphi,z) = \pi \mathrm{i}^{2n+3/2} \gamma^{n+1/2} \sqrt{\lambda} r^n \exp(in\varphi) R^{-n-1/2} H_{n+1/2}^{(1)}(i\gamma kR). \tag{11.9}$$

The function (11.9) satisfies the equation (11.1) and has been obtained under the condition that Re $\gamma > 0$. Further, we suppose in Eq. (11.9) that $\gamma = -i$. Then, we obtain the following function:

$$E(r,\varphi,z) = \pi \mathrm{i}^{n+1} \sqrt{\lambda} r^n \exp(in\varphi) \psi_{n+1/2}(R), \tag{11.10}$$

where $\psi_\nu(R) = H_\nu^{(1)}(kR) R^{-\nu}$. Note that by using $\gamma = -i$ we have broken the condition Re $\gamma > 0$. However, it can be verified (by substitution) that the function (11.10) satisfies the Helmholtz equation:

$$\frac{\partial^2 E}{\partial r^2} + \frac{1}{r}\frac{\partial E}{\partial r} + \frac{1}{r^2}\frac{\partial^2 E}{\partial \varphi^2} + \frac{\partial^2 E}{\partial z^2} + k^2 E = 0. \tag{11.11}$$

Indeed, after this substitution, the left part of Eq. (11.11) reads as

$$\nabla^2 E + k^2 E = \frac{k^2 r^n \mathrm{e}^{in\varphi}}{\xi^{n+1/2}} \left\{ \frac{\mathrm{d}^2}{\mathrm{d}\xi^2} + \frac{1}{\xi}\frac{\mathrm{d}}{\mathrm{d}\xi} + \left[1 - \left(n + \frac{1}{2} \right)^2 \xi^{-2} \right] \right\} H_{n+1/2}^{(1)}(\xi), \tag{11.12}$$

where $\xi = kR$. This equation is obviously equal to zero since the Hankel function is a solution of the Bessel differential equation. Note that according to Eq. (11.12), instead of the first Hankel function any other solution of the Bessel differential equation can be used. So, instead of $H^{(1)}_{n+1/2}(kR)$ in Eq. (11.12) we can use, for example, $J_{n+1/2}(kR)$. According to asymptotics of the cylindrical functions, at large propagation distances $H^{(1)}_{n+1/2}(kR) \sim \exp(ikR) \sim \exp(ikz)$. So, complex amplitude (11.10) describes outgoing waves in the positive direction of z-axis, while the same amplitude with $H^{(2)}_{n+1/2}(kR)$ and $J_{n+1/2}(kR)$ describe respectively incoming waves and standing waves.

The beams in Eq. (11.10) have been named in Ref. [342] as Type 1 Hankel beams (H1 beams). The H1 beam in Eq. (11.10) represents a linear optical vortex $(x + \mathrm{i}y)^n$ embedded into a generalized spherical wave $\Psi_{n+1/2}(R) = H^{(1)}_{n+1/2}(kR)R^{-n-1/2}$.

Angular Spectrum of the Hankel Beam

Further, we obtain the angular spectrum of plane waves of the complex amplitude (11.10). Note that it does not necessarily coincide with the function A from the Eq. (11.5) at $\gamma = -i$. Therefore, it should be derived directly from the Eq. (11.10). In the following we obtain the angular spectrum for the light field (11.10).

In the initial plane ($z = 0$) this beam (11.10) has the following complex amplitude:

$$E_{1,n}(r,\varphi,0) = \mathrm{i}^{n+1}\pi\sqrt{\lambda}\mathrm{e}^{\mathrm{i}n\varphi}\frac{H^{(1)}_{n+1/2}(kr)}{\sqrt{r}}. \tag{11.13}$$

Angular spectrum of plane waves is the Fourier transform of this amplitude and in polar coordinates reads as

$$A(\rho,\theta) = \frac{\mathrm{i}k^2}{2}\sqrt{\lambda}\exp(\mathrm{i}n\theta)(B_1 + \mathrm{i}B_2), \tag{11.14}$$

where:

$$\begin{cases} B_1 = \int\limits_0^\infty \sqrt{r}J_{n+1/2}(kr)J_n(k\rho r)dr, \\ B_2 = \int\limits_0^\infty \sqrt{r}Y_{n+1/2}(kr)J_n(k\rho r)dr. \end{cases} \tag{11.15}$$

In (11.15) the Hankel function has been split into the Bessel and Neumann functions.

For B_1, in Ref. [38] there is the following reference integral (expression 2.12.31.1):

$$\int\limits_0^\infty x^{\alpha-1}J_\mu(bx)J_\nu(cx)dx = A^\alpha_{\mu,\nu}, \tag{11.16}$$

where $b, c, \mathrm{Re}(\alpha+\mu+\nu)>0; \mathrm{Re}\,\alpha<2$ and

$$A_{\mu,\nu}^{\alpha}=2^{\alpha-1}\frac{c^{\nu}}{b^{\nu+\alpha}}\Gamma\left[\begin{matrix}\frac{\nu+\mu+\alpha}{2}\\ \frac{\mu-\nu-\alpha}{2}+1,\nu+1\end{matrix}\right]\times{}_2F_1\left(\frac{\nu+\mu+\alpha}{2},\frac{\nu-\mu+\alpha}{2};\nu+1;\frac{c^2}{b^2}\right),[0<c<b], \tag{11.17}$$

$$A_{\mu,\nu}^{\alpha}=2^{\alpha-1}\frac{b^{\mu}}{c^{\mu+\alpha}}\Gamma\left[\begin{matrix}\frac{\nu+\mu+\alpha}{2}\\ \frac{\nu-\mu-\alpha}{2}+1,\mu+1\end{matrix}\right]\times{}_2F_1\left(\frac{\nu+\mu+\alpha}{2},\frac{\mu-\nu+\alpha}{2};\mu+1;\frac{b^2}{c^2}\right),[0<b<c], \tag{11.18}$$

where ${}_2F_1$ is the hypergeometric function and

$$\Gamma\left[\begin{matrix}a_1,a_2,\ldots,a_m\\ b_1,b_2,\ldots,b_n\end{matrix}\right]=\prod_{k=1}^{m}\Gamma(a_k)\Big/\prod_{k=1}^{n}\Gamma(b_k). \tag{11.19}$$

For B_2, in Ref. [38] there is another reference integral (expression 2.13.15.4):

$$\int_0^{\infty}x^{\alpha-1}J_{\mu}(bx)Y_{\nu}(cx)dx=A_{\mu,\nu}^{\alpha}, \tag{11.20}$$

where $b,c>0, |\mathrm{Re}\,\nu|-\mathrm{Re}\,\mu<\mathrm{Re}\,\alpha<2$ and

$$A_{\mu,\nu}^{\alpha}=-\frac{2^{\alpha-1}b^{\mu}}{\pi c^{\alpha+\mu}}\cos\left[\frac{\pi(\alpha+\mu-\nu)}{2}\right]\times\frac{\Gamma\left(\frac{\alpha+\mu+\nu}{2}\right)\Gamma\left(\frac{\alpha+\mu-\nu}{2}\right)}{\Gamma(\mu+1)}{}_2F_1\left(\frac{\alpha+\mu+\nu}{2},\frac{\alpha+\mu-\nu}{2};\mu+1;\frac{b^2}{c^2}\right),[0<b<c] \tag{11.21}$$

$$
\begin{aligned}
A_{\mu,\nu}^{\alpha} = & -\frac{2^{\alpha-1}c^{\nu}}{\pi b^{\alpha+\nu}}\cos(\nu\pi)\frac{\Gamma(-\nu)\Gamma\left(\frac{\alpha+\mu+\nu}{2}\right)}{\Gamma\left(1-\frac{\alpha+\nu-\mu}{2}\right)} \\
& \times {}_2F_1\left(\frac{\alpha-\mu+\nu}{2},\frac{\alpha+\mu+\nu}{2};1+\nu;\frac{c^2}{b^2}\right) \\
& -\frac{2^{\alpha-1}b^{\nu-\alpha}}{\pi c^{\nu}}\frac{\Gamma(\nu)\Gamma\left(\frac{\alpha+\mu-\nu}{2}\right)}{\Gamma\left(1+\frac{\nu+\mu-\alpha}{2}\right)} \\
& \times {}_2F_1\left(\frac{\alpha-\mu-\nu}{2},\frac{\alpha+\mu-\nu}{2};1-\nu;\frac{c^2}{b^2}\right), [0<c<b]
\end{aligned}
\tag{11.22}
$$

In our case $\alpha = 3/2$, $\mu = n$, $\nu = n + 1/2$, $b = k\rho$, $c = k$. For such parameters, the hypergeometric function ${}_2F_1$ is simplified to the hypergeometric function ${}_1F_0$. All the conditions of the reference integrals are fulfilled. Using these integrals, we obtain:

$$
B_1 = \begin{cases} \dfrac{\sqrt{2}}{\sqrt{\pi}k^{3/2}}\dfrac{\rho^n}{\sqrt{1-\rho^2}}, \text{ if } \rho < 1, \\ 0, \text{ if } \rho > 1. \end{cases}
\tag{11.23}
$$

and

$$
B_2 = \begin{cases} 0, \text{ if } \rho < 1, \\ -\dfrac{\sqrt{2}\rho^n}{\sqrt{\pi}k^{\frac{3}{2}}}\dfrac{1}{\sqrt{\rho^2-1}}, \text{ if } \rho > 1. \end{cases}
\tag{11.24}
$$

Substituting these equations into the expression for the angular spectrum, it finally gets:

$$
A(\rho,\theta) = \begin{cases} \mathrm{i}\dfrac{\rho^n}{\sqrt{1-\rho^2}}\exp(\mathrm{i}n\theta), \rho < 1, \\ \dfrac{\rho^n}{\sqrt{\rho^2-1}}\exp(\mathrm{i}n\theta), \rho > 1. \end{cases}
\tag{11.25}
$$

It turned out that at $\gamma = -i$ this angular spectrum coincides with the function $A(\rho)$ in Eq. (11.5). At $n = 0$ it coincides with the well-known angular spectrum of a spherical wave [345]:

$$A(\rho,\theta) = \begin{cases} \dfrac{i}{\sqrt{1-\rho^2}}, \rho < 1, \\ \dfrac{1}{\sqrt{\rho^2-1}}, \rho > 1. \end{cases} \tag{11.26}$$

The intensity of the angular spectrum of Eq. (11.13) is shown in Figure 11.1. It is seen that this spectrum consists of two parts: propagating waves ($\rho < 1$) and evanescent waves ($\rho > 1$). Propagating waves are concentrated near the ring $\rho = 1$. So, they are similar to the standing (but not yet decaying) Bessel beam with the complex amplitude $E(r, \varphi, z) = J_n(kr)\exp(in\varphi)$. At $n > 1$, the contribution of evanescent constituent waves of the H1 beam increases with the spectral radial coordinate ρ (Figure 11.1b). However, according to Eq. (11.13), the larger is the contribution of particular evanescent wave, the faster is its decay in the longitudinal direction z. That's why the field amplitude at $z > 0$ does not tend to infinity.

With the $\Psi_{n+\nu}(R)$ function depending only on the distance R to the origin, it can be termed as a generalized spherical wave. At $n = 0$, the amplitude of the H1 beam can be shown to be related with the amplitude of a spherical wave. The fundamental H1 beam (at $n = 0$) describes a spherical wave outgoing from the origin:

$$E_{1,0}(r,\varphi,z) = i\pi\sqrt{\lambda}\psi_{1/2}(R) = 2\pi(kR)^{-1}e^{ikR}. \tag{11.27}$$

Type 2 Hankel beams (H2 beams) can be derived by differentiating the complex amplitude in Eq. (11.10) with respect to kz [342]:

$$E_{2,n}(r,\varphi,z) = i^{n-1}\pi\sqrt{\lambda}zr^n e^{in\varphi}\psi_{n+3/2}(R), \tag{11.28}$$

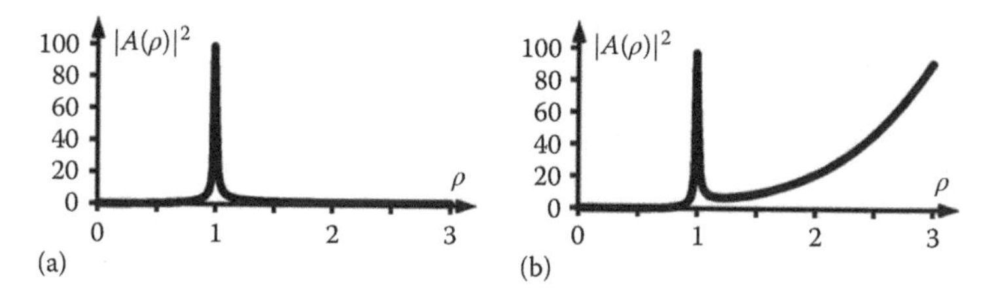

FIGURE 11.1 Intensity of the angular spectrum (11.13) versus the radial coordinate: (a) $n = 1$, (b) $n = 3$.

Similarly to the H1 beams, the H2 beams are structurally generalized spherical waves with the optical vortices embedded. The only difference is that the $\Psi_{n+\nu}(R)$ function takes a different form $\Psi_{n+3/2}(R) = H^{(1)}_{n+3/2}(kR)R^{-n-3/2}$ and amplitude (11.16) takes a nonzero value only at the origin of the initial plane ($z = 0$).

Different types of Hankel beams diverge in different ways. Actually, at large z, the intensity of the H1 beam of Eq. (11.10) is $I_1(r,z) \approx \lambda^2 r^{2n} R^{-2n-2}$ [342], while the intensity of the H2 beam of Eq. (11.16) is $I_2(r,z) \approx \lambda^2 z^2 r^{2n} R^{-2n-4}$. Differentiating these values with respect to the radial variable r and equating the result to zero yields:

$$r_{1,\max} \approx z\sqrt{n},\ r_{2,\max} = z\sqrt{n/2}. \tag{11.29}$$

From (11.29), the radii of intensity maxima of the vortex H1 and H2 beams ($n > 0$) are seen to linearly depend on the longitudinal coordinate z, also varying as a square root of the optical vortex topological charge. The divergence of the H2 beams is $\sqrt{2}$ times lower than that of the H1 beams.

Figure 11.2 depicts the (a, d) intensity, (b, e) phase, and (c, f) radial intensity profile for an H1 beam with topological charge $n = 1$ at distances (a, b, c) $z = \lambda$ and (d, e, f) $z = 5\lambda$. Plotted on the x-axis in Figure 11.2 is the radial coordinate in wavelengths.

In Figure 11.2c and f, vertical lines mark the intensity peaks on the profile plot for which an approximate value of radii can be calculated from (11.29). For instance, for the intensity profile in Figure 11.2c the exact radius of the intensity peak is $r = 0.994\lambda$. Thus, we can infer that it is not only in the far-field, Eq. (11.29), but also in the near-field that the H1 beam ($n = 1$) has a 45° divergence.

Note that both types of the Hankel beams under analysis are devoid of intensity side-lobes. Special techniques intended to suppress the side-lobes of laser optical vortices have been described [36,105,346], making them suitable for highly effective optical micromanipulation. With the Hankel beams of both types propagating without side-lobes, there is no need to employ such techniques.

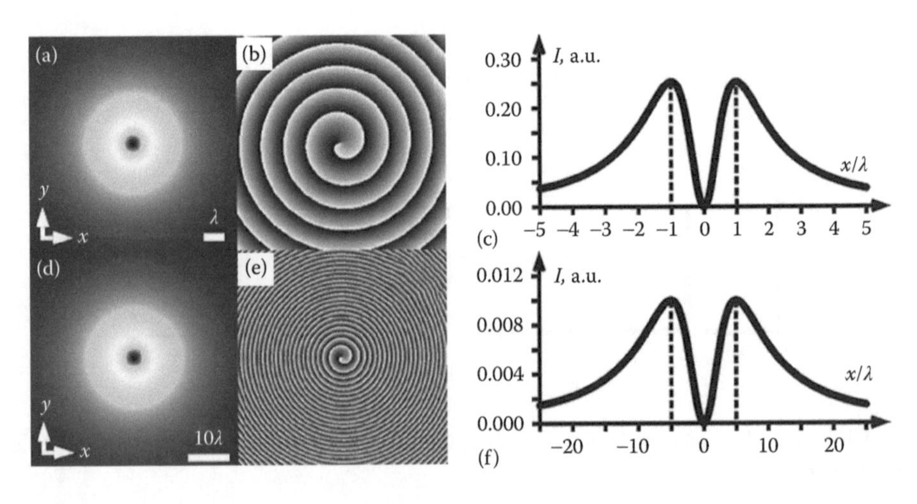

FIGURE 11.2 (a, d) Intensity, (b, e) phase, and (c, f) radial intensity profile for an H1 beam with topological charge $n = 1$ at distances (a, b, c) $z = \lambda$ and (d, e, f) $z = 5\lambda$.

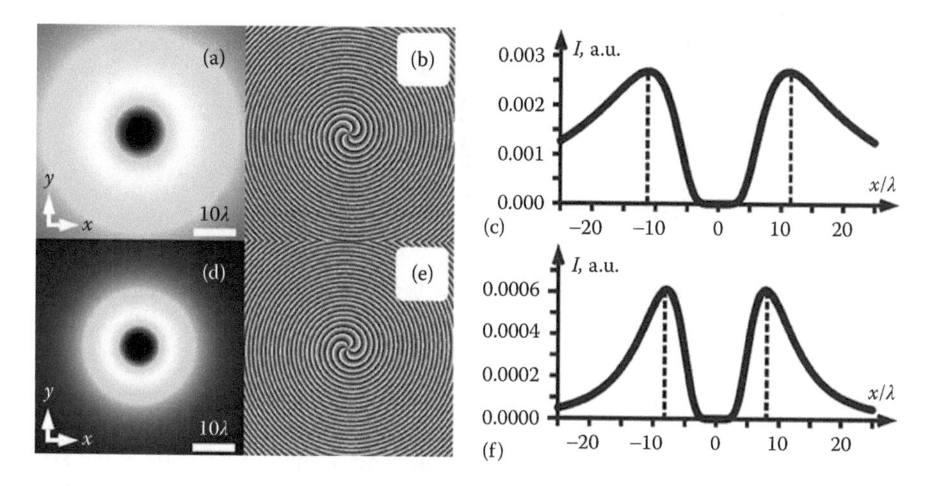

FIGURE 11.3 (a, d) Intensity, (b, e) phase, and (c, f) radial intensity profile for an (a, b, c) H1 beam and (d, e, f) H2 beam with topological charge $n = 5$ at distance $z = 5\lambda$.

Figure 11.3 depicts the (a, d) intensity, (b, e) phase, and (c, f) radial intensity profile for an (a, b, c) H1 beam and an (d, e, f) H2 beam with topological charge $n = 5$ at distance $z = 5\lambda$. From Figure 11.3c and f, the divergence of the H1 beam is seen to be higher than that of the H2 beam at the same topological charge. Radii of the intensity peaks are, respectively, 11λ (Figure 11.3c) and 7.9λ (Figure 11.3f).

Shown in Figure 11.4 are intensity distributions in the xz-plane for the Hankel beams of both types with $n = 5$, as in Figures 11.2 and 11.3. Calculation area in Figure 11.4 is $-25\lambda \leq x \leq 25\lambda$, $\lambda \leq z \leq 5\lambda$.

Figure 11.5 shows a radial intensity section of the H1 beam at a very close distance from the initial plane ($z << \lambda$). It is seen in Figure 11.5 that near the initial plane the intensity distribution has the form of a ring with a diameter much smaller than the wavelength. There is zero intensity on the axis, and a maximal intensity on the ring tends to infinity. The intensity pattern similar to that shown in Figure 11.5 will take place for any small value of z. If a spherical wave converges to (or diverges from) the point, the H1 beam converges to the "point with zero intensity in its center." So, it follows from the

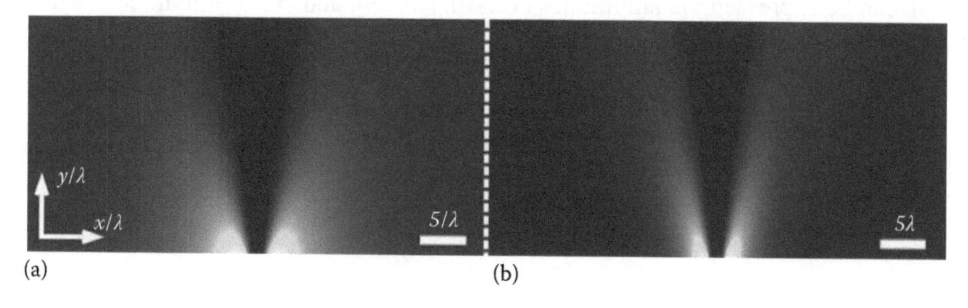

FIGURE 11.4 Intensity distribution of the Hankel beam with $n = 5$ in the xz-plane: (a) H1 beam, (b) H2 beam.

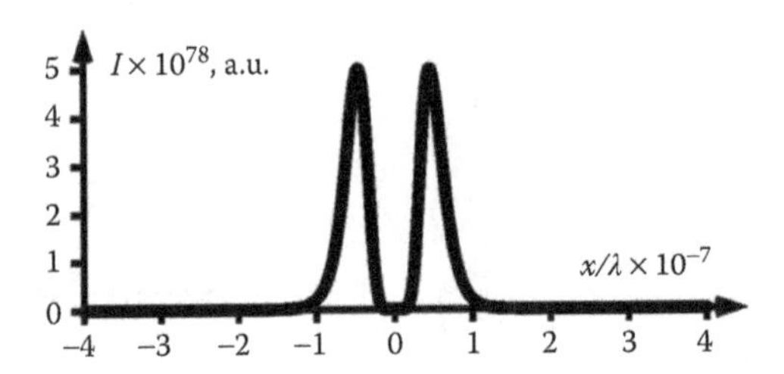

FIGURE 11.5 Intensity of the H1 beam at $n = 5$, $z = 10^{-7}\lambda$, size of calculation domain $8 \times 10^{-7}\lambda$.

written above that the source of H1 beams is an infinitesimally narrow light ring with infinitely small radius and with infinitely high maximal intensity.

Similarly to how the H2 beam (11.28) was derived from the amplitude of the H1 beam (11.10) by differentiation (11.10), it is possible to derive a H3 beam from a H2 beam by differentiation with respect to kz:

$$E_{3,n}(r,\varphi,z) = \mathrm{i}^{n+1}\pi\sqrt{\lambda}r^n e^{in\varphi}\left[z^2\psi_{n+5/2}(R) - k^{-1}\psi_{n+3/2}(R)\right]. \tag{11.30}$$

Similarly to the H1 beam and H2 beam, it can be shown that at large z, the intensity of the H3 beam of Eq. (11.18) is $I_3(r, z) \approx \lambda^2 r^{2n} z^4 R^{-2n-6}$, and, therefore, $r_{3,\max} = z\sqrt{n/3}$, that is, the divergence of the H3 beams is $\sqrt{3}$ times lower than that of the H1 beams.

It is worth noting that all types of Hankel beams in Eqs. (11.10), (11.16), and (11.18) have the same topological charge n, while their radial structure is described by different order Hankel functions. This means that while the normalized orbital angular momentum (OAM) (per unit power or per photon) is the same for all Hankel beams, being equal to the topological charge, the unnormalized OAM will be different for a Hankel beam of the different type.

Linearly Polarized Hankel Beams

Using complex amplitudes in Eqs. (11.10), (11.28), and (11.30), the electromagnetic field components of a vector Hankel beam can be expressed. Following Ref. [344], it can be shown that projections of the E field and H field of a linearly polarized Hankel beam can be expressed via amplitudes (11.10), (11.28), and (11.30) in the following way ($E_y = 0$):

$$\begin{gathered}
E_x = E_{2,n}(r,\varphi,z), \quad E_z = -\frac{1}{k}\frac{\partial E_{1,n}(r,\varphi,z)}{\partial x}, \\
H_x = \frac{-\mathrm{i}}{\mu k^2}\frac{\partial^2 E_{1,n}(r,\varphi,z)}{\partial y \partial x}, \quad H_z = \frac{-\mathrm{i}}{\mu k}\frac{\partial E_{2,n}(r,\varphi,z)}{\partial y}, \\
H_y = \frac{\mathrm{i}}{\mu}E_{3,n}(r,\varphi,z) + \frac{\mathrm{i}}{\mu k^2}\frac{\partial^2 E_{1,n}(r,\varphi,z)}{\partial x^2},
\end{gathered} \tag{11.31}$$

where μ is magnetic permeability. Therefore, explicit expressions for the amplitudes of the vectorial vortex Hankel beam, satisfying the Maxwell equations, read as ($E_y = 0$):

$$E_x(r,\varphi,z) = \mathrm{i}^{n-1}\pi\lambda^{1/2}zr^n \mathrm{e}^{\mathrm{i}n\varphi}\Psi_{n+3/2}(R), \tag{11.32}$$

$$E_z(r,\varphi,z) = \frac{\mathrm{i}^{n-1}\lambda^{3/2}r^{n-1}\mathrm{e}^{\mathrm{i}n\varphi}}{2}\left[n\mathrm{e}^{-\mathrm{i}\varphi}\psi_{n+1/2}(R) - kr^2\cos\varphi\,\psi_{n+3/2}(R)\right], \tag{11.33}$$

$$H_x(r,\varphi,z) = \frac{\mathrm{i}^n\lambda^{5/2}r^{n-2}\mathrm{e}^{\mathrm{i}n\varphi}}{4\pi\mu} \times\left[\mathrm{i}n(n-1)\mathrm{e}^{-\mathrm{i}2\varphi}\psi_{n+1/2}(R) - \mathrm{i}nkr^2\mathrm{e}^{-\mathrm{i}2\varphi}\psi_{n+3/2}(R) + k^2r^4\sin\varphi\cos\varphi\,\psi_{n+5/2}(R)\right], \tag{11.34}$$

$$\begin{aligned} H_y(r,\varphi,z) = -\frac{i^n\lambda^{5/2}r^{n-2}e^{in\varphi}}{4\pi\mu} & \\ \times\Big[n(n-1)\mathrm{e}^{-\mathrm{i}2\varphi}\Psi_{n+1/2}(R) & \\ -2kr^2\left(1+n\mathrm{e}^{-\mathrm{i}\varphi}\cos\varphi\right)\Psi_{n+3/2}(R) & \\ +k^2r^2\left(z^2+r^2\cos^2\varphi\right)\Psi_{n+5/2}(R)\Big], & \end{aligned} \tag{11.35}$$

$$H_z(r,\varphi,z) = -\frac{\mathrm{i}^n\lambda^{3/2}zr^{n-1}\mathrm{e}^{\mathrm{i}n\varphi}}{2\mu}\left[\mathrm{i}n\mathrm{e}^{-\mathrm{i}\varphi}\Psi_{n+3/2}(R) - kr^2\sin\varphi\,\Psi_{n+5/2}(R)\right]. \tag{11.36}$$

It is seen in Eqs. (11.32) to (11.36) that despite the radial symmetry of the intensity of the transverse electrical component E_x of the linearly polarized Hankel beam, intensities of other projections E_z, H_x, H_y, and H_z do not have radial symmetry. From Eqs. (11.20) and (11.21) it follows that at $n = 1$ on the optical axis $E_z \neq 0$, whereas at $n = 0$ on the optical axis $E_x \neq 0$. At $n > 1$, on the optical axis there will be isolated zero of the total intensity $I = I_x + I_z$ at any z. Shown in Figure 11.6 are distributions of the transverse and longitudinal projections of the intensity (i.e., $I_x = |E_x|^2$, $I_z = |E_z|^2$, $I = I_x + I_z$) and all three Cartesian components of the Poynting vector ($S_x = E_y^*H_z - E_z^*H_y$, $S_y = E_z^*H_x - E_x^*H_z$, $S_y = E_x^*H_y - E_y^*H_x$) of the vectorial Hankel beam with linear polarization along the x-axis and topological charge of $n = 1$, calculated in the transverse planes $z = 5\lambda$ and $z = 10\lambda$.

Furthermore, we can still see the symmetry of the intensity distributions in Figure 11.6 with respect to the horizontal and vertical axes. But it is not always the case. There are two terms in square brackets in Eq. (11.21) for the longitudinal E_z component. Modulus of the first term is radially symmetrical, that is, all asymmetry arises because of the second term, which begins to give a significant contribution when variable R decreases (i.e., closer to the origin). So, at small propagation distance z the

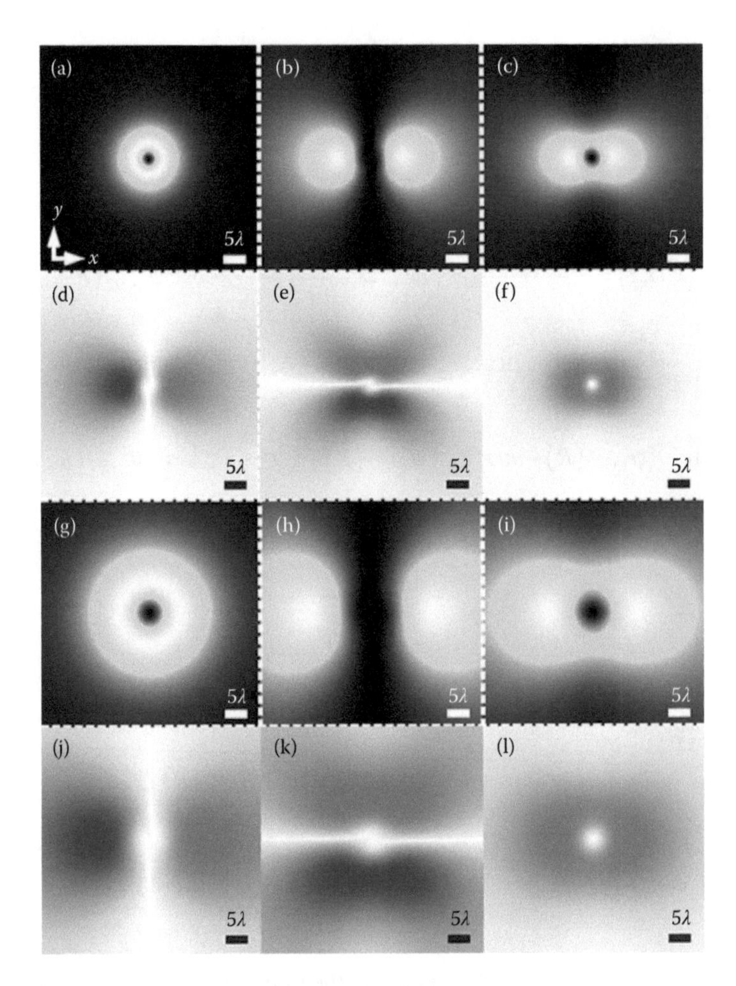

FIGURE 11.6 Distributions of the (a, g) transverse I_x, (b, h) longitudinal I_z, and (c, i) total I intensity and components (real part) of Poynting vector (d, j) S_x, (e, k) S_y, (f, l) S_z of the vectoral Hankel beam (wavelength $\lambda = 532$ nm, topological charge $n = 1$) in planes (a–f) $z = 5\lambda$ and (g–l) $z = 10\lambda$. In figures of the Poynting vector area 1 means positive values, area 2 means negative values.

both terms interfere with each other, and intensity distribution (both longitudinal I_z and total I) appears distorted and rotated in the transverse plane. Figure 11.7 shows the same beam as in Figure 11.6, but in the planes $z = 0.05\lambda$, $z = 0.4\lambda$, and $z = 0.5\lambda$. It is seen in Figure 11.7b and c that near the initial plane the longitudinal and the total intensity distributions are elongated vertically. Then, this elongated spot is rotated upon propagation and split into two spots (Figure 11.7h and i). There is no symmetry, neither radial nor with respect to the Cartesian axes. There is still symmetry with respect to the origin. Further propagation leads to the intensity distribution in Figure 11.6. So, the vectorial Hankel beam is seen to be rotated in space by 90° during the propagation on a distance of few wavelengths. Note that the intensity patterns rotate counterclockwise at $n > 0$ and clockwise at $n < 0$ (Figure 11.7).

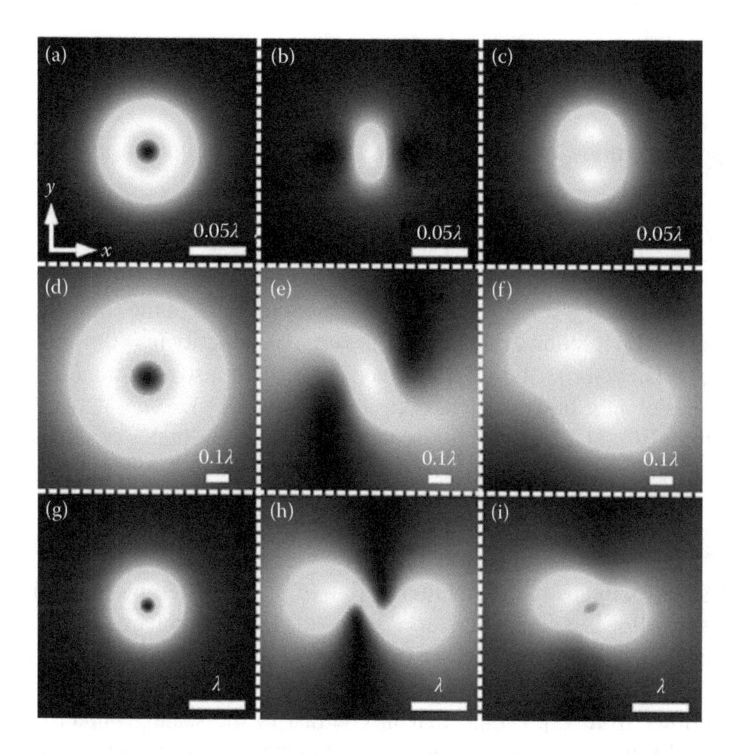

FIGURE 11.7 Distributions of the (a, d, g) transverse I_x, (b, e, h) longitudinal I_z, and (c, f, i) total I intensity of the vectorial Hankel beam (wavelength $\lambda = 532$ nm, topological charge $n = 1$) in planes (a, b, c) $z = 0.05\lambda$, (d, e, f) $z = 0.3\lambda$, and (g, h, i) $z = 0.5\lambda$. Calculation area is (a–c) $-0.1\lambda \le x, y \le 0.1\lambda$, (d–f) $-0.5\lambda \le x, y \le 0.5\lambda$, (g–i) $-2\lambda \le x, y \le 2\lambda$.

Linearly Polarized Hankel Beam in Far Field

Using the asymptotic expansion of the Hankel function, we can obtain the approximate values of all four components in far field, that is, $z >> \lambda$:

$$E_x(r,\varphi,z) = \frac{\mathrm{i}\lambda z r^n \mathrm{e}^{\mathrm{i}n\varphi+\mathrm{i}kR}}{R^{n+2}}, \tag{11.37}$$

$$E_z(r,\varphi,z) = -\frac{\lambda^2 r^{n-1}\mathrm{e}^{\mathrm{i}n\varphi+\mathrm{i}kR}}{2\pi R^{n+2}}\left(nR\mathrm{e}^{-\mathrm{i}\varphi} + \mathrm{i}kr^2\cos\varphi\right), \tag{11.38}$$

$$H_x(r,\varphi,z) = -\frac{\mathrm{i}\lambda^3 r^{n-2}\mathrm{e}^{\mathrm{i}n\varphi+\mathrm{i}kR}}{4\pi^2 R^{n+3}\mu}\left[\mathrm{i}n(n-1)R^2\mathrm{e}^{-\mathrm{i}2\varphi} - nkr^2R\mathrm{e}^{-\mathrm{i}2\varphi} - k^2r^4\sin\varphi\cos\varphi\right], \tag{11.39}$$

$$H_y(r,\varphi,z) = -\frac{\lambda^3 r^{n-2}\mathrm{e}^{\mathrm{i}n\varphi+\mathrm{i}kR}}{4\pi^2 R^{n+3}\mu} \times\left[-\mathrm{i}n(n-1)R^2\mathrm{e}^{-\mathrm{i}2\varphi} + 2kr^2R\left(1+n\mathrm{e}^{-\mathrm{i}\varphi}\cos\varphi\right) + \mathrm{i}k^2r^2\left(z^2 + r^2\cos^2\varphi\right)\right], \tag{11.40}$$

$$H_z(r,\varphi,z)=\frac{\mathrm{i}\lambda^2 zr^{n-1}\mathrm{e}^{\mathrm{i}n\varphi+\mathrm{i}kR}}{2\pi R^{n+3}\mu}\left(nR\mathrm{e}^{-\mathrm{i}\varphi}+kr^2\sin\varphi\right). \tag{11.41}$$

These expressions allow obtaining an expression for the total intensity:

$$I(r,\varphi,z)=\left|E_x\right|^2+\left|E_z\right|^2=\frac{\lambda^4 r^{2n-2}}{4\pi^2 R^{2n+4}}\left[(kzr)^2+(nR-kxy)^2+k^2x^4\right], \tag{11.42}$$

where $x = r\cos\varphi$ and $y = r\sin\varphi$ are Cartesian coordinates in the transverse plane.

From Eq. (11.30) it follows that at $n = 1$ the on-axis intensity ($r = 0$) is nonzero and equals $I(0, z) = (\lambda/z)^4/(4\pi^2)$. At $n = 0$, the on-axis intensity is also nonzero: $I(0, z) = (\lambda/z)^2$. At other values of n the on-axis intensity is always zero: $I(0, z) = 0$.

Asymptotic Eq. (11.42) makes it possible to obtain coordinates of intensity maxima on the horizontal x-axis and vertical y-axis:

$$x_{\max}=\pm\sqrt{n}z,\qquad y_{\max}=\pm\sqrt{n/2}z. \tag{11.43}$$

Figure 11.8 shows intensity distribution $|E_x|^2 + |E_z|^2$, calculated by Eqs. (11.20) and (11.21) for the following parameters: $n = 3$, $z = 30\lambda$ (calculation area is $-120\lambda \le x, y \le 120\lambda$). Black dots in Figure 11.8 are in the points of intensity maxima, calculated by Eq. (11.31).

So, it follows from Eq. (11.43) that in contrast to the scalar Hankel beam, the vectorial Hankel beam has different divergence in horizontal and vertical directions. From comparison of Eqs. (11.29) and (11.43) it can be concluded that the vectorial Hankel beam propagates as the scalar first-type Hankel beam in the xz-plane and as the scalar second-type Hankel beam in the yz-plane.

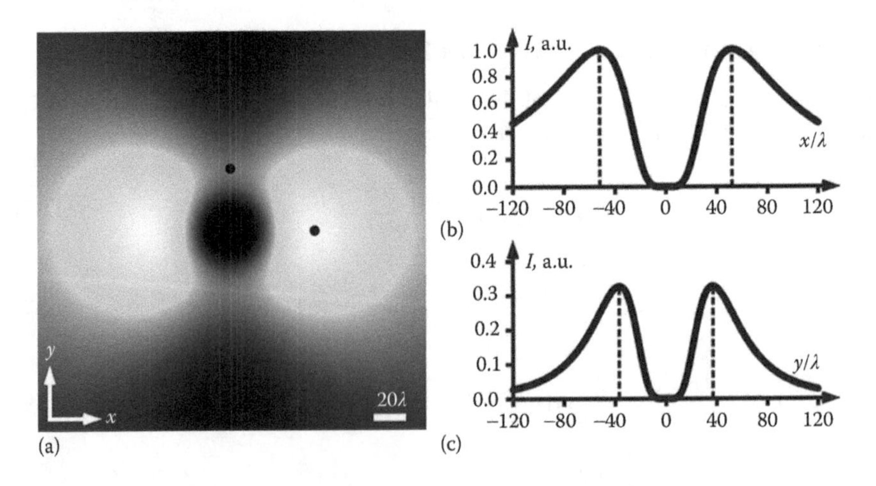

FIGURE 11.8 Distribution of the total intensity of the vectorial Hankel beam (wavelength $\lambda = 532$ nm, topological charge $n = 3$) in plane $z = 30\lambda$. (a) Black dots are in the intensity maxima along axes x and y. Normalized radial profiles of the total intensity along the axes (b) x and (c) y. Dashed lines mark intensity maxima obtained by Eq. (11.43).

Expressions (11.37) through (11.42) also allow obtaining an expression for axial projection of the vector of the OAM density:

$$j_z(r,\varphi,z) = \text{Re}\{xS_y - yS_x\}$$
$$= -\frac{\lambda^5 n r^{2n-2}}{8\pi^3 \mu R^{2n+4}}\left[n(n-1)R^2 - k(2n+1)Rxy + k^2r^2(z^2+x^2)\right]. \tag{11.44}$$

It is seen in Eq. (11.44) that, as in paraxial case, axial projection of the OAM density vector is proportional to the topological charge n. Therefore, $j_z \equiv 0$ at $n = 0$ in every point in space.

Thus, we have considered here a generalization of spherical waves in a form of linearly polarized beams with embedded optical vortices. These beams are vectorial counterparts of nonparaxial, strongly diverging scalar vortex Hankel beams of the first (Eq. 11.10), second (Eq. 11.28), and third (Eq. 11.30) types, considered recently in Ref. [342]. We have derived expressions for the amplitudes of all components of electric and magnetic vectors of the linearly polarized vortex Hankel beam. In all three types of Hankel beams, the intensity of transverse electric component is radially symmetric (doughnut-shaped) and devoid of side-lobes, while the distribution of the total beam intensity of the vectorial Hankel beam does not have radial symmetry. Based on the derived closed-form expressions, explicit expression has been derived for the intensity and for the axial projection of the OAM density have been obtained. From the expression for intensity it follows that the beams divergence in horizontal plane is $\sqrt{2}$ times larger than that in vertical plane (the horizontal plane is the plane of initial polarization). Rotation of the intensity pattern near initial plane (at a distance of few wavelengths) has been numerically demonstrated.

11.2 CIRCULARLY POLARIZED HANKEL VORTICES

There have been few vector laser beams that satisfy Maxwell's equations and for which explicit analytic relations to describe projections of the E and H field vectors have been derived. For such beams, exact calculation of energy flow and angular momentum (AM) is possible at any point in space. In this group, the most widely known are Bessel beams. For a symmetric zero-order Bessel beam, relations to express all six projections of the E and H field vectors in each point of space have been reported [347]. In Refs. [251,259,348,349] rather than considering the proper Bessel beam, projections of the E and H field vectors of its TE and TM modes were obtained. Expressions for an arbitrary-order vector Bessel beam with linear and circular polarization were deduced in Ref. [350]. A vector Bessel beam was derived in Ref. [350] using Hertz vector potentials. Along with the Bessel beam there is a well-known exact solution of Maxwell's equation—spherical wave. Spherical waves are highly diverging or converging, and they do not have the AM. Bessel beams are, conversely, neither diverging nor converging, but they do have AM. To the best of our knowledge, there are no known optical fields that are simultaneously: (1) described by an exact and relatively simple solution of the Maxwell equations, (2) having AM, and (3) diverging or converging like spherical wave. Most recently, we have introduced a novel vector beam (alongside the familiar Bessel beams) that can be fully defined by analytical relations. Nonparaxial scalar [342] and vector [351,352] Hankel beams with linear polarization were discussed. The beams were derived by using an expansion in terms of plane waves [344]. Vector cylindrical beams with radial and azimuthal polarization have also been proposed [353]. Note, however, that no analytical relations to express projections of the E and H field vectors of such beams have been derived so far. The cylindrical beams are analyzed using Richards-Wolf formulae and used for tightly focusing laser light [354].

Explicit formulae for projections of the E and H field vectors of a laser beam enable obtaining analytical relations to describe the Poynting vector and AM of the light field [17,246,350,352].

In this section, we expand the results we reported in Ref. [352] by deriving explicit relations to define the projections of the E and H field vectors of Hankel beams with clockwise and anticlockwise circular polarization. The relations derived suggest that in a Hankel beam with topological charge n, the amplitude of the longitudinal E and H field components is characterized by the topological charge of $n + 1$ and $n - 1$, respectively, for clockwise and anticlockwise polarization. We show that for non-negative topological charges Hankel beams with clockwise and anticlockwise circular polarization propagate in free space differently. The Hankel beam with the clockwise circular polarization has radial divergence (ratio between the radial and longitudinal projections of the Poynting vector) similar to that of the spherical wave, while the beam with the anticlockwise circular polarization has greater radial divergence. The beam with anticlockwise circular polarization has stronger longitudinal component of the electric vector than the beam with clockwise circular polarization. Laser beams under analysis can be generated by using high-resolution diffractive optical elements illuminated by circularly polarized light, showing promise for applications in optically trapping and rotating dielectric microobjects due to their nonparaxiality and absence of sidelobes.

Projections of the Electromagnetic Field Vectors for Clockwise and Anticlockwise Circular Polarization

In this section, we start with Maxwell's equations for monochromatic light propagating in free space:

$$\begin{cases} \text{rot } \mathbf{H} = -i\omega\varepsilon_0 \mathbf{E}, \\ \text{rot } \mathbf{E} = i\omega\mu_0 \mathbf{H}, \\ \text{div } \mathbf{E} = 0, \\ \text{div } \mathbf{H} = 0, \end{cases} \tag{11.45}$$

where:

w is the frequency of light

E and **H** are respectively electric and magnetic field strengths

ε_0 and μ_0 are the permittivity and permeability of free space

We suppose that the functions **E** and **H** depend only on spatial coordinates, while the time dependence factor exp($-iwt$) is omitted for brevity. From Eq. (11.45), the scalar Helmholtz equation follows

$$\nabla^2 P + k^2 P = 0, \tag{11.46}$$

where:

P is any Cartesian component of the vectors **E** or **H**
k is the wavenumber of light with a wavelength of λ

It is known [96] that if some function P is an exact solution of the Helmholtz equation (11.46), then its derivative with respect to the longitudinal coordinate $\partial P/\partial z$ is also a solution of this equation. So, if the transverse components of the electric strength read as $\partial P/\partial z$, that is, $E_x = \alpha_x \partial P/\partial z$ and $E_y = \alpha_y \partial P/\partial z$, where α_x and α_y are arbitrary constants, then the other components can be obtained by Maxwell's equations (11.45):

$$E_z = -\alpha_x \frac{\partial P}{\partial x} - \alpha_y \frac{\partial P}{\partial y}, \tag{11.47}$$

$$\mathbf{H} = -\frac{i}{k}\sqrt{\frac{\varepsilon_0}{\mu_0}} \text{rot } \mathbf{E}. \tag{11.48}$$

From now on, for obtaining the vector **H** by using Eq. (11.48) we will omit the factor $(\varepsilon_0/\mu_0)^{1/2}$ for brevity. In Ref. [352], an exact solution of the scalar Helmholtz equation was obtained (up to constant multipliers):

$$P(r,\varphi,z) = \frac{1}{2} i^{n+1} \lambda^{3/2} \left(re^{i\varphi}\right)^n \psi_{n+1/2}(R), \tag{11.49}$$

where (r, φ, z) are cylindrical coordinates, $R = (r^2 + z^2)^{1/2}$ and

$$\psi_\nu(R) = \frac{H_\nu^{(1)}(kR)}{R^\nu}, \tag{11.50}$$

where $H_\nu^{(1)}(x)$ is the νth order Hankel function of the 1-st kind. It can be verified by direct substitution of the function P into the Helmholtz equation (11.46), which leads to the following identity:

$$\left\{\frac{d^2}{d\xi^2} + \frac{1}{\xi}\frac{d}{d\xi} + \left[1 - \left(n + \frac{1}{2}\right)^2 \frac{1}{\xi^2}\right]\right\} H_{n+1/2}^{(1)}(\xi) = 0, \tag{11.51}$$

where $\zeta = kR$. This identity is obviously true since it is the Bessel differential equation. Earlier, we supposed that $n \geq 0$, but it can be shown that if a function $E(r, \varphi, z) = A(r, z)\exp(im\varphi)$ ($A(r, z)$ is an arbitrary complex function) is a solution of the Helmholtz equation (11.46), then a function $E(r, \varphi, z) = A(r, z)\exp(-im\varphi)$ is also a solution of this equation. Thus, the solution (11.49) can be generalized for the negative topological charges:

$$P(r,\varphi,z) = \frac{1}{2} i^{n+1} \lambda^{3/2} r^{|n|} e^{in\varphi} \psi_{|n|+1/2}(R). \tag{11.52}$$

Using the procedure (11.47), (11.48), and the function P of Eq. (11.49), linearly polarized Hankel beams were studied in Ref. [352]. Following the same calculation procedure, we can derive all projections of the electric $\mathbf{E} = (E_x, E_y, E_z)$ and magnetic $\mathbf{H} = (H_x, H_y, H_z)$ vectors of a Hankel beam with clockwise and anticlockwise circular polarization. To distinguish between the clockwise and anticlockwise circular polarizations, the amplitudes of the projections of the E and H field components will be marked with superscript "+" and "–". Then, the amplitudes of transverse components of electric vector of the Hankel beam with clockwise polarization are:

$$E^{+}_{ny}(r,\varphi,z) = iE_{nx}(r,\varphi,z) = i\frac{\partial P}{\partial z} = i^n \pi \sqrt{\lambda} z r^{|n|} e^{in\varphi} \psi_{|n|+3/2}(R). \tag{11.53}$$

Therefore, other components of the electromagnetic field read as ($\alpha_x = 1$ and $\alpha_y = i$ in [11.47])

$$\begin{aligned} E^{+}_{nz}(r,\varphi,z) &= -\frac{\partial P}{\partial x} - i\frac{\partial P}{\partial y} = -e^{i\varphi}\left(\frac{\partial P}{\partial r} + \frac{i}{r}\frac{\partial P}{\partial \varphi}\right) \\ &= \frac{1}{2} i^{n-1} \lambda^{3/2} r^{|n|-1} e^{i(n+1)\varphi} \left[\left(|n| - n\right)\psi_{|n|+1/2}(R) - kr^2 \psi_{|n|+3/2}(R)\right], \end{aligned} \tag{11.54}$$

$$\begin{aligned} H^{+}_{nx}(r,\varphi,z) &= \frac{-i}{k}\left(\frac{\partial E^{+}_{nz}}{\partial y} - i\frac{\partial E_{nx}}{\partial z}\right) \\ &= \frac{i^{n+1}\lambda^{3/2}}{2} r^{|n|} e^{in\varphi} \Big\{-\left(|n| - n\right)(n+1)k^{-1} r^{-2} e^{2i\varphi} \psi_{|n|+1/2}(R) \\ &+ \left[|n| + 2 - \left(|n| - n\right)e^{2i\varphi}\right]\psi_{|n|+3/2}(R) - k\left(z^2 - ir^2 e^{i\varphi}\sin\varphi\right)\psi_{|n|+5/2}(R)\Big\}, \end{aligned} \tag{11.55}$$

$$\begin{aligned} H^{+}_{ny}(r,\varphi,z) &= \frac{-i}{k}\left(\frac{\partial E_{nx}}{\partial z} - \frac{\partial E^{+}_{nz}}{\partial x}\right) \\ &= \frac{i^{n}\lambda^{3/2}}{2} r^{|n|} e^{in\varphi} \Big\{-\left(|n| - n\right)(n+1)k^{-1} r^{-2} e^{2i\varphi} \psi_{|n|+1/2}(R) \\ &- \left[|n| + 2 + \left(|n| - n\right)e^{2i\varphi}\right]\psi_{|n|+3/2}(R) + k\left(z^2 + r^2 e^{i\varphi}\cos\varphi\right)\psi_{|n|+5/2}(R)\Big\}, \end{aligned} \tag{11.56}$$

$$H_{nz}^{+}(r,\varphi,z)=\frac{-i}{k}\left(i\frac{\partial}{\partial x}-\frac{\partial}{\partial y}\right)E_{nx}=\frac{-i}{k}e^{i\varphi}\left(i\frac{\partial}{\partial r}-\frac{1}{r}\frac{\partial}{\partial\varphi}\right)E_{nx}$$
$$=i^{n+1}\pi\lambda^{1/2}zr^{|n|+1}e^{i(n+1)\varphi}\left[\psi_{|n|+5/2}(R)-(|n|-n)k^{-1}r^{-2}\psi_{|n|+3/2}(R)\right]. \tag{11.57}$$

For the Hankel beam with anticlockwise polarization, we have ($\alpha_x = 1$ and $\alpha_y = -i$ in (11.47)):

$$E_{ny}^{-}(r,\varphi,z)=-iE_{nx}(r,\varphi,z)=-i\frac{\partial P}{\partial z}=-i^{n}\pi\lambda^{1/2}zr^{|n|}e^{in\varphi}\psi_{|n|+3/2}(R), \tag{11.58}$$

$$E_{nz}^{-}(r,\varphi,z)=-\frac{\partial P}{\partial x}+i\frac{\partial P}{\partial y}=-e^{-i\varphi}\left(\frac{\partial P}{\partial r}-\frac{i}{r}\frac{\partial P}{\partial\varphi}\right)$$
$$=\frac{1}{2}i^{n-1}\lambda^{3/2}r^{|n|-1}e^{i(n-1)\varphi}\left[(|n|+n)\psi_{|n|+1/2}(R)-kr^{2}\psi_{|n|+3/2}(R)\right], \tag{11.59}$$

$$H_{nx}^{-}(r,\varphi,z)=\frac{-i}{k}\left(\frac{\partial E_{nz}^{-}}{\partial y}+i\frac{\partial E_{nx}}{\partial z}\right)$$
$$=\frac{i^{n-1}\lambda^{3/2}}{2}r^{|n|}e^{in\varphi}\left\{(|n|+n)(n-1)k^{-1}r^{-2}e^{-2i\varphi}\psi_{|n|+1/2}(R)\right. \tag{11.60}$$
$$+\left[|n|+2-(|n|+n)e^{-2i\varphi}\right]$$
$$\left.\psi_{|n|+3/2}(R)-k\left(z^{2}+ir^{2}e^{-i\varphi}\sin\varphi\right)\psi_{|n|+5/2}(R)\right\},$$

$$H_{ny}^{-}(r,\varphi,z)=\frac{-i}{k}\left(\frac{\partial E_{nx}}{\partial z}-\frac{\partial E_{nz}^{-}}{\partial x}\right)$$
$$=\frac{i^{n}\lambda^{3/2}}{2}r^{|n|}e^{in\varphi}\left\{(|n|+n)(n-1)k^{-1}r^{-2}e^{-2i\varphi}\psi_{|n|+1/2}(R)\right. \tag{11.61}$$
$$-\left[|n|+2+(|n|+n)e^{-2i\varphi}\right]$$
$$\left.\psi_{|n|+3/2}(R)+k\left(z^{2}+r^{2}e^{-i\varphi}\cos\varphi\right)\psi_{|n|+5/2}(R)\right\},$$

$$H_{nz}^{-}(r,\varphi,z)=\frac{-i}{k}\left(-i\frac{\partial}{\partial x}-\frac{\partial}{\partial y}\right)E_{nx}=\frac{i}{k}e^{-i\varphi}\left(i\frac{\partial}{\partial r}+\frac{1}{r}\frac{\partial}{\partial\varphi}\right)E_{nx}$$
$$=i^{n+1}\pi\lambda^{1/2}zr^{|n|+1}e^{i(n-1)\varphi}\left[(|n|+n)k^{-1}r^{-2}\psi_{|n|+3/2}(R)-\psi_{|n|+5/2}(R)\right]. \tag{11.62}$$

All these expressions (11.53) through (11.62) were verified by substitution into all eight scalar identities of Maxwell's equations (11.45) and by using finite-difference calculation of the partial derivatives.

From the comparison between (11.53) through (11.57) and (11.59) through (11.62) it is seen that if a Hankel beam with the positive topological charge n is circularly polarized clockwise, the amplitudes of the longitudinal projections E_{nz}^{+} and H_{nz}^{+} of the E and H field vectors have the topological charge $n + 1$, and vice versa, in the anticlockwise circularly polarized Hankel beam with positive topological charge n the amplitudes of the longitudinal projections E_{nz}^{-} and H_{nz}^{-} of the E and H field vectors have the topological charge $n - 1$. Similarly, if a Hankel beam with the negative topological charge n is circularly polarized anticlockwise, the amplitudes of the longitudinal projections E_{nz}^{-} and H_{nz}^{-} of the E and H field vectors have the negative topological charge $-|n| - 1$, and vice versa, in the clockwise circularly polarized Hankel beam with negative topological charge n the amplitudes of the longitudinal projections E_{nz}^{+} and H_{nz}^{+} of the E and H field vectors have the negative topological charge $-|n| + 1$. The topological charge of the rest projections remains unchanged and equal to n. As a result, in the absence of an optical vortex ($n = 0$), the longitudinal components of the Hankel beam with clockwise and anticlockwise circular polarization are optical vortices with the topological charge of 1:

$$E_{0z}^{\pm}(r,\varphi,z) = i\pi\lambda^{1/2} r e^{\pm i\varphi} \psi_{3/2}(R) = -i\lambda R^{-2}\left(1 + ik^{-1}R^{-1}\right) r e^{ikR \pm i\varphi}, \tag{11.63}$$

$$H_{0z}^{\pm}(r,\varphi,z) = \pm i\pi\lambda^{1/2} z r e^{\pm i\varphi} \psi_{5/2}(R), \tag{11.64}$$

while on the optical axis ($r = 0$) only transverse components of the electromagnetic field take non-zero values ($E_{0z}^{\pm}(r = 0,\varphi,z) = H_{0z}^{\pm}(r = 0,\varphi,z) = 0$):

$$E_{0y}^{+}(r = 0,\varphi,z) = iE_{0x}(r = 0,\varphi,z) = -E_{0y}^{-}(r = 0,\varphi,z) = \pi\lambda^{1/2} z \psi_{3/2}(z), \tag{11.65}$$

$$H_{0x}^{+}(r = 0,\varphi,z) = -H_{0x}^{-}(r = 0,\varphi,z) = \frac{i\lambda^{3/2}}{2}\left[2\psi_{3/2}(z) - kz^2\psi_{5/2}(z)\right], \tag{11.66}$$

$$H_{0y}^{+}(r = 0,\varphi,z) = H_{0y}^{-}(r = 0,\varphi,z) = -\frac{\lambda^{3/2}}{2}\left[2\psi_{3/2}(z) - kz^2\psi_{5/2}(z)\right], \tag{11.67}$$

At $n > 1$, all on-axis ($r = 0$) projections of the Hankel beam field equal zero. At $n = 1$, only longitudinal on-axis ($r = 0$) Hankel beam components with anticlockwise polarization are non-zero

$$E_{1z}^{-}(r = 0,\varphi,z) = \lambda^{3/2}\psi_{3/2}(z), \tag{11.68}$$

$$H_{1z}^{-}(r = 0,\varphi,z) = \lambda^{3/2} z \psi_{5/2}(z). \tag{11.69}$$

As a result, at $n = 1$, instead of generating a doughnut intensity at small distances z, the Hankel beam with anticlockwise circular polarization at first generates a focal spot, which is then transformed into a ring, although in the center of the ring the intensity is small, but not zero. This can be seen in Figure 11.9, which depicts

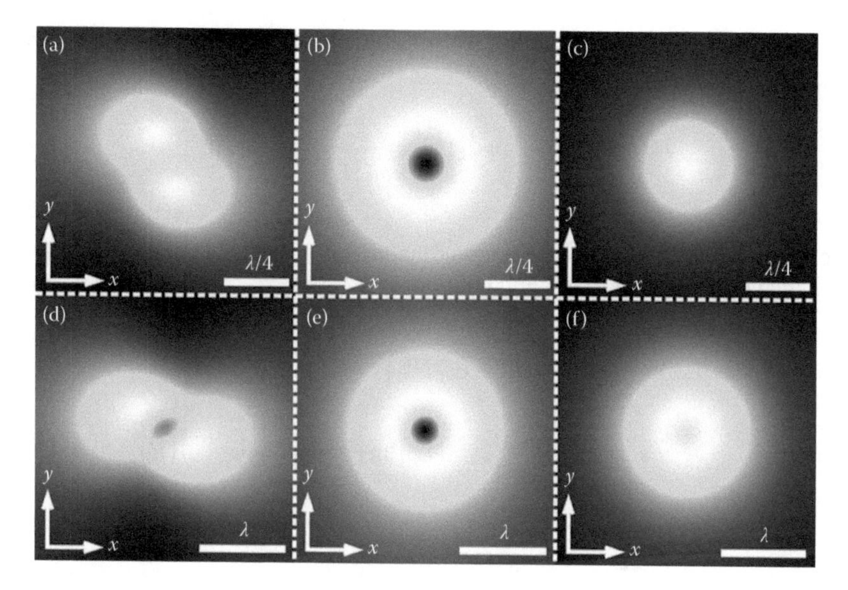

FIGURE 11.9 Intensity patterns $I = |E_x|^2 + |E_y|^2 + |E_z|^2$ from a Hankel beam (n = 1) with (a, d) linear, (b, e) clockwise circular, and (c, f) anticlockwise circular polarization at distances (a–c) $z = \lambda/4$, and (d–f) $z = \lambda/2$.

intensity patterns for a Hankel beam with n = 1, characterized by linear (E_y = 0) (Figure 11.9a and d), clockwise circular (Figure 11.9b and e), and anticlockwise circular (Figure 11.9c and f) polarization at distances $z = \lambda/4$ (Figure 11.9a–c) and $z = \lambda/2$ (Figure 11.9d–f). Shown in Figure 11.9a–c is the region $-\lambda/2 \le x, y \le \lambda/2$, while Figure 11.9(d–f) depicts the region $-3\lambda/2 \le x, y \le 3\lambda/2$.

Figure 11.9 suggests that while linearly polarized Hankel beams are devoid of circular symmetry, experiencing rotation in the near field upon propagation, circularly polarized Hankel beams are circularly symmetric, having a doughnut or circular intensity cross-section. The latter is also seen from the relation of the intensity for clockwise polarization, which is independent of the azimuthal angle:

$$I_n^+(r,z) = \pi^2\lambda r^{2|n|}\left(2z^2 + r^2\right)\left|\psi_{|n|+3/2}(R)\right|^2 + \frac{|n|-n}{2}\lambda^3 r^{2|n|-2}\left\{\frac{|n|-n}{2}\left|\psi_{|n|+1/2}(R)\right|^2 - kr^2\,\mathrm{Re}\left[\psi^*_{|n|+1/2}(R)\psi_{|n|+3/2}(R)\right]\right\}, \tag{11.70}$$

where Re(...) is the real part of the complex number. The intensity of a Hankel beam with anticlockwise circular polarization is also independent of the azimuthal angle:

$$I_n^-(r,z) = \pi^2\lambda r^{2|n|}\left(2z^2 + r^2\right)\left|\psi_{|n|+3/2}(R)\right|^2 + \frac{|n|+n}{2}\lambda^3 r^{2|n|-2}\left\{\frac{|n|+n}{2}\left|\psi_{|n|+1/2}(R)\right|^2 - kr^2\,\mathrm{Re}\left[\psi^*_{|n|+1/2}(R)\psi_{|n|+3/2}(R)\right]\right\}. \tag{11.71}$$

From (11.71), at $n = 1$, the on-axis intensity ($r = 0$) is seen to decrease with increasing z:

$$I_1^-(r=0,z) = \lambda^3 \left|\psi_{3/2}(z)\right|^2 = \frac{\lambda^4}{\pi^2 z^4}\left[1 + \frac{1}{(kz)^2}\right]. \tag{11.72}$$

Time-averaged power flow in the longitudinal direction is the same for both clockwise and anticlockwise circular polarizations:

$$\begin{aligned} S_{nz}^{\pm}(r,z) &= \frac{1}{2}\operatorname{Re}\left\{E_{nx}^{*}H_{ny}^{\pm} - E_{ny}^{\pm *}H_{nx}^{\pm}\right\} \\ &= -\frac{1}{2}\pi^2\lambda r^{2|n|}z\left(2z^2 + r^2\right)\operatorname{Im}\left\{\psi_{|n|+3/2}^{*}(R)\psi_{|n|+5/2}(R)\right\} \\ &= \frac{1}{2}\lambda^2 r^{2|n|}z\frac{2z^2 + r^2}{R^{2|n|+5}}. \end{aligned} \tag{11.73}$$

It seems interesting that according to Eq. (11.72) the on-axis intensity is always non-zero ($I_1^- \neq 0$), while from Eq. (11.73) it follows that there is no power flow along the optical axis ($S_{1z}^- = 0$). It can be explained since on the optical axis the Hankel beam with $n = 1$ has only longitudinal non-zero component (11.68), (11.69) of the light field. These longitudinal components do not affect the longitudinal power flow (11.73).

Comparison of Eqs. (11.70) and (11.71) at $n > 0$ shows that intensity of the Hankel beams with the clockwise and anticlockwise circular polarization is different. As it is shown in the following, this is due to the higher far-field divergence of the Hankel beam with the anticlockwise circular polarization compared to that with the clockwise circular polarization (Figure 11.10), although the power flow (11.73) along the optical axis is the same for the beams of both polarizations. The Hankel beam (11.71)

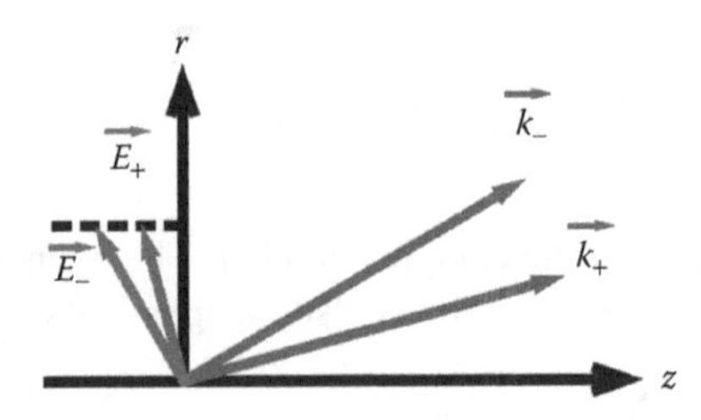

FIGURE 11.10 Higher divergence and greater longitudinal component of the electric vector of the Hankel beam with anticlockwise circular polarization compared to that with the clockwise circular polarization (k_+ and k_- are the wavevectors).

is diverging faster since the direction of rotation of the polarization plane coincides with the direction of increasing phase of the optical vortex (at $n > 0$), while the axial power flow (11.73) is the same since the transverse field components are the same in modulus: $|E_x^+| = |E_x^-| = |E_y^+| = |E_y^-|$.

It is seen in Eq. (11.73) that this power flow is always non-negative, that is, the Hankel beams do not possess the properties of tractor beams [355,356].

Vector Vortex Hankel Beams with Circular Polarization in the Far-Field

All of the above-obtained expressions are similar for similar components. For example, from Eqs. (11.70) and (11.71) the following rule follows, which we can use for intensity distributions of Hankel beams with negative topological charge: $I_{-|n|}^{\pm}(r,z) = I_{|n|}^{\mp}(r,z)$.

Therefore, from now on, without loss of generality we will assume that $n > 0$ since equations for complex amplitudes are cumbersome if they are written for both positive and negative topological charges.

Although we imply here that the Hankel beam propagates in the positive direction of the optical axis z, this beam is highly non-paraxial and under the far-field we mean the area where $R \gg \lambda$ (area B in Figure 11.11) instead of the area, where $z \gg r, \lambda$, as it is usually meant for paraxial beams (area A in Figure 11.11).

At $R \gg \lambda$ and for $n \geq 0$ the relations are simplified. Making use of an approximate Hankel relation ($x \gg 1$) [101] (Eq. 9.2.3)

$$H_{\nu}^{(1)}(x) \approx \sqrt{\frac{2}{\pi x}} (-i)^{\nu+1/2} e^{ix}, \tag{11.74}$$

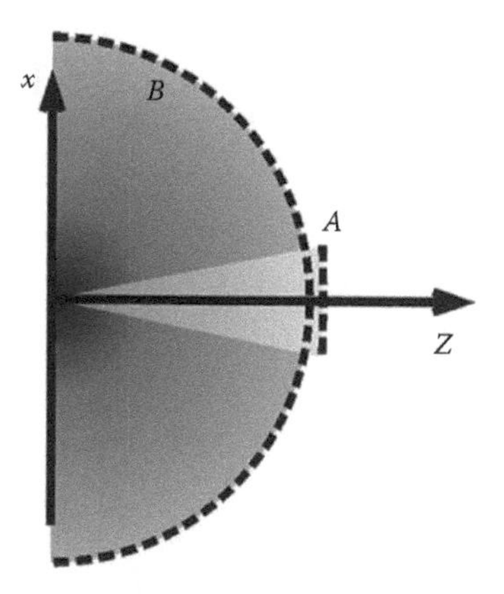

FIGURE 11.11 Far field of the Hankel beam (area B) and for arbitrary paraxial beam (area A).

Eq. (11.50) is rearranged to

$$\psi_\nu(R) = \frac{H_\nu^{(1)}(kR)}{R^\nu} \approx (-i)^{\nu+1/2} \frac{\lambda^{1/2} e^{ikR}}{\pi R^{\nu+1/2}}. \tag{11.75}$$

Substituting (11.75) into (11.53) through (11.57), we obtain for a Hankel beam with clockwise polarization:

$$E_{ny}^{+}(r,\varphi,z) = iE_{nx}(r,\varphi,z) = -\lambda \frac{r^n z}{R^{n+2}} e^{in\varphi+ikR}, \tag{11.76}$$

$$E_{nz}^{+}(r,\varphi,z) = -i\lambda \frac{r^{n+1}}{R^{n+2}} e^{i(n+1)\varphi+ikR}, \tag{11.77}$$

$$H_{nx}^{+}(r,\varphi,z) = -\lambda\left(i\frac{n+2}{kR^{n+2}} + \frac{ir^2 e^{i\varphi}\sin\varphi - z^2}{R^{n+3}}\right) r^n e^{in\varphi+ikR}, \tag{11.78}$$

$$H_{ny}^{+}(r,\varphi,z) = i\lambda\left(\frac{z^2 + r^2 e^{i\varphi}\cos\varphi}{R^{n+3}} - i\frac{n+2}{kR^{n+2}}\right) r^n e^{in\varphi+ikR}, \tag{11.79}$$

$$H_{nz}^{+}(r,\varphi,z) = -\lambda \frac{r^{n+1} z}{R^{n+3}} e^{i(n+1)\varphi+ikR}. \tag{11.80}$$

In a similar way, relations (11.59) through (11.62) to describe a Hankel beam with anticlockwise polarization in the far-field are:

$$E_{ny}^{-}(r,\varphi,z) = -iE_{nx}(r,\varphi,z) = \lambda \frac{r^n z}{R^{n+2}} e^{in\varphi+ikR}, \tag{11.81}$$

$$E_{nz}^{-}(r,\varphi,z) = -i\lambda \frac{r^{n+1}}{R^{n+2}}\left(1 - \frac{2inR}{kr^2}\right) e^{i(n-1)\varphi+ikR}, \tag{11.82}$$

$$\begin{aligned} H_{nx}^{-}(r,\varphi,z) = -\frac{\lambda}{k}\frac{r^n}{R^{n+3}}\Bigg\{ & \frac{2n(n-1)e^{-2i\varphi}}{kr^2}R^2 \\ & + i\left(2ne^{-2i\varphi} - n - 2\right)R + k\left(z^2 + ir^2 e^{-i\varphi}\sin\varphi\right)\Bigg\} e^{in\varphi+ikR}, \end{aligned} \tag{11.83}$$

$$\begin{aligned} H_{ny}^{-}(r,\varphi,z) = -\frac{\lambda}{k}\frac{r^n}{R^{n+3}}\Bigg\{ & i\frac{2n(n-1)e^{-2i\varphi}}{kr^2}R^2 \\ & - \left(2ne^{-2i\varphi} + n + 2\right)R - ik\left(z^2 + r^2 e^{-i\varphi}\cos\varphi\right)\Bigg\} e^{in\varphi+ikR}, \end{aligned} \tag{11.84}$$

$$H_{nz}^{-}(r,\varphi,z)=\lambda\frac{r^{n+1}z}{R^{n+3}}\left(1-\frac{2inR}{kr^2}\right)e^{i(n-1)\varphi+ikR}. \tag{11.85}$$

Thus, instead of Eqs. (11.70) through (11.72), the far-field intensity distributions for clockwise and anticlockwise polarizations are given by

$$\begin{aligned}I_n^{+}(r,z)&=\left|E_{nx}(r,\varphi,z)\right|^2+\left|E_{ny}^{+}(r,\varphi,z)\right|^2+\left|E_{nz}^{+}(r,\varphi,z)\right|^2\\&=\lambda^2\frac{r^{2n}}{R^{2n+4}}\left(r^2+2z^2\right),\end{aligned} \tag{11.86}$$

$$\begin{aligned}I_n^{-}(r,z)&=\left|E_{nx}(r,\varphi,z)\right|^2+\left|E_{ny}^{-}(r,\varphi,z)\right|^2+\left|E_{nz}^{-}(r,\varphi,z)\right|^2\\&=\lambda^2\frac{r^{2n}}{R^{2n+4}}\left(r^2+2z^2+\frac{4n^2R^2}{k^2r^2}\right).\end{aligned} \tag{11.87}$$

It is seen in Eqs. (11.86) and (11.87) that for the anticlockwise polarization the far-field intensity is always greater than that for the clockwise polarization. Since the transverse components have the similar intensity, it follows that the intensity of the longitudinal field component is greater for the anticlockwise polarization. If $n = 0$ then $I_0^{+}(r,z)=I_0^{-}(r,z)=\lambda^2\left(r^2+2z^2\right)R^{-4}$.

For clockwise polarization, the intensity maximum is found on a ring of radius

$$r_{\max}^{+}=z\sqrt{\frac{n-3+\sqrt{n^2+2n+9}}{2}}. \tag{11.88}$$

The ring radius only equals zero (resulting in an on-axis intensity peak) when $n = 0$. At large topological charges ($n \gg 1$), with increasing z, the radius of the intensity maximum increases as $r_{\max}^{+}=z\sqrt{n}$. For anticlockwise polarization, the intensity maximum is defined by a more complicated relation:

$$(n-3)\frac{r^4}{z^4}+2n\frac{r^2}{z^2}-\frac{r^6}{z^6}+\frac{4n^2(n-3)}{(kz)^2}\frac{r^2}{z^2}+\frac{4n^2(n-1)}{(kz)^2}-\frac{8n^2}{(kz)^2}\frac{r^4}{z^4}=0, \tag{11.89}$$

but putting $z \gg \lambda$, the denominators in the three last terms are much larger than the numerators, allowing us to retain just the first three terms. In this case, (11.89) is rearranged, resulting in the intensity maximum relation identical to that for clockwise circular polarization:

$$r_{\max}^{-}\approx z\sqrt{\frac{n-3+\sqrt{n^2+2n+9}}{2}}. \tag{11.90}$$

For circularly polarized Hankel beams, the Poynting vector $\mathbf{S}=\mathrm{Re}\{\mathbf{E}^{*}\times\mathbf{H}\}/2$ could be deduced from the relations for projections of the E and H fields in (11.53) through (11.62). However, the resulting cumbersome relations would be difficult to analyze.

Only the longitudinal projection of the Poynting vector (11.73) is described by the simple expression for any value of z. Meanwhile, far-field projections of the Poynting vector of a circularly polarized Hankel beam are described by less sophisticated relations. The exact expression for the longitudinal projection of vector **S** is written earlier (Eq. 11.73) and in the following we only give relations for transverse projections of vector **S**, which contribute to the on-axis projection of the AM vector. Then, for clockwise circular polarization:

$$\begin{aligned} S_{nx}^{+}(r,\varphi,z) &= \frac{1}{2}\operatorname{Re}\left\{E_{ny}^{+*}H_{nz}^{+} - E_{nz}^{+*}H_{ny}^{+}\right\} \\ &= \frac{1}{2}\lambda^2 \frac{r^{2n+1}}{R^{2n+5}}\left[\left(r^2+2z^2\right)\cos\varphi - (n+2)\frac{R}{k}\sin\varphi\right], \end{aligned} \tag{11.91}$$

$$\begin{aligned} S_{ny}^{+}(r,\varphi,z) &= \frac{1}{2}\operatorname{Re}\left\{E_{nz}^{+*}H_{nx}^{+} - E_{nx}^{*}H_{nz}^{+}\right\} \\ &= \frac{1}{2}\lambda^2 \frac{r^{2n+1}}{R^{2n+5}}\left[\left(r^2+2z^2\right)\sin\varphi + (n+2)\frac{R}{k}\cos\varphi\right], \end{aligned} \tag{11.92}$$

and for anticlockwise circular polarization:

$$\begin{aligned} S_{nx}^{-}(r,\varphi,z) &= \frac{1}{2}\operatorname{Re}\left\{E_{ny}^{-*}H_{nz}^{-} - E_{nz}^{-*}H_{ny}^{-}\right\} \\ &= \frac{1}{2}\lambda^2 \frac{r^{2n+1}}{R^{2n+5}}\left\{\left(2z^2+r^2\right)\cos\varphi - \left(4n\frac{z^2}{r^2}+n-2\right)\left(\frac{R}{k}\right)\sin\varphi\right. \\ &\quad \left. + \frac{2n}{r^2}(2n+3)\left(\frac{R}{k}\right)^2\cos\varphi - \frac{4n^2(n-1)}{r^4}\left(\frac{R}{k}\right)^3\sin\varphi\right\}, \end{aligned} \tag{11.93}$$

$$\begin{aligned} S_{ny}^{-}(r,\varphi,z) &= E_{nz}^{-*}(r,\varphi,z)H_{nx}^{-}(r,\varphi,z) - E_{nx}^{*}(r,\varphi,z)H_{nz}^{-}(r,\varphi,z) \\ &= \frac{1}{2}\lambda^2 \frac{r^{2n+1}}{R^{2n+5}}\left\{\left(2z^2+r^2\right)\sin\varphi + \left(4n\frac{z^2}{r^2}+n-2\right)\left(\frac{R}{k}\right)\cos\varphi\right. \\ &\quad \left. + \frac{2n}{r^2}(2n+3)\left(\frac{R}{k}\right)^2\sin\varphi + \frac{4n^2(n-1)}{r^4}\left(\frac{R}{k}\right)^3\cos\varphi\right\}. \end{aligned} \tag{11.94}$$

These expressions are cumbersome, but they allow obtaining of radial and azimuthal projections of the Poynting vector which are simpler and have clearer physical meaning since they are related with the beam divergence and with the beam AM. For the clockwise circular polarization, we obtain:

$$S_{nr}^{+}(r,z) = S_{nx}^{+}\cos\varphi + S_{ny}^{+}\sin\varphi = \frac{1}{2}\lambda^2 \frac{r^{2n+1}}{R^{2n+5}}\left(r^2 + 2z^2\right) \equiv \frac{r}{z}S_{nz}^{+},$$
$$S_{n\varphi}^{+}(r,z) = S_{ny}^{+}\cos\varphi - S_{nx}^{+}\sin\varphi = \frac{1}{2}\lambda^2 (n+2)\frac{r^{2n+1}}{kR^{2n+4}}, \tag{11.95}$$

while for the anticlockwise circular polarization similar expressions are longer:

$$S_{nr}^{-}(r,z) = \frac{1}{2}\lambda^2 \frac{r^{2n+1}}{R^{2n+5}}\left[r^2 + 2z^2 + \frac{2n}{r^2}(2n+3)\left(\frac{R}{k}\right)^2\right],$$
$$S_{n\varphi}^{-}(r,z) = \frac{1}{2}\lambda^2 \frac{r^{2n+1}}{kR^{2n+4}}\left[4n\frac{z^2}{r^2} + n - 2 + \frac{4n^2(n-1)}{r^4}\left(\frac{R}{k}\right)^2\right]. \tag{11.96}$$

It is seen in these equations that radial divergence S_{nr}^{+}/S_{nz}^{+} of the Hankel beam with clockwise circular polarization is independent on the topological charge n and is equal to z/r, that is, it is similar to that of the spherical wave (Figure 11.12). In addition, the Hankel beam with zero topological charge has non-zero azimuthal component of the Poynting vector: $S_{0\varphi}^{+} = -S_{0\varphi}^{-} = \lambda^2 r/(kR^4)$. This means, that for the clockwise polarization $S_{0\varphi}^{+} > 0$, that is, the power flow rotates anticlockwise, while for the anticlockwise polarization, vice versa, $S_{0\varphi}^{-} < 0$ and the power flow rotates clockwise. The radial divergence S_{nr}^{-}/S_{nz}^{-} of the Hankel beam with the anticlockwise circular polarization always exceeds the radial divergence S_{nr}^{+}/S_{nz}^{+} of the Hankel beam with the clockwise circular polarization. This explains why the intensity of the longitudinal field component is greater for the anticlockwise circular polarization.

Along with the forward-propagating Hankel beams, which travel predominantly in the positive direction of the optical axis z, we can consider back-propagating Hankel beams by changing variable $z \to f - z$ (f is the focal distance). According to (11.95) and (11.96), such back-propagating Hankel vortices are converging as sharply as converging spherical wave. Therefore, Hankel vortices have the potential to generate tight focus with highly concentrated light energy and with the doughnut intensity distribution, typical to the optical vortices.

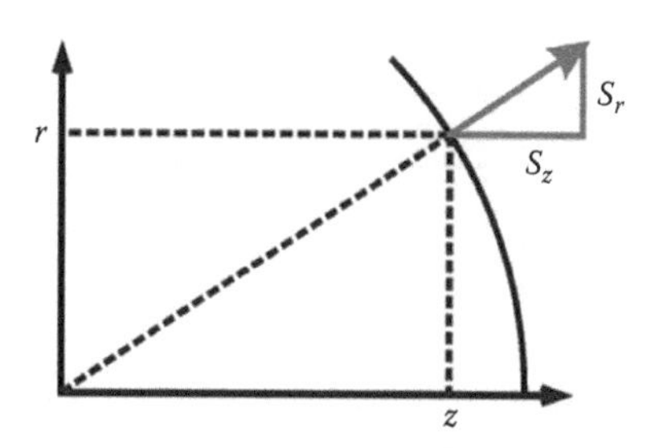

FIGURE 11.12 Radial divergence of the Hankel beam with clockwise circular polarization.

The far-field z-projections of the AM density vector $\mathbf{j} = \mathbf{r} \times \mathbf{S}$ for Hankel beams with clockwise and anticlockwise circular polarizations are given by

$$j_{nz}^{+}\left(r,\varphi,z\right) = rS_{n\varphi}^{+} = \frac{1}{2}\lambda^2\left(n+2\right)\frac{r^{2n+2}}{kR^{2n+4}}, \tag{11.97}$$

$$j_{nz}^{-}\left(r,\varphi,z\right) = rS_{n\varphi}^{-} = \frac{1}{2}\lambda^2\frac{r^{2n+2}}{kR^{2n+4}}\left[4n\frac{z^2}{r^2}+n-2+\frac{4n^2\left(n-1\right)}{r^4}\left(\frac{R}{k}\right)^2\right] \tag{11.98}$$

It can be shown that at $r = 0$, rather than tending to infinity, the AM density in (11.97) and (11.98) equals zero irrespective of polarization state. From (11.97) the total AM of the Hankel beam is seen to be infinite. Note that at $n = 0$, the Hankel beam still carries the spin AM:

$$j_{0z}^{+}\left(r,\varphi,z\right) = -j_{0z}^{-}\left(r,\varphi,z\right) = \frac{\lambda^2}{k}\frac{r^2}{R^4}. \tag{11.99}$$

The non-zero spin AM explains why there is a non-zero azimuthal component of the Poynting vector at $n = 0$.

For any $n > 0$, Hankel beam with the clockwise polarization has a positive axial AM (11.97), that is, the power flow in such beam propagates along a right-handed helix. Hankel beam with the anticlockwise polarization propagates as a left helix at $n = 0$ and as a right helix at $n \geq 2$. At $n = 1$ instead of Eq. (11.98) we obtain

$$j_{1z}^{-}\left(r,\varphi,z\right) = \frac{\lambda^2 r^4}{2kR^6}\left(\frac{4z^2}{r^2}-1\right). \tag{11.100}$$

It is seen in Eq. (11.100) that $j_{1z}^{-} > 0$ at $z > r/2$ and, vice versa, $j_{1z}^{-} \leq 0$ at $z \leq r/2$ (Figure 11.13), that is, the power flow in the Hankel beam with anticlockwise

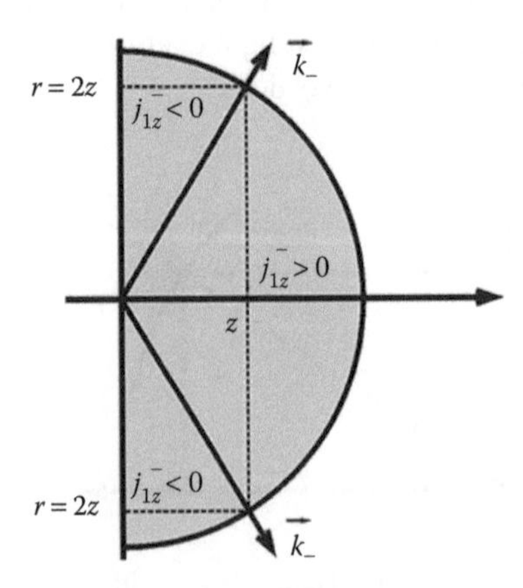

FIGURE 11.13 Direction of longitudinal AM for the Hankel beam with anticlockwise polarization with unitary topological charge $n = 1$.

polarization looks like two helices—right-handed near-axis helix nested into left-handed peripheral helix. This means that a Rayleigh microscopic particle, whose size is less than the wavelength, placed into different areas of the Hankel beam, rotates in different directions.

It is seen in Eqs. (11.97) through (11.99) that the longitudinal projection of the AM density is always positive with exception of the anticlockwise circularly polarized Hankel beam with zero or unitary topological charge. For $n = 0$, AM density coincides with the spin AM density and it is negative in every point in space.

From Eqs. (11.97) through (11.99) it follows that for both circular polarizations maximal density of the AM of the zero-order ($n = 0$) Hankel beam is on a circle $r = z$. When the topological charge is $n = 1$, the circle of maximal AM density becomes dependent on the polarization. For clockwise polarization, this circle widens and its radius is $r = \sqrt{2}z$, while for the anticlockwise polarization this circle, conversely, shrinks and its radius is $r = \sqrt{25-\sqrt{21}}\,z \approx 0.65z$. Figure 11.14 shows distributions of the longitudinal projection of the AM density for Hankel beams at $z = 50\lambda$ with clockwise and anticlockwise circular polarization with zero and unitary topological charge. Calculation area $-200\lambda \leq x, y \leq 200\lambda$. Widening and shrinking of the AM density distribution for $n = 1$ (compared to $n = 0$) is seen in Figure 11.14c and d. In addition, it is clearly seen in Figure 11.14d that at $r = 2z = 100\lambda$ the longitudinal AM projection changes, that is, at this value of radius $\bar{j}_{1z} = 0$ and a Rayleigh particle placed here would not rotate at all.

Summing up, we have derived relations to define all six projections of the E and H field vectors for circularly polarized vector vortex Hankel beams. These relations fully satisfy Maxwell's equations. For the non-negative topological charge, Hankel beams with clockwise and anticlockwise circular polarization have been shown to propagate in free space in a different manner. Thus, for a Hankel beam with the topological charge $n \geq 0$ the amplitude of the longitudinal E and H field components has the singularity order of $n + 1$ and $n - 1$, respectively, for clockwise and anticlockwise circular polarization. Expressions to describe the total E field in the entire space have been derived. The said relations have been found to differ for Hankel beams with clockwise and anticlockwise circular polarization, also demonstrating that the cross-section of circularly polarized Hankel beams shows circular symmetry. In the

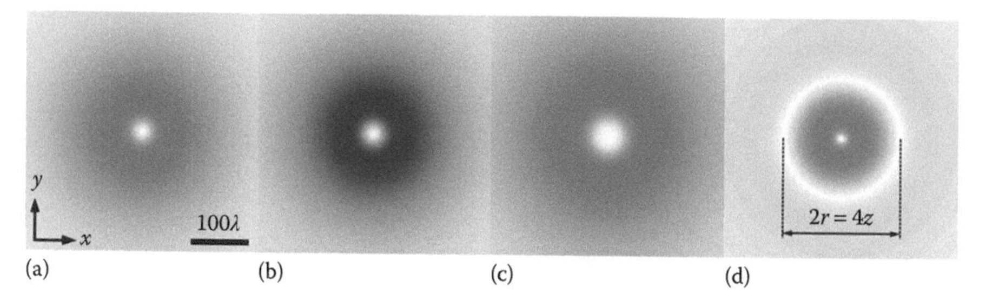

FIGURE 11.14 Longitudinal projection of the AM density for Hankel beams at $z = 50\lambda$ with (a, c) clockwise and (b, d) anticlockwise circular polarization with topological charge (a, b) $n = 0$ and (c, d) $n = 1$. Area 1—positive values, Area 2—negative values.

meantime, linearly polarized Hankel beams are devoid of circular symmetry [9]. Relations to describe projections of the Poynting vector and the AM for the far-field diffraction have also been derived. It is shown that a Hankel beam with clockwise circular polarization has radial divergence (ratio between the radial and longitudinal projections of the Poynting vector) similar to that of the spherical wave, while the beam with the anticlockwise circular polarization has greater radial divergence. At $n = 0$, the circularly polarized Hankel beam has non-zero spin AM. At $n = 1$, power flow of the Hankel beam with anticlockwise polarization consists of two parts: right-handed helical flow near the optical axis and left-handed helical flow in periphery. At $n \geq 2$, power flow is directed along the right-handed helix regardless of direction of the circular polarization. Power flow along the optical axis is the same for the Hankel beams of both circular polarizations if they have the same topological charge.

12 CONCLUSION

There is a problem of sustainable wireless communication through atmospheric turbulence. To address this problem, it is necessary to find laser beams which are resistant to the atmosphere distortions. For example, it is known that, unlike the Gaussian beam, the Bessel [1] and Pearcey [8] beams have a self-healing property. When these beams meet a small-sized obstacle, they restore its shape after some propagation distance. However, in a strong turbulence the self-healing property does not work, since the beam meets hundreds of phase inhomogeneities, comparable to the beam diameter. Therefore, there is a need in new beams and new methods of registering the information. In this book, we address this problem by researching new vortex laser beams with orbital angular momentum (OAM), which are resistant to the turbulence distortion. At this, instead of intensity distribution in the beam cross section, the phase value should be registered around the phase-singularity point since the turbulence distorts the intensity and phase of the vortex beam but conserves its topological charge.

The relevance of the studied problems for physics is due to the fact that theoretical results were obtained by using the analytically derived new solutions of the Helmholtz equation and its paraxial approximation (the Schrödinger-type equation), and therefore these results are universal for wave theories that occur in different areas of physics.

The relevance to optics is that light fields with OAM are and continue to be found useful for a wide class of practical problems. For example, in the area of fiber communications one prominent work was Ref. [357] where the authors demonstrate the viability of using the OAM of light to create orthogonal streams of data-transmitting channels that are multiplexed in a single fiber. Using over 1.1 kilometres of a specially designed optical fiber that minimizes mode coupling, 400-gigabits-per-second data transmission was achieved using four angular momentum modes at a single wavelength, and 1.6 terabits per second using two OAM modes over 10 wavelengths. In addition to fiber communications, more and more works appear with study of light beams with the OAM for wireless communications. These beams have the property to restore themselves after passing small-sized obstacles. Due to this property, such beams attracted an interest for the wireless transmission of information through a turbulent atmosphere. Today, research in this area is continuing. For example, in Ref. [358], the effects are modeled of turbulence on the spread and crosstalk of the spiral spectrum of the OAM states for Hankel-Bessel-Schell beams in paraxial and weak non-Kolmogorov turbulence channels. It is shown that such a beam is a better light source which has the ability to weaken turbulence spreading of the beams and can mitigate the effects of turbulence on the detection probability of the signal OAM state. In Ref. [359], OAM modes in a vortex Gaussian beam are studied theoretically after passing through weak-to-strong atmospheric turbulence. Numerically, the roles of the vortex Gaussian beam index, initial beam radius, and turbulence strength are explored, and the validity of the pure-phase-perturbation approximation employed in existing theoretical studies is examined. In Ref. [360], by using an adaptive feedback correction technique, turbulence compensation was experimentally demonstrated for free-space OAM links. The performance of OAM beams through emulated atmospheric turbulence and adaptive optics assisted compensation loop is investigated. The experimental results show that the scheme can efficiently compensate for the atmospheric turbulence induced distortions, that is, reducing power fluctuation of OAM channels, suppressing inter-channel crosstalk, and improving the bit-error rate performance. In Ref. [361], there is discussed an experimental demonstration of a free-space optical communication system where a light beam carrying different OAM modes and affected by turbulence is coupled to the multimode fiber link. In Ref. [362], the authors use a Gaussian beacon on a separate wavelength as a means to provide the information necessary to compensate for the effects of atmospheric turbulence on OAM and polarization-multiplexed beams in a free-space optical link. The influence of the Gaussian beacon's wavelength on the compensation of the OAM beams at 1560 nm is experimentally studied. It is found that the compensation performance degrades slowly with the increase in the beacon's wavelength offset, in the 1520–1590 nm band, from the OAM beams. Using this scheme, a 1 Tbit/s OAM and polarization-multiplexed link through emulated dynamic turbulence with a data-carrying beacon at 1550 nm is experimentally demonstrated. The experimental results show that the turbulence effects on all 10 data channels, each carrying a 100 Gbit/s signal, are mitigated efficiently, and the power penalties after compensation are below 5.9 dB for all channels.

Theoretically, there is an infinite number of orthogonal states with different OAM. Therefore, spatial modes of light can potentially carry a vast amount of information, making them promising candidates for both classical and quantum communication. However, the distribution of such modes over large distances remains difficult. Intermodal coupling complicates their use with common fibers, whereas free-space transmission is thought to be strongly influenced by atmospheric turbulence. In Ref. [363], the authors experimentally show the transmission of OAM modes of light over a distance of 143 km between two Canary Islands, which is 50× greater than the maximum distance achieved previously. As a demonstration of the transmission quality, the authors use superpositions of these modes to encode a short message. At the receiver, an artificial neural network is used for distinguishing between the different twisted light superpositions. The algorithm is able to identify different mode superpositions with an error rate of 8.33%. The quality of this free-space link can be further improved by the use of state-of-the-art adaptive optics systems.

Along with the over-ground atmosphere, there are studies of using the light beams with the OAM for communications in the marine [364] and even underwater [365,366] environments. So, in Ref. [364], there is a study of the effect of the anisotropic non-Kolmogorov turbulence of the marine atmosphere on propagation of OAM modes carried by partially coherent modified Bessel-Gaussian beams. It is concluded that such beams are robust for the OAM mode propagation in turbulence with long wavelength, low quantum number of the OAM mode, and a high spectral degree of coherence. In Ref. [365] an underwater wireless optical communications link over a 2.96 m distance with two 445-nm fiber-pigtailed laser diodes employing OAM to allow for spatial multiplexing is experimentally demonstrated. A data rate of 3 Gbit/s was achieved in water. To increase system capacity of underwater optical communications, in Ref. [366] the spatial domain is also used to simultaneously transmit multiple orthogonal spatial beams, each carrying an independent data channel. In this paper, it is shown up to a 40-Gbit/s link by multiplexing and transmitting four green OAM beams. Moreover, the degrading effects are investigated of scattering/turbidity, water current, and thermal gradient-induced turbulence, and it is found that thermal gradients cause the most distortions and turbidity causes the most loss. Finally, it is shown that inter-channel crosstalk induced by thermal gradients can be mitigated using multi-channel equalization processing.

It should be noted that in all these works the standard rotationally symmetric optical vortices were used. The proposed methods of compensation of turbulence-induced distortions can improve the quality of wireless communication; however, the possibility of many new types of light beams with OAM has not been investigated for data transmission. In addition, a well-known problem of using optical vortices for optical communications is the increase in their divergence with the increase of the topological charge. While the transverse dimensions of a fiber mode remain on propagation, using the higher-order modes in free-space requires increasing the size of the receiver and consequently its cost. Therefore, searching and investigation of alternative optical beams with OAM seems a relevant task.

Along with the classical optical communications, optical fields with the OAM are also investigated for quantum communications. For this purpose, it is necessary to generate OAM-entangled photons. Detection of such photons by using a fork-hologram was studied in 2001 in Ref. [22]. In a recent paper [325], the authors report a novel experimental scheme to generate a perfect vortex of any ring radius using a convex lens and an axicon by adjusting the separation between them. Using such vortices, in this paper the authors study a non-collinear interaction of photons having the OAM in spontaneous parametric down-conversion (SPDC) process and observed that the angular spectrum of the SPDC photons is independent of OAM of the pump photons but, rather, depends on spatial profile of the pump beam. In the presence of spatial walk-off effect in nonlinear crystals, the SPDC photons have an asymmetric angular spectrum with reducing asymmetry at increasing vortex radius.

Along with fiber and wireless communications, light fields with the OAM are used for optical trapping of microobjects. The problem of simulation of manipulating microscopic objects by laser radiation was considered in many ways. For the first time, a theoretical problem of light pressure on a sphere by using the Mie theory was considered by Debye in 1909 [367]. Generally, force of light pressure on a microparticle must be considered by using the Maxwell stress tensor of the electromagnetic (EM) field [368]. In Ref. [369], by using the stress tensor an expression was obtained for the force of light acting on a spherical particle with the Kerr nonlinearity. To calculate the forces acting on a microobject by an EM field, they usually use the finite difference solution of Maxwell's equations [370] or analytical expression in the case of spherical particles [371]. For example, in Ref. [372] the authors used the EM field stress tensor to calculate forces of light from the taper on small particles in the near-focus area in the presence of an AFM cantilever. Calculation of the field was done by the finite-difference FDTD method. In Ref. [373], the authors calculated the force and the torque onto the Rayleigh elliptical particles. It is also of interest to study the forces on microparticles where the incident light leads to resonance effects. For example, in Ref. [374], forces of light on the Rayleigh particles were experimentally examined in the presence of optical resonance. The resonance in the particles was due to the addition of dyes and the increase of light absorption at the resonant wavelength. Approximate methods for calculating the forces exerted by the beam on the particle are also popular: the method of geometrical optics with the Fresnel reflection and refraction taken into account [375,376].

In order to increase the efficiency of trapping of microscopic objects, a high concentration of laser beam power is necessary, which can be achieved by its sharp focusing. In the case of sharp focusing, rigorous calculation of optical force and torque on the object, as well as the Maxwell stress tensor, requires knowing all the components of the electric and magnetic field. The complex amplitude of many paraxial beams, used for optical trapping, is not a rigorous solution of Maxwell's equations, and therefore the distribution of their amplitudes in the focus of the lens with a high numerical aperture has no analytical expression. This means that there is a need to find new exact solutions of Maxwell's equations. If the medium is homogeneous, an exact solution of Maxwell's equations can be derived from the exact solution of the nonparaxial Helmholtz equation, but not many solutions are known. Known solutions are usually obtained by separating the variables in some coordinate systems. Thus, separation of variables in Cartesian coordinates leads to a plane wave, though it does not suit for optical trapping. Separation of variables in cylindrical coordinates leads to the diffraction-free Bessel modes [1]. In Ref. [246], the OAM density of Bessel beams with linear, radial, and azimuthal polarization is calculated explicitly within a rigorous vectorial treatment. Mechanical transfer of OAM to trapped particles in optical tweezers using a high-order Bessel beam was demonstrated experimentally. In Ref. [251] there is a description of vectorial Bessel beams, obtained by using vectorial Hertzian potential. Expressions for all components are derived as well as for the energy density and Poynting vector. However, there are no expressions for the OAM. In Ref. [377], nondiffracting Bessel beams are generated whose polarization state varies with propagation distance, although there are no equations for the complex amplitude of such fields. In a recent paper [378], an analytical description of optical force on a Rayleigh particle by a vector Bessel beam is investigated. Linearly, radially, azimuthally, and circularly polarized Bessel beams are considered. The radial, azimuthal, and axial forces by a vector Bessel beam are numerically simulated. Numerical calculations show that optical forces, especially azimuthal forces, are very sensitive to the polarization of beams. In a 2017 paper [379], axial and transverse radiation force cross-sections of optical tractor Bessel polarized beams are theoretically investigated for a dielectric sphere with particular emphasis on the beam topological charge (or order), half-cone angle, and polarization. The angular spectrum decomposition method (ASDM) is used to derive the non-paraxial EM field components of the Bessel beams. The focus of this investigation is to identify some of the tractor beam conditions so as to achieve retrograde motion of a dielectric sphere. In Refs. [377–379], there are studies of one of the most known Bessel beams. However, it has a rotational symmetry and thus turns microobjects to rotate along a circular trajectory and, in addition, has a quantized OAM. Therefore, in the following we consider briefly the alternative light fields, using which new types of light traps can be created that cause micro-objects to move on other paths.

Based on the separability of the Helmholtz equation into elliptical cylindrical coordinates, Ref. [98] describes invariant optical fields that may have a highly localized distribution along one of the transverse directions and a sharply peaked quasi-periodic structure along the other. These fields are described by the radial and angular Mathieu functions. In Ref. [380], experimental generation of high-order Mathieu beams is described. There are also equations for the OAM, although they are obtained for the paraxial case. Ref. [381] describes the transfer of OAM to trapped particles in a helical Mathieu beam. The average rotation rate and instantaneous angular displacement of the trapped particles are measured experimentally. This is an experimental work and there are no expressions for the OAM. Ref. [97] demonstrates nondiffracting parabolic beams and determines their associated angular spectrum. Their transverse structure is described by parabolic cylinder functions and looks like set of light parabolas. Experimental generation of parabolic beams and using them for trapping and manipulation of microscopic particles is reported in Ref. [382]. It is written that the particles acquire OAM and exhibit an open trajectory following the parabolic fringes of the beam. This work is also purely experimental and there are no expressions for the OAM and optical force. Among the other well-known exact solutions of the Helmholtz equation we can mention our work [257] in which we described nonparaxial Hankel-Bessel laser beams. Now, their properties for optical communications are studied [258,383], but their vector counterparts and applicability to trap microscopic objects have not been studied.

Thus, it seems a relevant problem of finding new exact solutions of the Maxwell and Helmholtz equations, as well as studying their optical properties. The scientific significance of these solutions is the analytical description of new-type optical traps and the possibility of calculating the forces acting on the trapped microparticles.

The other applications of laser beams with OAM include measurements of the frequency of the rotating object and the vorticity of the fluid flow by using the rotational Doppler shift [384,385], sample concentration measurements [386], and signal processing [387].

References

1. Durnin, J., J. Miceli Jr, and J.H. Eberly. 1987. Diffraction-free beams. *Phys. Rev. Lett.* 58:1499–1501.
2. Siegman, A.E. 1986. *Lasers*. Mill Valley, CA: University Science Books.
3. Abramochkin, E.G., and V.G. Volostnikov. 2004. Generalized Gaussian beams. *J. Opt. A: Pure Appl. Opt.* 6:5157–5161.
4. Gutiérrez-Vega, J.C., M.D. Iturbe-Castillo, G.A. Ramırez, E. Tepichın, R.M. Rodrıguez-Dagnino, S. Chávez-Cerda, and G.H.C. New. 2001. Experimental demonstration of optical Mathieu beams. *Opt. Commun.* 195(1):35–40.
5. Bandres, M.A., and J.C. Gutierr-Vega. 2004. Ince-Gaussian beams. *Opt. Lett.* 29(2):144–146.
6. Kotlyar, V.V., S.N. Khonina, R.V. Skidanov, and V.A. Soifer. 2007. Hypergeometric modes. *Opt. Lett.* 32(7):742–744.
7. Siviloglou, G.A., and D.N. Christodoulides. 2007. Accelerating finite energy Airy beams. *Opt. Lett.* 32:979–981.
8. Ring, J., J. Lindberg, A. Mourka, M. Mazilu, K. Dholakia, and M. Dennis. 2012. Auto-focusing and self-healing of Pearcey beams. *Opt. Express* 20:18955–18966.
9. Seshardi, S.R. 2006. Self-interation and mutual interaction of complex-argument Laguerre-Gauss beams. *Opt. Lett.* 31(5):619–621.
10. Kostenbauder, A., Y. Sun, and A.E. Siegman. 2006. Eigenmode expansions using biorthogonal functions: Complex-valued Hermite-Gaussians: Reply to comment. *J. Opt. Soc. Am. A* 23(6):1528–1529.
11. Kotlyar, V.V., S.N. Khonina, A.A. Almazov, V.A. Soifer, K. Jefimovs, and J. Turunen. 2006. Elliptic Laguerre-Gaussian beams. *J. Opt. Soc. Am. A* 23(1):43–56.
12. Mei, Z., J. Gu, and D. Zhao. 2007. The elliptical Laguerre-Gaussian beam and its propagation. *Optik* 118:9–12.
13. Karimi, E., G. Zito, B. Piccirillo, L. Marrucci, and E. Santamato. 2007. Hypergeometric-Gaussian modes. *Opt. Lett.* 32:3053–3055.
14. Bandres, M.A., and J.C. Gutiérrez-Vega. 2008. Circular beams. *Opt. Lett.* 33:177–179.
15. Caron, C.F.R., and R.M. Potvliege. 1999. Bessel-modulated Gaussian beams with quadratic radial dependence. *Opt. Commun.* 164:83–93.
16. Rozas, D., C.T. Law, and G.A. Swartzlander. 1997. Propagation dynamics of optical vortices. *J. Opt. Soc. Am. B* 14(11):3054–3065.
17. Allen, L., M.W. Beijersergen, R.J.C. Spreeuw, and J.P. Woerdman. 1992. Orbital angular momentum of light and the transformation of Laguerre-Gaussian laser modes. *Phys. Rev. A* 45:8185–8189.
18. Zhu, K., G. Zhou, X. Li, X. Zhen, and H. Tang. 2008. Propagation of Bessel-Gaussian beams with optical vortices in turbulent atmosphere. *Opt. Express* 16(26):12315–12320.
19. Krenn, M., R. Fickler, M. Fink, J. Handsteiner, M. Malik, T. Scheidl, R. Ursin, and A. Zeilinger. 2014. Communication with spatially modulated light through turbulent air across Vienna. *New J. Phys.* 16:113028.
20. Hadzievski, L., A. Maluckov, A.M. Rubenchik, and S. Turitsyn. 2015. Stable optical vortices in nonlinear multicore fibers. *Light: Science Appl.* 4:e314.
21. Foo, G., D.M. Palacios, and G.A. Swartzlander Jr. 2005. Optical vortex coronagraph. *Opt. Lett.* 30(24):3308–3310.
22. Mair, A., A. Vaziri, G. Weihs, and A. Zeilinger. 2001. Entanglement of the orbital angular momentum states of photons. *Nature* 412(6844):313–316.

23. Otsu, T., T. Ando, Y. Takiguchi, Y. Ohtake, H. Toyoda, and H. Itoh. 2014. Direct vidence for three-dimensional off-axis trapping with single Laguerre-Gaussian beam. *Scient. Rep.* 4:4579.
24. Khonina, S.N., V.V. Kotlyar, M.V. Shinkarev, V.A. Soifer, and G.V. Uspleniev. 1992. The phase rotor filter. *J. Mod. Opt.* 39(5):1147–1154.
25. Beijersbergen, M.W., R.P.C. Coerwinkel, M. Kristensen, and J.P. Woerdman. 1994. Helical-wavefront laser beams produced with a spiral phaseplate. *Opt. Commun.* 112(5–6):321–327.
26. Curtis, J.E., and D.G. Grier. 2003. Structure of optical vortices. *Phys. Rev. Lett.* 90(13):133901.
27. Oemrawsingh, S.S.R., J.A.W. van Houwelinger, E.R. Eliel, J.R. Woerdman, E.J.K. Vestegen, J.G. Kloosterboer, and G.W. Hooft. 2004. Production and characterization of spiral phase plates for optical wavelengths. *Appl. Opt.* 43(3):688–694.
28. Cheong, W.G., W.M. Lee, X.-C. Yuan, L.-S. Zhang, K. Dholakia, and H. Wang. 2004. Direct electron-beam writing of continuous spiral phase plates in negative resist with high power efficiency for optical manipulation. *Appl. Phys. Lett.* 85(23):5784–5786.
29. Lee, W.M., B.P.S. Ahluwalia, X.-C. Yuan, W.C. Cheong, and K. Dholakia. 2005. Optical steering of high and low index microparticles by manipulating an off-axis optical vortex. *J. Opt. A: Pure Appl. Opt.* 7:1–6.
30. Sueda, K., G. Miyaji, N. Miyanaga, and M. Nakatsura. 2004. Laguerre-Gaussian beam generated with a multilevel spiral phase plate for high intensity laser pulses. *Opt. Express* 12(15):3548–3553.
31. Khonina, S.N., V.V. Kotlyar, R.V. Skidanov, V.A. Soifer, K. Jefimovs, J. Simonen, and J. Turunen. 2004. Rotation of microparticles with Bessel beams generated by diffractive elements. *J. Mod. Opt.* 51(14):2167–2184.
32. Oemrawsingh S.S.R., A. Aiello, E.R. Eliel, G. Nienhuis, and J.P. Woerdman. 2004. How to observe high-dimensional two-photon entanglement with only two detectors. *Phys. Rev. Lett.* 92(21):217901.
33. Kotlyar, V.V., A.A. Almazov, S.N. Khonina, V.A. Soifer, H. Elfstrom, and J. Turunen. 2005. Generation of phase singularity through diffracting a plane or Gaussian beam by a spiral phase plate. *J. Opt. Soc. Am. A* 22(5):849–861.
34. Berry, M.V. 2004. Optical vortices evolving from helicoidal integer and fractional phase steps. *J. Opt. A: Pure Appl. Opt.* 6:259–268.
35. Sacks, Z.S., D. Rozas, and G.A. Swartzlander Jr. 1998. Holographic formation of optical-vortex filaments. *J. Opt. Soc. Am. B* 15:2226–2234.
36. Guo, C., X. Liu, J. He, and H. Wang. 2004. Optimal annulus structures of optical vortices. *Opt. Express* 12(19):4625–4634.
37. Sundbeck, S., I. Gruzberg, and D.G. Grier. 2005. Structure and scaling of helical modes of light. *Opt. Lett.* 30(5):477–479.
38. Prudnikov, A.P., Y.A. Brichkov, and O.I. Marichev. 1983. *Integrals and Series. Special Functions.* Moscow, Russia: Nauka.
39. Furhapter, S., A. Jesacher, S. Bernet, and M. Ritsch-Marte. 2005. Spiral phase contrast imaging in microscopy. *Opt. Express* 13(3):689–694.
40. Ganic, D., X. Gan, and M. Gu. 2003. Focusing of doughnut laser beams by a high numerical-aperture objective in free space. *Opt. Express* 11(21):2747–2752.
41. Khonina, S.N., V.V. Kotlyar, V.A. Soifer, M.V. Shinkarev, and G.V. Uspleniev. 1992. Trochoson. *Opt. Commun.* 91(3–4):158–162.
42. Paterson, C., and R. Smith. 1996. Higher-order Bessel waves produced by axicon-type computer-generated holograms. *Opt. Commun.* 124:123–130.
43. Alonzo, C.A., P.J. Rodrigo, and J. Gluckstad. 2005. Helico-conical optical beams: A product of helical and conical phase fronts. *Opt. Express* 13(5):1749–1760.

44. Davis, J.A., E. McNamara, D.M. Cottrell, and J. Campos. 2000. Image processing with the radial Hilbert transform: Theory and experiments. *Opt. Lett.* 25:99–101.
45. Guo, C., Y. Han, J. Xu, and J. Ding. 2006. Radial Hilbert transform with Laguerre-Gaussian spatial filters. *Opt. Lett.* 31:1394–1396.
46. Furhapter, S., A. Jesacher, S. Bernet, and M. Ritsch-Marte. 2006. Spiral interferometry. *Opt. Lett.* 30:1953–1955.
47. Bernet, S., A. Jesacher, S. Furhapter, C. Maurer, and M. Ritsch-Marte. 2006. Quantitative imaging of complex samples by spiral phase contrast microscopy. *Opt. Express* 14:3792–3805.
48. Lin, J., X. Yuan, S.H. Tao, and R.E. Burge. 2006. Synthesis of multiple collinear helical modes generated by a phase-only element. *J. Opt. Soc. Am. A* 23:1214–1218.
49. Davis, J.A., J. Guertin, and D.M. Cottrell. 1993. Diffraction-free beam generated with programmable spatial light modulators. *Appl. Opt.* 32:6368–6370.
50. Davis, J.A., E. Carcole, and D.M. Cottrell. 1996. Intensity and phase measurements of nondiffracting beams generated with the magneto-optic spatial light modulator. *Appl. Opt.* 35:593–598.
51. Davis, J.A., E. Carcole, and D.M. Cottrell. 1996. Nondiffracting interference patterns generated with programmable spatial light modulators. *App. Opt.* 35:599–602.
52. Chattrapiban, N., E.A. Rogers, D. Cofield, W.T. Hill, and R. Roy. 2003. Generation of non-diffracting Bessel beams by use of a spatial light modulator. *Opt. Lett.* 28:2183–2185.
53. Hakola, A., A. Shevchenko, S.C. Buchter, M. Kaivola, and N.V. Tabiryan. 2006. Creation of a narrow Bessel-like laser beam using a nematic liquid crystal. *J. Opt. Soc. Am. B* 23:637–641.
54. Chakraborty, R., and A. Ghosh. 2006. Generation of an elliptic Bessel beam. *Opt. Lett.* 31:38–40.
55. Bentley, J.B., J.A. Davis, M.A. Bandres, and J.C. Gutiérrez-Vega. 2006. Generation of helical Ince-Gaussian beams with a liquid-crystal display. *Opt. Lett.* 31:649–651.
56. Fatemi, F.K., and M. Bashkansky. 2006. Generation of hollow beams by using a binary spatial light modulator. *Opt. Lett.* 31:864–866.
57. Kotlyar, V.V., S.N. Khonina, A.A. Kovalev, V.A. Soifer, H. Elfstrom, and J. Turunen. 2006. Diffraction of a plane, finite-radius wave by a spiral phase plate. *Opt. Lett.* 31:1597–1599.
58. Kotlyar, V.V., A.A. Kovalev, S.N. Khonina, R.V. Skidanov, V.A. Soifer, H. Elfstrom, N. Tossavainen, and J. Turunen. 2006. Diffraction of conic and Gaussian beams by a spiral phase plate. *Appl. Opt.* 45:2656–2665.
59. Ahluwalia, B.P.S., W.C. Cheong, X.-C. Yuan, L.-S. Zhang, S.-H. Tao, J. Bu, and H. Wang. 2006. Design and fabrication of a double-axicon for generation of tailorable self-imaged three-dimensional intensity voids. *Opt. Lett.* 31:987–989.
60. M.R. Dennis. 2006. Rows of optical vortices from elliptically perturbing a high-order beam. *Opt. Lett.* 31:1325–1327.
61. Ling, D., J. Li, and J. Chen. 2006. Analysis of eigenfields in the axicon-based Bessel-Gauss resonator by the transfer-matrix method. *J. Opt. Soc. Am. A* 23:912–918.
62. Lin, J., X. Yuan, S.H. Tao, and R.E. Burge. 2006. Variable-radius focused optical vortex with suppressed sidelobes. *Opt. Lett.* 31:1600–1602.
63. Courtial, J., G. Whyte, Z. Bouchel, and J. Wagner. 2006. Iterative algorithm for holographic shaping of non-diffracting and self-imaging light beams. *Opt. Express* 14:2108–2116.
64. Whyte, G., and J. Courtial. 2005. Experimental demonstration of holographic three-dimensional light shaping using a Gerchberg-Saxton algorithm. *New J. Phys.* 7:1–12.
65. Wang, Q., X.W. Sun, P. Shum, and X.J. Yin. 2005. Dynamic switching of optical vortices with dynamic gamma-correction liquid-crystal spiral phase plate. *Opt. Express* 13:10285–10291.

66. Lin, J., X. Yuan, S.H. Tao, X. Peng, and H.B. Nin. 2005. Deterministic approach to the generation of modified helical beams for optical manipulation. *Opt. Express* 13:3862–3867.
67. Hahn, J., H. Kim, K. Choi, and B. Lee. 2006. Real-time digital holographic beam-shaping system with a genetic feedback tuning loop. *Appl. Opt.* 45:915–924.
68. Cojoc, D., E. Di Fabrizio, L. Businaro, S. Carbini, F. Romanato, L. Vaccari, and M. Altissimo. 2002. Design and fabrication of diffractive optical elements for optical tweezer arrays by means of e-beam lithography. *Microelectr. Engineer.* 61–62:963–969.
69. Swartzlander, G.A. 2005. Broadband nulling of a vortex phase mask. *Opt. Lett.* 30:2876–2878.
70. Soifer, V.A. 2002. *Method for Computer Design of Diffractive Optical Elements.* New York: John Wiley & Sons.
71. Menon, R., and H.I. Smith. 2006. Absorbance-modulation optical lithography. *J. Opt. Soc. Am. A* 23:2290–2294.
72. Jesacher, A., S. Fürhapter, S. Bernet, and M. Ritsch-Marte. 2006. Spiral interferogram analysis. *J. Opt. Soc. Am. A* 23:1400–1409.
73. Palacios, D.M., and S.L. Hunyadi. 2006. Low-order aberration sensitivity of an optical vortex coronagraph. *Opt. Lett.* 31:2981–2983.
74. Oron, R., N. Davidson, A.A. Friesem, and E. Hasman. 2000. Efficient formation of pure helical laser beams. *Opt. Commun.* 182:205–208.
75. Peele, A.G., P.J. McMahon, D. Paterson, C.Q. Tran, A.P. Mancuso, K.A. Nugent, J.P. Hayes, E. Harvey, B. Lai, and I. McNulty. 2002. Observation of an x-ray vortex. *Opt. Lett.* 27:1752–1754.
76. Swartzlander Jr, G.A. 2006. Achromatic optical vortex lens. *Opt. Lett.* 31:2042–2044.
77. Guo, C., D. Xue, Y. Han, and J. Ding. 2006. Optimal phase steps of multi-level spiral phase plates. *Opt. Commun.* 268(2):235–239.
78. Kim, G.-H., J.-H. Jeon, K.-H. Ko, H.-J. Moon, J.-H. Lee, and J.-S. Chang. 1997. Optical vortices produced with a nonspiral phase plate. *Appl. Opt.* 36:8614–8621.
79. Saga, N. 1987. New line integral expressions for Fraunhofer diffraction. *Opt. Commun.* 64:4–8.
80. Kotlyar, V.V., A.A. Kovalev, R.V. Skidanov, O.Y. Moiseev, and V.A. Soifer. 2007. Diffraction of a finite-radius plane wave and a Gaussian beam by a helical axicon and a spiral phase plate. *J. Opt. Soc. Am. A* 24:1955–1964.
81. Cai, Y., and Q. Lin. 2002. Decentered elliptical Gaussian beam. *Appl. Opt.* 41(21):4336–4340.
82. Cai, Y., and Q. Lin. 2003. Decentered elliptical Hermite-Gaussian beam. *J. Opt. Soc. Am. A* 20(6):1111–1119.
83. Cai, Y., and Q. Lin. 2004. A partially coherent elliptical flattened Gaussian beam and its propagation. *J. Opt. A: Pure and Appl. Opt.* 6:1061–1066.
84. Mitreska, Z. 1994. Diffraction of elliptical Gaussian light beams on rectangular profile grating of transmittance. *J. Opt. A: Pure and Appl. Opt.* 3:995–1004.
85. Seshadri, S. 2003. Basic elliptical Gaussian wave and beam in a uniaxial crystal. *J. Opt. Soc. Am. A* 20(9):1818–1826.
86. Steinbach, A., M. Ranner, F.C. Crnz, and J.C. Bergquist. 1996. CW second harmonic generation with elliptical Gaussian beam. *Opt. Commun.* 123:207–214.
87. Cai, Y., and Q. Lin. 2004. Light beams with elliptical flat-topped profiles. *J. Opt. A: Pure and Appl. Opt.* 6:390–395.
88. Cai, Y., and Q. Lin. 2004. Hollow elliptical Gaussian beam and its propagation through aligned and misaligned paraxial optical systems. *J. Opt. Soc. Am. A* 21(6):1058–1065.
89. Bandres, M.A., and J. Gutierrez-Vega. 2004. Ince-Gaussian modes of the paraxial wave equation and stable resonators. *J. Opt. Soc. Am. A* 21(5):873–880.

90. Bandres, M.A., and J. Gutierrez-Vega. 2004. Elegant Ince-Gaussian beams. *Opt. Lett.* 29(15):1724–1726.
91. Bandres, M.A., and J. Gutierrez-Vega. 2004. Higher-order complex source for elegant Laguerre-Gaussian waves. *Opt. Lett.* 29(19):2213–2215.
92. Schwarz, U.T., M.A. Bandres, and J. Gutierrez-Vega. 2004. Observation of Ince-Gaussian modes in stable resonators. *Opt. Lett.* 29(16):1870–1872.
93. Bin, Z., and L. Zhu. 1998. Diffraction property of an axicon in oblique illumination. *Appl. Opt.* 37(13):2563–2568.
94. Thaning, A., Z. Jaroszewicz, and A.T. Friberg. 2003. Diffractive axicons in oblique illumination: Analysis and experiments and comparison with elliptical axicons. *Appl. Opt.* 42(1):9–17.
95. Khonina, S.N., V.V. Kotlyar, V.A. Soifer, K. Jefimovs, P. Paakkonen, and J. Turunen. 2004. Astigmatic Bessel laser beams. *J. Mod. Opt.* 51(5):677–686.
96. Miller Jr, W. 1977. *Symmetry and Separation of Variables.* Reading, MA: Addison-Wesley.
97. Bandres, M.A., J.C. Gutierrez-Vega, and S. Chavez-Cerda. 2004. Parabolic nondiffracting optical wave fields. *Opt. Lett.* 29(1):44–46.
98. Gutierrez-Vega, J.C., M.D. Iturbe-Castillo, and S. Chavez-Cerda. 2000. Alternative formulation for invariant optical fields: Mathieu beams. *Opt. Lett.* 25(20):1493–1495.
99. Gutierrez-Vega, J.C., and M.A. Bandres. 2005. Helmholtz-Gauss waves. *J. Opt. Soc. Am. A* 22(2):289–298.
100. Bandres, M.A., and J.C. Gutierrez-Vega. 2005. Vector Helmholtz-Gauss and vector Laplace-Gauss beams. *Opt. Lett.* 30(16):2155–2157.
101. Abramowitz, M., and I.A. Stegun. 1964. *Handbook of Mathematical Function.* Washington D.C.: National Bureau of Standards, Applied Mathematics Series 55.
102. Gao, M., C. Gao, and Z. Lin. 2007. Generation and application of the twisted beam with orbital angular momentum. *Chin. Opt. Lett.* 5:89–92.
103. Duan, K., and B. Lu. 2007. Application of the Wiener distribution function to complex-argument Hermite-and Laguerre-Gaussian beams beyond the paraxial approximation. *Opt. Las. Tech.* 39:110–115.
104. Burvall, A., K. Kolacz, A.V. Goncharov, Z. Jaroszewicz, and C. Dainty. 2007. Lens axicons in oblique illumination. *Appl. Opt.* 46:312–318.
105. Kotlyar, V., A. Kovalev, V. Soifer, C. Tuvey, and J. Devis. 2007. Sidelobes contrast reduction for optical vortex beams using a helical axicon. *Opt. Lett.* 32:921–923.
106. Singh, R.K., P. Senthilkumaran, and K. Singh. 2007. Effect of coma on the focusing of an apertured singular beam. *Opt. Las. Eng.* 45:488–494.
107. Singh, R.K., P. Senthilkumaran, and K. Singh. 2007. Focusing of a vortex carrying beam with Gaussian background by a lens in the presence of spherical aberration and defocusing. *Opt. Las. Eng.* 45:773–782.
108. Singh, R.K., P. Senthilkumaran, and K. Singh. 2007. Influence of astigmatism and defocusing on the focusing of a singular beam. *Opt. Commun.* 270:128–138.
109. Senthilkumaran, P., K. Singh, and R.K. Singh. 2007. The effect of astigmatism on the diffraction of a vortex carrying beam with a Gaussian background. *J. Opt. A: Pure and Appl. Opt.* 9:543–554.
110. Jaroszewicz, Z., J. Sochacki, A. Kolodziejczyk, and L.R. Staronski. 1993. Apodized annular-aperture logarithmic axicon: Smoothness and uniformity of intensity distributions. *Opt. Lett.* 18:1893–1895.
111. Sochacki, J., Z. Jaroszewicz, L.R. Staronski, and A. Kolodziejczyk. 1993. Annular-aperture logarithmic axicon. *J. Opt. Soc. Am. A* 10:1765–1768.
112. Khonina, S.N., V.V. Kotlyar, V.A. Soifer, P. Pääkkönen, J. Simonen, and J. Turunen. 2001. An analysis of the angular momentum of a light field in terms of angular harmonics. *J. Mod. Opt.* 48:1543–1557.

113. Kotlyar, V.V., and A.A. Kovalev. 2008. Family of hypergeometric laser beams. *J. Opt. Soc. Am. A* 25:262–270.
114. Kotlyar, V.V., A.A. Kovalev, R.V. Skidanov, S.N. Khonina, and J. Turunen. 2008. Generating hypergeometric laser beams with a diffractive optical element. *Appl. Opt.* 47:6124–6133.
115. Chen, J., G. Wang, and Q. Xu. 2011. Production of confluent hypergeometric beam by computer-generated hologram. *Opt. Eng.* 50:024201.
116. Bernardo, B., and F. Moraes. 2011. Data transmission by hypergeometric modes through a hyperbolic-index medium. *Opt. Express* 19:11264–11270.
117. Li, J., and Y. Chen. 2012. Propagation of confluent hypergeometric beam through uniaxial crystals orthogonal to the optical axis. *Opt. Las. Technol.* 44:1603–1610.
118. Levy, U., M. Nezhad, H. Kim, C. Tsai, L. Pang, and Y. Fainman. 2005. Implementation of a graded-index medium by use of subwavelength structures with graded fill factor. *J. Opt. Soc. Am. A* 22:724–733.
119. Tien, P.K., J.P. Gordon, and J.R. Whinnery. 1965. Focusing of light beam of Gaussian field distribution in continuous and periodic lenslike media. *Proc. IEEE* 53:129–136.
120. Newstein, M., and B. Rudman. 1987. Laguerre-Gaussian periodically focusing beams in a quadratic index medium. *IEEE J. Quantum Electron.* 23:481–482.
121. Gutierrez-Vega, J.C., and M.A. Bandres. 2005. Ince-Gaussian beams in a quadratic-index medium. *J. Opt. Soc. Am. A* 22:306–309.
122. Bandres, M.A., and J.C. Gutierrez-Vega. 2007. Airy-Gauss beams and their transformation by paraxial optical systems. *Opt. Express* 15:16719–16728.
123. Deng, D.M. 2011. Propagation of Airy-Gaussian beams in a quadratic-index medium. *Eur. Phys. J. D* 65:553–556.
124. Saleh, B.E.A., and M.C. Teich. 1991. *Fundamentals of Photonics*. New Jersey: John Wiley & Sons.
125. Di Falco, A., S.C. Kehr, and U. Leonhardt. 2011. Luneburg lens in silicon photonics. *Opt. Express* 19:5156–5162.
126. Zentgrat, T., Y. Liu, M.N. Mikkelsen, J. Valentine, and X. Zhang. 2011. Plasmonic Luneburg and Eaton lenses. *Nat. Nanotechn.* 6:151–155.
127. Dyachenko, P.N., V.S. Pavelyev, and V.A. Soifer. 2012. Graded-index photonic quasi-crystals. *Opt. Lett.* 37:2178–2180.
128. Triandaphilov, Y.R., and V.V. Kotlyar. 2008. Photonic Crystal Mikaelian Lens. *Opt. Mem. Neur. Netw. (Inform. Opt.)* 17:1–7.
129. Kotlyar, V.V., A.A. Kovalev, and V.A. Soifer. 2010. Subwavelength focusing with a Mikaelian planar lens. *Opt. Mem. Neur. Netw. (Inform. Opt.)* 19:273–278.
130. Kotlyar, V.V., and S.S. Stafeev. 2009. Sharply focusing a radially polarized laser beam using a gradient Mikaelian's microlens. *Opt. Commun.* 282:459–464.
131. Kotlyar, V.V., A.A. Kovalev, and A.G. Nalimov. 2012. Planar gradient-index hyperbolic secant lens for subwavelength focusing and superresolution imaging. *Optics (SciencePG)* 1:1–10.
132. Mendlovic, D., and H.M. Ozaktas. 1993. Fractional Fourier transform and their optical implementation: I. *J. Opt. Soc. Am. A* 10:1875–1881.
133. Lohmann, A.W. 1993. Image rotation, Wigner rotation, and the fractional Fourier transform. *J. Opt. Soc. Am. A* 10:2181–2186.
134. Khonina, S.N., A.S. Striletz, A.A. Kovalev, and V.V. Kotlyar. 2009. Propagation of laser vortex beams in a parabolic-index optical fiber. *Proc. SPIE* 7523:75230B.
135. Gradshteyn, I.S., and I.M. Ryzhik. 2007. *Table of Integrals, Series, and Products*. Amsterdam: Elsevier.
136. Hensler, J.R. 1975. Method of Producing a Refractive Index Gradient in Glass. *U.S. Patent* 3,873,408 (March 25, 1975).
137. Keck, D.B., and R. Olshansky. 1975. Optical Waveguide Having Optimal Index Gradient. *U.S. Patent* 3,904,268 (September 9, 1975).

138. Dupuis, A. 2007. Guiding in the visible with "colorful" solid-core Bragg fiber. *Opt. Lett.* 32:2882–2884.
139. Liu, L., G. Liu, Y. Xiong, J. Chen, W. Li, and Y. Tian. 2010. Fabrication of X-ray imaging zone plates by e-beam and X-ray lithography. *Microsyst. Technol.* 16:1315–1321.
140. Kamijo, N., Y. Suzuki, H. Takau, S. Tamura, M. Yasumoto, A. Takeuchi, and M. Awaji. 2003. Microbeam of 100keV X-ray with sputtered-sliced Fresnel zone plate. *Rev. Sci. Instrum.* 74:5101–5104.
141. Takenaka, T., M. Yokota, and O. Fukumitsu. 1985. Propagation of light beams beyond the paraxial approximation. *J. Opt. Soc. Am. A* 2:826–829.
142. Bomzon, Z., V. Kleiner, and E. Hasman. 2001. Formation of radially and azimuthally polarized light using spacevariant subwavelength metal strip grating. *Appl. Phys. Lett.* 79:1587–1589.
143. Armstrong, D.J., M.C. Philips, and A.V. Smith. 2003. Generation of radially polarized beams with an image rotating resonator. *Appl. Opt.* 42:3550–3554.
144. Passilly, N., R. de S. Denis, K. Aït-Ameur, F. Treussart, R. Hierle, and J.-F. Roch. 2005. Simple interferometric technique for generation of a radially polarized light beam. *J. Opt. Soc. Am. A* 22:984–991.
145. Volpe, G., and D. Petrov. 2004. Generation of cylindrical vector beams with few-mode fibers by Laguerre-Gaussian beams. *Opt. Commun.* 237:89–95.
146. Niziev, V.G., and A.V. Nesterov. 1999. Influence of beam polarization on laser cutting efficiency. *J. Phys. D* 32:1455–1461.
147. Gahagan, K.T., and G.A. Swartzlander Jr. 1999. Simultaneous trapping of low-index and high-index microparticles observed with an optical-vortex trap. *J. Opt. Soc. Am. B* 16:533–537.
148. Wu, G., Q. Lou, J. Zhou, J. Dong, and Y. Wei. 2007. Focal shift in focused radially polarized ultrashort pulsed laser beams. *Appl. Opt.* 46:6251–6255.
149. Deng, D. 2006. Nonparaxial propagation of radially polarized light beams. *J. Opt. Soc. Am. B* 23:1228–1234.
150. Banerjee, P., G. Cook, and D. Evans. 2009. A q-parameter approach to analysis of propagation, focusing, and waveguiding of radially polarized Gaussian beams. *J. Opt. Soc. Am. A* 26:1366–1374.
151. Kotlyar, V.V., and A.A. Kovalev. 2009. Nonparaxial hypergeometric beams. *J. Opt. A: Pure Appl. Opt.* 11:045711.
152. Rashid, M., O.M. Marago, and P.H. Jones. 2009. Focusing of high order cylindrical vector beams. *J. Opt. A: Pure Appl. Opt.* 11:065204.
153. Chen, B., and J. Pu. 2009. Tight focusing of elliptically polarized vortex beams. *Appl. Opt.* 48:1288–1294.
154. Luneburg, R.K. 1966. *Mathematical Theory of Optics*. Berkeley, CA: University of California Press.
155. Mei, Z., and D. Zhao. 2007. Nonparaxial analysis of vectorial Laguerre-Bessel-Gaussian beams. *Opt. Express* 15:11942–11951.
156. Niv, A., G. Biener, V. Kleiner, and E. Hasman. 2006. Manipulation of the Pancharatnam phase in vectorial vortices. *Opt. Express* 14:4208–4220.
157. Tidwell, S.C., D.H. Ford, and W.D. Kimura. 1990. Generating radially polarized beams interferometrically. *Appl. Opt.* 29:2234–2239.
158. Yirmiyahu, Y., A. Niv, G. Biener, V. Kleiner, and E. Hasman. 2007. Excitation of a single hollow waveguide mode using inhomogeneous anisotropic subwavelength structures. *Opt. Express* 15:13404–13414.
159. Li, Y., and E. Wolf. 1981. Focal shifts in diffracted converging spherical waves. *Opt. Commun.* 39:211–215.
160. Carter, W.H. 1982. Focal shift and concept of effective Fresnel number for a Gaussian laser beam. *Appl. Opt.* 21:1989–1994.

161. Salamin, Y.I. 2006. Electron acceleration from rest in vacuum by an axicon Gaussian laser beam. *Phys. Rev. A* 73:043402.
162. Zhu, Q. 2009. Description of the propagation of a radially polarized beam with the scalar Kirchhoff diffraction. *J. Mod. Opt.* 56(14):1621–1625.
163. Born, M., and E. Wolf. 1986. *Principles of Optics*, 6th ed. New York: Pergamon.
164. Kalnins, E.G., and W. Miller Jr. 1974. Lie theory and separation of variables. *J. Math. Phys.* 15:1728–1737.
165. Berry, M.V., and N.L. Balazs. 1979. Nonspreading wave packets. *Am. J. Phys.* 47:264–267.
166. Besieris, I.M., A.M. Shaarawi, and R.W. Ziolkowski. 1994. Nondispersive accelerating wave packets. *Am. J. Phys.* 62:519–521.
167. Siviloglou, G.A., J. Broky, A. Dogariu, D.N. Christodoulides. 2007. Observation of accelerating Airy beams. *Phys. Rev. Lett.* 99:213901.
168. Siviloglou, G.A., J. Broky, A. Dogariu, D.N. Christodoulides. 2008. Ballistic dynamics of Airy beams. *Opt. Lett.* 33:207–209.
169. Besieris, I.M., and A.M. Shaarawi. 2007. A note on an accelerating finite energy Airy beam. *Opt. Lett.* 32:2447–2449.
170. Bandres, M.A. 2008. Accelerating parabolic beams. *Opt. Lett.* 33:1678–1680.
171. Bandres, M.A. 2009. Accelerating beams. *Opt. Lett.* 34:3791–3793.
172. Greenfield, E., M. Segev, W. Walasik, and O. Raz. 2011. Accelerating light beams along arbitrary convex trajectories. *Phys. Rev. Lett.* 106:213902.
173. Froehly, L., F. Courvoisier, A. Mathis, M. Jacquot, L. Furfaro, R. Giust, P.A. Lacourt, and J.M. Dudley. 2011. Arbitrary accelerating micron-scale caustic beams in two and three dimensions. *Opt. Express* 19:16455–16465.
174. Courvoisier, F., A. Mathis, L. Froehly, R. Giust, R. Furfaro, P.A. Lacourt, M. Jacquot, and J.M. Dudley. 2012. Sending femtosecond pulses in circles: Highly nonparaxial accelerating beams. *Opt. Lett.* 37:1736–1738.
175. Cottrell, D.M., J.A. Devis, and T.M. Hazard. 2009. Direct generation of accelerating Airy beams using a 3/2 phase-only pattern. *Opt. Lett.* 34:2634–2636.
176. Kaminer, I., R. Bekenstein, J. Nemirovsky, and M. Segev. 2012. Nondiffracting accelerating wave packets of Maxwell's equations. *Phys. Rev. Lett.* 108:163901.
177. Zhang, P., Y. Hu, D. Cannan, A. Salandrino, T. Li, R. Morandotti, X. Zhang, and Z. Chen. 2012. Generation of linear and nonlinear nonparaxial accelerating beams. *Opt. Lett.* 37:2820–2822.
178. Aleahmad, P., M. Miri, M.S. Mills, and I. Kaminer. 2012. Fully vectoral accelerating diffraction-free Helmholtz beams. *Phys. Rev. Lett.* 109:203902.
179. Bandres, M.A., and B.M. Rodiguez-Lara. 2013. Nondiffracting accelerating waves: Weber waves and parabolic momentum. *New J. Phys.* 15:013054.
180. Torre, A. 2008. A note on the general solution of the paraxial wave equation: A Lie algebra view. *J. Opt. A: Pure Appl. Opt.* 10:055006.
181. Bandres, M.A., M.A. Alonso, I. Kaminer, and M. Segev. 2013. Three-dimensional accelerating electromagnetic waves. *Opt. Express* 21:13917–13929.
182. Kotlyar, V.V., and A.A. Kovalev. 2014. Airy beam with a hyperbolic trajectory. *Opt. Commun.* 313:290–293.
183. Kotlyar, V.V., S.S. Stafeev, and A.A. Kovalev. 2013. Curved laser microjet in near field. *Appl. Opt.* 52:4131–4136.
184. Torre, A. 2009. A note on the Airy beams in the light of the symmetry algebra based approach. *J. Opt. A: Pure Appl. Opt.* 11:125701.
185. Torre, A. 2011. The Appell transformation for the paraxial wave equation. *J. Opt.* 13:015701.
186. Kogelnik, H., and T. Li. 1966. Laser beams and resonators. *Proc. IEEE* 54:1312–1329.

187. Pratesi, R., and L. Ronchi. 1977. Generalized Gaussian beams in free space. *J. Opt. Soc. Am.* 67:1274–1276.
188. Kotlyar, V.V., and A.A. Kovalev. 2014. Hermite–Gaussian modal laser beams with orbital angular momentum. *J. Opt. Soc. Am. A* 31:274–282.
189. Kotlyar, V.V., A.A. Kovalev, and A.G. Nalimov. 2013. Propagation of hypergeometric laser beams in a medium with the parabolic refractive index. *J. Opt.* 313:290–293.
190. Papazoglou, D.G., N.K. Efremidis, D.N. Christodoulides, and S. Tzortakis. 2011. Observation of abruptly autofocusing waves. *Opt. Lett.* 36:1842–1844.
191. Berry, M., J. Nye, and F. Wright. 1979. The elliptic umbilic diffraction catastrophe. *Phil. Trans. R. Soc. Lond.* 291:453–484.
192. Vaughan, J.M., and D. Willetts. 1979. Interference properties of a light-beam having a helical wave surface. *Opt. Commun.* 30:263–267.
193. Coullet, P., G. Gil, and F. Rocca. 1989. Optical vortices. *Opt. Commun.* 73:403–408.
194. Bazhenov, V., M.V. Vasnetsov, and M.S. Soskin. 1990. Laser-beam with screw dislocations in the wavefront. *JETP Lett.* 52:429–431.
195. Abramochkin, E.G., and V.G. Volostnikov. 1991. Beam transformation and nontransformed beams. *Opt. Commun.* 83:123–125.
196. Beijersbergen, M.W., L. Allen, H.E. Van der Veen, and J.P. Woerdman. 1993. Astigmatic laser mode converters and transfer of orbital angular momentum. *Opt. Commun.* 96:123–132.
197. Yao, A.M., and M.J. Padgett. 2011. Orbital angular momentum: Origins, behavior and applications. *Adv. Opt. Photon.* 3:161–204.
198. Kotlyar, V.V., and A.A. Kovalev. 2012. *Vortex Laser Beams*. Samara, Russia: Novaya Tekhnika (in Russian).
199. Vyas, S., and P. Senthilkumaran. 2007. Interferometric optical vortex array generator. *Appl. Opt.* 46:2893–2898.
200. Fraczek, E., and G. Budzyn. 2009. An analysis of an optical vortices interferometer with focused beam. *Opt. Applicata*. XXXIX:91–99.
201. Singh, B.K., G. Singh, P. Senthilkumaran, and D.S. Metha. 2012. Generation of optical vortex array using single-element reversed-wavefront folding interferometer. *Int. J. Opt.* 2012:689612.
202. Vaity, P., A. Aadhi, and R. Singh. 2013. Formation of optical vortices through superposition of two Gaussian beams. *Appl. Opt.* 52:6652–6656.
203. Shen, Y., G.T. Campbell, B. Hage, H. Zou, B.C. Buchler, and P.K. Lam. 2013. Generation and interferometric analysis of high charge optical vortices. *J. Opt.* 15:044005.
204. Gotte, J.B., K. O'Holleran, D. Precce, F. Flossman, S. Franke-Arnold, S.M. Barnett, and M.J. Padgett. 2008. Light beams with fractional orbital angular momentum and their vortex structure. *Opt. Express* 16:993–1006.
205. O'Dwyer, D.P., C.F. Phelan, Y.P. Rakovich, P.R. Eastham, J.C. Lunney, and J.F. Donegan. 2010. Generation of continuously tunable fractional orbital angular momentum using internal conical diffraction. *Opt. Express* 18:16480–16485.
206. Siegman, A.E. 1973. Hermite-Gaussian functions of complex argument as optical beam eigenfunction. *J. Opt. Soc. Am.* 63:1093–1094.
207. Humblet, J. 1943. Sur le moment d'impulsion d'une onde electromagnetique. *Physica (Utrecht)* 10(7):585–603.
208. Kotlyar, V.V., S.N. Khonina, R.V. Skidanov, and V.A. Soifer. 2007. Rotation of laser beams with zero of the orbital angular momentum. *Opt. Commun.* 274:8–14.
209. Cai, Y., and Q. Lin. 2002. The elliptical Hermite-Gaussian beam and its propagation through paraxial systems. *Opt. Commun.* 207(1–6):139–147.
210. Durnin, J. 1987. Exact solutions for nondiffracting beams. I. The scalar theory. *J. Opt. Soc. Am. A* 4:651–654.

211. Kotlyar, V.V., S.N. Khonina, and V.A. Soifer. 1992. Algorithm for the generation of non-diffracting Bessel beams. *J. Mod. Opt.* 42:1231–1239.
212. Chavez-Cedra, S., J.C. Gutierrez-Vega, and G.H.C. New. 2001. Elliptic vortices of electromagnetic wave fields. *Opt. Lett.* 26:1803–1805.
213. Dennis, M.R., and J.D. Ring. 2013. Propagation-invariant beams with quantum pendulum spectra: From Bessel beams to Gaussian beam-beams. *Opt. Lett.* 17:3325–3328.
214. Gori, F., G. Guattari, and C. Padovani. 1987. Bessel-Gauss beams. *Opt. Commun.* 64:491–495.
215. Li, Y., H. Lee, and E. Wolf. 2004. New generalized Bessel-Gauss beams. *J. Opt. Soc. Am. A* 21:640–646.
216. Kisilev, A.P. 2004. New structures in paraxial Gaussian beams. *Opt. Spectrosc.* 96:479–481.
217. Bagini, V., F. Frezza, M. Santarsiero, G. Schettini, and G. Shirripa Spagnolo. 1996. Generalized Bessel-Gauss beams. *J. Mod. Opt.* 43:1155–1166.
218. Sheppard, C.J.R. 2009. Beam duality, with application to generalized Bessel-Gaussian, and Hermite- and Laguerre-Gaussian beams. *Opt. Express* 17:3690–3697.
219. Kotlyar, V.V., A.A. Kovalev, and V.A. Soifer. 2014. Assimetric Bessel modes. *Opt. Lett.* 38:2395–2398.
220. Lu, J., and J.F. Greenleaf. 1992. Diffraction-limited beams and their applications for ultrasonic imaging and tissue characterization. *Proc. SPIE* 1733:92–119.
221. Kotlyar, V.V., A.A. Kovalev, and V.A. Soifer. 2014. Nondiffracting asymmetric elegant Bessel beams with fractional orbital angular momentum. *Comput. Opt.* 38:4–10.
222. Zamboni-Rached, M. 2004. Stationary optical wave fields with arbitrary longitudinal shape by superposing equal frequency Bessel beams: Frozen Wave. *Opt. Express* 12:4001–406.
223. Zamboni-Rached, M., E. Recami, and H. Figueroa. 2005. Theory of Frozen Wave: Modelling the shape of stationary wave field. *J. Opt. Soc. Am. A* 22:2465–2475.
224. Zamboni-Rached, M. 2006. Diffraction-attenuation resistant beams in absorbing media. *Opt. Express* 14:1804–1809.
225. Nelson, W., J.P. Palastro, C.C. Davis, and P. Sprangle. 2014. Propagation of Bessel and Airy beams through atmospheric turbulence. *J. Opt. Soc. Am. A* 31:603–609.
226. Froehly, L., M. Jacquot, P.A. Lacourt, J.M. Dudley, and F. Courvoisier. 2014. Spatiotemporal structure of femtosecond bessel beams from spatial light modulators. *J. Opt. Soc. Am. A* 31:790–793.
227. Watson, G.N. 1966. *A Treatise on the Theory of Bessel Functions*, 2nd ed. Cambridge, UK: Cambridge University Press (§16.5–16.59, pp. 537–550).
228. Vengsarkar, A.M., A.M. Shaarawi, R.W. Ziolkowski, and I.M. Besieris. 1992. Closed-form localized wave solution in optical fiber waveguides. *J. Opt. Soc. Am. A* 9:937–949.
229. Zamboni-Rached, M., and H.H. Figueroa. 2001. A rigorous analysis of localized wave propagation in optical fiber. *Opt. Commun.* 191:49–54.
230. Sheppard, C.J.R., and P. Saari. 2008. Lommel pulses: An analytic form for localized wave of the focus wave mode type with bandlimited spectrum. *Opt. Express* 16:150–160.
231. Sheppard, C.J.R. 2014. Focusing of vortex beams: Lommel treatment. *J. Opt. Soc. Am. A* 31:644–651.
232. Kovalev, A.A., and V.V. Kotlyar. 2014. Diffraction-free Lommel beams. *Comput. Opt.* 28:188–192.
233. Rykov, M.A., and R.V. Skidanov. 2014. Modifying the laser beam intensity distribution for obtaining improved strength characteristics of an optical trap. *Appl. Opt.* 53:156–164.
234. Wang, J., J. Yang, I.M. Fazal, N. Ahmed, Y. Yan, H. Huang, Y. Ren et al. 2012. Terabit free-space data transmission employing orbital angular momentum multiplexing. *Nat. Photonics* 6:488–496.

235. Kotlyar, V.V., A.A. Kovalev, R.V. Skidanov, and V.A. Soifer. 2014. Rotating elegant Bessel-Gaussian beams. *Comput. Opt.* 38(2):162–170.
236. Kotlyar, V.V., A.A. Kovalev, R.V. Skidanov, and V.A. Soifer. 2014. Assymetric Bessel-Gauss beams. *J. Opt. Soc. Am. A* 31(9):1977–1983.
237. Gong, L., X. Qiu, Y. Ren, H. Zhu, W. Liu, J. Zhou, M. Zhong, X. Chu, and Y. Li. 2014. Observation of the asymmetric Bessel beams with arbitrary orientation using a digital micromirror device. *Opt. Express* 22(22):26763–26776.
238. Sheppard, C.J.R., S.S. Kou, and J. Lin. 2014. Two-dimensional complex source point solutions: Application to propagationally invariant beams, optical fiber modes, planar waveguides, and plasmonic devices. *J. Opt. Soc. Am. A* 31(12):2674–2679.
239. Mitri, F.G. 2014. Partial-wave series expansion in spherical coordinates for the acoustic field of vortex beams generated from finite circular aperture. *IEEE Trans.* 61(12):2089–2097.
240. Mendez, G., A. Fernando-Vazquez, and R.P. Lopez. 2015. Orbital angular momentum and highly efficient holographic generation of nondiffractive TE and TM vector beams. *Opt. Commun.* 334:174–183.
241. Turunen, J., A. Vasara, and A.T. Friberg. 1988. Holographic generation of diffractive-free beams. *Appl. Opt.* 27(19):3959–3962.
242. Vasara, A., J. Turunen, and A.T. Friberg. 1989. Realization of general nondiffracting beams with computer-generated holograms. *J. Opt. Soc. Am. A* 6(11):1748–1754.
243. MacDonald, R.P., S.A. Boothroyd, T. Okamato, J. Chrostowski, and B.A. Syrett. 1996. Interboard optical data distribution by Bessel beam shadowing. *Opt. Commun.* 122:169–177.
244. McQueen, C.A., J. Arlt, and K. Dholakia. 1999. An experiment to study a "non-diffracting" light beam. *Am. J. Phys.* 67:912–915.
245. Barnett, S.M., and L. Allen. 1994. Orbital angular momentum and nonparaxial light-beams. *Opt. Commun.* 110:670–678.
246. Volke-Sepulveda, K., V. Garces-Chavez, S. Chavez-Cedra, J. Arlt, and K. Dholakia. 2002. Orbital angular momentum of a high-order Bessel light beam. *J. Opt. B: Quantum Semiclass. Opt.* 4:S82–S89.
247. Kotlyar, V.V., S.N. Khonina, and V.A. Soifer. 1997. An algorithm for the generation of laser beams with longitudinal periodicity: Rotating images. *J. Mod. Opt.* 44(7):1409–1416.
248. Paakkonen, P., J. Lautanen, M. Honkanen, M. Kuittinen, J. Turunen, S.N. Khonina, V.V. Kotlyar, V.A. Soifer, and A.T. Friberg. 1998. Rotating optical fields: Experimental demonstration with diffractive optics. *J. Mod. Opt.* 45(11):2355–2369.
249. Khonina, S.N., V.V. Kotlyar, V.A. Soifer, J. Lautanen, M. Honkanen, and J. Turunen. 1999. Generating a couple of rotating nondiffracting beams using a binary-phase DOE. *Optik* 110(3):137–144.
250. Lee, H.S., B.W. Stewart, K. Choi, and H. Fenichel. 1994. Holographic nondiverging hollow beam. *Phys. Rev. A* 49(6):4922–4927.
251. Herman, R.M., and T.A. Wiggins. 1991. Production and uses of diffractionless beams. *J. Opt. Soc. Am. A* 8(6):932–942.
252. Arlt, J., and K. Dholakia. 2000. Generation of high-order Bessel beams by use of an axicon. *Opt. Commun.* 177:297–301.
253. MacDonald, M.P., L. Paterson, K. Volke-Sepulveda, J. Arlt, W. Sibbett, and K. Dholakia. 2002. Creation and manipulation of three-dimensional optically trapped structures. *Science* 296:1101–1103.
254. Garces-Chavez, V., D. McGloin, H. Melville, W. Sibbett, and K. Dholakia. 2002. Simultaneous micromanipulation in multiple planes using a self-reconstructing light beam. *Nature* 419:145–147.
255. Arlt, J., T. Hitomi, and K. Dholakia. 2000. Atom guiding long Laguerre-Gaussian and Bessel beams. *Appl. Phys. B* 71:549–554.

256. Arlt, J., K. Dholakia, J. Soneson, and E.M. Wright. 2001. Optical dipole traps and atomic waveguides based on Bessel light beams. *Phys. Rev. A* 63:063602.
257. Kotlyar, V.V., A.A. Kovalev, and V.A. Soifer. 2012. Hankel-Bessel laser beams. *J. Opt. Soc. Am. A* 29(5):741–747.
258. Zhu, Y., X. Liu, J. Gao, Y. Zhang, and F. Zhao. 2014. Probability density of the orbital angular momentum mode of Hankel-Bessel beams in an atmospheric turbulence. *Opt. Express* 22(7):7765–7772.
259. Bouchal, Z., and M. Olivik. 1995. Non-diffractive vector Bessel beams. *J. Mod. Opt.* 42(8):1555–1566.
260. Yu, Y.Z., and W.B. Dou. 2008. Vector analysis of nondiffracting Bessel beams. *Prog. In Electrom. Res. Lett.* 5:57–71.
261. Litvin, I.A., A. Dudley, and A. Forbes. 2011. Poynting vector and orbital angular momentum density of superpositions of Bessel beams. *Opt. Express* 19(18):16760–16771.
262. Chen, Y.F., Y.C. Lin, W.Z. Zhuang, H.C. Liang, K.W. Su, and K.F. Huang. 2012. Generation of large orbital angular momentum from superposed Bessel beams corresponding to resonant geometric modes. *Phys. Rev. A* 85:043833.
263. Goodman, J.W. 1968. *Introduction to Fourier Optics*. New York: McGraw-Hill.
264. Pearcey, T. 1946. The structure of an electromagnetic field in the neighbourhood of a cusp of a caustic. *Phil. Mag. S* 7:311–317.
265. Berry, M.V., and C.J. Howls. 2012. Integrals with coalescing saddles. Digital Library of Mathematical Functions, National Institute of Standards and Technology. http://dlmf.nist.gov/36.2.
266. Deng, D., C. Chen, X. Zhao, B. Chen, X. Peng, and Y. Zheng. 2014. Virtual Source of a Pearcey beam. *Opt. Lett.* 39:2703–2706.
267. Rogel-Salazar, J., H.A. Jiménez-Romero, and S. Chávez-Cerda. 2014. Full characterization of Airy beams under physical principles. *Phys. Rev. A* 89:023807.
268. Anguiano-Morales, M., A. Martínez, M. Iturbe-Castillo, S. Chávez-Cerda, and N. Alcalá-Ochoa. 2007. Self-healing property of a caustic optical beam. *Appl. Opt.* 46:8284–8290.
269. Kovalev, A.A., V.V. Kotlyar, and S.G. Zaskanov. 2014. Structurally stable three-dimensional and two-dimensional laser half Pearcey beams. *Comput. Opt.* 38(2):193–197.
270. Kovalev, A.A., V.V. Kotlyar, S.G. Zaskanov, and A.P. Profirev. 2015. Half Pearcey laser beams. *J. Opt.* 17:035604.
271. Kotlyar, V.V., A.A. Kovalev, and V.A. Soifer. 2014. Transformation of decelerating laser beams into accelerating ones. *J. Opt.* 16(8):085701.
272. Vaziri, A., G. Weihs, and A. Zeilinger. 2002. Superpositions of the orbital angular momentum for applications in quantum experiments. *J. Opt. B: Quant. Semicl. Opt.* 4(2):S47–S51.
273. Chen, Q., B. Shi, Y. Zhang, and G. Guo. 2008. Entanglement of the orbital angular momentum states of the photon pairs generated in a hot atomic ensemble. *Phys. Rev. A* 78:053810.
274. Janicijevic, L., and S. Topuzoski. 2016. Gaussian laser beam transformation into an optical vortex beam by helical lens. *J. Mod. Opt.* 63(2):164–176.
275. Ricci, F., W. Loffler, and M.P. van Exter. 2012. Instability of higher-order optical vortices analyzed with a multi-pinhole interferometer. *Opt. Express* 20(20):22961–22975.
276. Kumar, A., P. Vaity, and R.P. Singh. 2011. Crafting the core asymmetry to lift the degeneracy of optical vortices. *Opt. Express* 19(7):6182–6190.
277. Li, Z., M. Zhang, G. Liang, X. Li, X. Chen, and C. Cheng. 2013. Generation of high-order optical vortices with asymmetrical pinhole plates under plane wave illumination. *Opt. Express* 21(13):15755–15764.
278. Kovalev, A.A., V.V. Kotlyar, and A.P. Porirev. 2016. Asymmetric Laguerre-Gaussian beams. *Phys. Rev. A* 93:063858.

279. Kotlyar, V.V., A.A. Kovalev, and A.P. Porfirev. 2015. Vortex Hermite-Gaussian laser beams. *Opt. Lett.* 40(5):701–704.
280. Kotlyar, V.V., A.A. Kovalev, and A.P. Porfirev. 2017. Asymmetric Gaussian optical vortex. *Opt. Lett.* 42:139–142.
281. Nugrowati, A., W. Stam, and J. Woerdman. 2012. Position measurement of non-integer OAM beams with structurally invariant propagation. *Opt. Express* 20:27429–27441.
282. Oemrawsingh, S.S.R., X. Ma, D. Voigt, A. Aiello, E.R. Eliel, G.W. 'tHooft, and J.P. Woerdman. 2005. Experimental demonstration of fractional orbital angular momentum entanglement of two photons. *Phys. Rev. Lett.* 95:240501.
283. Götte, J., S. Franke-Arnold, R. Zambrini, and S.M. Barnett. 2007. Quantum formulation of fractional orbital angular momentum. *J. Mod. Opt.* 54:1723–1738.
284. Gutiérrez-Vega, J.C., and C. López-Mariscal. 2008. Nondiffracting vortex beams with continuous orbital angular momentum order dependence. *J. Opt. A: Pure Appl. Opt.* 10:015009.
285. Tao, S., W. Lee, and X. Yuan. 2004. Experimental study of holographic generation of fractional Bessel beams. *Appl. Opt.* 43:122–126.
286. Kotlyar, V.V., A.A. Kovalev, and A.P. Porirev. 2017. Elliptic Gaussian optical vortices. *Phys. Rev. A* 95:053805.
287. Bekshaev, A.Y., and A.I. Karamoch. 2008. Astigmatic telescopic transformation of a high-order optical vortex. *Opt. Commun.* 281(23):5687–5696.
288. Xu, D., X. Song, H. Li, D. Zhang, H. Wang, J. Xiong, and K. Wang. 2015. Experimental observation of sub-Rayleigh quantum imaging with a two-photon entangled source. *Appl. Phys. Lett.* 106:171104.
289. Huang, S., Z. Miao, C. He, F. Pang, Y. Li, and T. Wang. 2016. Composite vortex beams by coaxial superposition of Laguerre-Gaussian beams. *Opt. Lasers Eng.* 78:132–139.
290. Plick, W.N., and M. Krenn. 2015. Physical meaning of the radial index of Laguerre-Gauss beams. *Phys. Rev. A* 92(6):063841.
291. Savelyev, D.A., and S.N. Khonina. 2015. Characteristics of sharp focusing of vortex Laguerre-Gaussian beams. *Comput. Opt.* 39(5):654–662.
292. Stilgoe, A.B., T.A. Nieminen, and H. Rubinsztein-Dunlop. 2015. Energy, momentum and propagation of non-paraxial high-order Gaussian beams in the presence of an aperture. *J. Opt.* 17(12):125601.
293. Zhang, Y., X. Liu, M. Belić, W. Zhong, F. Wen, and Y. Zhang. 2015. Anharmonic propagation of two-dimensional beams carrying orbital angular momentum in a harmonic potential. *Opt. Lett.* 40:3786–3789.
294. Kim, D.J., and J.W. Kim. 2015. High-power TEM00 and Laguerre–Gaussian mode generation in double resonator configuration. *Appl. Phys. B* 121(3):401–405.
295. Lin, D., J. Daniel, and W. Clarkson. 2014. Controlling the handedness of directly excited Laguerre-Gaussian modes in a solid-state laser. *Opt. Lett.* 39:3903–3906.
296. Ruffato, G., M. Massari, and F. Romanato. 2014. Generation of high-order Laguerre-Gaussian modes by means of spiral phase plates. *Opt. Lett.* 39:5094–5097.
297. Karimi, E., R.W. Boyd, P. de la Hoz, H. de Guise, J. Rehacek, Z. Hradil, A. Aiello, G. Leuchs, and L.L. Sanchez-Soto. 2014. Radial quantum number of Laguerre-Gauss modes. *Phys. Rev. A* 89:063813.
298. Das, B.C., D. Bhattacharyya, and S. De. 2016. Narrowing of Doppler and hyperfine line shapes of Rb – D2 transition using a Vortex beam. *Chem. Phys. Lett.* 644:212–218.
299. Allocca, A., A. Gatto, M. Tacca, R.A. Day, M. Barsuglia, G. Pillant, C. Buy, and G. Vajente. 2015. Higher-order Laguerre-Gauss interferometry for gravitational-wave detectors with in situ mirror defects compensation. *Phys. Rev. D* 92(10):102002.
300. Sun, K., C. Qu, and C. Zhang. 2015. Spin-orbital-angular-momentum coupling in Bose-Einstein condensates. *Phys. Rev. A* 91(6):063627.

301. Mondal, P.K., B. Deb, and S. Majumder. 2014. Angular momentum transfer in interaction of Laguerre-Gaussian beams with atoms and molecules. *Phys. Rev. A* 89(6):063418.
302. Krenn, M., J. Handsteiner, M. Fink, R. Fickler, and A. Zeilinger. 2015. Twisted photon entanglement through turbulent air across Vienna. *Proc. Natl. Acad. Sci.* 112:14197–14201.
303. Grillo, V., G.C. Gazzadi, E. Mafakheri, S. Frabboni, E. Karimi, and R.W. Boyd. 2015. Holographic generation of highly twisted electron beams. *Phys. Rev. Lett.* 114:034801.
304. Clark, C.W., R. Barankov, M.G. Huber, M. Arif, D. Cory, and D.A. Pushin. 2015. Controlling neutron orbital angular momentum. *Nature* 525:504–506.
305. Kravtsov, Y.A. 1967. Complex ray and complex caustics. *Radiophys. Quantum Electron.* 10:719–730.
306. Bandres, M.A., and J.C. Gutierrez-Vega. 2008. Elliptical beams. *Opt. Express* 16:21087–21092.
307. Kim, H.C., and Y.H. Lee. 1999. Hermite–Gaussian and Laguerre–Gaussian beams beyond the paraxial approximation. *Opt. Commun.* 169:9–16.
308. Kovalev, A.A., V.V. Kotlyar, and A.P. Porfirev. 2015. Shifted nondiffractive Bessel beams. *Phys. Rev. A* 91(5):053840.
309. Grier, D. 2003. A revolution in optical manipulation. *Nature* 424:810–816.
310. Kuga, T., Y. Torii, N. Shiokawa, and T. Hirano. 1997. Novel optical trap of atoms with a doughnut beam. *Phys. Rev. Lett.* 78:4713–4716.
311. Willig, K.I., S.O. Rizzoli, V. Westphal, R. Jahn, and S.W. Hell. 2006. STED microscopy reveals that synaptotagmin remains clustered after synaptic vesicle exocytosis. *Nature* 440:935–939.
312. Courtial, J., K. Dholakia, L. Allen, and M.J. Padgett. 1997. Gaussian beams with very high orbital angular momentum. *Opt. Commun.* 144:210–213.
313. Izdebskaya, Y., T. Fadeyeva, V. Shvedov, and A. Volyar. 2006. Vortex-bearing array of singular beams with very high orbital angular momentum. *Opt. Lett.* 31(17):2523–2525.
314. Krenn, M., N. Tischler, and A. Zeilinger. 2016. On small beams with large topological charge. *New. J. Phys.* 18:033012.
315. Zheng, S., and J. Wang. 2017. Measuring orbital angular momentum (OAM) states of vortex beams with annular gratings. *Sci. Rep.* 7:40781.
316. Vieira, J., R.M.G.M. Trines, E.P. Alves, R.A. Fonseca, J.T. Mendonca, R. Bingham, P. Norreys, and L.O. Silva. 2016. High orbital angular momentum harmonic generation. *Phys. Rev. Lett.* 117:265001.
317. Chen, Y., Z. Fang, Y. Ren, L. Gong, and R. Lu. 2015. Generation and characterization of a perfect vortex beam with a large topological charge through a digital micromirror device. *Appl. Opt.* 54(27):8030–8035.
318. Jesacher, A., S. Furhapter, C. Maurer, S. Bernet, and M. Ritsch-Marte. 2006. Holographic optical tweezers for object manipulations at an air-liquid surface. *Opt. Express* 14(13):6342–6352.
319. Fickler, R., R. Lapkiewicz, W.N. Plick, M. Krenn, C. Schaeff, S. Ramelow, and A. Zeilinger. 2012. Quantum entanglement of high angular momenta. *Science* 338:640–643.
320. Campbell, G., B. Hage, B. Buchler, and P.K. Lam. 2012. Generation of high-order optical vortices using directly machined spiral phase mirrors. *Appl. Opt.* 51(7):873–876.
321. Mafakheri, E., A.H. Tavabi, P. Lu, R. Balboni, F. Venturi, C. Menozzi, G.C. Gazzadi, S. Frabboni, and A. Sit. 2017. Dunin-Borkowski RE, Karimi E. Realization of electron vortices with large orbital angular momentum using miniature holograms fabricated by electron beam lithography. *Appl. Phys. Lett.* 110:093113.
322. Fickler, R., G. Campbell, B. Buchler, P.K. Lam, and A. Zeilinger. 2016. Quantum entanglement of angular momentum states with quantum number up to 10010. *Proc. Natl. Acad. Sci. USA* 113(48):13642–13647.

323. Ostrovsky, A.S., C. Rickenstorff-Parrao, and V. Arrizon. 2013. Generation of the "perfect" optical vortex using a liquid-crystal spatial light modulator. *Opt. Lett.* 38(4):534–536.
324. Chen, M., M. Mazilu, Y. Arita, E.M. Wright, and K. Dholakia. 2013. Dynamics of microparticles trapped in a perfect vortex beam. *Opt. Lett.* 38(22):4919–4922.
325. Jabir, M.V., N. Apurv Chaitanya, A. Aadhi, and G.K. Samanta. 2016. Generation of "perfect" vortex of variable size and its effect in angular spectrum of the down-converted photons. *Sci. Rep.* 6:21877.
326. Li, P., Y. Zhang, S. Liu, C. Ma, L. Han, H. Cheng, and J. Zhao. 2016. Generation of perfect vectorial vortex beams. *Opt. Lett.* 41(10):2205–2208.
327. Vaity, P., and L. Rusch. 2015. Perfect vortex beam: Fourier transformation of a Bessel beam. *Opt. Lett.* 40:597–600.
328. García-García, J., C. Rickenstorff-Parrao, R. Ramos-García, V. Arrizón, and A. Ostrovsky. 2014. Simple technique for generating the perfect optical vortex. *Opt. Lett.* 39(18):5305–5308.
329. Fedotowsky, A., and K. Lehovec. 1974. Optimal filter design for annular imaging. *Appl. Opt.* 13(12):2919–2923.
330. Korn, G.A., and T.M. Korn. 1968. *Mathematical Handbook for Scientists and Engineers.* New York: McGraw-Hill.
331. Reddy, S.G., P. Chithrabhanu, P. Vaity, A. Aadhi, S. Prabhakar, and R.P. Singh. 2016. Non-diffracting speckles of a perfect vortex beam. *J. Opt.* 18(5):055602.
332. Zhang, C., C. Min, L. Du, and X.-C. Yuan. 2016. Perfect optical vortex enhanced surface plasmon excitation for plasmonic structured illumination microscopy imaging. *Appl. Phys. Lett.* 108(20):201601.
333. Sun, L., Y. Zhang, Y. Wang, C. Zhang, C. Min, Y. Yang, S. Zhu, and X. Yuan. 2017. Refractive index mapping of single cells with a graphene-based optical sensor. *Sens. Actuators B* 242:41–46.
334. Ma, H., X. Li, Y. Tai, H. Li, J. Wang, M. Tang, Y. Wang, J. Tang, and Z. Nie. 2017. In situ measurement of the topological charge of a perfect vortex using the phase shift method. *Opt. Lett.* 42:135–138.
335. Kotlyar, V.V., A.A. Kovalev, and A.P. Porfirev. 2016. Optimal phase element for generating a perfect optical vortex. *J. Opt. Soc. Am. A* 33:2376–2384.
336. Ye, F., L. Dong, B. Malomed, D. Mihalache, and B. Hu. 2010. Elliptic vortices in optical waveguides and self-attractive Bose-Einstein condensates. *J. Opt. Soc. Am. B* 27:757–762.
337. Fadeyeva, T., C. Alexeyev, B. Sokolenko, M. Kudryavtseva, and A. Volyar. 2011. Non-canonical propagation of high-order elliptic vortex beams in a uniaxially anisotropic medium. *Ukr. J. Phys. Opt.* 12(2):62–82.
338. Ye, F., D. Mihalache, and B. Hu. 2009. Elliptic vortices in composite Mathieu lattices. *Phys. Rev. A* 79(5):053852.
339. Tao, H., Y. Liu, Z. Chen, and J. Pu. 2012. Measuring the topological charge of vortex beams by using an annular ellipse aperture. *Appl. Phys. B* 106:927–932.
340. Miret, J., and C. Zapata-Rodríguez. 2008. Diffraction-free beams with elliptic Bessel envelope in periodic media. *J. Opt. Soc. Am. B* 25, 1–6.
341. Jin, J., J. Luo, X. Zhang, H. Gao, X. Li, M. Pu, P. Gao, Z. Zhao, and X. Luo. 2016. Generation and detection of orbital angular momentum via metasurface. *Sci. Rep.* 6:24286.
342. Kotlyar, V.V., A.A. Kovalev, and V.A. Soifer. 2015. Nonparaxial Hankel vortex beams of the first and second types *Comput. Opt.* 39:299–304.
343. Kovalev, A.A., V.V. Kotlyar, and A.G. Nalimov. 2013. Nonparaxial optical vortices and Kummer laser beams. *Opt. Eng.* 52:091716.
344. Cerjan, A., and C. Cerjan. 2011. Orbital angular momentum of Laguerre-Gaussian beams beyond the paraxial approximation. *J. Opt. Soc. Am. A* 28:2253–2260.

345. Carter, W. 1975. Band-limited angular-spectrum approximation to a spherical scalar wave field. *J. Opt. Soc. Am.* 65:1054–1058.
346. Guo, J., Z. Wei, Y. Liu, and A. Huang. 2015. Analysis of optical vortices with suppressed sidelobes using modified Bessel-like function and trapezoid annulus modulation structures. *J. Opt. Soc. Am.* A 32:195–203.
347. Mishra, S.R. 1991. A vector wave analysis of a Bessel beam. *Opt. Commun.* 85:159–161.
348. Horak, R., Z. Bouchal, and J. Bajer. 1997. Nondiffracting stationary electromagnetic fields. *Opt. Commun.* 133:314–327.
349. Jauequi, R., and S. Hacyan. 2005. Quantum-mechanical properties of Bessel beams. *Phys. Rev. A* 71:033411.
350. Wang, Y., W. Dou, and H. Meng. 2014. Vector analyses of linearly and circularly polarized Bessel beams using Hertz vector potentials. *Opt. Express* 22:7821–7830.
351. Kotlyar, V.V., and A.A. Kovalev. 2015. Vectorial Hankel laser beams carrying orbital angular momentum. *Comput. Opt.* 39:449–452.
352. Kotlyar, V.V., A.A. Kovalev, and V.A. Soifer. 2016. Vectorial rotating vortex Hankel laser beams. *J. Opt.* 18:095602.
353. Youngworth, K.S., and T.G. Brown. 2000. Focusing of high numerical aperture cylindrical-vector beams. *Opt. Express* 7:77–87.
354. Zhan, Q., and J.R. Leger. 2002. Focus shaping using cylindrical vector beams. *Opt. Express* 10:324–331.
355. Novitsky, A., and D. Novitsky. 2007. Negative propagation of vector Bessel beams. *J. Opt. Soc. Am. A* 24:2844–2849.
356. Sukhov, S., and A. Dogariu. 2010. On the concept of "tractor beams". *Opt. Lett.* 35:3847–3849.
357. Bozinovic, N., Y. Yue, Y. Ren, M. Tur, P. Kristensen, H. Huang, A.E. Willner, and S. Ramachandran. 2013. Terabit-scale orbital angular momentum mode division multiplexing in fibers. *Science* 340(6140):1545–1548.
358. Cheng, M., Y. Zhang, Y. Zhu, J. Gao, W. Dan, Z. Hu, and F. Zhao. 2015. Effects of non-Kolmogorov turbulence on the orbital angular momentum of Hankel-Bessel-Schell beams. *Opt. Laser Technol.* 67:20–24.
359. Chen, C., H. Yang, S. Tong, and Y. Lou. 2016. Changes in orbital-angular-momentum modes of a propagated vortex Gaussian beam through weak-to-strong atmospheric turbulence. *Opt. Express* 24:6959–6975.
360. Li, S., and J. Wang. 2016. Compensation of a distorted N-fold orbital angular momentum multicasting link using adaptive optics. *Opt. Lett.* 41:1482–1485.
361. Jurado-Navas, A., A. Tatarczak, X. Lu, J. Olmos, J. Garrido-Balsells, and I. Monroy. 2015. 850-nm hybrid fiber/free-space optical communications using orbital angular momentum modes. *Opt. Express* 23:33721–33732.
362. Ren, Y., G. Xie, H. Huang, L. Li, N. Ahmed, Y. Yan, M. Lavery et al. 2015. Turbulence compensation of an orbital angular momentum and polarization-multiplexed link using a data-carrying beacon on a separate wavelength. *Opt. Lett.* 40:2249–2252.
363. Krenn, M., J. Handsteiner, M. Fink, R. Fickler, R. Ursin, M. Malik, and A. Zeilinger. 2016. Twisted light transmission over 143 km. *Proc. Natl. Acad. Sci. USA* 113(48):13648–13653.
364. Zhu, Y., M. Chen, Y. Zhang, and Y. Li. 2016. Propagation of the OAM mode carried by partially coherent modified Bessel–Gaussian beams in an anisotropic non-Kolmogorov marine atmosphere. *J. Opt. Soc. Am. A* 33:2277–2283.
365. Baghdady, J., K. Miller, K. Morgan, M. Byrd, S. Osler, R. Ragusa, W. Li, B. Cochenour, and E. Johnson. 2016. Multi-gigabit/s underwater optical communication link using orbital angular momentum multiplexing. *Opt. Express* 24:9794–9805.

366. Ren, Y., L. Li, Z. Wang, S.M. Kamali, E. Arbabi, A. Arbabi, Z. Zhao et al. 2016. Orbital angular momentum-based space division multiplexing for high-capacity underwater optical communications. *Sci. Rep.* 6:33306.
367. Debye, P. 1909. Der Lichtdruck and Kugeln von beliebige Material. *Ann. Phys.* 30:57–136.
368. Rohrbach, A., and E.H. Stelzer. 2001. Optical trapping of dielectric particles in arbitrary fields. *J. Opt. Soc. Am. A* 18(4):813–839.
369. Pobre, R., and C. Salome. 2002. Radiation force on a nonlinear microsphere by a lightly focused Gaussian beam. *Appl. Opt.* 41(36):7694–7701.
370. Gauthier, R.C. 2005. Computation of the optical trapping force using an FDTD based technique. *Opt. Express* 13(10):3707–3718.
371. Moine, O., and B. Stout. 2005. Optical force calculations in arbitrary beams by use of the vector addition theorem. *J. Opt. Soc. Am. B* 22(8):1620–1631.
372. Liu, B., and Y. Wang. 2011. Optical trapping force combining an optical fiber probe and an AFM metallic probe. *Opt. Express* 19(4):3703–3714.
373. Li, M., S. Yan, B. Yao, Y. Liang, G. Han, and P. Zhang. 2016. Optical trapping force and torque on spheroidal Rayleigh particles with arbitrary spatial orientations. *J. Opt. Soc. Am. A* 33(7):1341–1347.
374. Kendrick, M.J., D.H. McIntyre, and O. Ostroverkhova. 2009. Wavelength dependence of optical tweezer trapping forces on dye-doped polystyrene microspheres. *J. Opt. Soc. Am. B* 26(11):2189–2198.
375. Ashkin, A. 1992. Forces of a single-beam gradient laser trap on a dielectric sphere in the ray-optics region. *Biophys. J.* 61:569–582.
376. Callegari, A., M. Mijalkov, A. Gokoz, and G. Volpe. 2015. Computational toolbox for optical tweezers in geometrical optics. *J. Opt. Soc. Am. B* 32(5):B11–B19.
377. Moreno, I., J. Davis, M. Sánchez-López, K. Badham, and D. Cottrell. 2015. Nondiffracting Bessel beams with polarization state that varies with propagation distance. *Opt. Lett.* 40:5451–5454.
378. Yang, R., and R. Li. 2016. Optical force exerted on a Rayleigh particle by a vector arbitrary-order Bessel beam. *J. Quant. Spectrosc. Radiat. Transfer* 178:230–243.
379. Mitri, F.G., R.X. Li, L.X. Guo, and C.Y. Ding. 2017. Optical tractor Bessel polarized beams. *J. Quant. Spectrosc. Radiat. Transfer* 187:97–115.
380. Chávez-Cerda, S., M.J. Padgett, I. Allison, G.H.C. New, J.C. Gutiérrez-Vega, A.T. O'Neil, I. MacVicar, and J. Courtial. 2002. Holographic generation and orbital angular momentum of high-order Mathieu beams. *J. Opt. B: Quant. Semicl. Opt.* 4(2):S52–S57.
381. Lóxpez-Mariscal, C., J. Gutiérrez-Vega, G. Milne, and K. Dholakia. 2006. Orbital angular momentum transfer in helical Mathieu beams. *Opt. Express* 14:4182–4187.
382. Ortiz-Ambriz, A., J. Gutiérrez-Vega, and D. Petrov. 2014. Manipulation of dielectric particles with nondiffracting parabolic beams. *J. Opt. Soc. Am. A* 31:2759–2762.
383. Wu, G., C. Tong, M. Cheng, and P. Peng. 2016. Superimposed orbital angular momentum mode of multiple Hankel–Bessel beam propagation in anisotropic non-Kolmogorov turbulence. *Chin. Opt. Lett.* 14(8):080102.
384. Lavery, M.P.J., F.C. Speirits, S.M. Barnett, and M.J. Padgett. 2013. Detection of a spinning object using light's orbital angular momentum. *Science* 341(6145):537–540.
385. Ryabtsev, A., S. Pouya, A. Safaripour, M. Koochesfahani, and M. Dantus. 2016. Fluid flow vorticity measurement using laser beams with orbital angular momentum. *Opt. Express* 24:11762–11767.
386. Ashrafi, S., R. Linquist, and N. Ashrafi. 2015. Sample concentration measurements using orbital angular momentum. *US Patent* US20160169799.
387. Schmitt, H.A., D.E. Waagen, N.N. Shah, D.L. Barker, and A.D. Greentree. 2010. System and method of orbital angular momentum (OAM) diverse signal processing using classical beams. *US Patent* US20100013696.

Index